Heme, Chlorophyll, and Bilins

Heme, Chlorophyll, and Bilins

Methods and Protocols

Edited by

Alison G. Smith

Department of Plant Sciences, University of Cambridge, Cambridge, UK

and

Michael Witty

Department of Biochemistry, University of Cambridge, Cambridge, UK

Humana Press **Totowa, New Jersey**

999 Riverview Drive, Suite 208
Totowa, New Jersey 07512

www.humanapress.com

This publication is printed on acid-free paper. ∞
ANSI Z39.48-1984 (American Standards Institute)
Permanence of Paper for Printed Library Materials.

Cover design by Patricia F. Cleary.
For additional copies, pricing for bulk purchases, and/or information about other Humana titles, contact Humana at the above address or at any of the following numbers: Tel.: 973-256-1699; Fax: 973-256-8341; E-mail: humana@humanapr.com; or visit our Website: www.humanapress.com

Printed in the United States of America. 10 9 8 7 6 5 4 3 2 1

Library of Congress Cataloging in Publication Data

Heme, chlorophyll, and bilins: methods and protocols / edited by Alison G. Smith and Michael Witty
p. cm.
Includes bibliographical references (p.)
ISBN 1-58829-111-1 (alk. paper)
1. Chlorophyll. 2. Heme. 3. Tetrapyrroles. 4. Plant pigments. I. Smith, Alison G. II. Witty, Michael.
QK898.C5 H46 2001
572'.46–dc21 2001039604

Preface

> The men of experiment are like the ant, they only collect and use; the reasoners resemble spiders, who make cobwebs out of their own substance. But the bee takes the middle course: it gathers its material from the flowers of the garden and field, but transforms and digests it by a power of its own. Not unlike this is the true business of philosophy [science]; for it neither relies solely or chiefly on the powers of the mind, nor does it take the matter which it gathers from natural history and mechanical experiments and lay up in the memory whole, as it finds it, but lays it up in the understanding altered and digested. Therefore, from a closer and purer league between these two faculties, the experimental and the rational (such as has never been made), much may be hoped.
>
> **Francis Bacon**, *Novum Organum*, 1620
> (Republished in 1960 by Liberal Arts Press, New York, p. 93)

Each time a new researcher joins a laboratory, there is a passing on of methods and technical know-how from existing members, so that expertise is maintained and refined. As long as the procedures are current, then the information remains easily accessible, and can be transferred to other research groups by exchange visits, or when a researcher moves labs. But it is seldom that the methods are published in anything other than an abbreviated form, or with the inclusion of technical tips that can make the difference between a method working or failing. With the handling and manipulation of tetrapyrroles, a discipline that has been carried out over the last hundred years or so, there have been a number of excellent handbooks published over the years that detail the characteristics of these important compounds, and provide methods for their preparation, analysis, and use. However, these books are now mostly out-of-print, and in many cases had a theoretical rather than practical orientation. In the experience of one of us (MW), as someone new moving into the area of tetrapyrrole research, despite collecting all the methods from publications and colleagues, the knowledge was disjointed and hard to put into practice. Furthermore, it seemed that although many modern and state-of-the-art procedures were practiced, the simpler, more traditional methods had been forgotten about, or lost with the retirement of older scientists.

Our goal in producing this book, therefore, was to ask scientists who routinely carry out the experiments, to describe their basic protocols and technology for the study of chlorophyll, heme, and related molecules, including technical tips and ways to avoid common pitfalls. In the editing process, we have worked hard to ensure that the contributions from each author provided a coherent and accessible introduction to their topic, be it chemical, biophysical, or molecular biological, and that the protocols were comprehensible to novices (us!). We are extremely grateful to all the contributors for their willingness to modify their chapters as we requested, and for their forbearance in the length of time it has taken to complete the project. We would also like to thank Tom Lanigan at Humana Press Inc., for being prepared to take the project on, and Christine McAndrew for all her help at a difficult time.

Alison G. Smith
Michael Witty

Contents

Contributors

JERRY C. BOMMER • *Frontier Scientific/Porphyrin Products, Logan, UT, USA*
DONALD A. BRYANT • *Department of Biochemistry and Molecular Biology, The Pennsylvania State University, University Park, PA, USA*
DANIEL FORCINITI • *Chemical Engineering Department, University of Missouri-Rolla, Rolla, MO, USA*
TERRENCE G. FREY • *San Diego State University, San Diego, CA, USA*
PETER HAMBRIGHT • *Department of Chemistry, Howard University, Washington, DC, USA*
J. CLARK LAGARIAS • *University of California–Davis, Davis, CA, USA*
CHANG KEE LIM • *MRC Bioanalytical Science Group, School of Biological and Chemical Sciences, Birbeck College, University of London, London, UK*
MICHAEL T. MCDOWELL • *University of California–Davis, Davis, CA, USA*
HARALD PAULSEN • *Institut für Allgemeine Botanik der Johannes-Gutenberg, Univerität Mainz, Mainz, Germany*
CONSTANTIN A. REBEIZ • *University of Illinois, Urbana, IL, USA*
WENDY M. SCHLUCHTER • *Department of Biological Sciences, University of New Orleans, New Orleans, LA, USA*
VOLKMAR H. R. SCHMID • *Institut für Allgemeine Botanik der Johannes-Gutenberg, Univerität Mainz, Mainz, Germany*
PETER M. SHOOLINGIN-JORDAN • *School of Biological Sciences, University of Southampton, Southamton, UK*
ALISON G. SMITH • *Department of Plant Sciences, University of Cambridge, Cambridge, UK*
KEVIN M. SMITH • *Department of Chemistry, University of California–Davis, Davis, CA, USA*
MATTHEW J. TERRY • *University of Southampton, Southampton, UK*
GEORGIOS TSIOTIS • *Department of Chemistry, University of Crete, Heraklion, Greece*
MARTIN J. WARREN • *School of Biological Sciences, Queen Mary Westfield College, London, UK*
ANGELA WILKS • *University of Maryland, Baltimore, MD, USA*
MICHAEL WITTY • *Department of Biochemistry, University of Cambridge, Cambridge, UK*

1 Laboratory Methods for the Study of Tetrapyrroles

Alison G. Smith and Michael Witty
University of Cambridge, Cambridge, CB2 3EA, UK

1. TETRAPYRROLE STRUCTURE AND FUNCTION

1.1. Structure of Tetrapyrroles

Tetrapyrroles are a group of organic molecules that includes chlorophyll (Figure 1), hemes (Figure 2), bilins (Figure 3), and corrins, such as vitamin B_{12} (37). These molecules are also often referred to as porphyrins, although strictly, these are only those compounds with the same oxidation state as heme. Chlorophyll, for example, has one more saturated bond and is therefore a chlorin (30).

A pyrrole is a 5-membered ring containing one nitrogen, which is colorless, but when four pyrroles are linked by unsaturated methine groups, the properties of the tetrapyrrole macrocycle are changed dramatically, and two extremely important characteristics emerge. Tetrapyrroles contain a ring rich in conjugated double bonds that absorb light strongly, and they have four nitrogens oriented towards a cavity that may accommodate metal ions and allow coordination of the metal ion above or below the plane of the macrocycle. These metals have stabilized oxidation states and solubility. Aside from these two important properties, tetrapyrroles also have a subtly substituted ring structure which alters the light absorbance properties of the conjugated double bond system, the geometry of metal ion binding (and therefore the type of metal bound), and mediates interactions of the tetrapyrrole with proteins.

Most metals and metalloids in the periodic table have been incorporated into complexes with tetrapyrroles (27), and many metals are observed in mineral porphyrins (10). However, because of the differences in abundance and differential stability of the complexes, nickel and vanadium are the most common ions in natural abiotic porphyrins, whereas the following seven have been seen in living systems: Mg, Fe, Mn, Co, Zn, Ni, and V (6).

1.2. Distribution of Tetrapyrroles

Porphyrins are spontaneous products of organic chemical reactions which can be synthesized in Urey-Miller type experiments that mimic prebiotic atmospheric conditions: UV irradiation of 5-aminole-

Heme, Chlorophyll, and Bilins: ***Methods and Protocols***
Edited by A.G. Smith and M. Witty

vulinic acid (ALA) has produced pyrroles (33), while electrical discharge in the presence of pyrrole and formaldehyde has produced porphyrins (14), and they have been detected in sterile meteorites (14,15).

Porphyrins are chemically stable (30) and can persist in the environment for many millions of years. Porphyrins are found in large fossils such as mollusk shells (17) and also in molecular fossil forms in geological strata. The best examples of these are coal and oil deposits, where they are found as mostly nickel(II) and vanadyl complexes (9). Mineral porphyrins have been detected in sedimentary deposits with high organic content laid down as early as precambrian times (8). They may precipitate to form distinct bedding planes and, although most deposits contain only a few parts per million, some contain significant amounts of free or complexed porphyrins, for example the Gibellina sedimentary deposits, which contain 24 mg/g copper and nickel porphyrins (29).

Although they are found in abiotic systems, most tetrapyrroles are biological, and indeed they are the most conspicuous living molecule on earth. Chlorophylls can be seen from satellites in space, where vegetation types can be identified and used to predict underlying geology (31). Even when viewed from outside, the Earth looks enticing because of tetrapyrroles. If there are Men from Mars, they would pick on Earth for special interest, and they would be right to do so (25).

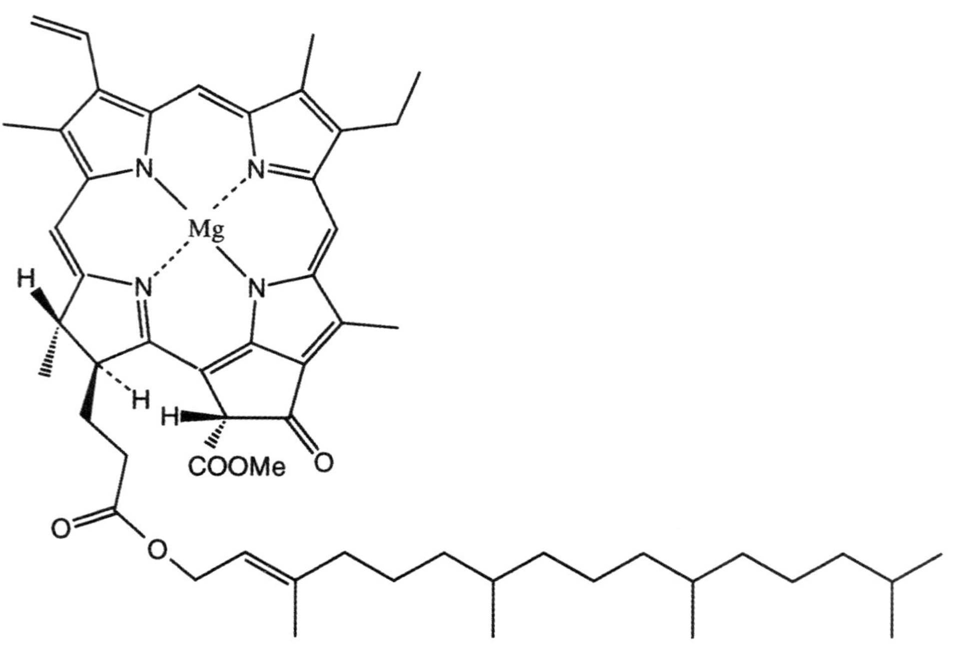

Figure 1. The structure of chlorophyll *a*. Chlorophylls are present in protein complexes in the membrane of photosynthetic bacteria and the thylakoid membrane of chloroplasts, where they harvest and trap light energy during photosynthesis (Chapters 10 and 11).

1.3. Importance of Tetrapyrroles in Nature

Although there are a large number of chemical types and ionic conjugates of tetrapyrroles, only a few species and their derivatives are very abundant in nature: chlorophylls, hemes, and linear tetrapyrroles, the bilins. Tetrapyrroles are important in living cells because of their physical properties. The tetrapyrrole macrocycle can be highly conjugated and absorb visible light strongly, therefore many tetrapyrroles are photochemically active, the most important interaction with light being the capture of energy by chlorophyll in photosynthesis.

Chlorophylls are an essential part of the photosynthetic apparatus, and the heme of cytochromes is an essential part of electron transfer chains in both respiration and photosynthesis. These two tetrapyrrole types are essential for the most significant reduction and oxidation processes in nature. Tetrapyrroles are also essential in many other biochemical processes. They form the prosthetic groups of metalloenzymes such as sulfite reductase, nitrite reductase, peroxidase, and catalase, which carry out a wide range of oxidation and reduction reactions. Vitamin B_{12} is a cobalt tetrapyrrole complex that acts as a cofactor in methyltransferases, and factor F_{430} is a nickel tetrapyrrole that is involved in methane formation in certain bacteria. Bilins are linear tetrapyrroles with no tightly bound metal and are important as the accessory pigments in algae and as phytochromobilin, the red-light receptor of higher plants (Chapters 12–14) (21).

2. A COMMON BIOCHEMICAL PATHWAY

As might be expected from their common structure, all cellular tetrapyrroles are

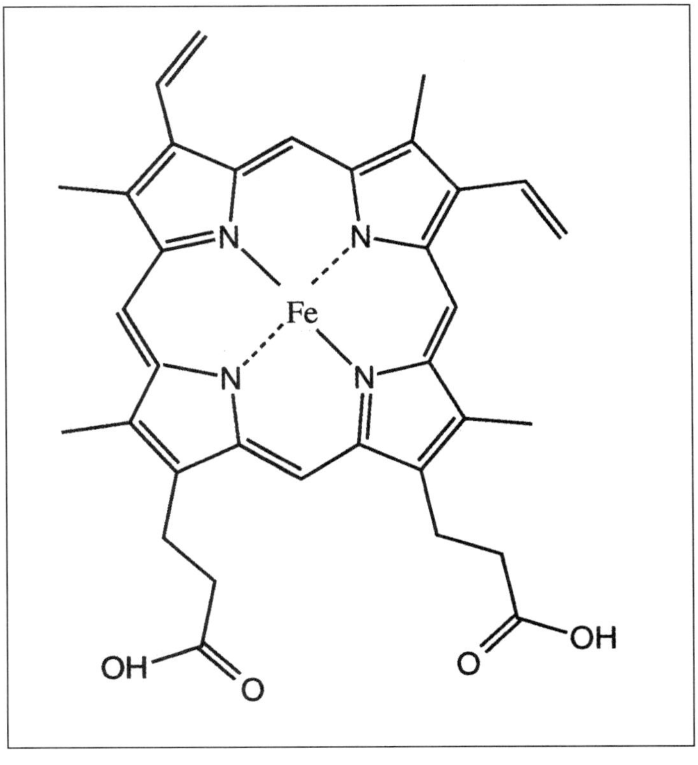

Figure 2. The structure of protoheme IX. Hemes are found in a wide range of different proteins, including photosynthetic and respiratory cytochromes involved in electron transfer, the oxidative enzymes catalase and peroxidase, cytochrome P450s, which catalyze mono-oxygenase reactions, and oxygen-carrying proteins such as hemoglobin and myoglobin (Chapters 7, 8, and 9).

made by a common biochemical pathway (Figure 4) from the central intermediate uroporphyrinogen III (for a review see Reference 37).

The first committed precursor is ALA, which contains all of the carbon and nitrogen atoms required by the tetrapyrrole nucleus. Two biosynthetic pathways that lead to ALA have evolved (Chapter 4). The first pathway to be discovered was the so-called Shemin pathway, in which ALA is formed from glycine and succinyl CoA by ALA synthase (ALAS). This occurs in animals, fungi, and some bacteria. However, the ancestral pathway, characteristic of the majority of bacteria, algae, and plants, is the C_5 pathway, in which ALA is formed from glutamate in three steps involving glutamyl-tRNA as an intermediate. Monomers are formed by condensation of ALA by ALA dehydratase (Figure 5) to form porphobilinogen (PBG), which is in turn tetramerized by PBG deaminase (Figure 6) to form the linear intermediate 1-hydroxymethylbilane (or preuroporphyrinogen). This is cyclized and isomerized by uroporphyrinogen III synthase, to produce the common intermediate to all cellular tetrapyrroles.

Reduced uroporphyrinogen III is formed with methylene rather than methine bridges to prevent photoactivity, production of singlet oxygen, and similar damaging species. The porphyrinogen form is maintained until the step preceding metal ion insertion. Uroporphyrinogen III has two possible fates. On the corrin pathway, it is methylated and used to produce siroheme, the cofactor of sulfite and nitrite reductases, or vitamin B_{12}, after the insertion of ferrous iron or cobalt, respectively. Alternatively, uroporphyrinogen III is oxidatively decarboxylated in three steps to form protoporphyrin IX, the last common intermediate of heme and chlorophyll (Chapters 4 and 5).

Ferrochelatase catalyzes insertion of iron into protoporphyrin IX for heme biosynthesis, which is followed by insertion into protein complexes. Heme may be metabolized further to form bilins (Chapters 12–14) by linearization and the loss of the iron atom, catalyzed by heme oxygenase. The insertion of magnesium into protoporphyrin IX by magnesium chelatase is the first step of chlorophyll biosynthesis and is followed by further modification of the tetrapyrrole nucleus by esterification, methylation,

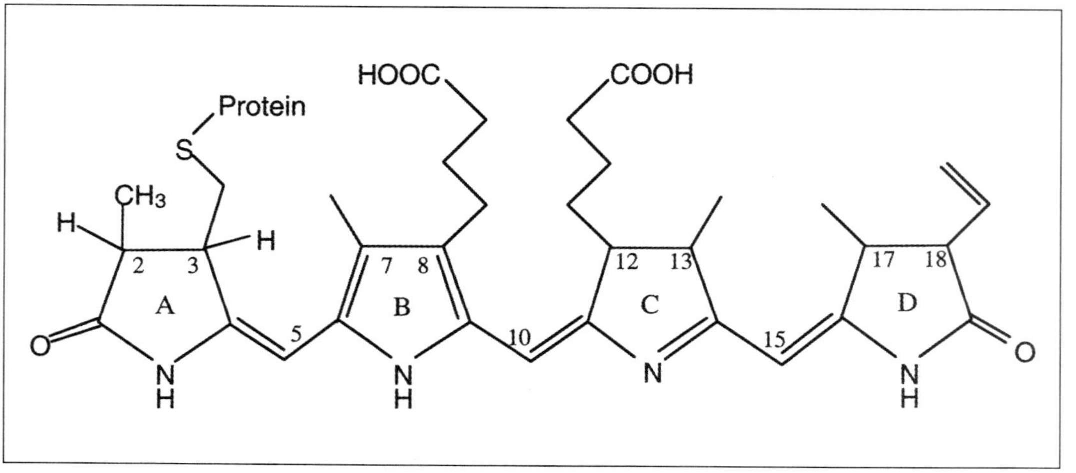

Figure 3. The structure of phytochromobilin. This is the chromophore of phytochromobilin, which is the red-light receptor of higher plants (Chapter 13). Linear tetrapyrroles are also found as accessory light-harvesting pigments in cyanobacteria and many algae (Chapter 14).

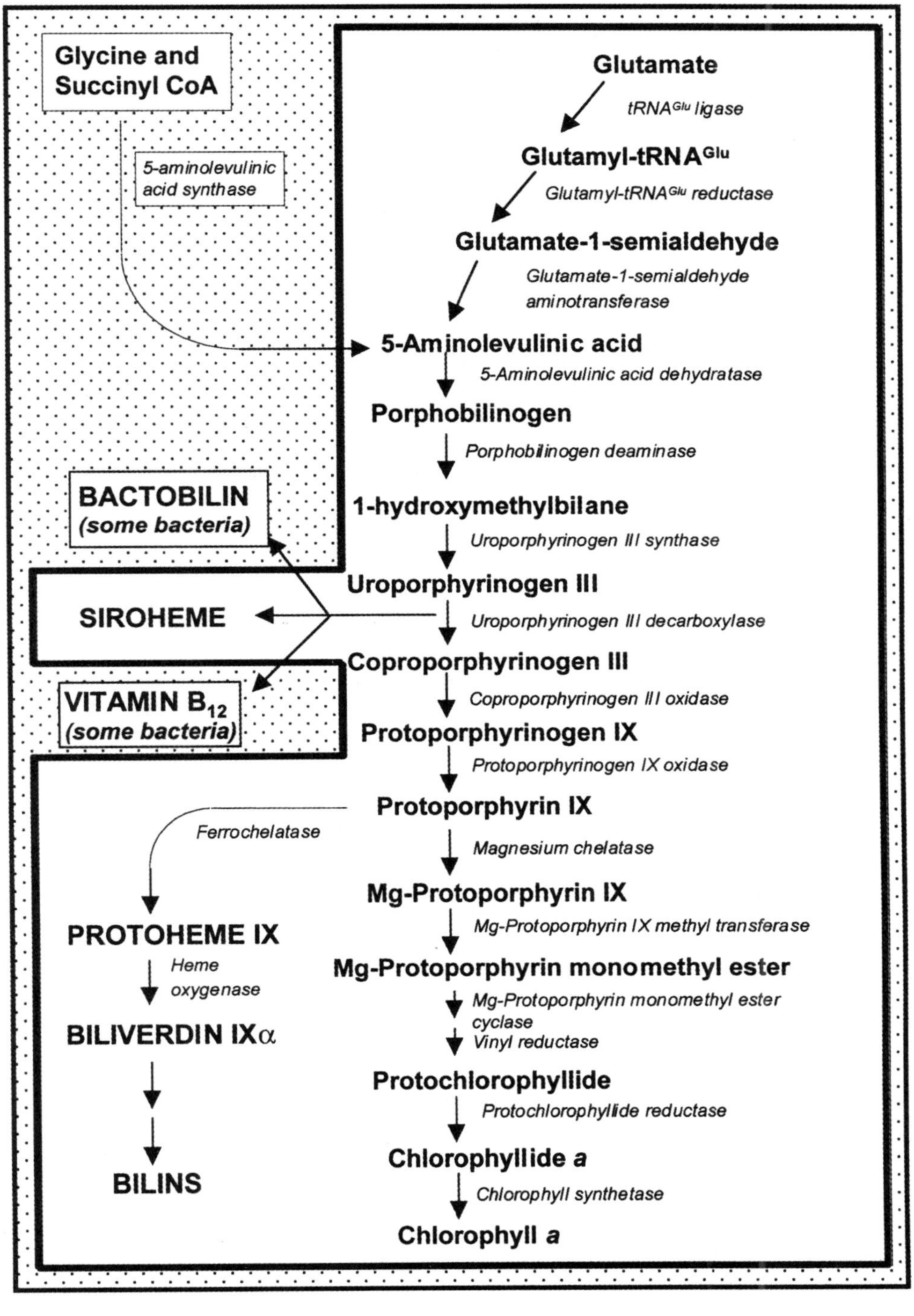

Figure 4. The tetrapyrrole biosynthetic pathway, showing the different endproducts and the major intermediates (Chapters 4, 5, 6, and 12). Enzymes are shown in italics.

reduction of vinyl group, and formation of a fifth ring to produce protochlorophyllide. In the presence of light, protochlorophyllide is reduced to form chlorophyllides, which undergo esterification by phytyl diphosphate or geranylgeranyldiphosphate to produce chlorophyll (see Chapter 6).

3. ROLES OF TETRAPYRROLES

3.1. Light Harvesting

Photosynthetic organisms contain a sophisticated system of several hundred chlorophylls (or bacteriochlorophylls) and other accessory pigments, which act as antennae to absorb light and pass the energy to special chlorophylls in reaction centers. Here the light energy is trapped as excited electrons, which are then transferred through an electron transfer chain to generate ATP. In higher plants, algae, and cyanobacteria, this process results in the oxidation of water to evolve molecular oxygen and the production of reduced nicotinamide adenine dinucleotide phosphate (NADPH) (see Chapters 10 and 11 for more detail). The ATP and NADPH generated by the light-dependent reactions are used to fix CO_2 into organic combination via the Calvin cycle. Photosynthesis not only provides the means for photosynthetic organisms to live, but also indirectly supports almost all life on earth with carbohydrates and oxygen.

3.2. Oxidation of Carbohydrates to Produce Usable Energy

Nonphotosynthetic cells obtain their energy by the oxidation of carbohydrates, which in aerobic organisms results in the formation of CO_2. This process involves a series of reduction-oxidation (redox) reactions whereby the large gap in oxidation state between carbohydrate and carbon dioxide is released in a series of gentle and efficient steps, with oxygen as the final electron acceptor. Transition metals, such as the iron found in heme (Figure 2), are well suited to catalyze these reactions because they contain *d*-electron orbital systems with small differences in energy levels, thus allowing a range of oxidation states so that energy can be released in a controlled and useful way (cf section 3.4).

In the bacterial membrane, and the mitochondria of eukaryotes, a series of protein complexes containing cofactors, which include heme (see Chapter 7), undergo a series of reversible redox reactions that generates ATP. In this respect, the process of

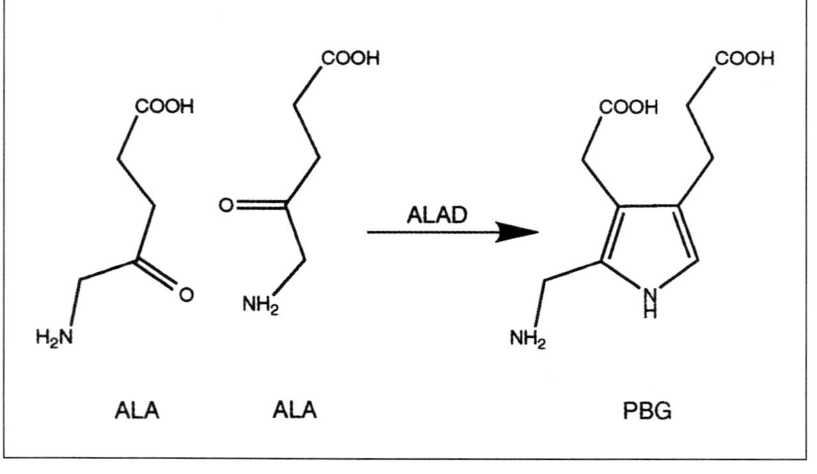

Figure 5. The formation of a pyrrole. The reaction catalyzed by ALA dehydratase.

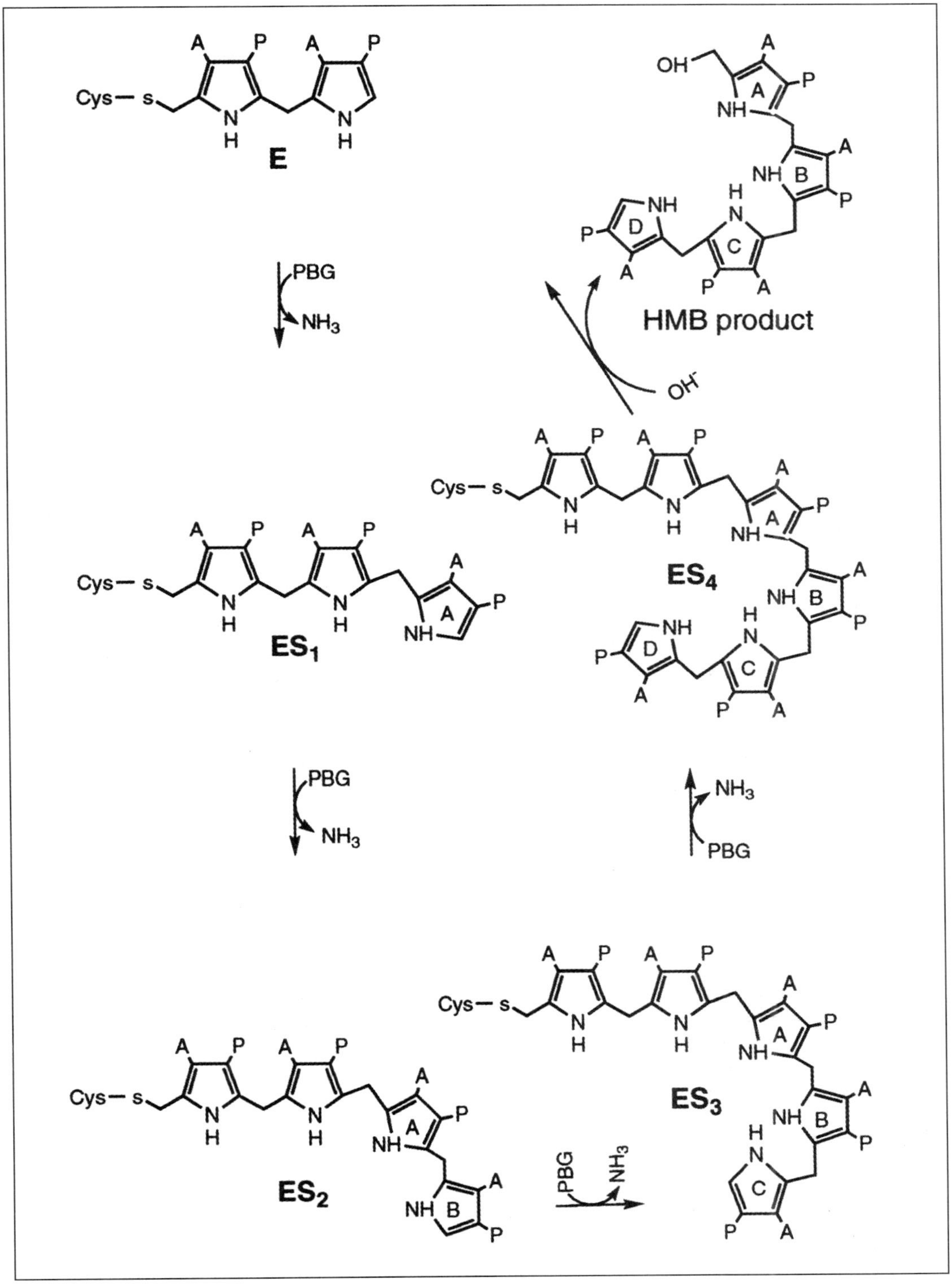

Figure 6. The formation of a tetrapyrrole. The reaction catalyzed by PBG deaminase. The holoenzyme (E) contains an active site dipyrromethane cofactor. This is used to accept PBG monomers and form enzyme substrate complexes. A, acetate; P, propionate moiety. Pyrrole rings are labeled A, B, C, and D.

oxidative phosphorylation is similar to that in photosynthesis (section 3.1), reflecting common evolutionary ancestors of some of the components involved.

3.3. Transport and Homeostasis of Oxygen

Metabolism requires the consumption of large amounts of oxygen, and therefore transport from the atmosphere to cells buried deep in animal bodies is necessary to allow a rapid rate of metabolism. While several kinds of oxygen transport proteins exist, including hemerythrins (non-heme iron proteins) and hemocyanins (copper proteins), globins are the most abundant and widespread oxygen transport protein (12). They use reversible oxygen coordination to protoheme IX iron (Fe^{II}) for transport of oxygen from respiratory organs throughout the animal body.

Hemoglobins transport oxygen in many animal groups, and most have similar structures and functions (35). In circulating red blood cells, hemoglobins consist of 17 kDa polypeptides, each containing a heme group that can bind one oxygen molecule. Hemoglobins are typically tetramers that allow for cooperative binding and release of oxygen, depending on the P_{O_2} of the surrounding tissue.

In addition to oxygen transport, hemoglobins are associated with other important functions. For example, circulating hemoglobin I of the swamp clam *Lucina pectinata* is involved in sulfide transport. Sulfides are bound with high affinity within a cage of heme and three phenylalanine residues and are released to symbiotic bacteria upon reduction (3).

Myoglobin is a monomeric protein that contributes to the transport of oxygen by diffusion, and large amounts are found in skeletal and cardiac muscles of mammals. In cardiac muscle, myoglobin acts as a short-term oxygen buffer, smoothing supply from one beat to the next. Intracellular myoglobin is also found in bacteria, protozoans, plants, and invertebrates (32).

Plants also contain leghemoglobin used for the homeostasis of oxygen. The best studied example is in the legume root nodule, where symbiotic bacteria consume large amounts of energy during the reduction of atmospheric nitrogen to ammonia, which is then available to the host plant. Leghemoglobin facilitates the maintenance of high levels of oxygen needed for bacterial respiration, while preventing poisoning of the nitrogen fixing machinery by molecular oxygen (2). Unlike in animals, the protein is not circulated, but rather simple diffusion down an oxygen gradient created by bacterial respiration promotes transport of oxygen from the outside.

3.4. Protection of Cellular Processes from Reactive Oxygen Species

As well as the desired biological reaction of oxygen as a terminal acceptor of electrons in the controlled oxidation of carbohydrates in respiration, oxygen will also react with electrons encountered at random, to produce reactive oxygen species (ROS) such as oxygen radicals and superoxide. These are very harmful to living systems, causing lipid peroxidation, membrane damage, and genetic mutation. Cells contain a number of enzymes that act to remove these intermediates quickly. Many of these oxidases are heme-containing proteins, including peroxidases and catalase (7).

3.5. Taking Advantage of Reactive Oxygen Species

Although ROS can be harmful, organisms have evolved systems in which they are useful. One important example of this is the biosynthesis and degradation of lignin. Lignocellulose is a composite of lignin and cellulose and probably the most abundant organic molecule in the biosphere, functioning as material for mechanical strength in wood. Rather than a regular polymer such as cellulose, lignin is synthesized by peroxidases, which oxidize phenolpropane units (such as coniferyl alcohol) to form reactive radical species. These polymerize in an irregular fashion to form lignin, which thereby contains a wide array of chemical linkages (38). Because of this unsymmetrical arrangement, lignin is unusually resistant to enzymatic breakdown, and only a few microbes, such as the white rot fungi, have enzymes capable of doing so (20). Lignin is also an important byproduct of the paper industry, which generates about 30 million tons of unused lignin per year, in a process involving harsh chemical treatments (13). Genetic modification of tree species to reduce lignin content is being explored as a means of avoiding this costly and polluting process (34).

A more general, and essential, role of these reactive species is found in both plants and animals, where ROS have been shown to modulate a wide range of cellular and physiological processes, acting as part of the signal transduction pathway. For example, one of the earliest responses to pathogen attack is a marked increase in ROS in the infected tissue, produced in part by the activity of plasma membrane-bound NADPH oxidase, a hemoprotein complex. The ROS, in turn, act as a trigger for defense responses, such as modification of membrane permeability and ion fluxes, and systemic acquired resistance in plants (1). Similarly in animals, ROS influence signaling cascades and transcriptional–posttranscriptional control of gene expression, thereby playing an essential role in processes such as apoptosis (23).

3.6. Control of Metabolic and Cellular Processes by Signaling

As well as functioning as an enzymic prosthetic group, tetrapyrroles also function in key regulatory processes. These include control of gene expression, cellular signaling pathways, and control of protein transport within the cell.

An example is heme-mediated feedback control of its own synthesis, which appears to occur in all groups of organisms, most importantly at the production of the first committed precursor ALA. In plants, there is evidence that heme inhibits the enzyme glutamyl-tRNA reductase (36). In animals, although heme inhibits the activity of ALAS, control is exerted at several other points as well. In mammals, there are two forms of the enzyme, constitutive and erythroid-specific. Liver ALAS is inhibited by heme in a negative feedback loop (11) to maintain levels of heme production for the maintenance of cellular processes. This feedback regulation is achieved by a combination of effects including inhibition of ALAS gene expression, increased ALAS mRNA degradation, and inhibition of pre-ALAS protein transport to the mitochondrion (19), with only a minor contribution by inhibition of ALAS catalytic activity (28). In contrast to the liver, in erythroid cells, transcription of the ALAS gene, together with genes for later enzymes in the pathway and for globins, is stimulated by heme to produce the large amounts of heme needed for hemoglobin in red blood cells.

In yeast, expression of the ALAS gene is controlled by heme and mediated through the transcription factor HAP1, which binds heme for activity. The binding domain contains multiple copies of a short motif, which is also found in the mitochondrial transit peptide of mammalian ALAS. This motif, involved in transient binding of heme, is quite different to the more stable heme-binding domain of cytochromes and globins (39).

There is accumulating evidence that tetrapyrrole intermediates play a role in signaling. In plants, there is coordination between the chloroplast and the nucleus, such that nuclear-encoded genes for chloroplast-targeted proteins are transcribed only in cells with functional chloroplasts. Although the exact identity of the so-called "plastid-factor" remains elusive, plant mutants with defects in certain steps of the tetrapyrrole biosynthetic pathway have altered plastid-nuclear signaling (24).

3.7. Subtle Pigmentation

While plants are green because of the presence of chlorophyll and animal tissues are largely red due to heme, some of the more subtle animal colors are also conferred by tetrapyrroles. The cuticle of birds' eggs with colored shells contain tetrapyrroles which contribute to their camouflage. Most markings and pigmentation are due to protoporphyrin IX, which is associated with brown and black coloring. Blue eggshells are associated with biliverdin IXα, and green eggshells are associated with zinc-biliverdin IXα with traces of coproporphryin III (18). The feathers of some birds also contain tetrapyrrole pigments. The feathers of *Turocos* contain red turacin (copper-uroporphyrin III) and green turacoverdin (22). Uroporphryin I is found in many calcified mollusk tissues such as shells (17) and pearls (16), though the function of the tetrapyrrole is unknown.

3.8. Artificial Uses of Tetrapyrroles

In addition to their importance in biology, tetrapyrroles are increasingly of interest to a much wider range of researchers. For example, chemists are able to create synthetic molecules which mimic the recognition and catalytic properties of enzymes. A particular aspect of this work is catalysis of reactions for which there are no known natural enzymes, such as Diels-Alder reactions (4). Porphyrins have proved very useful for this sort of study because of the rigid structures that they are able to form and the fact that they can coordinate a number of metal ions which are involved in the catalysis. For example, using porphyrin molecular boxes and zinc coordination, it has been possible to influence the stereospecificity of reactions by the geometrical constraints of a host cavity (26).

Other novel uses of tetrapyrroles have been established in clinical medicine, in particular for the treatment of cancer cells, in a technique called photodynamic therapy (PDT). The rationale behind the method is to load the cancerous cells with photosensitizing porphyrin mixtures, which, upon irradiation with visible light, cause the production of singlet oxygen, thereby leading to the destruction of the cells as described in section 3.4. Porphyrins are ideal compounds for this technique, not only because of their light absorption properties, but also because there is some preferential uptake of these molecules by tumor cells. Initially, in the 1960s and 1970s, the major photosensitizers used were hematoporphyrins and related preparations derived from acid extraction of blood (or hemoglobin), and therefore, are not chemically defined compounds. However, since 1980, new sensitizers have been developed, including chlorins and phthalocyanines, which have been chemically synthesized (5).

4. LABORATORY METHODS FOR THE STUDY OF TETRAPYRROLES

Tetrapyrroles are clearly a diverse and important group of molecules, and researchers from a wide range of different fields may wish to study them, whether it be a clinician using them for PDT, an ecologist studying the chlorophyll composition of leaves in a tropical forest, or a cell biologist investigating the function of specific hemoproteins. However, as we know from our own laboratory experience, there are certain "tricks-of-the-trade" which are necessary to use in order to carry out successful experiments. In this volume, we have selected articles written by people who actually carry out these procedures on a daily basis in their own laboratories. Each chapter provides an overview of the topic with general information on the experimental approach, as well as a number of step-by-step procedures, which should provide the basis for any novice tetrapyrrologist taking their first steps into this field.

ABBREVIATIONS

ALA, 5-aminolevulinic acid; ALAS, ALA synthase; PBG, porphobilinogen; PDT, photodynamic therapy; ROS, reactive oxygen species.

REFERENCES

1. **Alvarez, M.E., R.I. Pennell, P.J. Meijer, A. Ishikawa, R.A. Dixon, and C. Lamb.** 1998. Reactive oxygen intermediates mediate a systemic signal network in the establishment of plant immunity. Cell *98*:773-784.
2. **Appleby, C.A.** 1984. Leghemoglobin and rhizobium respiration. Annu. Rev. Plant Physiol. *35*:443-478.
3. **Bolognesi, M., C. Rosano, R. Losso, A. Borassi, M. Rizzi, J.B. Wittenberg, A. Boffi, and P. Ascenzi.** 1999. Cyanide binding to *Lucina pectinata* hemoglobin I and to sperm whale myoglobin: an X-ray crystallographic study. Biophys. J. *77*:1093-1099.
4. **Bonarlaw, R.P., L.G. Mackay, C.J. Walter, V. Marvaud, and J.K.M. Sanders.** 1994. Towards synthetic enzymes based on porphyrins and steroids. Pure Appl. Chem. *66*:803-810.
5. **Bonnett, R.** 1999. Photodynamic therapy in historical perspective. Rev. Contemp. Pharmacother. *10*:1-17.
6. **Buchler, J.W.** 1975. Static coordination chemistry of metalloporphyrins, p. 157-231. *In* K.M. Smith (Ed.), Porphyrins and Metalloporphyrins. Elsevier, Amsterdam.
7. **Cadenas, E.** 1989. Biochemistry of oxygen toxicity. Annu. Rev. Biochem. *58*:79-100.
8. **Callot, H.J.** 1991. Geochemistry of chlorophylls, p. 339-364. *In* H. Scheer (Ed.) Chlorophylls. CRC Press, Boca Raton.
9. **Czernuszewicz, R.S., J.G. Rankin, and T.D. Lash.** 1996. Fingerprinting petroporphyrin structures with vibrational spectroscopy. 4. Resonance raman spectra of nickel(II) cycloalkanoporphyrins: structural effects due to exocyclic ring size. Inorg. Chem. *35*:199-209.
10. **Dailey, K.K. and T.B. Rauchfuss.** 1997. π-Complexes of metalloporphyrins as model intermediates in hydrodemetallation (HDM) catalysis. Polyhedron *16*:3129-3136.
11. **Granick, S.** 1966. The induction in vitro of the synthesis of δ-aminolevulinic acid synthase in chemical porphyria: a response to certain drugs, sex hormones, and foreign chemicals. J. Biol. Chem. *241*:1359-1375.
12. **Hardison, R.** 1998. Hemoglobins from bacteria to man: evolution of different patterns of gene expression. J. Exp. Biol. *201*:1099-1117.
13. **Hartley, B.S., P.M.A. Broda, and P.J. Senior.** 1987. Technology in the 1990s: Utilization of Lignocellulosic Wastes. The Royal Society, London.
14. **Hodgson, G.W. and B.L. Baker.** 1964. Evidence for porphyrins in the orgueil meteorite. Nature *202*:125-131.
15. **Hodgson, G.W. and B.L. Baker.** 1967. Porphyrin abiogenesis from pyrrole and formaldehyde under simulated geochemical conditions. Nature *216*:29-32.
16. **Iwahashi, Y. and S. Akamatsu.** 1994. Porphyrin pigment in black-lip pearls and its application to pearl identification. Fisheries Sci. *60*:69-71.
17. **Kennedy, G.Y.** 1975. Porphyrins in invertebrates. Ann. NY Acad. Sci. *244*:662-673.
18. **Kennedy, G.Y. and H.G. Vevers.** 1976. A survey of avian eggshell pigments. Comp. Biochem. Physiol. B *55*:117-123.
19. **Lathrop, J.T. and M.P. Timko.** 1993. Regulation by heme of mitochondrial protein-transport through a conserved amino-acid motif. Science *259*:522-525.
20. **Leonowicz, A., A. Matuszewska, J. Luterek, D. Ziegenhagen, M. Wojtas-Wasilewska, N.S. Cho, M. Hofrichte, and J. Rogalski.** 1999. Biodegradation of lignin by white rot fungi. Fungal Genet. Biol. *27*:175-185.
21. **McDonagh, A.F.** 1979. Bile pigments: bilatrienes and 5,15-biladienes, p. 293-491. *In* D. Dolphin (Ed.), The Porphyrins, Vol. 1. Academic Press, London.
22. **Nicholas, R.E.H. and C. Rimington.** 1952. Isolation of unequivocal uroporphyrin III, a further study of turacin. Biochem. J. *50*:194-201.
23. **Nose K.** 2000. Role of reactive oxygen species in the regulation of physiological functions. Biol. Pharmacol. Bull. *23*:897-903.
24. **Oster, U., H. Brunner, and W. Rudiger.** 1996. The greening process in cress seedlings. 5. Possible interference of chlorophyll precursors, accumulated after thujaplicin treatment, with light-regulated expression of Lhc genes. J. Photochem. Photobiol. B *36*:255-261.
25. **Sagan, C., W.R. Thompson, R. Carlson, D. Gurnett, and C. Hord.** 1993. A search for life on earth from the Galileo spacecraft. Nature *365*:715-721.
26. **Sanders, J.K.M.** 1998. Supramolecular catalysis in transition. Chem. Eur. J. *4*:1378-1383.
27. **Sanders, J.K.M., N. Bampos, Z. Clyde-Watson, S.L. Darling, J.C. Hawley, H.J. Kim, C.C. Mak, and S.J. Webb.** 2000. Axial coordination chemistry of metalloporphyrins, p. 349-390. *In* K.M. Kadish, K.M. Smith, and R. Guilard (Eds.), The Porphyrin Handbook. Academic Press, London.

28. **Sassa, S. and T. Nagai.** 1996. The role of heme in gene expression. Int. J. Hematol. *63*:167-178.
29. **Schaeffer, P., R. Ocampo, H.J. Callot, and P. Albrecht.** 1993. Extraction of bound porphyrins from suphur-rich sediments and their use for reconstruction of palaeoenvironments. Nature *364*:133-136.
30. **Smith, K.M.** 1975. Porphyrins and Metalloporphyrins, p. 829-836. Elsevier, Amsterdam.
31. **Smith, M.O., S. Jacquemond, M. Verstraete, and Y. Govaerts.** 1999. Geobotany: vegetation mapping for earth sciences, p. 189-248. *In* Remote Sensing for the Earth Sciences, Manual of Remote Sensing, Vol. 3. John Wiley & Sons, New York.
32. **Suzuki, T. and K. Imai.** 1998. Evolution of myoglobin. Cell. Mol. Life Sci. *54*:979-1004.
33. **Szutka, A.** 1966. Formation of pyrrolic compounds by ultra-violet irradiation of δ-aminolevulinic acid. Nature *212*:401-402.
34. **Tamagnone, L., A. Merida, A. Parr, S. Mackay, F.A. Culianez-Macia, K. Roberts, and C. Martin.** 1998. The AmMYB308 and AmMYB330 transcription factors from *Antirrhinum* regulate phenylpropanoid and lignin biosynthesis in transgenic tobacco. Plant Cell *10*:135-154.
35. **Terwilliger, N.B.** 1998. Functional adaptations of oxygen-transport proteins. J. Exp. Biol. *201*:1085-1098.
36. **Vothknecht, U.C., C.G. Kannangara, and D. Wettstein.** 1998. Barley glutamyl tRNA(Glu) reductase: mutations affecting haem inhibition and enzyme activity. Phytochemistry *47*:513-519.
37. **Warren, M.J. and A.I. Scott.** 1990. Tetrapyrrole assembly and modification into the ligands of biologically functional cofactors. Trends Biochem. Sci. *15*:486-491.
38. **Whetten, R.W., J.J. MacKay, and R.R. Sedoroff.** 1998. Recent advances in understanding lignin biosynthesis. Annu. Rev. Plant Physiol. Plant Mol. Biol. *49*:585-609.
39. **Zhang, L. and L. Guarente.** 1995. Heme binds to a short sequence that serves a regulatory function in diverse proteins. EMBO J. *14*:313-320.

2 Syntheses of Tetrapyrroles

Kevin M. Smith
Department of Chemistry, University of California–Davis, Davis, CA, USA

1. INTRODUCTION

This chapter addresses basic methodology that can be used to obtain tetrapyrrole macrocycles in the porphyrin and chlorin series from natural materials and some simple methods for the total chemical synthesis of typical pyrroles and porphyrins. The aim is to provide investigators with enough information to decide whether to take on the task of preparing samples of useful porphyrin and chlorophyll derivatives or whether to simply purchase them or collaborate with other individuals more expert in the established synthetic procedures. The procedures reported herein are usually those which are easiest for the nonexpert to perform, while at the same time being sufficient to provide pure samples of the required product.

The porphyrin field has a very rich history; Hans Fischer's books present a laboratory approach to synthesis of porphyrin compounds dating back from the 1930s (20,22,24). In 1975, *Porphyrins and Metalloporphyrins* was published (64); this contained a fairly detailed laboratory methods section, which was useful at that time and is probably still useful to many investigators. An up-to-date and highly detailed description of the synthetic art of porphyrin chemistry can be found in *The Porphyrin Handbook* (39).

At the outset it must be mentioned that a certain degree of expertise in experimental organic chemistry is essential for success in the endeavors described herein; also essential are the appropriate laboratory equipment (fume hoods, rotary evaporators, temperature controlled reaction monitors, chromatographic equipment, etc.) and glassware. Since hazardous waste chemicals and solvents will also need to be disposed of, approved facilities for these responsibilities must also be available.

In terms of chemical technique and procedures, pyrrole and porphyrin derivatives tend to be easy to work with. With the exception of porphyrinogens, they usually do not require stringent exclusion of oxygen and water vapor (as is the case with much of the rest of organometallic chemistry), they are stable at room temperature and higher temperatures, and they can be purified by recrystallization and chromatography in the air at room temperature. As might be expected with any colored compound (which will be absorbing

Heme, Chlorophyll, and Bilins: *Methods and Protocols*
Edited by A.G. Smith and M. Witty

light of various wavelengths and therefore will be accessing excited electronic states—porphyrins fluoresce strongly), attempts should be made routinely to keep porphyrin and chlorin compounds out of the light; this is not difficult, and aluminum foil wrapped around a sample flask or around a chromatography column usually suffices. In the particular case of protoporphyrin IX [**1**] or its dimethyl ester [**2**], a well-characterized so-called Diels-Alder reaction is known to take place in the presence of oxygen and light to afford a mixture of photoprotoporphyrin and isophotoprotoporphyrin IX dimethyl ester [**3** and **4**, respectively] (7,34); this represents the extreme of normal porphyrin photolability and is caused by the presence of the 3- or 8-vinyl groups. If you can successfully handle protoporphyrin IX without continually

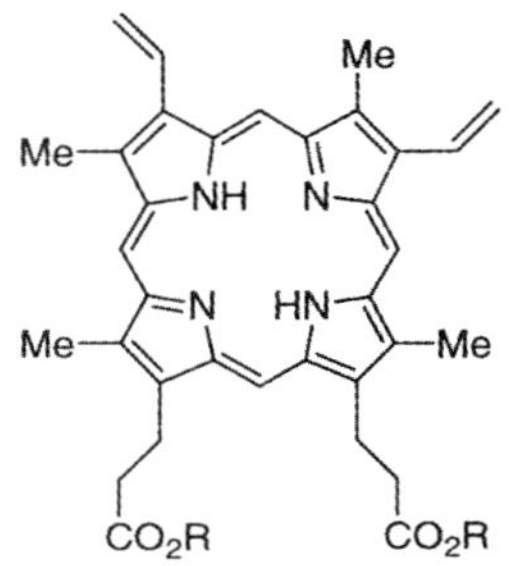

(**1**) R = H (protoporphyrin IX)
(**2**) R = Me (protoporphyrin IX dimethyl ester)

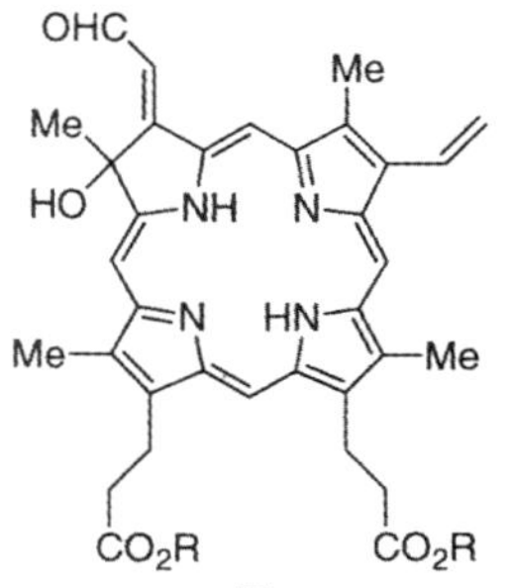

(**3**)

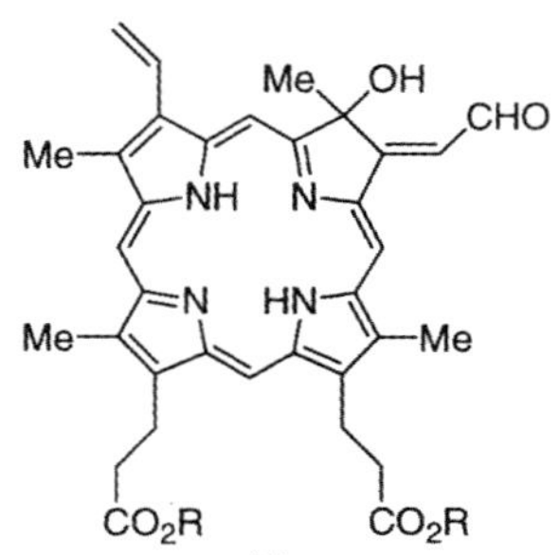

(**4**)

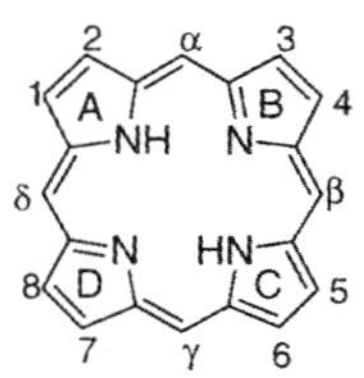

Fischer nomenclature

(**5**)

generating two polar green bands upon chromatography, you should do just fine. Further advice on the specific requirements for handling these molecules can be found in Chapter 3 by Bommer and Hambright.

2. NOMENCLATURE

Over the years, two different schemes for nomenclature of porphyrin systems have been used. The Fischer system for porphyrin nomenclature [structure **5**] provides a link back to the rich history of porphyrin chemistry mentioned above—many trivial names were generated which, particularly in the field of chlorophyll chemistry, are almost impossible to do without. Likewise, in the porphyrin field, there are some names that are indispensable (e.g., protoporphyrin IX, the "first" porphyrin, and deuteroporphyrin IX, the "second" porphyrin); the "IX" given after the porphyrin name refers to the (secondary) type-IX arrangement of the porphyrin substituents. When there are only two types of substituent, for example methyl and ethyl, with one methyl and one ethyl on each pyrrole ring, only four "primary type-isomers" **[6–9]** of the so-called "etio"porphyrins are possible. When there are three kinds of substituent (as with the methyl, vinyl, and propionic substituents in protoporphyrin IX), no less than fifteen "secondary type-isomers" are possible (provided there is one methyl on each pyrrole subunit), and the type-IX isomer is the biologically significant one. In the primary type isomer series, type-III is the biologically significant arrangement. But all that said, and given the near impossibility of naming some porphyrin and chlorophyll derivatives without the use of Fischer's trivial names, the International Union of Pure and Applied Chemistry (IUPAC) system of nomenclature [structure **10**] is the officially adopted nomenclature system, and this will have to be used in this chapter.

3. PREPARATION OF PORPHYRINS AND CHLORINS BY DEGRADATION OF NATURAL PIGMENTS

It is truly fortunate that massive amounts of natural products containing both hemin **[11]** and chlorophylls *a* and *b* **[12,13]** can be accessed. Fischer's three volumes (20,22,24), *Die Chemie des Pyrrols*, report an astonishing array of procedures for obtaining tetrapyrrole compounds from natural sources. Thus, large volumes of blood can be processed to provide hemin in kilogram quantities. From hemin, a large number of porphyrins and derivatives can be obtained (see later). Similarly, chlorophyll derivatives in the *a* and *b* series can be obtained by extraction of leaves, usually spinach. But if only chlorophyll *a* derivatives are desired, one can take advantage of the fact that certain algae, such as *Spirulina*, produce only chlorophyll *a*; thus, a laborious separation of the chlorophyll *a* and *b* series can be avoided. If chlorophyll *b* derivatives are required, there used to be no option but to extract plant chlorophylls and perform the chromatographic separation, either by preparative scale high-performance liquid chromatography (HPLC) or by gravity column chromatography on sucrose. Some years ago, *a* chemical derivatization approach was developed to make the chromatographic separations more palatable, and that will be discussed later.

3.1. Porphyrins from Hemoglobin

3.1.1. Hemin [11]

Because of the relative ease with which hemin can be obtained from blood, it can be purchased from a number of chemical companies at costs around a few dollars per gram. The method of choice (19) for preparation of hemin from blood involves addition of heparinized, citrated, or defibrinated blood to hot acetic acid containing sodium chloride. After cooling and removal of coagulated protein (usually with a wooden stick), the hemin separates and can be collected by filtration. Alternatively, the messy protein can be precipitated by addition of a solution of strontium chloride, followed by concentration to give hemin as above (16,44). Hemes [iron(II) porphyrins] can be obtained from hemins [iron(III) chloride porphyrins] most commonly by reduction with sodium dithionite under nitrogen or argon. Since autoxidation of iron(II) to iron(III) porphyrins is very facile in air, use of nitrogen or (preferably the heavier) argon

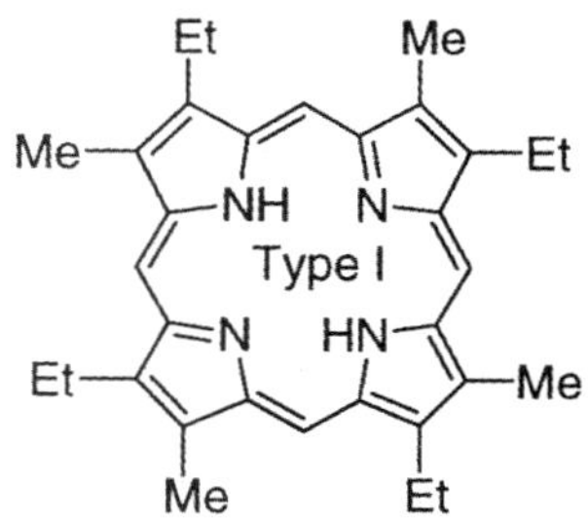

(6)

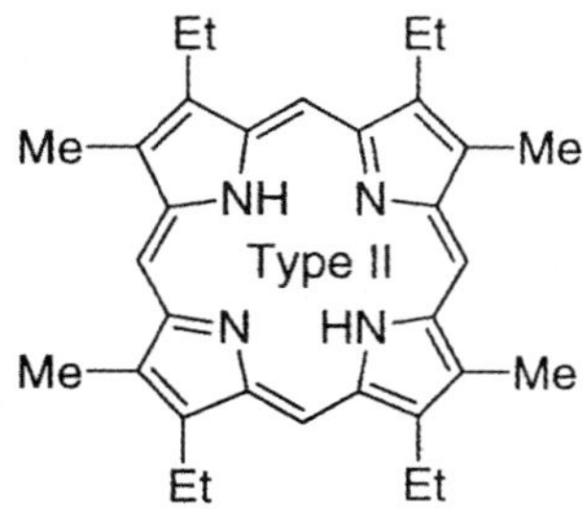

(7)

The four etioporphyrin type-isomers

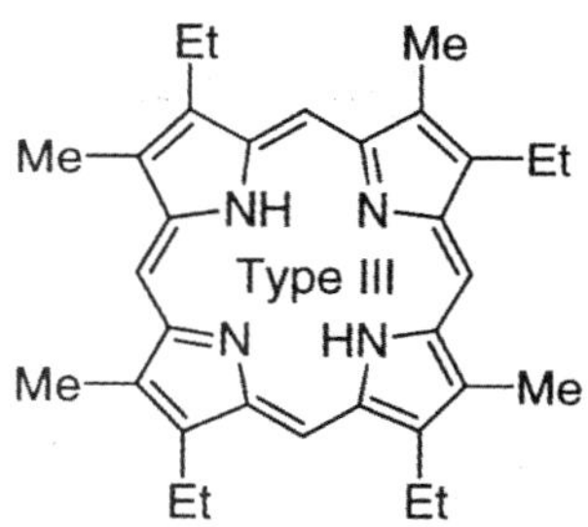

(8)

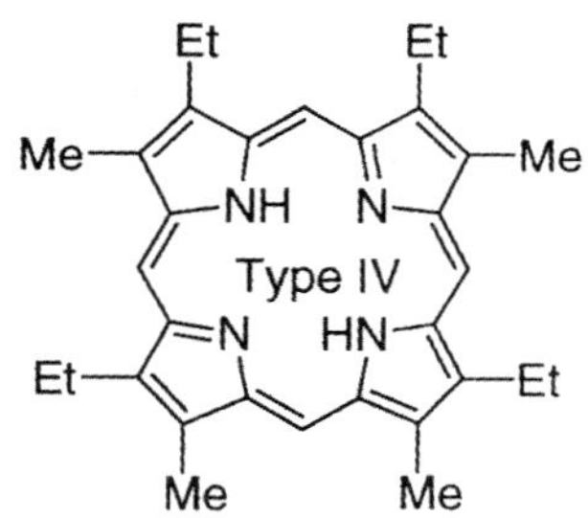

(9)

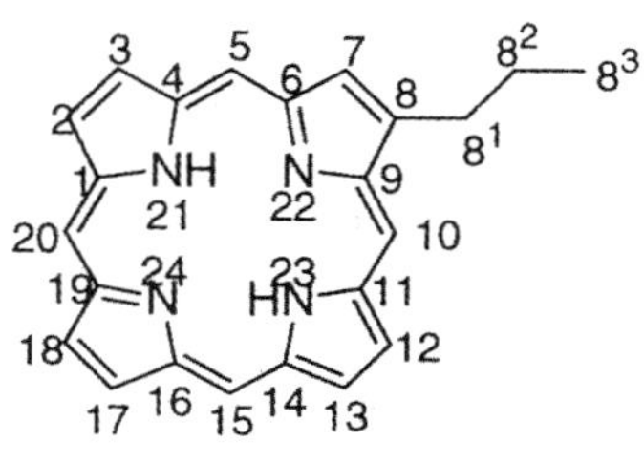

IUPAC nomenclature

(10)

is absolutely essential. Chromatographic purification of hemins is best accomplished on the corresponding (usually methyl) esters; but hemins [e.g., **11**] bearing carboxylic acid groups should not be esterified with diazomethane—a side-reaction takes place with the iron atom. For methyl esters (the simplest and best ester to use under normal circumstances), 5% sulfuric acid in methanol is the best mixture to use (CAUTION: take care to gently add the acid to the stirred and cooled alcohol) (66). Hemin esters can be hydrolyzed to the corresponding free carboxylic acids using base (66).

3.1.2. Protoporphyrin IX [1]

Protoporphyrin IX [**1**] is the product obtained by removal of iron from hemin [**11**], but acid alone does not accomplish this result because iron(III) is very difficult to eject from a porphyrin. Commercial samples of protoporphyrin IX are usually not very pure because of the sensitivity of protoporphyrin to photo-oxygenation at the vinyl groups (see above). The best method for obtaining protoporphyrin IX is to treat hemin [**11**] with ferrous sulfate in hydrochloric acid (46,51,52); the hemin is reduced to heme, and the iron(II), in strict contrast to iron (III), is readily removed by the acid. Commercial hematoporphyrin IX [**14**] is often very pure (unlike protoporphyrin IX), so a method for the preparation of [**1**] by double dehydration of hematoporphyrin IX [**14**] has been reported (40). This involves brief heating of [**14**] with toluene *p*-sulfonic acid in 1,2-dichlorobenzene. The dimethyl ester [**2**] of protoporphyrin IX can be obtained by esterification with either diazomethane (CAUTION: diazomethane can be explosive under certain circumstances) or with methanol–sulfuric acid (CAUTION) as mentioned above for hemin. The very useful Grinstein method (33) can be used to prepare protoporphyrin IX dimethyl ester [**2**] in one step from hemin [**11**].

3.1.3. Mesoporphyrin IX [15]

Mesoporphyrin IX [**15**] is related to protoporphyrin IX [**1**] with the important difference that the sensitive 3- and 8-vinyl groups in [**1**] are replaced with durable ethyl groups—hence, mesoporphyrin IX does not undergo the photo-oxygenation reaction mentioned above for protoporphyrin. Early biosynthetic investigations of the metabolism of protoporphyrin IX often used the easy to handle mesoporphyrin IX [**15**], and so incorporated a hydrogenation step to accomplish reduction of the 3- and 8-vinyl groups in protoporphyrin IX (9); the method of choice (22) is catalytic hydrogenation over palladium in formic acid. Either protoporphyrin IX, its ester, or hemin are used, and the iron in [**11**] is removed concomitantly during the reaction. Mesohemin IX [**16**], the iron(III) chloride of mesoporphyrin IX, is best obtained by the introduction of iron into [**15**] rather than by reduction of hemin [**11**]. Esterification of mesoporphyrin IX can be carried out using either diazomethane or sulfuric acid acid–alcohol.

3.1.4. Hematoporphyrin IX [14]

Hematoporphyrin IX [**14**] was the first porphyrin to be isolated (in 1867) (69); it was obtained by treatment of blood with concentrated sulfuric acid. Nominally, hematoporphyrin IX [**14**] is obtained chemically from protoporphyrin by hydration of both of the 3- and 8-vinyl groups. Since the 3^1- and 8^1-carbon atoms are chiral in [**14**], a mixture of four optical isomers (enantiomers and diastereomers) is obtained, and these can be separated by HPLC. Porphyrin [**14**] can also be purchased from commercial sources.

Using protoporphyrin IX [**1**] as the starting material, hematoporphyrin IX is best prepared by treatment with hydrogen bromide in acetic acid, followed by hydrolysis of the resulting 3,8-di(1-bromoethyl)-derivative [**17**] with water (22). If a common alcohol (R^1OH) such as methanol ($R^1 = CH_3$) is used in this last stage, then the 3,8-di(1-alkoxyethyl) analogue [**18**] is produced. Alternatively, reduction of 3,8-diacetyldeuteroporphyrin IX dimethyl ester [**19**] with sodium borohydride affords hematoporphyrin IX dimethyl ester [**20**] [e.g., Reference 66]. 3,8-Diacetyldeuteroporphyrin IX [**21**] can be prepared by oxidation of hematoporphyrin IX (62), or by Friedel-Crafts acetylation of deuterohemin IX [**22**] using acetic anhydride and pyridine, followed by removal of the iron (66). Use of sulfuric acid and methanol to esterify the propionic acids in [**14**] is not advised because acid-catalyzed dehydra-

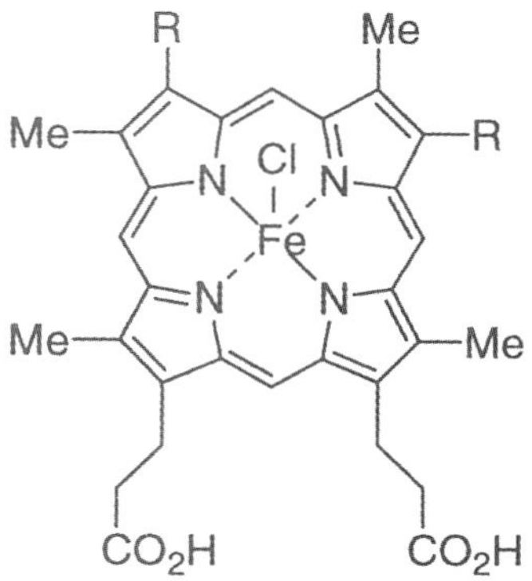

(**11**) R = CH=CH$_2$ (hemin)
(**16**) R = Et (mesohemin)

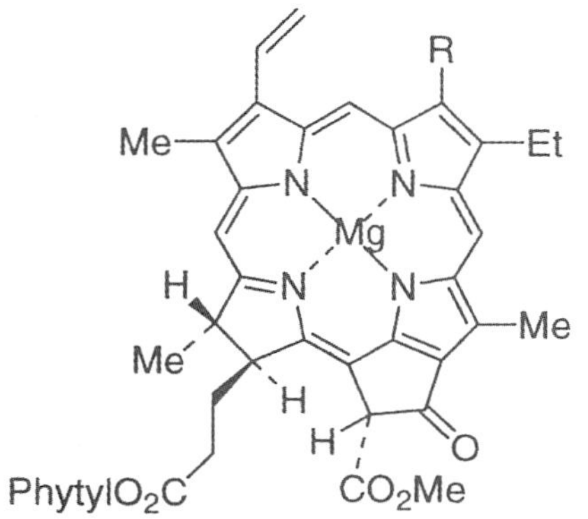

(**12**) R = Me (chlorophyll *a*)
(**13**) R = CHO (chlorophyll *b*)

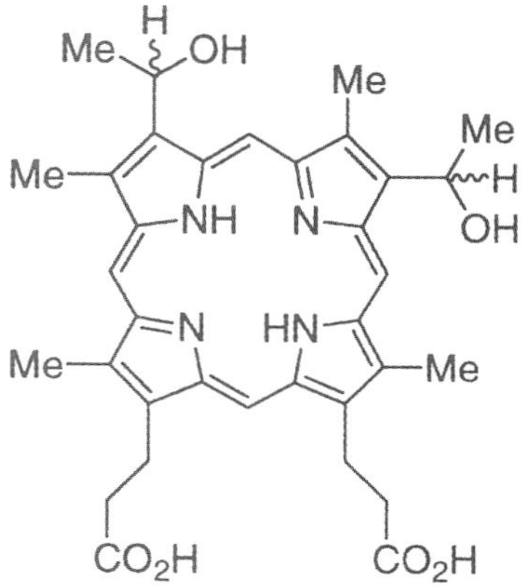

(**14**) (hematoporphyrin IX)

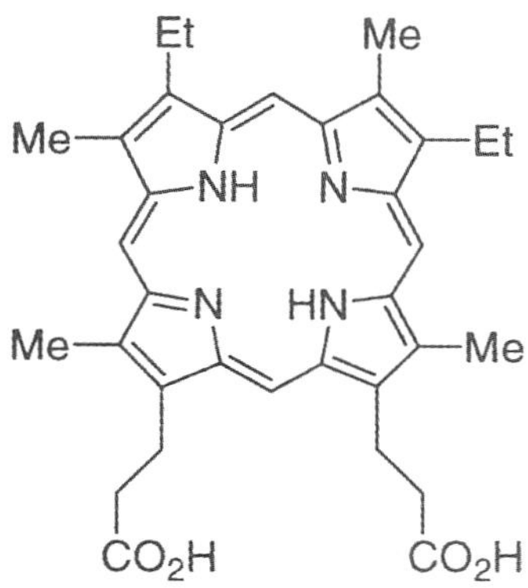

(**15**) (mesoporphyrin IX)

tion, or ether formation, at the 3,8-(1-hydroxyethyl) groups is a problem; it is best to use diazomethane in methanol to obtain the dimethyl ester [**20**] (CAUTION).

3.1.5. Deuteroporphyrin IX [23]

Deuteroporphyrin IX [**23**] is of significant historical importance because it was the first porphyrin isolated in Fischer's Nobel Prize winning synthesis of hemin [**11**] (29). Deuterohemin [**22**] can be obtained from "proto" hemin by brief heating of [**11**] in a resorcinol melt (60), via the so-called Schumm reaction in which the vinyl groups are replaced by hydrogen atoms (10,12,17,42). Demetalation, as reported above for the preparation of protoporphyrin IX from hemin, then affords deuteroporphyrin IX [**23**].

Numerous 3,8-disubstitution products (and 3- or 8-monosubstitution analogues)

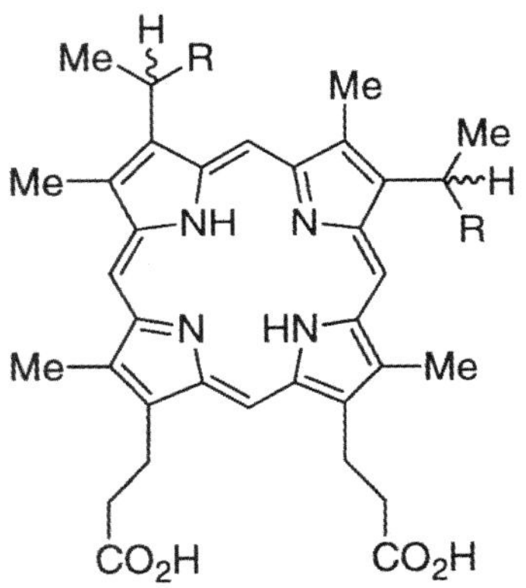

(**17**) R = Br
(**18**) R = OR[1]

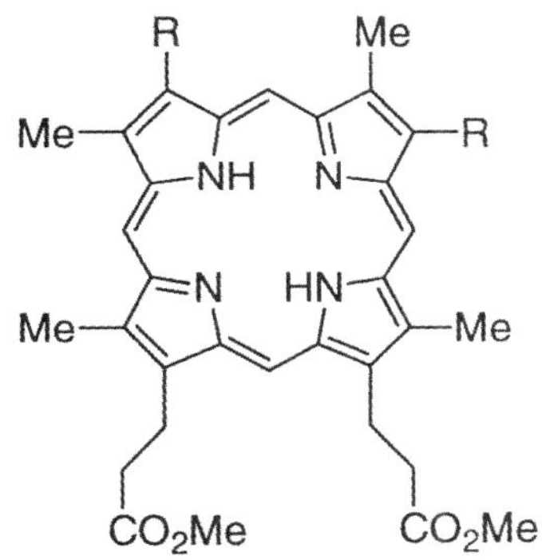

(**19**) R = COMe
(**20**) R = CH(OH)Me

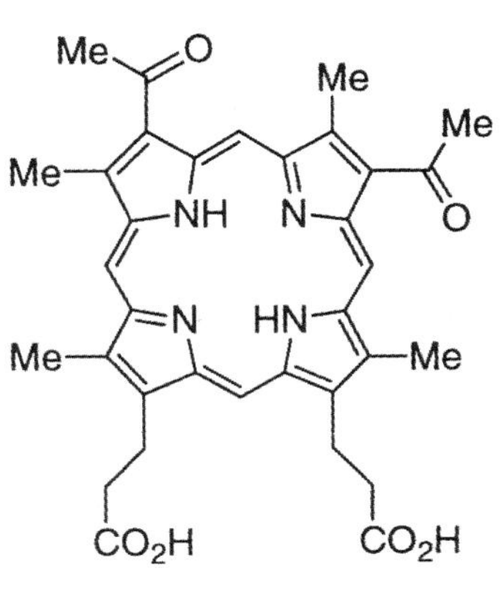

(**21**) (diacetyldeuteroporphyrin IX)

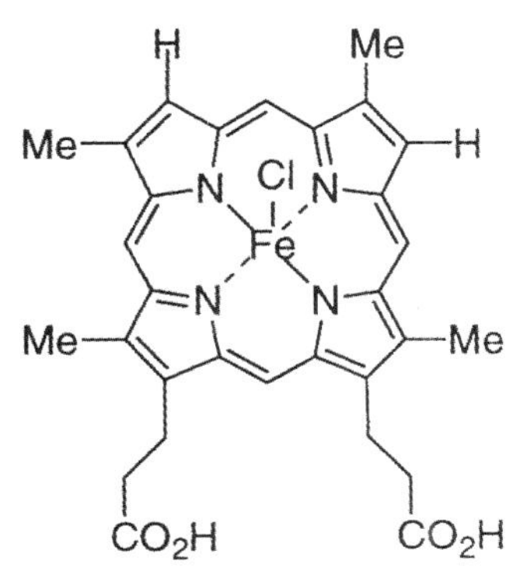

(**22**) (deuterohemin)

of deuteroporphyrin IX and its esters can be prepared, usually by way of aromatic electrophilic substitution on the hemin or its copper(II) complex. A typical example is 3,8-diacetyldeuteroporphyrin IX [**21**] (see above), which was also an intermediate in the Fischer's hemin total synthesis.

3.1.6. Coproporphyrin III [24]

Coproporphyrin III [**24**] is a biologically significant porphyrin because its hexahydroderivative, coproporphyrinogen III [**25**], is a colorless intermediate on the pathway between uroporphyrinogen III [**26**], protoporphyrinogen IX [**27**], and protoporphyrin IX [**1**] in normal porphyrin metabolism. Under normal circumstances, the amount of [**25**] present at steady state is small. However, biological oxidation of coproporphyrinogen III yields the colored coproporphyrin III, which takes it out of the normal metabolic sequence. Hence, certain diseases of porphyrin metabolism can result in a buildup of excess photochemically active porphyrins in tissues; such diseases are known collectively as porphyrias. Chemically, porphyrinogens can be oxidized very efficiently to porphyrins by use of 2,3-dichloro-5,6-dicyanobenzoquinone (DDQ). If biosynthetic work using porphyrinogens is to be carried out, the corresponding porphyrin can usually be reduced to porphyrinogen using sodium amalgam or catalytic hydrogenation (15). When vinyl groups are present on the porphyrin macrocycle, of course, only the sodium amalgam route is recommended—catalytic hydrogenation will probably reduce the vinyls to ethyls. It must be kept in mind when handling porphyrinogens, that oxygen and light can efficiently oxidize the hexahydro material to the porphyrin level, which will make it inactive in biosynthetic investigations—the first true porphyrin in porphyrin biosynthesis is protoporphyrin IX itself.

3.2 Porphyrins and Chlorins from Plants and Algae

In this section, some simple degradation reactions, which furnish porphyrins and chlorins in useful quantities from plants and algae, will be described.

The traditional source for chlorophylls *a* [**12**] and *b* [**13**], usually present in a ratio of about 3:1, was leaf tissue, usually spinach (25,68). A very useful chemical adjunct for simplification of the mandatory chromatographic separation of the chlorophyll *a* and *b* pigments has been reported (41); it employs the Girard reagent T as a means of dramatically increasing the polarity of the series *b* component in the mixture. For example, reaction of methyl pheophorbide *a* [**28**] and *b* [**29**] mixture (see above) with Girard's reagent T gives a mixture consisting of unreacted *a* series compound, i.e., methyl pheophorbide *a* [**28**], and the salt [**30**] from the *b* series. Column chromatography then achieves a very simple separation in which [**30**] remains adsorbed to the top of the column, whereas the relatively nonpolar *a* series compound [**28**] is eluted quickly. Use of a polar solvent then elutes the *b* series salt, which can be hydrolyzed to give pure methyl pheophorbide *b* [**29**].

Investigators wishing only to deal with chlorophyll derivatives in the *a* series were advantaged when it was shown that *Spirulina maxima* (from Mexico) or *S. pacifica* (from Hawaii) contain only the chlorophyll *a* series of pigments. On account of the fairly drastic extraction conditions, chlorophyll *a* itself is usually not obtained directly from the alga, but large quantities of pheophytin *a* [**31**] and methyl pheophorbide *a* [**28**] (up to 0.4% measured by dry weight) can be obtained (67).

Treatment of the plant chlorophylls (either separately or as a mixture) with mild acid gives the metal-free pheophytins *a* [**31**] and *b* [**32**]; this, as a dried paste, is usually the form in which the pigments are stored prior to further degradation to useful materials. Hydrolysis of the pheophytins gives the corresponding pheophorbides *a* [**33**] and *b* [**34**]; (note that the pheophorbides still contain one ester, and that hydrolysis of this ester will cause concomitant decarboxylation on ring E). Alternatively, and preferably (for ease of handling), methanolysis of pheophytin *a*

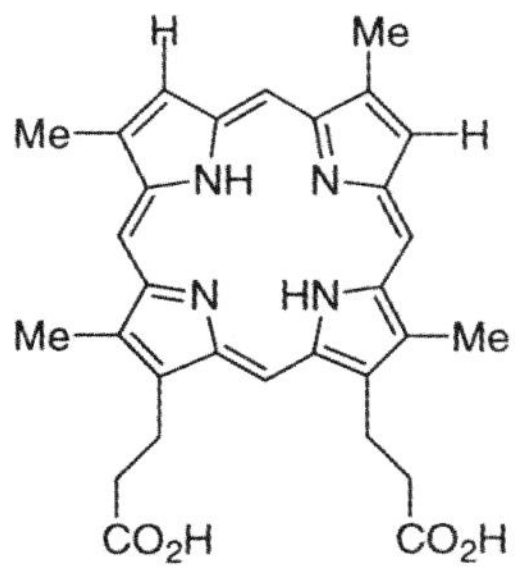

(**23**) (deuteroporphyrin IX)

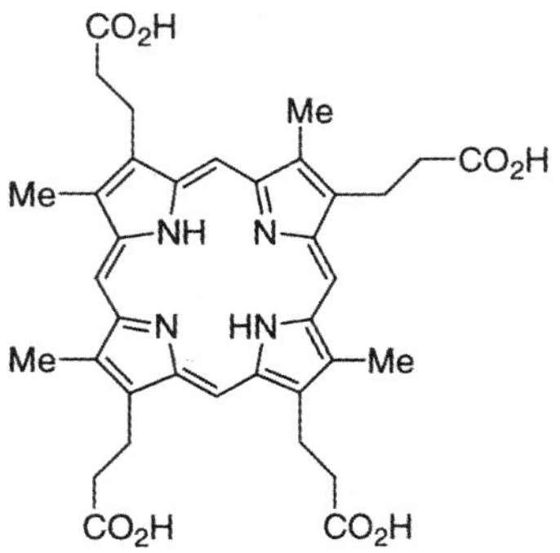

(**24**) (coproporphyrin III)

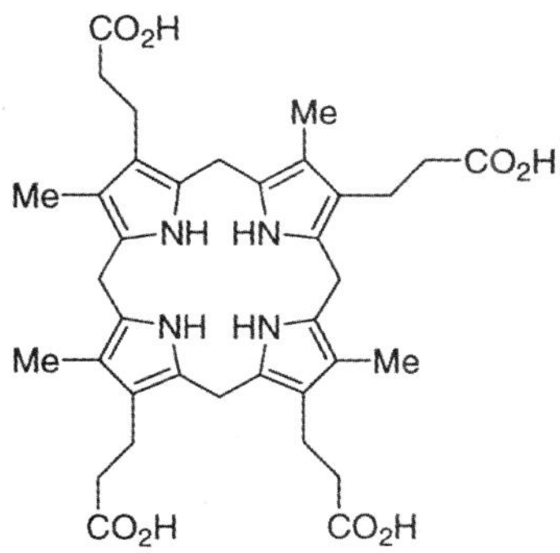

(**25**) (coproporphyrinogen III)

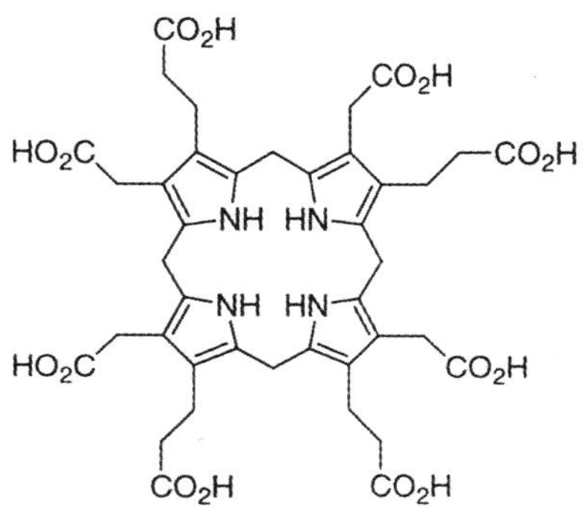

(**26**) (uroporphyrinogen III)

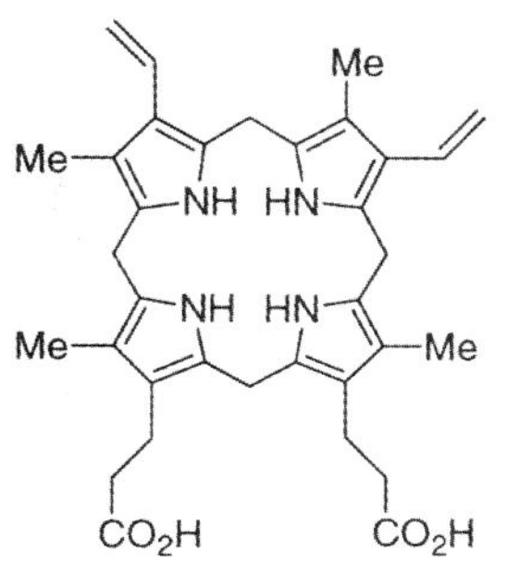

(**27**) (protoporphyrinogen IX)

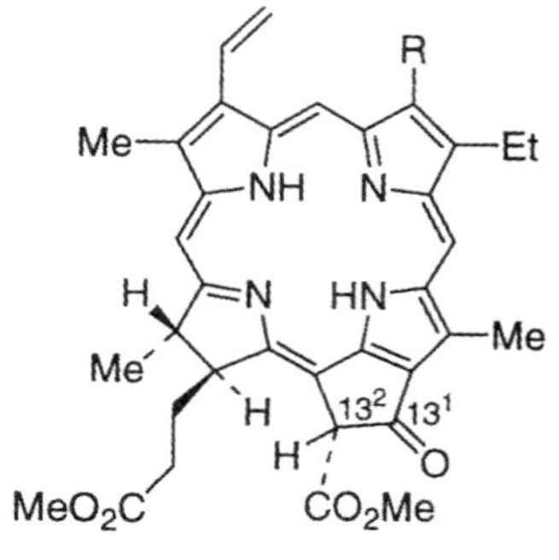

(**28**) R = Me (methyl pheophorbide *a*)
(**29**) R = CHO (methyl pheophorbide *b*)

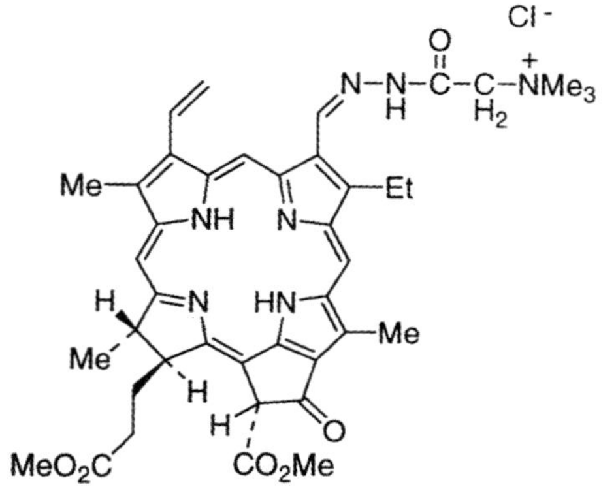

(**30**)

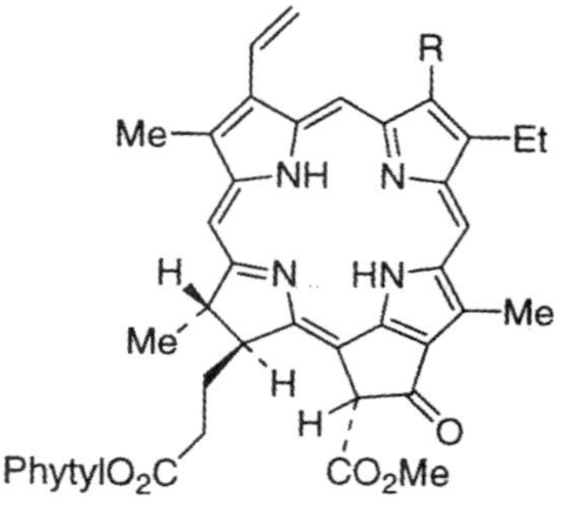

(**31**) R = Me (pheophytin *a*)
(**32**) R = CHO (pheophytin *b*)

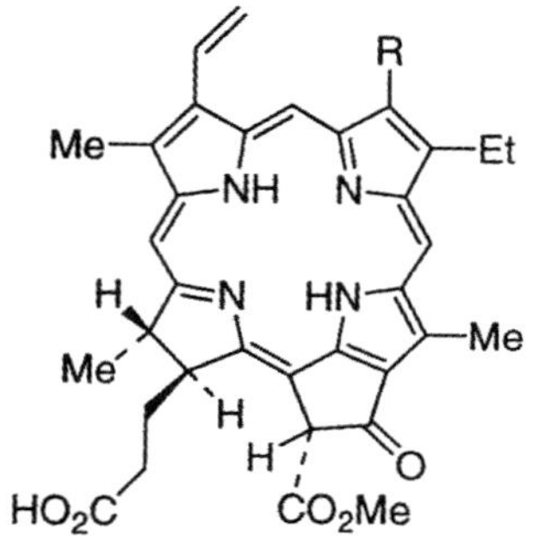

(**33**) R = Me (pheophorbide *a*)
(**34**) R = CHO (pheophorbide *b*)

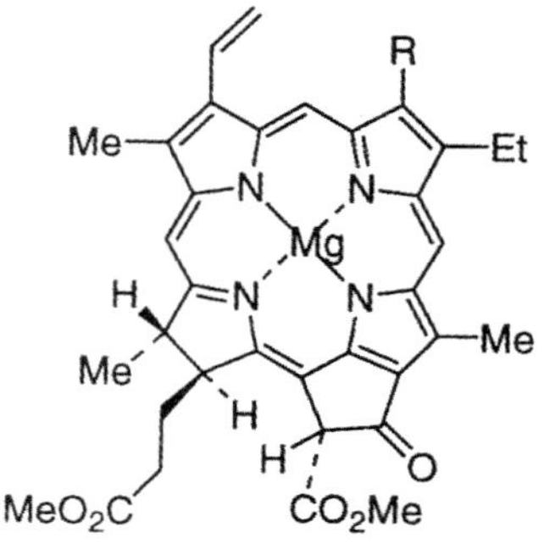

(**35**) R = Me
(**36**) R = CHO

or *b* provides the corresponding methyl pheophorbides *a* or *b* [**28** or **29**, respectively]—these contain two methyl esters. Transesterification of the phytyl ester for methyl, without removal of the magnesium atom, can be accomplished to afford the methyl chlorophyllides [**35**] and [**36**] (26).

A number of simple to perform but mechanistically complex reactions can be carried out on chlorophyll derivatives. For example, oxidation of pheophytin *a* [**31**] under highly alkaline conditions accomplishes cleavage the 13^1-13^2 bond in the β-ketoester ring E, with hydrolysis of the of phytyl ester, to give Fischer's "unstable chlorin" [**37**] (28). Simple evaporation of the solution affords the so-called purpurin 18 [**38**], which bears a very useful anhydride ring [**45**]. On the other hand, diazomethane esterification (CAUTION) yields purpurin 7 trimethyl ester [**39**] (26–28,45). Heating of [**39**] in collidine gives a diversely substituted porphyrin, 3-vinylrhodoporphyrin XV dimethyl ester [**40**] (28). If the so-called "meso" (i.e., 3-ethyl instead of 3-vinyl) series of pigments is used, another porphyrin, rhodoporphyrin XV dimethyl ester [**41**], is obtained.

The isocyclic ring (E) in chlorophylls and their derivatives contains a β-ketoester function which imparts a high degree of chemical reactivity upon the compounds containing it. Such lability is often a disadvantage in the use of chlorophyll derivatives

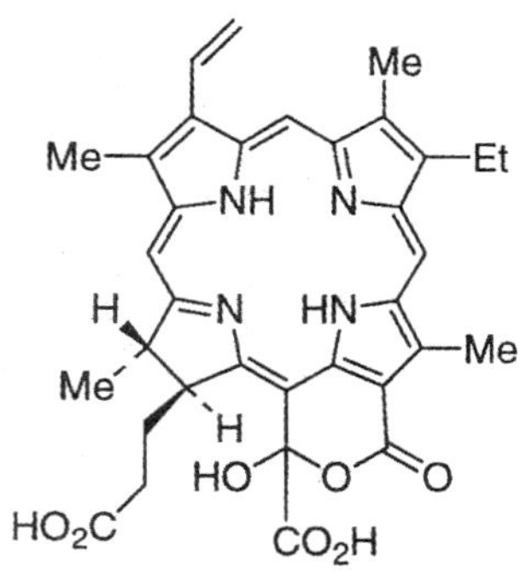

(**37**) ("unstable chlorin")

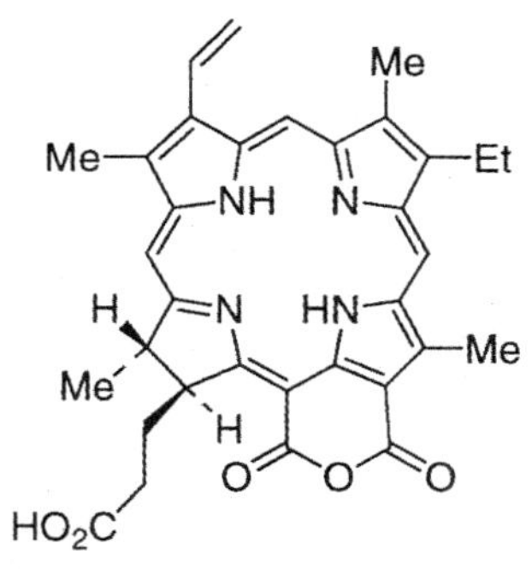

(**38**) (purpurin 18)

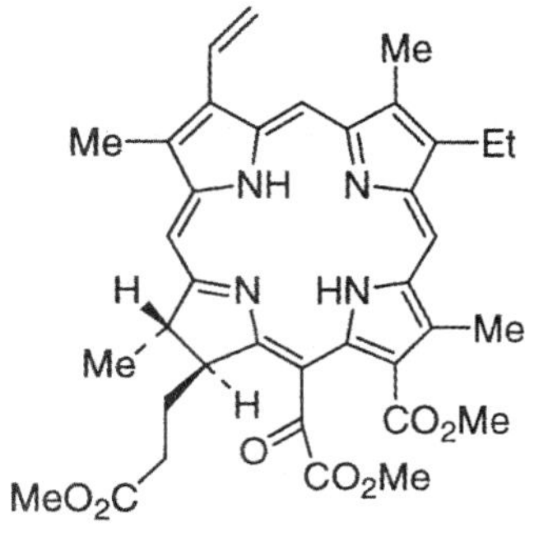

(**39**) (purpurin 7 trimethyl ester)

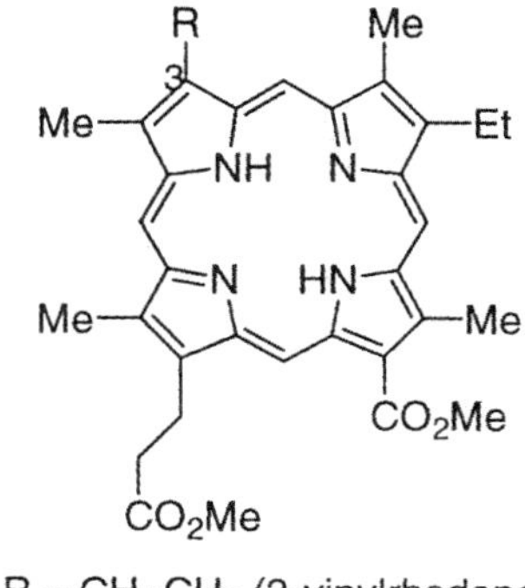

(**40**) R = $CH{=}CH_2$ (3-vinylrhodoporphyrin XV dimethyl ester)
(**41**) R = Et (rhodoporphyrin XV dimethyl ester)

for specific purposes; the spectrum of chemical reactivity in the ring E portion of the pigments can be dramatically decreased by removal of the 13^2-CO_2Me group. When the 13^2-CO_2Me group is removed, the so-called "pyro" series of chlorophyll derivatives are obtained. Basically, ketones are much less reactive than are conjugated ketoesters. Thus, heating of methyl pheophorbide *a* [**28**] (or *b* [**29**]) in collidine (30) gives methyl pyropheophorbide *a* [**42**] (or *b* [**43**]) in virtually quantitative yield; use of collidine is a yield-enhancing improvement upon the classical method (28) which uti-

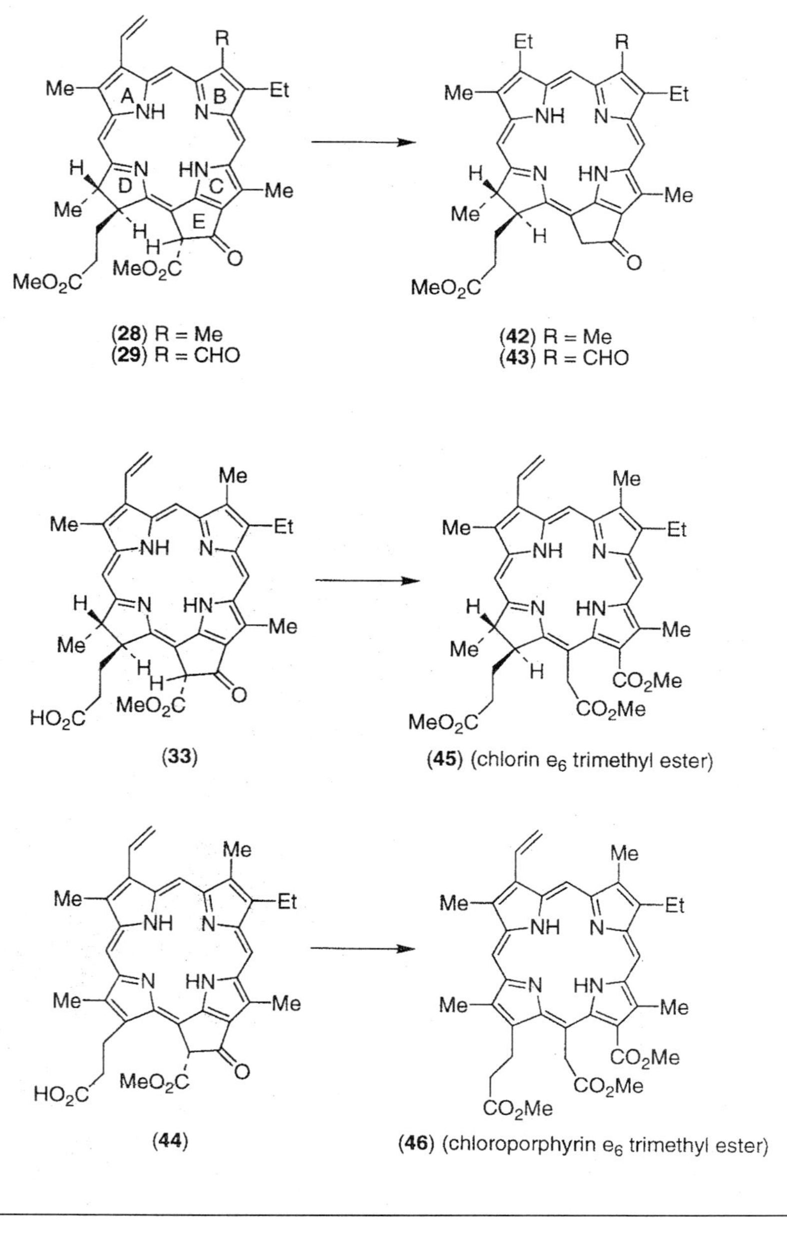

lized pyridine. Identical demethoxycarbonylation reactions take place with the so-called meso- (3-ethyl) series of compounds.

The 5-membered isocyclic ring in the pyro-series of chlorophylls cannot be cleaved, but the ring E in its β-ketoester form can be readily opened since the highly reactive conjugated functionality provides a handle for chemical elaboration of ring E. For example, pheophorbide a [**33**] and 3-vinylpheoporphyrin a_5 [**44**, vide infra] can be treated with alkali to give, after esterification with diazomethane, chlorin e_6 trimethyl ester [**45**] and chloroporphyrin-e_6 trimethyl ester [**46**], respectively (24). Methanolysis of pheophorbide *a* also affords [**45**]. This reaction can be reversed, and ring E is reformed either by treatment with methoxide (24), with *tert*-butoxide (65), or best of all using triphenylphosphine and bis(trimethylsilyl) amide (31).

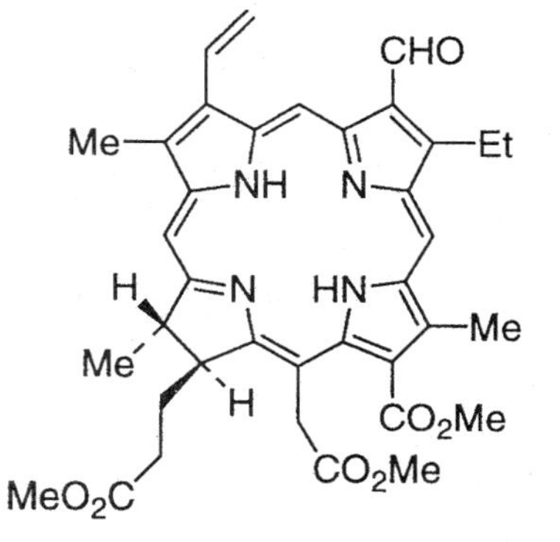

(**47**) (rhodin g_7 trimethyl ester)

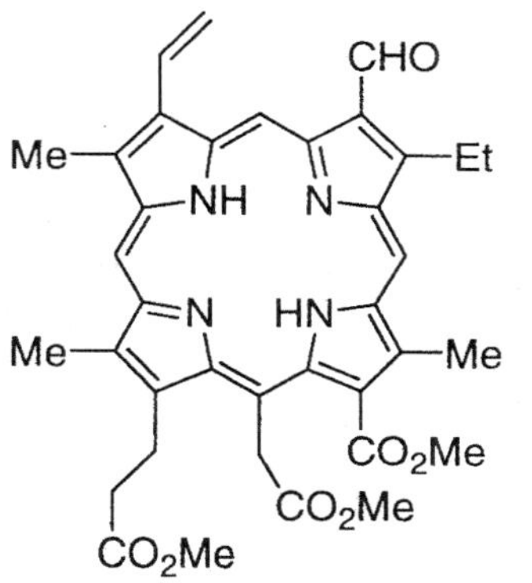

(**48**) (rhodinporphyrin g_7 trimethyl ester)

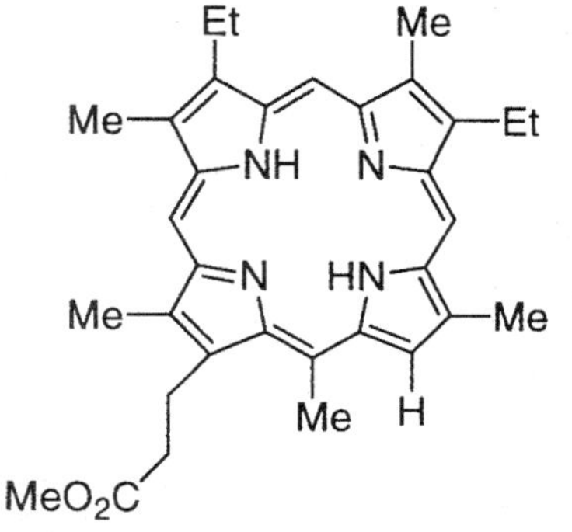

(**49**) (phylloporphyrin XV methyl ester)

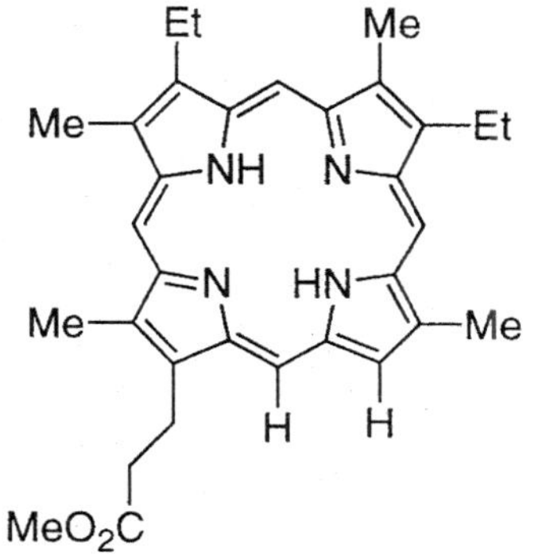

(**50**) (pyrroporphyrin XV methyl ester)

Although the chlorophyll *b* series of pigments is less accessible than those from chlorophyll *a* (and indeed, as mentioned above, *Spirulina* algae contains no chlorophyll *b*) a series of reactions parallel to those described above also occurs in the *b* series; the analogue of chlorin e_6 trimethyl ester in the *b* series is called rhodin g_7 trimethyl ester [**47**] and of chloroporphyrin e_6 trimethyl ester is rhodinporphyrin g_7 trimethyl ester [**48**].

Chlorins can be converted into porphyrins by using DDQ as a dehydrogenation agent. The β-ketoester functionality does not take kindly to oxidative stress, so methyl pheophorbide *a* [**28**] gives only a low yield of 3-vinylpheoporphyrin a_5 dimethyl ester [**44**]. Using a "sledgehammer" approach to preparation of porphyrins from chlorophyll derivatives, chlorophyll *a* under very vigorous basic conditions followed by esterification (methanolysis), affords phylloporphyrin XV methyl ester [**49**] and pyrroporphyrin XV methyl ester [**50**] (23).

❖ Procedure 1. Isolation of Methyl Pheophorbide *a* [28] from *S. maxima* (67)

1. About 500 g of dried *S. maxima* alga is slurried in 2 L of acetone, and then liquid nitrogen is added to form a frozen slush.
2. After transferring to a 5-L 3-necked round-bottom flask, the mixture is heated at reflux with mechanical stirring for 2 hours. The supernatant is filtered through a Whatman filter paper (Whatman, Clifton, NJ, USA) using a Buchner funnel, and more acetone is added to the solid debris.
3. The extraction process, with refluxing, is repeated twice more—note that the debris retains its deep green color, but amounts of additional chlorophyll obtained are marginal.
4. The green filtrate is evaporated and then purified by flash chromatography on Grade V neutral alumina, eluting first with n-hexane to remove a fast running yellow band, with dichlormethane to remove the major blue-gray pheophytin *a* band, and finally with 97:3 dichloromethane:tetrahydrofuran to remove some bright green magnesium-containing pigments.
5. The evaporated pheophytin *a* fraction is treated with 500 mL of 5% sulfuric acid in methanol (degassed by bubbling with nitrogen gas) for 12.5 hours at room temperature in the dark (aluminum foil) under nitrogen, followed by dilution with dichloromethane, and rinsing with water.
6. The mixture is rinsed with 10% saturated aqueous sodium bicarbonate, the organic layer is dried over anhydrous sodium sulfate, followed by evaporation and crystallization of the residue from dichloromethane:methanol. This gives methyl pheophorbide *a* [**28**] (average yield 1.8 g).

4. CHEMICAL SYNTHESES OF PORPHYRINS

Porphyrin chemical synthesis will be discussed here in connection with two series of compounds: *(i)* those porphyrins which have been most often used in connection with model studies, e.g., 5,10,15,20-tetraphenylporphyrin (TPP) [**51**] and 2,3,7,8,12,13,17,18-octaethylporphyrin (OEP) [**52**]; and *(ii)* those related to protoporphyrin IX [**1**]. Simply based on the symmetry in the substituent arrays of [**51**] and [**52**] and the lack of symmetry in [**1**], it is obvious that it would be a waste of time to approach the synthesis of both series of compounds using the same strategy. To attempt the synthesis of OEP [**52**] by laborious multistep construction of an open-chain tetrapyrrolic intermediate would be plainly unwise—such symmetrically substituted compounds are most efficiently obtained by tetramerization of a suitable monopyrrole (see below). On the other hand, there is no way (in the absence of enzymes) that protoporphyrin IX [**1**] can be synthesized chemically by monopyrrole chemical self-condensation, so a more sophisticated chemical approach is essential. As it happens, porphyrins [**51**] and [**52**] can be synthesized by self-condensation of monopyrroles, while protoporphyrin IX [**1**] can be accessed by a number of routes, the most simple (and the one to be used as an example in this chapter) being from dipyrroles.

4.1. Syntheses of Pyrroles

For both series of compounds mentioned above, it is first essential to synthesize monopyrroles. Pyrrole itself [**53**] is commercially available. Syntheses of two common examples of useful pyrroles (from the many dozens in the literature) (5,20,32,37,38) will be illustrated here.

Pyrroles bearing peripheral substituents are those which are most useful for application to dipyrrole and porphyrin synthesis. The Johnson–Kleinspehn synthesis (11,43) is perhaps the most useful for tetrasubstituted pyrroles. For example, pyrrole [**54**], bearing very useful methyl and propionate groups, is prepared by the reaction of dione [**55**] with benzyl oximinoacetoacetate [**56**]—compound [**56**] is in turn obtained by the reaction of benzyl acetoacetate [**57**] with sodium nitrite in the presence of acetic acid. Slow admixture of equimolar amounts of [**55**] and [**56**] and excess zinc powder and sodium acetate in hot acetic acid results in reduction of the oximinoderivative [**56**] to the amine, followed by in situ condensation with [**55**] to give pyrrole [**54**]. Simply pouring the cooled reaction mixture into iced water causes precipitation of the product pyrrole. The reaction works with a variety of substituents on the central (i.e., 2-) carbon of the 1,3-dione and with a variety of esters on the acetoacetate. The reaction described above (using acetoacetates) is the Johnson version, while the Kleinspehn modification employs oximinomalonic esters in place of the acetoacetates.

Compared with the above synthesis of pyrroles, methodology for preparing pyrroles such as [**58**] is relatively new. A major advance in the field was made when the Barton–Zard pyrrole synthesis was reported (8); the importance of this route was related to the substituent patterns which could be accessed using it. Thus, treatment of a nitroalkene [**59a**] or its synthetic precursor, an acetoxynitroalkane [e.g., **59b**], with an isocyanoacetate [e.g., **60**] affords excellent yields of pyrroles such as [**58**].

❖ Procedure 2. Synthesis of Ethyl 3,4-Diethylpyrrole-2-Carboxylate [58] (55)

1. A mixture of 3-acetoxy-4-nitrohexane [**59b**] (8)(16.3 g), ethyl isocyanoacetate [**60**] (9.8 g; Sigma, St. Louis, MO, USA), and 1,8-diazabicyclo[5.4.0]undec-7-ene (26.4 g; Sigma) in tetrahydrofuran (100 mL) is stirred at 20°C for 12 hours.
2. The mixture is poured into water containing 1 M HCl and then extracted with ethyl acetate.
3. The extracts are washed with water and dried over anhydrous magnesium sulfate.
4. After evaporation of the solvents, the residue is chromatographed on a column of silica gel eluted with hexane: dichloromethane mixtures.

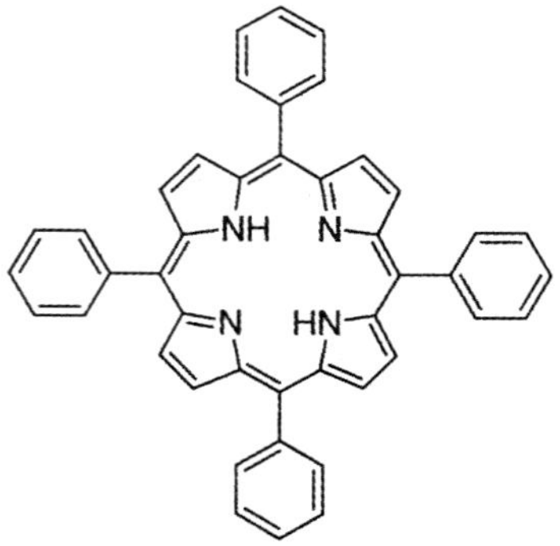

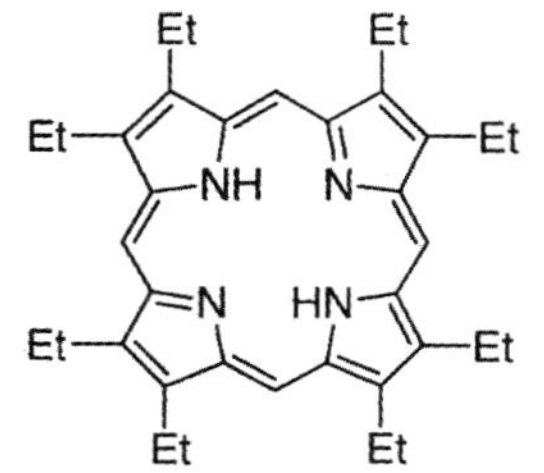

(**52**) (2,3,7,8,12,13,17,18-octaethylporphyrin "OEP")

(**51**) (5,10,15,20-tetraphenylporphyrin, "TPP")

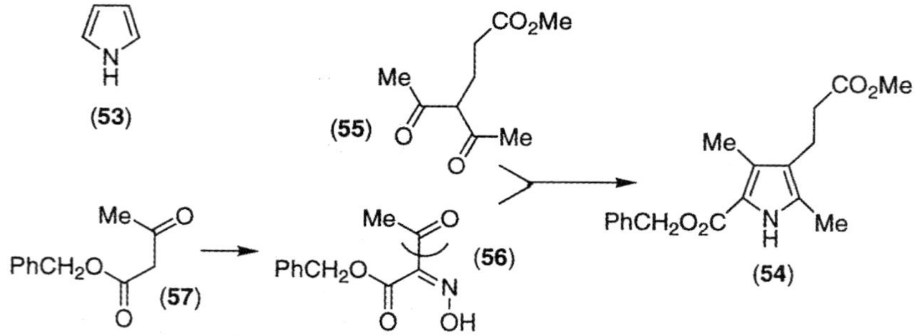

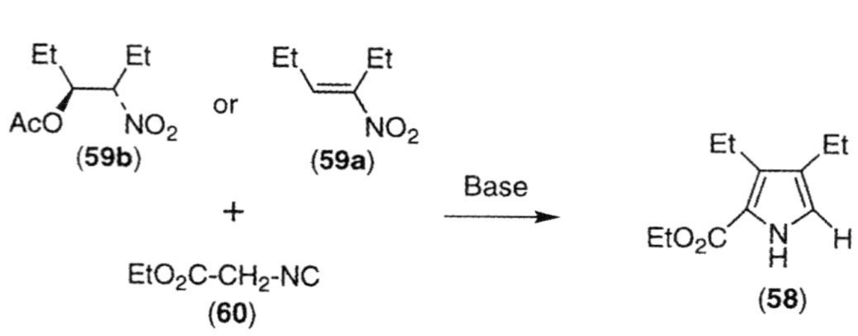

5. Evaporation of the eluates containing the red band will give the required pyrrole [**58**], with an average yield of 14.2 g.

4.2. Syntheses of Dipyrromethanes

Unsymmetrically substituted dipyrromethanes, [e.g., **61**], can be prepared by condensation of 2-acetoxymethylpyrroles [**62**] with 2-unsubstituted pyrroles [**63**] in acetic acid containing a catalytic amount (<0.1 equiv.) of toluene *p*-sulfonic acid (13). Montorillonite K-10 clay has also been shown to be a very useful acid catalyst in dipyrromethane syntheses (30,36); the advantage of using the clay is that it can be removed simply by filtration after the reaction is complete. Compound [**62**] is obtained from the corresponding methylpyrrole [**64**] simply by treatment with lead tetra-acetate. Note that pyrrole [**64**] possesses the same "symmetry pattern" as does pyrrole [**54**]; its synthesis is relatively straightforward (21). Pyrrole [**63**] can be obtained from [**64**] by following a sequence of reactions as shown in Scheme 1. The dipyrromethane [**61**] will form rings A and B of protoporphyrin IX dimethyl ester [**2**] in the synthesis, which will be described later. The 1- and 9-car-

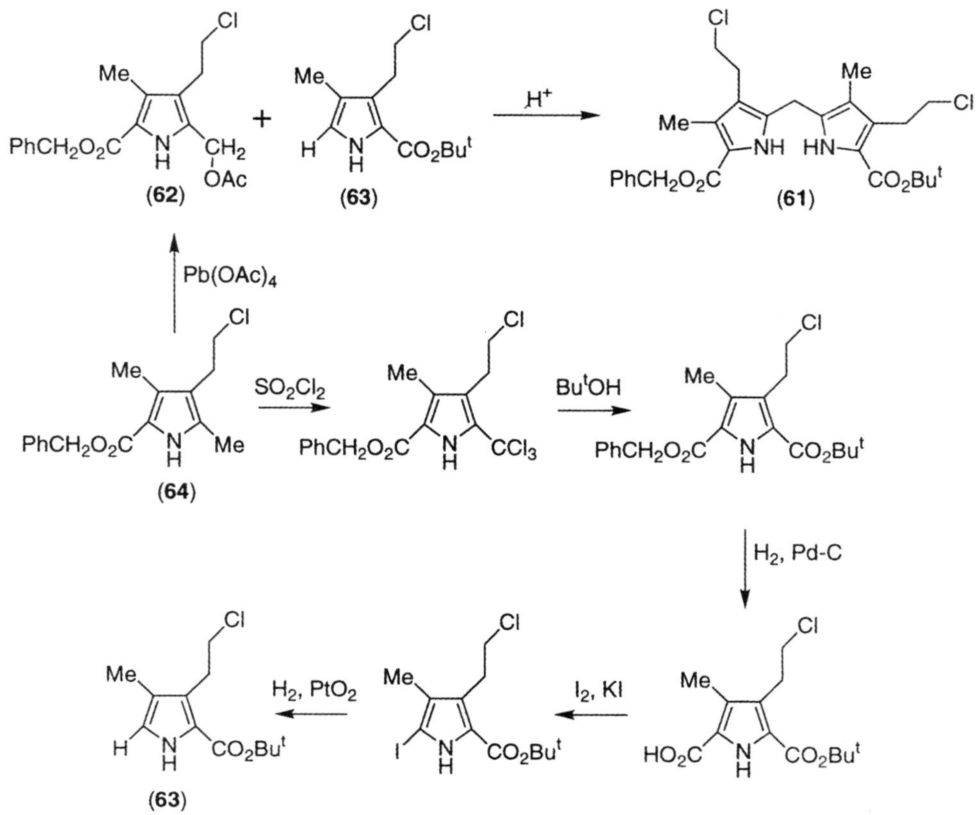

Scheme 1

boxylate substituents are differentially protected, and the future vinyl groups are protected as 2-chloroethyls.

Symmetrically substituted dipyrromethanes [e.g., **65**] are best prepared in one step by self-condensation of bromomethylpyrroles [e.g., **66**] in hot methanol (21), or by heating 2-acetoxymethylpyrroles [e.g., **67**] in methanol-hydrochloric acid (50). The 1- and 9-benzyl esters can be cleaved using catalytic hydrogenation with hydrogen gas and 5% (or 10%) palladium–carbon as catalyst. The resulting 1,9-dicarboxylic acid [**68**] can then be formylated using the Vilsmeier reagent (phosphoryl chloride or benzoyl chloride mixed with equimolar amounts of dimethylformamide) to give [**69**]. The 1- and 9-formyl groups serve as the bridging carbons in the MacDonald porphyrin macrocyclization, which will be described later.

4.3. Porphyrins via Monopyrrole Tetramerization

By definition, tetramerization of monopyrroles must result either in a single symmetrically substituted porphyrin (if the 3- and 4-substituents are identical) or in a mixture of porphyrins (if the 3- and 4-substituents are different—see later). By far, the easiest way to prepare a porphyrin involves the reaction of pyrrole [**53**] with benzaldehyde. The product is the almost legendary TPP [**51**]. This simple route was first reported by Rothemund (57,58) and, after modification by Adler, Longo and colleagues (involving use of refluxing propionic acid instead of sealed tube chemistry) (1), was finally optimized as a two-step procedure by Lindsey's group (48). The nonexpert procedure that is easiest to follow for the synthesis of [**51**] involves addition of equimolar amounts of crude (undis-

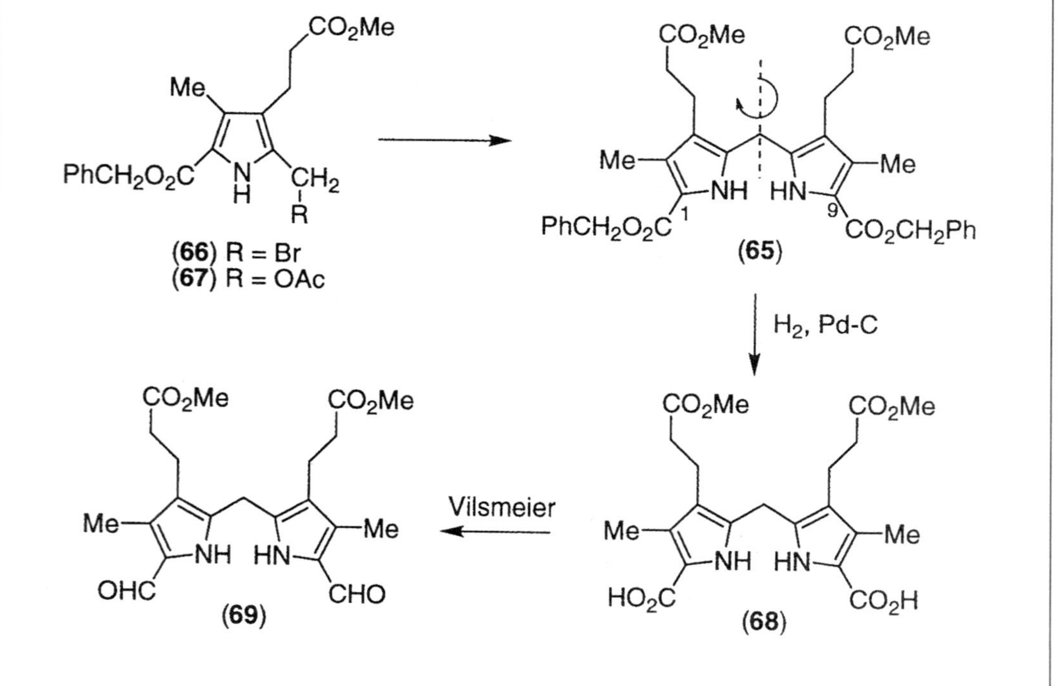

tilled) pyrrole [53] and benzaldehyde to refluxing propionic acid. After heating for about 30 minutes, the mixture is allowed to cool, and the TPP is filtered off, usually in 20% to 22% yield. The propionic acid can be recovered by distillation and then reused. Higher yields of TPP can be obtained by use of more elaborate and expensive chemistry, but, for TPP, quick and dirty seems to work well. The product

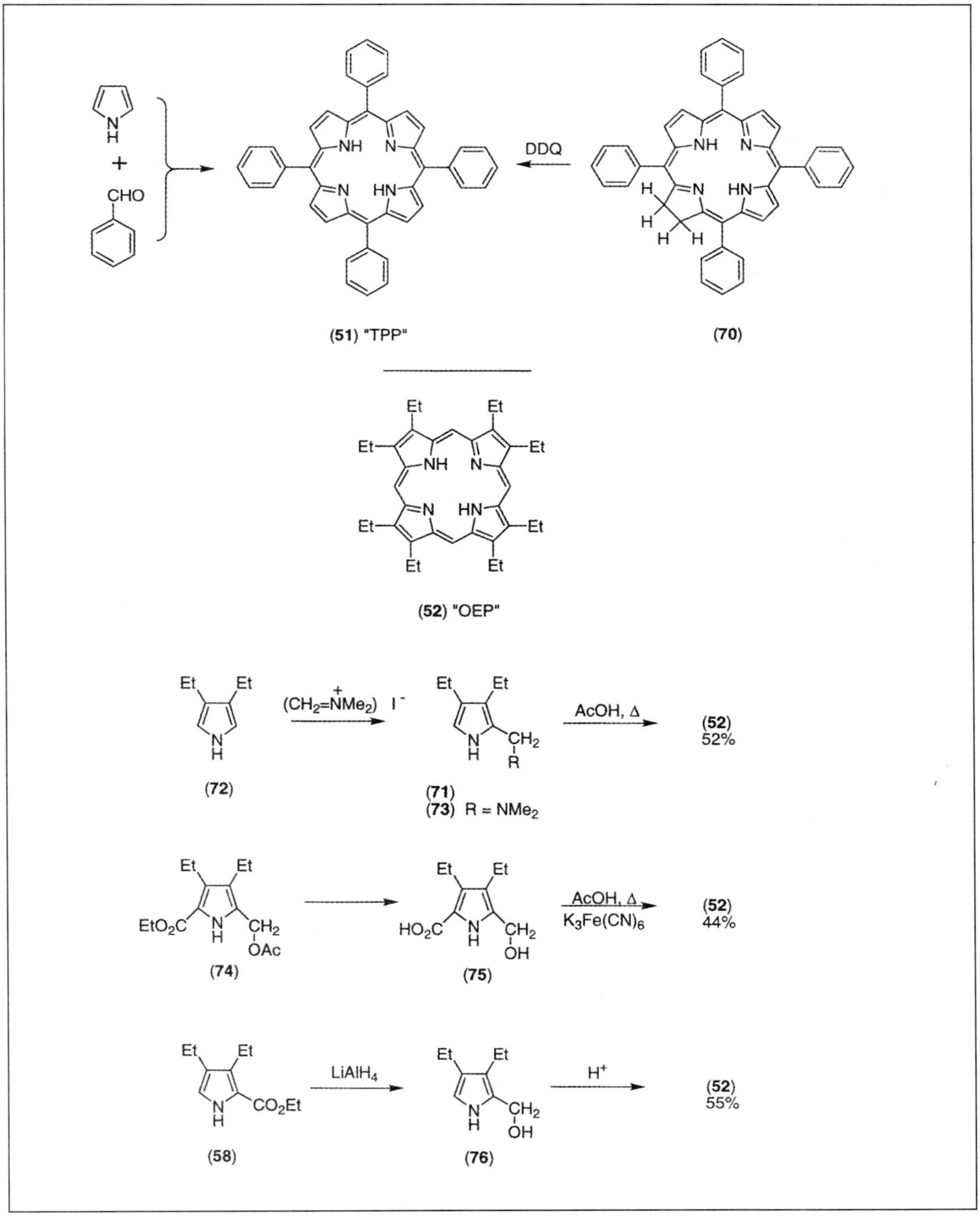

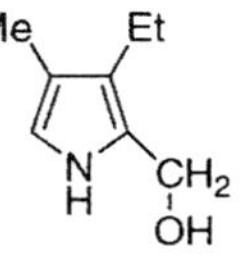

(77)

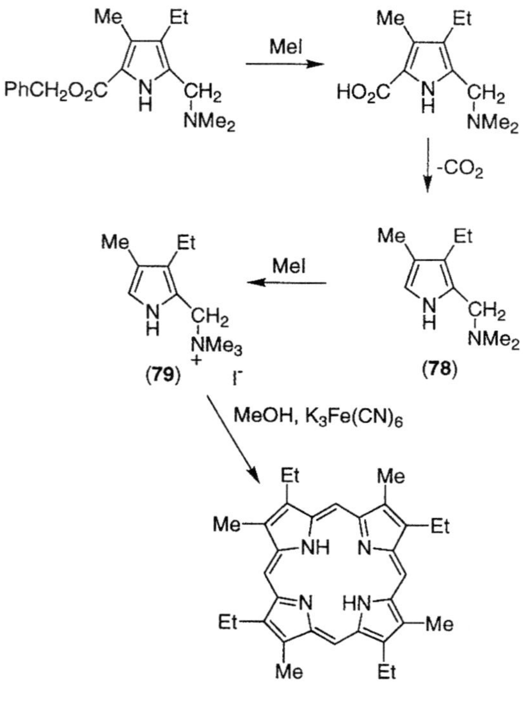

etioporphyrin I
(6)

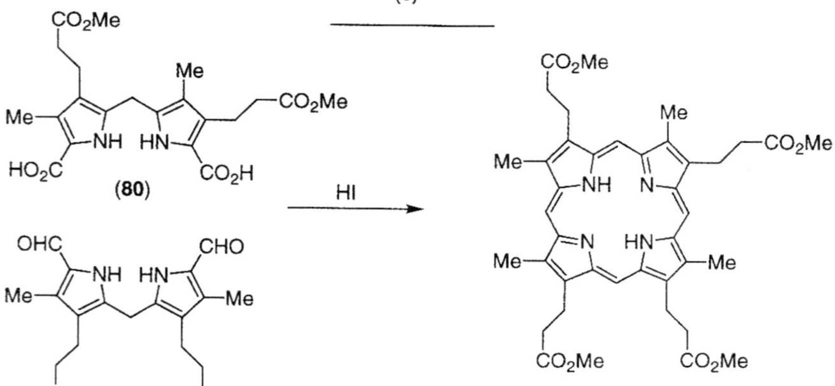

(69)

coproporphyrin III tetramethyl ester

(81)

from the Rothemund and Adler–Longo (propionic acid) approaches is somewhat impure (though highly crystalline) and contains (4) about 5% or less of meso-tetraphenylchlorin [**70**]. Brief treatment (6) of the crude product with DDQ accomplishes transformation of [**70**] into [**51**]; earlier methodology involved the separation of these two components on a chromatography column, but transformation of [**70**] into [**51**] instead of separation of [**51**] from [**70**] is much more sensible. Using this kind of methodology, literally kilograms of TPP can be prepared. TPP can, in any case, be purchased either "chlorin-free" or crude. Additionally, with only relatively few exceptions, the reaction tolerates substitution of other arylaldehydes for benzaldehyde, and good yields of a variety of tetra-arylporphyrins can be obtained (47).

❖ Procedure 3. Synthesis of Chlorin-Free TPP [51] (6)

1. Benzaldehyde (66.5 mL) and pyrrole (46.5 mL) are simultaneously added to refluxing propionic acid (2.5 L), and the mixture is refluxed for a further 30 minutes before being allowed to cool overnight to room temperature.
2. The crude TPP is filtered off, washed with hot water, and then washed with methanol until the filtrate is colorless, to give 20.4 g (20% yield) of purple glistening crystals.
3. Concentration of the propionic acid filtrate affords a second crop of crystals.
4. The crude TPP (20 g) is dissolved in refluxing ethanol-free chloroform (2.5 L) before addition of DDQ (5 g) in dry toluene (150 mL).
5. The mixture is refluxed for 3 hours before filtration of the yellowish solution, under suction, through a sintered glass funnel containing Grade I alumina (300 g).
6. The alumina is washed with dichloromethane (200 mL), and the combined filtrates are concentrated to approximately 200 mL before addition of 200 mL of methanol.
7. Filtration results in the chlorin-free product as glistening purple crystals, with an average yield of 19.2 g.

Approaches to so-called octaalkylporphyrins (such as OEP [**52**]) are a little more complicated, but only with regard to the difficulty of preparing the pyrrole starting materials. In this case, the future meso- (i.e., interpyrrolic) carbons can either be present already on the pyrrole, or as in the case with TPP (wherein the meso-carbons were provided by the formyl carbon in benzaldehyde), the meso-carbons can be added separately from the pyrrole. A primary stricture, as mentioned above, is that a monopyrrole tetramerization approach can only be used to give structurally unique product if the substituents at positions 3- and 4- in the monopyrrole are identical.

Thus, fully symmetrical porphyrins such as octaethylporphyrin [**52**] can be prepared easily using two major routes. The first approach, which chronologically was developed first, involves the tetramerization of pyrroles [**71**] bearing 2-CH_2R substituents; the "R" group must be a good leaving group, and the methylene carbon of the 2-substituent will eventually be the source of the 5,10,15, and 20-carbons of the product porphyrin. After the condensation reaction, an oxidation step is necessary to afford good yields of symmetrical porphyrin. Useful examples for attachment of CH_2R groups to pyrroles are *(i)* the Mannich reaction of pyrrole [**72**] with formaldehyde and dimethylamine [or better, with commercially available (N,N-dimethylmethylene)ammonium iodide, Eschenmoser's reagent (56,59)] to give the 2-(N,N-dimethylaminomethyl)pyrrole [**73**]; heating of this in acetic acid gives a 52% yield of [**52**]

(18,70); and *(ii)* hydrolysis of the pyrrole [**74**] to give pyrrole [**75**] which is tetramerized to give [**52**] in 44% yield by heating in acetic acid containing potassium ferricyanide (35,63). Most recently, the Barton–Zard pyrrole synthesis (see above) (8) has greatly simplified preparative approaches to monopyrroles of the type [**58**]; lithium aluminum hydride (CAUTION: reacts violently with moisture) reduction, followed by tetramerization of the resulting pyrrole-2-carbinol [**76**] under acidic conditions, gives [**52**] in 55% yield (2,55).

❖ Procedure 4. Synthesis of OEP [52] from Pyrrole [58] (54)

1. Ethyl 3,4-diethylpyrrole-2-carboxylate [**58**] (see above, 657 mg) is added dropwise at 0°–5°C to a stirred solution of lithium aluminum hydride (320 mg; Sigma) in dry tetrahydrofuran (15 mL). The mixture is stirred for 2 hours at 0°–5°C before destroying the excess lithium aluminum hydride by addition of ethyl acetate.
2. It is then poured into saturated aqueous ammonium chloride, extracted with ethyl acetate (3 times 10 mL), washed with aqueous sodium chloride, and dried over anhydrous magnesium sulfate.
3. The solution is evaporated to dryness under vacuum before addition of dichloromethane (15 mL).
4. To this solution is added dimethoxymethane (0.7 mL; Sigma) and toluene *p*-sulfonic acid (110 mg), and the mixture is stirred for 12 hours at room temperature. Aerial oxidation occurs under these conditions, but chloranil (Sigma) can also be used, although without any improvement in yield.
5. The mixture is washed with aqueous sodium bicarbonate, and the organic layer is dried over anhydrous magnesium sulfate.
6. Evaporation gives a residue which is chromatographed on silica gel, eluted with dichloromethane to give OEP [**52**], with an average 55% yield (240 mg).

Alternatively, tetramerization of 2,5-di-unsubstituted pyrroles [e.g., **72**] in the presence of reagents that can provide the four meso-methine carbons of the product can be used. Cyclization of 3,4-diethylpyrrole [**72**] with formaldehyde affords 55%–75% yields of OEP [**52**] (61).

If the 3- and 4-substituents on the monopyrrole component are not identical, mixtures will usually result due to acid catalyzed pyrrole ring scrambling reactions (49). Thus, acid catalyzed tetramerization of pyrrole [**77**] will result in production of a mixture of the four etioporphyrin type isomers [**6–9**]. However, a method has been devised which does produce only etioporphyrin I [**6**] from tetramerization of a pyrrole related to [**77**]; treatment of 2-(N,N-dimethylaminomethyl)pyrroles [e.g., **78**] with methyl iodide gives [**79**], which has a leaving group that is labile even under neutral conditions (i.e., no acid), which would cause pyrrole ring scrambling, is present (53,54). This, quaternized in methanol containing potassium ferricyanide (to accomplish rapid in situ oxidation of the labile porphyrinogen intermediate), gives a good yield of pure etioporphyrin I [**6**].

❖ Procedure 5. Synthesis of Etioporphyrin I [6] from Pyrrole [78]

1. Benzyl 4-ethyl-5-(N,N-dimethylaminomethyl)-3-methylpyrrole-2-carboxylate (53,54) (1.88 g) is dissolved in tetrahydrofuran (600 mL), and 10% palladium on carbon (500 mg; Sigma) is added.
2. The resulting mixture is stirred under hydrogen at room temperature for 12 hours before the catalyst is removed, and the solvent is evaporated to dryness.

3. Recrystallization from dichloromethane/hexane affords 4-ethyl-5-(N,N-dialkyl-aminomethyl)-3-methylpyrrole-2-carboxylic acid as an off-white powder in quantitative yield. Note that because of spontaneous decarboxylation at room temperature to give [**78**], this must be used immediately.
4. The pyrrole carboxylic acid (1.29 g) is dissolved in a solution of methanol (200 mL) and triethylamine (2 mL) and heated under reflux for 15 minutes.
5. Potassium ferricyanide (3.8 g) is added, and the reaction is continued at reflux for another 10 hours.
6. After removal of the solvent, the residue is redissolved in chloroform, the insoluble material filtered off, and the red solution is passed through a short column of silica gel (eluted with chloroform).
7. The solvent is evaporated, and the residual porphyrin is recrystallized from

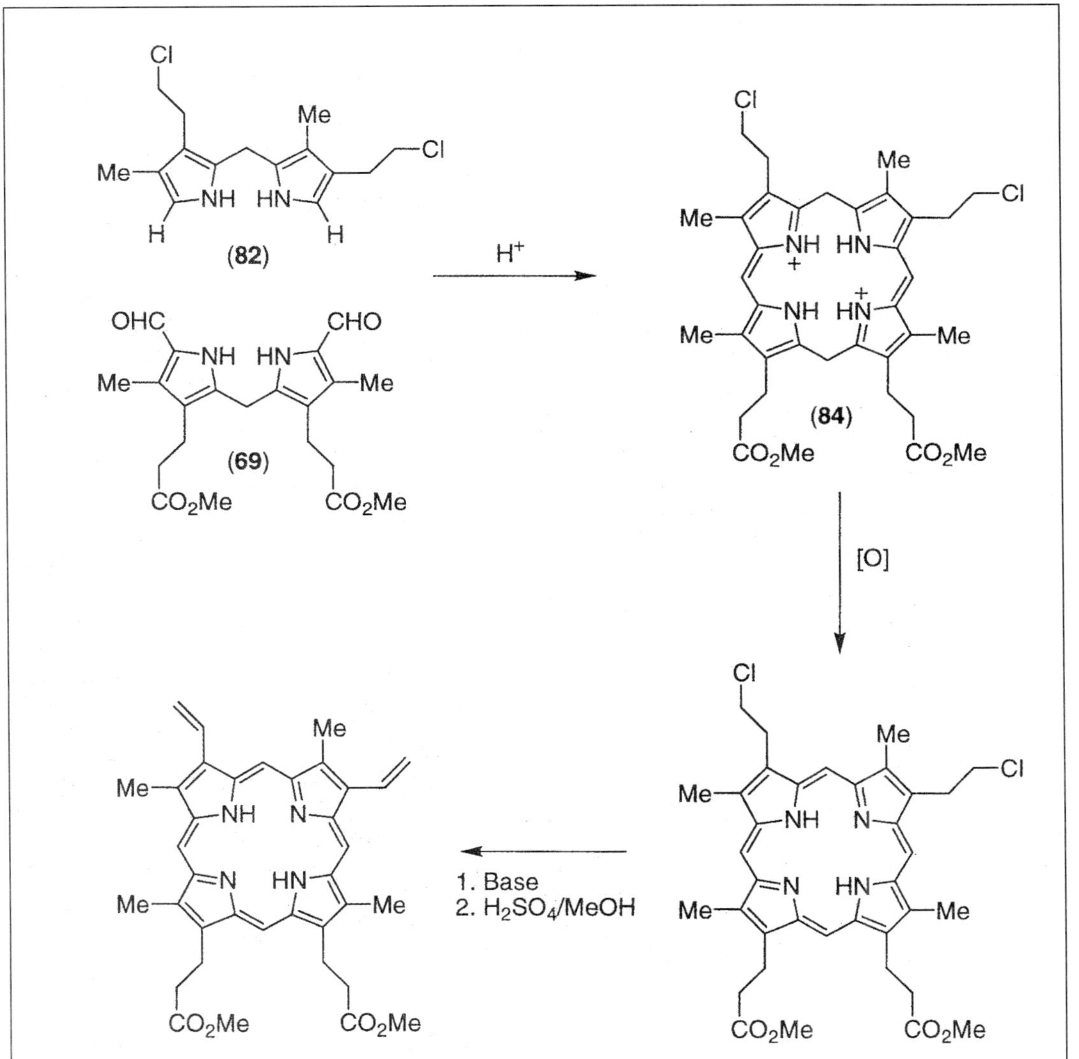

(**2**) (protoporphyrin IX dimethyl ester

dichloromethane/methanol to afford etioporphyrin-I in 36% yield (284 mg).

4.4. Porphyrins via Dipyrromethane Intermediates

If two dipyrromethane units with appropriate bridging carbons are condensed together, there are three possible products because the dipyrromethanes can either react with themselves or with each other. If the dipyrromethanes individually possess an unsymmetrical array of substituents, even greater mixtures can occur because there is no control over which end of one dipyrromethane reacts with the end of another. These symmetry limitations are common with all so-called "2 + 2" syntheses; in a porphyrin synthesis involving an A-B and a C-D dipyrromethane is to be condensed, the symmetry problems can be avoided if the A-B or C-D dipyrromethane unit is symmetrical about the interpyrrolic (5-) carbon atom. Arsenault, MacDonald, and coworkers showed (3) that a 1,9-diformyldipyrromethane, [e.g., **69**], can be condensed with a 1,9-di-unsubstituted dipyrromethane or its 1,9-dicarboxylic acid [**80**] in the presence of an acid catalyst to give pure porphyrin [e.g., **81**], often in high yields. MacDonald used hydriodic acid, but since that time, toluene *p*-sulfonic acid has been shown (14,15) to be a much better choice and more convenient too.

This 2 + 2 route using dipyrromethanes is probably the most widely used pathway to synthetic porphyrins. Thus, for example, treatment of dipyrromethane [**82**] (obtained from [**61**] by catalytic debenzylation followed by treatment with trifluoroacetic acid) with 1,9-diformyldipyrromethane [**69**] gives a good yield of porphyrin [**83**] after oxidation of the intermediate porphodimethene [**84**]; no mixtures are produced because both of the future linking meso-carbons are sited on the same dipyrromethane (preventing either of the two individual dipyrromethanes from reacting with themselves) and dipyrromethane [**69**] is symmetrical about its 5-carbon. Conversion of the 3,8-bis(2-chloroethyl)porphyrin [**83**] into protoporphyrin IX dimethyl ester [**2**] is accomplished simply by treatment with base (40)—just in case this base treatment also accomplished hydrolysis of the methyl esters, the product is then set aside in methanol containing 5% sulfuric acid (CAUTION: add the acid to the methanol, cooled and slowly). Workup and chromatography [NOTE: protoporphyrin is photolabile (see above), so the column should be run in the dark or with aluminum foil wrapped around it] then produces the product, [**2**].

It has been my intention to provide a summary of fairly simple procedures that can be used, given a certain competence in synthetic organic chemistry, to obtain by extraction or by total synthesis some useful porphyrins with defined symmetry and structural characteristics. But competence in organic chemistry is not easily earned. If the procedures still look too complex, or (more likely) if you do not have the basic laboratory equipment with which to carry out the procedures described, then the best bet is to collaborate. Just remember, most organic chemists would not know where to start if they needed to run a gel or if they needed to do a northern blot. They will want to collaborate also if they need these things.

ABBREVIATIONS

DDQ, 2,3-dichloro-5,6-dicyanobenzoquinone; OEP, 2,3,7,8,12,13,17,18-octaethylporphyrin; TPP, 5,10,15,20-tetraphenylporphyrin.

ACKNOWLEDGMENTS

Some of the work reported herein was

developed within my own research group. I would like to thank all of my collaborators, over the years, for their important contributions to our group effort; their names can be found in the various literature citations. I would also like to thank the National Institutes of Health (HL 22252) and the National Science Foundation (CHE 99-04076) for uninterrupted support of our efforts over the past 24 years.

REFERENCES

1. **Adler, A.D., F.R. Longo, J.D. Finarelli, J. Goldmacher, J. Assour, and L. Korsakoff.** 1967. A simplified synthesis for meso-tetraphenylporphyrin. J. Org. Chem. *32*:476.
2. **Aoyagi, K., T. Haga, H. Toi, Y. Aoyama, T. Mizutani, and H. Ogoshi.** 1997. Electron deficient porphyrins. III. Facile syntheses of perfluoroalkylporphyrins including water soluble porphyrin. Bull. Chem. Soc. Jpn. *70*:937-943.
3. **Arsenault, G.P., E. Bullock, and S.F. MacDonald.** 1960. Pyrromethanes and porphyrins therefrom. J. Am. Chem. Soc. *82*:4384-4389.
4. **Badger, G.M., R.A. Jones, and R.L. Laslett.** 1964. Porphyrins. VII. The synthesis of porphyrins by the Rothemund reaction. Aust. J. Chem. *17*:1028-1035.
5. **Baltazzi, E. and L.I. Krimen.** 1963. Recent advances in the chemistry of pyrrole. Chem. Rev. *63*:511.
6. **Barnett, G.H., M.F. Hudson, and K.M. Smith.** 1975. Concerning *meso*-tetraphenylporphyrin purification. J. Chem. Soc. [Perkin 1] 1401-1403.
7. **Barrett, J.** 1959. Detection of hydroxyl groups in porphyrins and chlorins. Nature *183*:1185-1186.
8. **Barton, D.H.R., J. Kervagoret, and S.Z. Zard.** 1990. A useful synthesis of pyrroles from nitroolefins. Tetrahedron *46*:7587-7598.
9. **Battersby, A.R. and E. McDonald.** 1975. Biosynthesis of porphyrins, chlorins, and corrins. *In* K.M. Smith (Ed.), Porphyrins and Metalloporphyrins. Elsevier, Amsterdam.
10. **Bonnett, R., I.H. Campion-Smith, and A.J. Page.** 1977. Protiodevinylation: the Schumm reaction of vinylporphyrins. J. Chem. Soc. [Perkin 1] 68-71.
11. **Bullock, E., A.W. Johnson, E. Markham, and K.B. Shaw.** 1958. A synthesis of coproporphyrin III. J. Chem. Soc. 1430-1440.
12. **Burbidge, P.A., G.L. Collier, A.H. Jackson, and G.W. Kenner.** 1967. Syntheses and spectra of some meso-methylated porphyrins. J. Chem. Soc. B 930-937.
13. **Cavaleiro, J.A.S., A.M.d'A.R. Gonsalves, G.W. Kenner, and K.M. Smith.** 1973. Synthesis of pyrromethanes and a tripyrrane. J. Chem. Soc. [Perkin 1] 2471-2478.
14. **Cavaleiro, J.A.S., A.M.d'A.R. Gonsalves, G.W. Kenner, and K.M. Smith.** 1974. Total synthesis of deuteriated derivatives of protoporphyrin-IX for NMR studies of haemoproteins. J. Chem. Soc. [Perkin 1] 1771-1781.
15. **Cavaleiro, J.A.S., G.W. Kenner, and K.M. Smith.** 1974. Biosynthesis of protoporphyrin-IX from coproporphyrinogen-III. J. Chem. Soc. [Perkin 1] 1188-1194.
16. **Chu, T.C. and E.J.-H. Chu.** 1955. Paper chromatography of iron complexes of porphyrins. J. Biol. Chem. *212*:1-7.
17. **DiNello, R.K. and D.H. Dolphin.** 1981. Evidence for a fast (major) and slow (minor) pathway in the Schumm devinylation reaction of vinylporphyrins. J. Org. Chem. *46*:3498-3502.
18. cf. **Eisner, U., A. Lichtarowicz, and R.P. Linstead.** 1957. Chlorophylls and related compounds. Part VI. The synthesis of octaethylchlorin. J. Chem. Soc. 733-739.
19. **Fischer, H.** 1955. Preparation of hemin. Org. Synth. *3*:442.
20. **Fischer, H. and H. Orth.** 1934. Die Chemie des Pyrrols. Vol. 1. Akademische Verlagsgesellschaft, Leipzig.
21. **Fischer, H. and H. Orth.** 1934. Die Chemie des Pyrrols. Vol. 1, p. 333. Akademische Verlagsgesellschaft, Leipzig.
22. **Fischer, H. and H. Orth.** 1937. Die Chemie des Pyrrols. Vol. 2, part 1. Akademische Verlagsgesellschaft, Leipzig.
23. **Fischer, H. and H. Orth.** 1937. Die Chemie des Pyrrols. Vol. 2, part 1, p 358. Akademische Verlagsgesellschaft, Leipzig.
24. **Fischer, H. and A. Stern.** 1940. Die Chemie des Pyrrols. Vol. 2, part 2. Akademische Verlagsgesellschaft, Leipzig.
25. **Fischer, H. and A. Stern.** 1940. Die Chemie des Pyrrols. Vol. 2, part 2, p. 48. Akademische Verlagsgesellschaft, Leipzig.
26. **Fischer, H. and A. Stern.** 1940. Die Chemie des Pyrrols. Vol. 2, part 2, p. 52. Akademische Verlagsgesellschaft, Leipzig.
27. **Fischer, H. and A. Stern.** 1940. Die Chemie des Pyrrols. Vol. 2, part 2, p. 73. Akademische Verlagsgesellschaft, Leipzig.
28. **Fischer, H. and A. Stern.** 1940. Die Chemie des Pyrrols. Vol. 2, part 2, p. 108. Akademische Verlagsgesellschaft, Leipzig.
29. **Fischer, H. and K. Zeile.** 1929. Synthese des haematoporphyrins, protoporphyrins und haemins. Liebigs Ann. Chem. *468*:98-116.
30. **Freeman, B.A. and K.M. Smith.** 1995. Novel uses of a naturally occurring heterogeneous catalyst to improve organic syntheses. Proc. NOBCChE *22*:227-242.
31. **Gerlach, B., S.E. Brantley, and K.M. Smith.** 1998. Novel synthetic routes to 8-vinyl chlorophyll derivatives. J. Org. Chem. *63*:2314-2320.
32. **Gossauer, A.** 1974. Die Chemie der Pyrrole. Springer, Berlin.
33. **Grinstein, M.** 1947. Studies of protoporphyrin. J. Biol. Chem. *167*:515-519.
34. **Iakovides, P. and K.M. Smith.** 1996. Syntheses of oxygen analogues of sulfhemes-A and -C. Tetrahedron *52*:1123-1148.

35.**Inhoffen, H.H., J.-H. Fuhrhop, H. Voigt, and H. Brockmann, Jr.** 1966. Formylierung der meso-Kohlenstoffeatome von alkylsubstituierten porphyrinen. Liebigs Ann. Chem. *695*:133.
36.**Jackson, A.H., R.K. Pandey, K.R.N. Rao, and E. Roberts.** 1985. Reactions on solid supports, part II: a convenient method for synthesis of pyrromethanes using a Montmorillonite clay as catalyst. Tetrahedron Lett. *26*:793-796.
37.**Jackson, A.H. and K.M. Smith.** 1973. *In* J.W. ApSimon (Ed.), Total Synthesis of Natural Products, Vol. 1, p. 143-278. John Wiley & Sons, New York.
38.**Jackson, A.H. and K.M. Smith.** 1984. *In* J.W. ApSimon (Ed.)Total Synthesis of Natural Products Vol. 6, p. 237-280. John Wiley & Sons, New York.
39.**Kadish, K., K.M. Smith, and R. Guilard. (Eds.).** 1999. The Porphyrin Handbook. Academic Press, Boston.
40.**Kenner, G.W., S.W. McCombie, and K.M. Smith.** 1973. Protection of porphyrin vinyl groups. A synthesis of coproporphyrin-III from protoporphyrin-IX. Liebigs Ann. Chem. 1329-1338.
41.**Kenner, G.W., S.W. McCombie, and K.M. Smith.** 1973. Separation and oxidative degradation of chlorophyll derivatives. J. Chem. Soc. [Perkin 1]. 2517-2523.
42.**Kenner, G.W., J.M.E. Quirke, and K.M. Smith.** 1976. Transformations of protoporphyrin-IX into harderoprophyrin, pemptoporphyrin, chlorocruoroporphyrin and their isomers. Tetrahedron *32*:2753-2756.
43.**Kleinspehn, G.G.** 1955. Pyrrole series XXVII. Novel route to certain 2-pyrrolecarboxylic esters and nitriles. J. Am. Chem. Soc. *77*:1546-1548.
44.**Labbe, R.F. and G. Nishida.** 1957. A new method of hemin isolation. Biochim. Biophys. Acta *26*:437.
45.**Lee, S.-J.H., N. Jagerovic, and K.M. Smith.** 1993. Use of the chlorophyll derivative, purpurin-18, for syntheses of sensitizers for photodynamic therapy. J. Chem. Soc. [Perkin 1] 2369-2377.
46.**Lemberg, R., B. Bloomfield, P. Caiger, and W. Lockwood.** 1955. Porphyrins with formyl groups. V. Isolation of porphyrin a from heart muscle and determination of its heme a content. Aust. J. Exp. Biol. Med. Sci. *33*:435-450.
47.**Lindsey, J.S.** 1999. Synthesis of meso-substituted porphyrins. *In* K. Kadish, K.M. Smith, and R. Guilard (Eds.), The Porphyrin Handbook, Chapter 2. Academic Press, Boston.
48.**Lindsey, J.S., I.C. Schreiman, H.C. Hsu, P.C. Kearney, and A.M. Marguerettaz.** 1987. Rothemund and Adler-Longo reactions revisited: synthesis of tetraphenylporphyrins under equilibrium conditions. J. Org. Chem. *52*:827-836.
49.**Mauzerall, D.** 1960. The thermodynamic stability of porphyrinogens. J. Am. Chem. Soc. *82*:2601-2605.
50.**Mironov, A.F., T.R. Ovsepyan, R.P. Evstigneeva, and N.A. Preobrazenskii.** 1965. Synthetic studies in the dipyrrylmethane series. Synthesis of 4,4′-dimethyl-3,3′-bis(β-carbomethoxyethyl)dipyrrylmethane. Zh. Obshch. Khim. *35*:324-328.
51.**Morell, D.B., J. Barrett, and P.S. Clezy.** 1961. The prosthetic groups of cytochrome oxidase. 1. Purification as porphyrin a and conversion into haemin a. Biochem. J. *78*:793-797.
52.**Morell, D.B. and M. Stewart.** 1956. Removal of iron from hemins. Aust. J. Exp. Biol. Med. Sci. *34*:211-218.
53.**Nguyen, L.T., M.O. Senge, and K.M. Smith.** 1996. Simple methodology for syntheses of porphyrins possessing multiple peripheral substituents with an element of symmetry. J. Org. Chem. *61*:998-1003.
54.**Nguyen, L.T. and K.M. Smith.** 1996. Syntheses of type-I porphyrins via monopyrrole tetramerization. Tetrahedron Lett. *37*:7177-7180.
55.**Ono, N., H. Kawamura, M. Bougauchi, and K. Maruyama.** 1990. Porphyrin synthesis from nitrocompounds. Tetrahedron *46*:7483-7496.
56.**Ono, M., R. Lattmann, K. Imomota, C. Lehmann, T. Früh, and, A. Eschenmoser.** 1985. Monopyrrole precursors for the synthesis of uroporphyrinogenoctanitrile. Croat. Chem. Acta *58*:627.
57.**Rothemund, P.** 1935. Formation of porphyrins from pyrrole and aldehydes. J. Am. Chem. Soc. *57*:2010-2011.
58.**Rothemund, P. and A.R. Menotti.** 1941. Porphyrin studies. IV. The synthesis of α,β,γ,δ-tetraphenylporphyrin. J. Am. Chem. Soc. *63*:267-270.
59.**Schreiber, J., H. Maag, N. Hashimoto, and A. Eschenmoser.** 1971. Dimethyl(methylene)ammonium iodide. Angew. Chem. Int. Ed. Engl. *10*:330-331.
60.**Schumm, O.Z.** 1928. Preparation of hemin derivatives by pyro reactions. Z. Physiol. Chem. *178*:1-18.
61.**Sessler, J.L., A. Mozaffari, and M.R. Johnson.** 1991. 3,4-Diethylpyrrole and 2,3,7,8,12,13,17,18-octaethylporphyrin. Org. Synth. *70*:68-78.
62.**Shiau, F.-Y., R.K. Pandey, S. Ramaprasad, T.J. Dougherty, and K.M. Smith.** 1990. The isomeric mono-acetyl-mono-(1-hydroxyethyl)-deuteroporphyrins: synthesis, characterization and use for syntheses of regioselectively methyl- and vinyl-deuterated hemins. J. Org. Chem. *55*:2190-2195.
63.cf. **Siedel, W. and F. Winkler**. 1943. Oxydation von pyrrolderivaten mit bleitetraacetat. Liebigs Ann. Chem. *554*:162-201.
64.**Smith, K.M. (Ed.).** 1975. Porphyrins and Metalloporphyrins. Elsevier, Amsterdam.
65.**Smith, K.M., G.M.F. Bisset, and M.J. Bushell.** 1980. Partial syntheses of optically pure methyl bacteriopheophorbides c and d from methyl pheophorbide a. J. Org. Chem. *45*:2218-2224.
66.**Smith, K.M., E.M. Fujinari, K.C. Langry, D.W. Parish, and H.D. Tabba.** 1983. Manipulation of vinyl groups in protoporphyrin-IX: introduction of deuterium and carbon-13 labels for spectroscopic studies. J. Am. Chem. Soc. *105*:6638-6646.
67.**Smith, K.M., D.A. Goff, and D.J.Simpson.** 1985. Meso-substitution of chlorophyll derivatives: direct route for transformation of bacteriopheophorbides-d into bacteriopheophorbides-c. J. Am. Chem. Soc. *107*:4946-4954.
68.**Strain, H.H. and W.A. Svec.** 1966. *In* L.P. Vernon and G.R. Seely (Eds.), p. 21. The Chlorophylls, Academic Press, New York.
69.**Thudichum, J.L.W.** 1867. Rpt. Med. Off. Privy Council, App. 7. *10*:152.
70.**Whitlock, H.W. and R. Hanauer.** 1968. Octaethylporphyrin. J. Org. Chem. *33*:2169-2171.

3 General Laboratory Methods for Tetrapyrroles

Jerry C. Bommer[1] **and Peter Hambright**[2]
[1]Frontier Scientific/Porphyrin Products, Logan, UT, and [2]Department of Chemistry, Howard University, Washington, DC, USA

1. INTRODUCTION

There are thousands of porphyrins and metalloporphyrins, and hundreds of new derivatives appear each year. This variety arises because the cyclic conjugated tetrapyrrole nucleus (Figure 1) can have different substituents at the eight β-pyrrole positions, at the four *meso* [5,10,15,20] carbon atoms, and N-alkyl groups can be added to the four central nitrogen atoms. Since its synthesis in 1972, over 8000 literature references have appeared on 5,10,15,20-tetrakis(N-methyl-4-pyridyl)porphyrin compounds, and a similar number on its precursor, 5,10,15,20-tetrakis(4-pyridyl) porphyrin and its derivatives. Most metals and metalloids in the periodic table form metalloporphyrins, and iron porphyrins have been prepared in oxidation states ranging from 0 to +5. In addition, the porphyrin ring itself can be oxidized, reduced, and cleaved. In solution, metalloporphyrins can exist as monomers, homo- and heteronuclear dimers and higher polymers, they can form molecular complexes and exist in various protonation and axial ligation states, and certain porphyrins form large supramolecular aggregates.

A number of books are important in the biological porphyrin area. The classics are *Die Chemie des Pyrroles* in 1934 (38) and 1937 (39) by Fischer and Orth, the 1940 Fischer and Stern (41), and the 1949 *Hematin Compounds and Bile Pigments* by Lemberg and Legge (74). The monograph by Lascelles on *Tetrapyrrole Biosynthesis and Its Regulation* appeared in 1964 (70). Following the publication of *Porphyrins and Metalloporphyrins* by Falk in 1964 (35), Smith edited a new expanded edition of the same title in 1974 (106). The last chapter on "Laboratory Methods" gives information on basic laboratory manipulations and transformations of porphyrin macrocycles (44). Dolphin edited an eight volume series entitled *The Porphyrins* in 1978 (30a), followed by *The Chemistry and Biochemistry of N-Substituted Porphyrins* by Lavallee in 1987 (71). Kalyansundaram's book *Photochemistry of Polypyridine and Porphyrin Complexes* appeared in 1991 (63). *The Colours of Life: An Introduction to the Chemistry of Porphyrins and Related*

Heme, Chlorophyll, and Bilins: ***Methods and Protocols***
Edited by A.G. Smith and M. Witty

Compounds by Milgrom in 1998 (84) is suggested reading both for beginners and experienced workers desiring a broader view of the field. *The Porphyrin Handbook*, a ten volume series edited by Kadish, Smith, and Guilard was published in 2000 (61a). Dupré maintains "Porphynet, the porphyrin site" (www. porphyrin.net), which lists 40 currently available monographs on porphyrins. This site is a source of general information to those working in all aspects of the porphyrin field, and lists vendors, e-mail addresses of porphyrin chemists, future conferences, etc. *The Journal of Porphyrins and Phthalocyanines* began publication in 1997. The Chemical Abstracts Service "CA Selects: Porphyrins" gives biweekly coverage of books, dissertations, reports, reviews, patents, and journal abstracts dealing with most areas of porphyrin chemistry. The free Medline (www.ncbi.nih.gov/PubMed) has over 23 000 listings under "porphyrins".

The chapter by Smith in this volume describes the nomenclature of the porphyrin nucleus, and the preparations of numerous 3,8-disubstituted deuteroporphyrins, chlorophylls, and recent advances in porphyrin synthesis. We intend to cover laboratory techniques useful for the synthesis, purification, and analysis of water-soluble and insoluble porphyrins and metalloporphyrins, and general aspects of the solution chemistry of such macrocyclic species. While only a few examples are mentioned, such procedures can usually be applied to whole classes of compounds.

2. CHROMATOGRAPHY

Most porphyrins are purified by some form of chromatography. Column chromatography is used for isolation of 100 mg to multigram quantities of material. Thin layer chromatography (TLC) on coated glass plates is employed in the 1 to 100 mg range and is valuable for the separation of compounds having similar retention factor (R_f) values. High-pressure liquid chromatography (HPLC) can separate and quantitate samples on the microgram scale. Many porphyrins and diamagnetic metalloporphyrins fluoresce under UV irradiation, and a hand-held long wavelength 366 nm UV lamp is an aid in monitoring the separations. Porphyrin solution should always be filtered before application to columns or plates, and one should avoid "overloading" the media with too high a concentration of pigment. A small amount of the column material often accompanies the purified porphyrin and tends to complicate subsequent CHN determinations. Filtering the solution through a 47 mm 0.2- or 0.45-μm nylon filter (Millipore, Bedford, MA, USA) aids in the removal of small particles. Such filters are also useful for the collection of small quantities of a valuable precipitate. Our seventy years experience at the bench indicates that accidents do happen, and we have tried to indicate throughout this chapter where accidental loss of a valuable porphyrin can be avoided. A clean heavy-wall empty flask should be between the water aspirator and the filtration flask, and the filter flask should always be securely clamped to a ring stand. It is worthwhile to date all sample vials containing solid porphyrins. This is an aid in rapidly locating in your laboratory book at a later time the synthetic or isolation method used for the particular preparation, and also because certain compounds have unexpectedly short shelf-lives. Columns and TLC work should be done under the hood to minimize exposure to the often hazardous solvents, and it is worthwhile to wear disposable gloves in all porphyrin manipulations. Lim has edited a volume on porphyrin chromatography to appear in 2000 (75a). White, Bachman, and Burnham (116) have an excellent chapter on chromatography of porphyrins and metalloporphyrins, and Varaldi, Longo and

Adler (114) have summarized nonchromatographic methods (electrophoresis, countercurrent distribution, sublimation, extraction, precipitation, and crystallization) for the purification of porphyrins.

2.1. Column Chromatography

Most column chromatography involves glass columns of varying lengths, although some prefer plastic columns that can be cut to isolate the porphyrin bands. Others claim that better separations occur in pear-shaped separating funnels. In contrast to gravity techniques, some groups prefer "flash chromatography", in which a low pressure (5–15 psi) of an inert gas is applied to force the sample and solvent rapidly through the sorbent. A glass-wool plug and sand are placed on the bottom of open columns, or commercial columns with glass frits overlaid with a layer of sand are employed. The "dry column" technique (116) in which the column is packed with dry absorbent is sometimes useful, but in general, better separations are achieved when the column is loaded with a slurry of the packing material, which is prepared by stirring the solid into a large volume of the liquid. A layer of sand is added to the top of the column so that the bed is not disturbed by solvent addition. It is best to dissolve the impure porphyrin mixture in a relatively nonpolar solvent, so that the majority of the porphyrin and impurities are initially adsorbed near the top of the column. This is followed by selective elution with a solvent, solvent mixture, or successive solvent mixtures of higher polarity, which move the desired porphyrin away from the adsorbed impurities, or separate a mixture of porphyrins one from another. One elutropic series of solvents useful in porphyrin work are n-hexane, petroleum ether, cyclohexane, carbon tetrachloride, toluene, benzene, diethyl ether, trichloroethylene, 1,2-dichloroethane, chloroform, methylene chloride, acetone, ethyl acetate, tetrahydrofuran, acetonitrile, pyridine, n-propanol, ethanol, methanol, acetic acid, and water. Numerous examples of useful solvent mixtures and sorbents are mentioned in later sections of this chapter for particular compounds. For silica gel or alumina chromatography, we find that ethyl acetate in various concentrations (1%–15%) mixed with $CHCl_3$ or CH_2Cl_2, in combination with increasing amounts of methanol, is effective for the separation of a wide range of synthetic porphyrins and for porphyrin esters in general.

Many groups prefer to evaporate a concentrated organic solution of the impure porphyrin(s) in the presence of the packing material. This solid is then placed on top of the column, and elution is begun with a relatively nonpolar solvent. One rationale for this procedure is that many porphyrins are relatively insoluble in solvents of low polarity. For example, the room temperature solubility of nickel(II) etioporphyrin I (μg/mL) in various solvents is: chloroform (2.2×10^3), methylene chloride (1.6×10^3), ethyl acetate (2.2×10^2), cyclohexane (approximately 7), followed by acetonitrile cyclohexane, ethanol, methanol, and hexane at approximately 2 μg/mL. Solubility depends on both the type of porphyrin ring and the identity of the coordinated metal; for example, the solubilities in chloroform are Ni-Etio I (2.2×10^3), Ni-octaethylporphyrin (5.1×10^3), VO-Etio I (6.8×10^3), and VO- octaethylporphyrin (OEP) (8.2×10^3). OEP has eight ethyl groups in the β- pyrrole positions, while etioporphyrin I has four methyl and four ethyl substituents.

Commercial $CHCl_3$ is often stabilized with approximately 0.75% ethanol, and the polar ethanol affects the separations. If necessary, the ethanol can be removed by shaking the $CHCl_3$ with an equal volume of water, drying with $CaCl_2$, and distilling over P_2O_5, collecting only the middle fractions. Some groups directly use $CHCl_3$ stabilized with amylenes. $CHCl_3$ and CH_2Cl_2 (both carcinogenic to laboratory animals) often contain HCl, which can protonate the porphyrin free base H_2-P into the charged diacid H_4-P^{2+}, leading to poor separations. To test for HCl, the inexpensive and readily available tetraphenylporphyrin (H_2-TPP) is dissolved in benzene, producing a red solution of the free base. Several milliliters of $CHCl_3$ or CH_2Cl_2 are added to a similar volume of the red reagent, and a green solution of the porphyrin diacid forms if HCl is present. The acid can be removed by shaking these solvents with an aqueous solution of 0.1 M NaOH or 0.1 M ammonia until the test solution remains red. The solvents are then dried over solid anhydrous $CaCl_2$, Na_2SO_4, or K_2CO_3 before use. For the separation of a metalloporphyrin that is stable in acid from small amounts of the unmetalated free base, the mixture is often dissolved in chloroform containing 2% trifluoroacetic acid to purposely protonate the undesired H_2-P into H_4-P^{2+}. The green diacid will tend to stick near the top of the column, resulting in a better separation than can be obtained simply with the metalloporphyrin and its free base.

A variety of packing materials have been used in column techniques. Silica gel (100–200 mesh) and alumina (80–200 mesh) are perhaps the most common choices. Activated alumina can be purchased in acidic, neutral, or basic forms, and these activated (Grade I) materials often adsorb the porphyrin so strongly that it becomes difficult to remove from the column. The alumina can be deactivated by adding water to the adsorbent (wt/wt) in a stoppered Erlenmeyer flask. The mixture is shaken to remove all lumps and allowed to cool to room temperature before use. Activity grade I has no added water, grade II contains 3% water, grade III contains 6% water, grade IV contains 10% water, and grade V contains 15% water.

Other adsorbents include Florisil (60–100 mesh), calcium carbonate (for protoporphyrin IX dimethyl ester [DME]), sugar (for chlorophylls), talc (to remove reduced porphyrin impurities), 80% magnesol-20% cellulose (for magnesium porphyrin esters) microcrystalline cellulose, Kieselguhr, Fullers earth, and diatomaceous earth (for sulfonated deuteroporphyrins). Bachmann and Burnham (8) separated, on the macro or micro scale, the methyl esters of meso, proto, copro I, and uro I by gel filtration on Sephadex® LH-20 (Amersham Pharmacia Biotech, Piscataway, NJ, USA), with 1/1 $CHCl_3$-MeOH (containing 1 g Tris base/L) as the eluant. Sephadex has been used to remove undesired salts from porphyrin solutions. Polyamide columns are useful for iron porphyrin carboxylic acids and acid-labile metalloporphyrins (29). Reverse phase octadecyl (C18) columns using methanol-buffered water mixtures can separate water-soluble anionic porphyrins by total porphyrin charge (110). Cationic porphyrins of differing charge can be resolved on silica gel plates developed with saturated KNO_3:water:acetonitrile in a 1:1:8 mixture (10). Cation exchange columns ($-SO_3^-H^+$ or $-SO_3^-Na^+$) are used to separate divalent metal cations from aqueous solutions of anionic water-soluble porphyrins. For cationic porphyrins, anion exchange columns [$-N(CH_3)_3^+Cl^-$] replace tosylates, or the precipitating agents iodide, or perchlorate in solution with chloride.

TLC or absorption spectrophotometry (350–800 nm) are used to monitor the fractions finally coming off of columns and also to determine what impurities are present prior to running a column. In many cases, a given fraction must be rechromatographed many times, filtered, and then recrystallized to obtain pure material. It is advantageous to use a long spatula or aspirator to remove the obvious impurity band(s) from the top of the column once development has begun. Sometimes, it is best to carefully empty the column and slice out the bands of interest for further treatment. It is prudent to empty the column of packing material after the separation, as some sorbents swell and will break the column upon prolonged standing. Chromatography is an art, and often trial and error is the only means of discovering an efficient separation technique for new and unusual compounds. The following procedure illustrates a typical column chromatographic purification of a water insoluble iron porphyrin, crude iron(III) OEP chloride (section 10.2, Procedure 4).

TLC of this iron porphyrin in HCl-free CH_2Cl_2 on an alumina plate showed one major band with an R_f value of 0.6 and a moderate amount of substance that did not move from the origin. The absorption spectra of the impure compound in the same solvent gave bands (and relative peak heights) at 633.5 nm (1.0), 535.5 nm (1.9), and 506 nm (1.8), indicating that the major form in solution was the Fe(OEP)Cl monomer. If present, the metal-free H_2-OEP would be the first species to elute, having bands in CH_2Cl_2 at 619.5 nm (1.0), 566.0 nm (1.5), 531.5 nm (2.7), and 497.5 nm (2.9). In this particular example, no H_2-OEP was found, and approximately 500 mL of the red metalloporphyrin solution was collected. An approximately 2-cm thick impurity band was present at the origin, which did not move in CH_2Cl_2. The addition of ethanol-stabilized chloroform eluted an unidentified minor band from the origin material, and thus chromatography in $CHCl_3$ would have been undesirable. The spectra of the eluted metalloporphyrin in CH_2Cl_2 showed peaks at 588.0 nm (1.0) and 560.0 nm (1.7) indicating the Fe(OEP)Cl monomer had been transformed into the u-oxo dimer (OEP)Fe^{III}-O-Fe^{III}(OEP), which is typical behavior for most iron(III) protoporphyrin type esters and tetra-aryl porphyrins on alumina. With sterically hindered compounds not allowing dimerization, such as tetrakis(2,4,6-trimethyl-phenyl)porphyrin, the monohydroxy Fe(III) porphyrin would be obtained.

To isolate the solid OEP dimer, the solution was filtered through a 0.45-µm filter, the solvent removed on a rotary evaporator at 60°C, and the purple solid dried overnight at 60°C in a vacuum oven. To transform the dimer into the Fe(OEP)Cl monomer, the CH_2Cl_2 solution was filtered, saturated with HCl(g) followed by N_2(g) to remove excess HCl(g), and the solvent was then removed under the hood on a steam bath. The monomer was dried overnight at 70°C in a vacuum oven. The HCl(g) could come from a lecture bottle with a trap between the cylinder and solution, or blown into the CH_2Cl_2 from the head-space of a reagent bottle of HCl.

❖ Procedure 1. Column Chromatography of Iron-Octaethylporphyrin

1. Wearing gloves and under the hood, a 3- × 50-cm glass column is fitted with glass wool, tamped down with a meter stick, and overlaid with approximately 2 cm of acid-washed sea sand.
2. The bottom of the column is placed in a small diameter circular metallic clamp (to prevent the heavy column from sliding into the collection beaker), and the upper portion is held vertical with a vinyl covered extension clamp (not overly tightened such that the column cracks).
3. A beaker is placed under the column, which is then loaded with 250 g of Grade IV 80 to 200 mesh alumina (A-540; Fisher Scientific, Pittsburgh, PA, USA) that has been slurried in 400 mL CH_2Cl_2. The slurry is poured into the column with the Teflon® stopcock open, taking care that solvent is always

present above the solid phase. If the upper layer runs dry, more solvent should be added and the upper layer stirred with a glass rod until no more bubbles percolate up through the slurry.

4. After settling, sand is added to a depth of approximately 2 cm. The alumina fills 64% of the column, leaving room such that a reasonable amount of solution or solvent can be added at one time.
5. Two grams of impure iron OEP is dissolved in 350 mL of CH_2Cl_2, filtered, and slowly poured onto the column with the stopcock in the open position.
6. Once all of the metalloporphyrin reaches the sand layer, small portions of solvent are added until the entire compound has been absorbed, and then the upper space is filled with methylene chloride.
7. The void volume of the solid phase is approximately 200 mL, and since the metalloporphyrin runs near the solvent front, about 200 mL of pure solvent is collected before the porphyrin enters the collection beaker.

2.2. Thin Layer Chromatography

To purify small amounts of porphyrins in the 1 to 100 mg range, or larger amounts if necessary, TLC plates usually give better resolutions than do columns, with the resolution inversely proportional to the thickness of the plate. Commercially, 200-, 250-, 500-, and 1000-µm silica gel plates are available, and 2000-µm or thicker preparative plates can be prepared in house. Alumina, microcrystalline cellulose, RP-C18-silica gel, Kieselguhr, polyamide and Florisil (Mg silicate) plates can be purchased. Fluorescent TLC plates (and fluorescent column material) are to be avoided. These plates often contain Zn^{2+} salts as a component of the phosphor, and the zinc invariably incorporates into a fraction or all of the H_2-P on the plate forming the unwanted Zn-P complex. For fairly basic porphyrins, shaking the organic solvent containing the Zn-P with an approximately 3 M solution of HCl, followed by a water wash, is often sufficient to remove the zinc from Zn-P. Another technique for less basic porphyrins involves adding trifluoroacetic acid to a chloroform solution containing the Zn-P and stirring until the absorption spectrum of neutralized samples indicate that the two bands between 500 and 700 nm, due to Zn-P, have been transformed into the four-banded spectra characteristic of many H_2-P derivatives. The acidic mixture should be extracted repeatedly with deionized water before shaking with a base. Immediate neutralization with a dilute ammonia or NaOH often forms reactive $Zn(NH_3)_x^{2+}$ or $Zn(OH)^+/Zn(OH)_3^-$ complexes, which then reintroduce zinc back into the porphyrin.

All TLC manipulations should be done under the hood and in dim light, as certain porphyrins and metalloporphyrins have a tendency to photodecompose on the plates. A line of closely spaced spots resolves better than material simply streaked onto the plates, and the walls of the developing chamber can be lined with absorbent paper towels to saturate the atmosphere in the enclosure. A single- edged razor blade is useful for scraping off the various fractions, which should then be ground into a fine powder. Methanol often must be added to deactivate the packing material before the porphyrin can be eluted from the powder with an organic solvent. With low R_f material, pretreatment with acetic acid is sometimes necessary.

An excellent TLC method for determining the purity of methyl esters of natural porphyrins is one originally developed by Chu and Chu (25) and modified by Elder (33). The system consists of chloroform, kerosene, and methanol in a volume ratio of 200:100:7 on silica plates. This TLC system with modifications in the methanol content is useful over a broad array of the more polar organic soluble porphyrins and metalloporphyrins including many synthetic derivatives. There is something about the kerosene that cannot be duplicated by substitution with any pure hydrocarbon.

If only two or three bands of slightly different R_f values are present, we have had success using radial chromatography on a commercial Chromatotron (Harrison Research, Palo Alto, CA, USA) with an automated sample collector to isolate 50 to 100 mg amounts of pure material. Circular silica gel 1-, 2-, and 4-mm thick disks are readily prepared. The same disk can be reused many times. A typical TLC procedure is outlined below for the separation of the atropisomers of tetrakis(2-aminophenyl)porphyrin.

In this example, the R_f values were calculated as 0.12 (0.60/5.0) for the α,α,α,α isomer, 0.6 for α,α,α,β, 0.8 for α,α,β,β, and 0.88 for the least polar α,β,α,β compound. The notation α,α,α,α is for the *cis*-isomer with all four amino groups on the same side of the porphyrin plane; α,α,α,β is for three amino groups on one side and the other in the opposite direction, and so on. The R_fs from the literature are 0.04, 0.43, 0.64, and 0.77, respectively (26), indicating the variability often found between laboratories. Statistically, the relative concentration of each isomer should be 4 for α,α,α,β, 2 for α,α,β,β, 1 for α,α,α,α, and 1 for α,β,α,β. The observed intensities of the spots were in qualitative agreement with predictions. In some cases, investigators isolate each band and use absorption spectrophotometry to determine the relative amounts of each compound. A small amount of unidentified material was found at the origin, which is usually the case for impure porphyrins.

❖ Procedure 2. TLC Separation of the Atropisomers of Tetrakis(2-aminophenyl)porphyrin

1. Under the hood, 100 mL of 1:1 (vol/vol) benzene-diethylether is placed in a 27.0 × 26.5 × 7.0 cm developing chamber, which is covered and allowed to stand for 20 minutes.
2. A saturated solution of tetrakis(2-aminophenyl)porphyrin (26) is prepared in benzene and filtered.
3. Eight individual spots are made approximately 2 cm above the bottom of a 5 × 20 cm 250 μm, 60 Å pore size general purpose TLC plate (Aldrich Chemical, Milwaukee, WI, USA), using an open capillary tube for spotting and allowing the solvent to evaporate.
4. The plate is placed in the chamber and developed until the solvent is 5.0 cm above the origin.
5. The plate is removed from the chamber, and a pencil is used to mark quickly the four band positions (0.60, 3.0, 4.0, and 4.4 cm) and the height of the solvent front (5.0 cm).

2.3. High-Pressure Liquid Chromatography

HPLC seems an especially attractive method for analysis of tetrapyrroles. However, like many analytical methods, it can be deceiving without input from other analytical techniques. Similar to analysis of just about any material by HPLC, truly accurate analytical results must rely on comparing the response in the chromatogram of the sample being tested to the response of a standard sample of known purity and concentration. There are some points to be aware of when doing qualitative or semiquantitative work by HPLC.

Many detection systems for tetrapyrrole

work are chosen for their sensitivity towards the compounds of interest, e.g., wavelengths near the Soret maxima for UV/VIS absorbance detection systems. This can give a good idea of purity relative to contamination by other tetrapyrroles, but tells one nothing about what other types of dissimilar organic or inorganic compounds may reside in the sample. The same can be said for fluorescence detection, except in this case, another class of contaminant, nonfluorescent or weakly fluorescent metallo-tetrapyrroles, may be missed or under-represented along with other nonfluorescent compounds. Also, only substances that are actually eluted from the column are detected. In the case of tetrapyrroles, there are a number of contaminants that will stick more or less irreversibly on the column, such as partially esterified porphyrins in the instance of silica chromatography of porphyrin carboxylic acid esters and black polypyrroles, etc., from synthesis of meso-substituted porphyrins. Thus, initially, TLC is useful in determining whether such contaminants are present in the HPLC samples.

HPLC chromatography of tetrapyrroles is particularly prone to artifacts. This probably arises from the relative insolubility of most of these compounds and their tendency to aggregate. This is especially pronounced in reverse phase chromatography of some of the less aqueous soluble porphyrins such as mono- or dicarboxylic acid porphyrins or chlorins derived from natural sources and some of the derivatives of meso-tetraphenylporphine. Systems that tend to minimize this problem usually involve ion-pairing reagents such as tetrabutylammonium ion and a relatively low pH. Sometimes, multiple peaks appear on the chromatogram, which seem to be due to dimers and higher aggregates that are relatively stable under the chromatographic conditions. Changing the concentration of injected sample, or the solvent in which it is injected, will often change the relative size of the eluted peaks to give a good indication that artifacts are indeed a problem.

Generally, we have found that normal phase HPLC of porphyrins tends to be less reproducible than reverse phase chromatography. Very precise conditioning of the column seems to be required, and therefore, we have moved to reverse phase systems exclusively. Systems that seem to work well for monocarboxylic through tetracarboxylic tetrapyrroles use C-18 columns with eluant consisting of tetrabutylammonium phosphate (0.002–0.01 M) at about pH 2.7 and methanol or acetonitrile as organic modifiers. Some systems for more polar tetrapyrroles rely on sodium or potassium phosphate buffers at near neutral pH with concentrations from 0.01 to 0.1 M containing methanol or acetonitrile. The higher salt concentrations are most useful for the most polar tetrapyrroles, such as highly sulfonated species or uroporphyrins. It is best to use some type of gradient system unless analyzing very pure samples, so that less polar contaminants are eluted from the column in a reasonable period of time.

As an example, the system we use to evaluate a mixture of porphyrin carboxylic acids, which contain from 2 to 8 carboxyl groups is as follows: Hamilton PRP-1 polymeric supported C-18 column with a gradient from 0% acetonitrile in 0.01 M sodium phosphate buffer at pH 6.85 and flow 2 mL/minute to 25% acetonitrile over 14 minutes, then a further gradient to 80% acetonitrile over 2 minutes.

3. RECRYSTALLIZATION

As with many organic compounds, recrystallization can be a powerful tool for the purification of tetrapyrroles. It is generally only practical however when more than a few milligrams of compound are available. It has been our experience that for most porphyrins, a fairly high level of purity, perhaps 80% or better for tetrapyrrole impurities and somewhat less if impurities are not tetrapyrroles, is necessary before satisfactory results can be obtained by recrystallization. It has also been our experience that certain types of porphyrins do not crystallize well, and recrystallization as a purification step is not productive. Type III isomers of the biologically important porphyrins for instance are very difficult to crystallize, but type I isomers crystallize nicely with improvement in isomeric purity as well as removal of other porphyrin contaminants. Some porphyrins can be recrystallized as the dihydrochlorides by dissolution in concentrated HCl and refluxing until constant boiling HCl is achieved then adding water slowly to the refluxing solution until a lower normality of HCl is achieved in which the porphyrins have reduced solubility. In general, dissolution in a minimum amount of low boiling solvent in which they have high solubility, then adding an equal or greater amount of a higher boiling solvent in which they have only slight solubility can recrystallize small quantities of porphyrins or other tetrapyrroles. The solution in an open container is then left to evaporate at room temperature or slightly warmed on a trivet until the desired degree of crystal formation is complete. Filtering and washing with the higher boiling solvent completes the operation. Larger amounts of tetrapyrroles can be handled more efficiently and rapidly by recrystallization on a rotary evaporator. In this case, the tetrapyrrole is dissolved in a low boiling solvent, placed in a round bottom flask on the rotary evaporator fitted with a continuous feed tube, and the solvent removed under reduced pressure while adding the higher boiling solvent in which the compound is less soluble, until the desired degree of crystallization is reached. A typical procedure is described below.

As noted in Section 4, tetrapyrroles are notorious for retaining solvents, especially if crystallized from a high boiling solvent like N,N-dimethylformamide (DMF) and a certain amount of DMF decomposes upon heating into dimethylamine, which binds to certain metals in metalloporphyrins. Sometimes these molecules can only be removed by heating the solid near the boiling point of the solvent, if the stability of the tetrapyrrole allows. The removal can be followed by monitoring the increase in absorbance at the Soret peak for a standard concentration by weight over a period of hours or days. Lower temperatures can be employed if the substance is placed under vacuum in a vacuum oven or drying pistol.

❖ Procedure 3. Recrystallization of Protoporphyrin IX Dimethyl Ester

1. Crude protoporphyrin IX DME (approximately 7 g) is obtained by removing the iron from 10 g of hemin with concurrent esterification [Grinstein method (29)] and partially purified by flash chromatography over silica or alumina eluting with 10% ethyl acetate in dichloromethane.
2. The solution containing the porphyrin (about 300–500 mL) is placed in a L-rotary evaporating flask, and the removal of solvent under vacuum is commenced using a 50° to 60°C bath with continuous addition of 500 mL of ethyl acetate. The rate of addition is controlled such that the volume in the flask approaches about 300 mL concurrently with complete addition of the ethyl acetate.
3. Completion of the crystallization occurs upon standing in the refrigerator for several hours, at which time the crystals are collected by filtration on a Buchner funnel and washed with ethyl acetate.

4. The product is analyzed by the Elder kerosene TLC technique (33), and the procedure can be repeated as desired if further purification is necessary by dissolving in minimum dichloromethane and again exchanging with ethyl acetate.
5. The yield is about 5 g, and an overall purity of about 97% to 98% can be expected after one or two recrystallizations.

4. ANALYSIS OF PORPHYRINS AND METALLOPORPHYRINS

It is often important to know the concentration in solution of a particular porphyrin or metalloporphyrin. Since this class of molecules often contains various kinds and amounts of nonporphyrin material that are isolated along with the porphyrin itself, simply weighing out a given amount of compound may give inaccurate results. With water-soluble porphyrins, water molecules are usually indicated by chemical analysis of the isolated solids. The anionic water-soluble H_2-$TPPS_4$ (Figure 1) analyzes as Na_4H_2-$TPPS_4 \cdot 10\ H_2O$ when oven dried at 110°C (83), the cationic pyridinium compound as H_2-TMPyP(4)$Cl_4 \cdot 4\ H_2O$, various phenyl/(4-sulfonatophenyl), and phenyl/(N-methyl-4-pyridyl) porphyrins contain from 2 to 11 moles of water, often in nonstoichiometric amounts, as Zn-$TPPS_4$ analyzes for 16.6 H_2O. Thermo-gravimetric work (53) indicated that the latter compound gradually lost weight up to 200°C, a constant-weight region was found in the 220° to 400°C range, while rapid decomposition was noted above 400°C. Compounds prepared at different times sometimes contain differing amounts of water. The Fe(III)-TMPyP(4) is a monomer in acid, but when precipitated from acid solution by the addition of 1 M $HClO_4$, Mossbauer spectra on the solid indicated 85% monomer and 15% of the μ-oxo-bridged P-Fe-O-FeP dimer (91). The CHN analyses and corresponding ratios are always helpful, but the moles of water are usually determined by calculation based on the observed values. Crystal structures of water-insoluble porphyrins usually contain several molecules of the crystallizing solvent, and tetra-arylporphyrin crystals have channels throughout the solid that occlude solvent molecules and act as “porphyrin sponges” (19). On new compounds, most authors report CHN and metal analyses, the absorption spectra and extinction coefficients, infrared (IR) band positions, the analyzed ^1H nuclear magnetic resonance (NMR) data, and the major peaks obtained by mass spectrometry.

It is better to determine solution concentrations based on known extinction coefficients, than to rely on calculations based on the weight of the sample. Due to aggregation effects, the spectra should be measured at concentration levels similar to the value quoted in the literature. DiNello and Chang (29) give band positions and extinction coefficients in $CHCl_3$ for the free bases of modified natural porphyrin derivatives, and for the corresponding Fe(II) pyridine hemochromes, measured in 4 M pyridine-0.2 M KOH with the addition of small amounts of the reducing agent sodium dithionite. Caughey and coworkers (23) have spectra on other free base 3,8-disubstituted deuteroporphyrin DMEs and metallo derivatives (22). Caughey et al. have compiled data on isomers of uroporphyrins and coproporphyrins and other biologically important molecules (21). Fuhrhop and Smith (44) have five tables of extinction coefficients of various porphyrins and metalloporphyrins in organic solvents and in aqueous sodium dodecyl sulfate solutions. The spectra of centrally N-alkylated porphyrins and metalloporphyrins are found in Lavallee's monograph (71), and James and coworkers

(83) give data on well-characterized water-soluble porphyrins and their precursors. Methods for the determination of porphyrins in biological material have been reviewed (103).

Finally, there are a number of laboratory procedures to determine porphyrin concentrations. In Drabkin's method, iron protoporphyrin IX is digested with hydrogen peroxide in basic solution, the liberat-

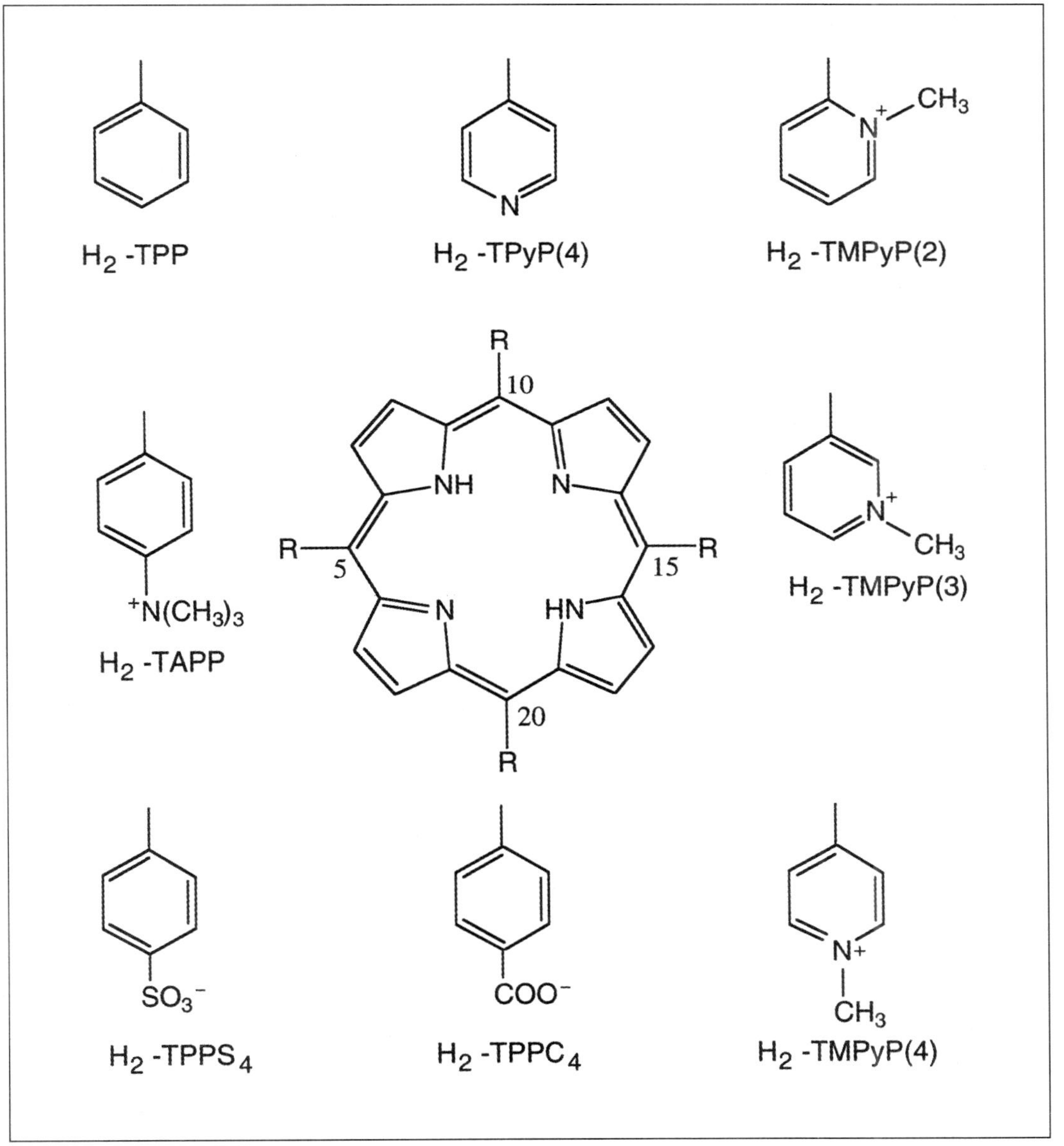

Figure 1. The structures of some water-soluble porphyrins and several of their precursors. Each compound is the porphyrin with the indicated substituents on the four meso (5, 10, 15 and 20) positions. H_2-TPP: meso-tetraphenylporphyrin; H_2-TPyP(4), meso-tetrakis(4-pyridyl)porphyrin; H_2-TMPyP(X): the ortho (2), meta (3) and para (4) isomers of meso-tetrakis(N-methyl-X-pyridyl)porphyrin; H_2-TPPC$_4$, meso-tetrakis(4-carboxyphenyl)porphyrin; H_2-TPPS$_4$: meso-tetrakis(4-sulfonatophenyl)porphyrin; H_2-TAPP, meso-tetrakis(4-N,N,N-trimethylanilinium)porphyrin. A fuller explanation of the nomenclature of porphyrins can be found in Chapter 2.

ed ferric iron reduced with ascorbate, and the Fe(II) spectrophotometrically analyzed with *o*-phenanthroline (32). With metallo-$TPPS_4$ complexes, approximately 10-mg samples were digested with a 3:1:1 mixture of H_2SO_4-HNO_3-$HClO_4$, and the metal then determined by titration with EDTA (53). Brisbin and coworkers (13) spectrophotometrically determined the concentration of protoporphyrin IX DME to ± 2%. To a constant amount of porphyrin (approximately 10^{-5} M) in acetic acid, higher and lower concentrations of standardized metal acetate solutions (divalent Zn, Co, Ni, Fe, and Cu) were added. After heating to completely form the 1:1 complex, appropriate plots of absorbance versus metal ion concentration indicate the point at which the concentration of the metal is equal to the concentration of the porphyrin. This method can also be applied to determine the concentration of the metals ($2–5 \times 10^{-5}$ M) by titration with a known concentration of porphyrin. Petro and Marzilli (93) determined the molar extinction coefficients of a series of cationic porphyrins by adding excess standardized Zn(II) to aqueous solutions of the porphyrins at pH 12.0. After the reaction was complete, the pH was brought to 9.0, and excess zinc was determined with the colorimetric zinc reagent, zincon. The precision of this method was approximately 3%.

5. WATER-SOLUBLE PORPHYRINS

There are a variety of synthetic porphyrins and metalloporphyrins with positive and negatively charged substituents that are of biochemical and biomedical interest. The *meso*-tetrakis(N-methyl-4-pyridyl)porphyrin, H_2-TMPyP$(4)^{4+}$ and its 3 and 2 isomers (Figure 1) are soluble in water the full pH range and monomeric below 10^{-4} M in aqueous solution (20,49,62). These cationic compounds aid in the delivery of oligonucleotides into the nucleus (43). Four coordinate Cu(II), Ni(II), and Au(III) TMPyP(4) complexes intercalate into certain DNAs at low ionic strengths, and show external groove binding at higher salt levels (89). The N-methyl groups in the ortho position of metallo TMPyP(2) complexes provide a steric barrier to intercalation, and the bulky tetracationic tetrakis(4-N,N,N-trimethylanilinium)porphyrin, H_2-$TAPP^{4+}$ also does not intercalate (81). Fe(III) and Mn(III)-TMPyP(X) compounds are excellent superoxide dismutase mimics (10–12), and they also transform the cytotoxic peroxynitrite ($ONOO^-$) into NO_2 and the O = Mn(IV) porphyrin (36,73). Mn(III)-tetra(N-ethyl-2-pyridyl)porphyrin is superior to oxyhemoglobin for the determination of NO concentrations. Water-soluble Fe(III), Mn(III), and Gd(III) porphyrins have been explored as liver and tumor imaging agents using magnetic resonance imaging (MRI) techniques (79), and radioactive sulfonated metalloporphyrins have potential as tumor localizing agents in nuclear medicine. Natural and synthetic water-soluble porphyrins act against the human immunodeficiency virus (30), and H_2-TMPyP(2) and H_2-TMPyP(4) both inhibit human telomerase (51,58). Porphyrins that produce singlet dioxygen within tumors are valuable in photodynamic therapy (108). The structures and common abbreviations used for certain water-soluble compounds are shown in Figure 1.

6. PRECURSORS FOR CATIONIC AND ANIONIC PORPHYRINS

From health considerations, we mention again that one should always wear inexpensive disposable gloves when working with porphyrins and the reagents and solvents involved in porphyrin synthesis. Most work should be done in a well-ventilated hood.

The H_2-TMPyP(X)$^{4+}$ compounds are made from the *meso*-tetrakis(4-pyridyl)porphyrin, H_2-TPyP(4), and its isomers. These are prepared by refluxing, for 45 minutes, a propionic acid solution which is 0.24 M in pyrrole and 0.24 M in the 4(3 or 2)-pyridine-carboxaldehyde (78). In contrast to tetraphenylporphyrin, H_2-TPP, which is prepared in the same manner and precipitates from solution (3), the pyridyl porphyrins are soluble in propionic acid. One method of isolation involves adding a large volume of water to the cooled mixture, and then stirring in solid sodium acetate to bring the pH to approximately 3.0 (48). The pyridyl groups are deprotonated at this pH, and the purple porphyrin filtered off and washed with methanol, DMF, and water. Another method is to evaporate off all of the propionic acid (78) and titurate the resulting tar with DMF. Most of the impurities are soluble in DMF, and the porphyrin is filtered from the cooled solution. Other groups neutralize the tar with NaOH, dissolve the material in CH_2Cl_2, and run column chromatography on alumina (slurried in acetone), eluting the porphyrin fraction with CH_2Cl_2 containing 5% to 10% pyridine (62). The yields approach 16% with the para- and meta-isomers, while less than 1% of H_2-TPyP(2) can be isolated under such conditions.

To obtain pure compounds, the crude porphyrin is dissolved in HCl-free chloroform, and refluxed with 2,3-dichloro-5,6-dicyano-1,4-quinone (DDQ) in order to oxidize the 5% to 10% chlorin impurity that is always formed from the porphyrin itself during the synthesis (9). The warm mixture is then chromatographed on a column of dry alumina, the porphyrin eluted with CH_2Cl_2, and finally recrystallized from $CHCl_3$-MeOH. To avoid chromatography on a prepurified porphyrin contaminated with its chlorin, the mixture can be refluxed in toluene containing DDQ (100). The cooled solution is extracted with a 1% NaOH solution containing sodium dithionite, and the organic phase then washed with water. The toluene is removed under vacuum, and the chlorin-free porphyrin is crystallized from CH_2Cl_2-MeOH. The solubilities in 17 solvents of the H_2-TPyP(X) compounds have been tabulated (109), and the observed order 3 >>2 >4, was also noted for other tetraaryl porphyrin isomers (42). The Adler-Longo propionic acid method (3) and the Smith DDQ oxidation procedure outlined above are general techniques for the preparation and purification of an array of meso-substituted porphyrins.

More complicated meso-substituted compounds can be prepared by the "mixed aldehyde" approach (6). For example, 0.1 moles each of benzaldehyde and 4-pyridine carboxaldehyde are mixed with 0.2 mole of pyrrole and refluxed in propionic acid (105). TLC on silica gel plates developed with 97.5/2.5 chloroform-methanol show six bands for the product, with R_f values of 0.97 for H_2-TPP, 0.94 for the monopyridyl, 0.86 for *trans*, 0.75 for *cis*, 0.66 for the tri(4-pyridyl)-mono-phenyl, and 0.60 for H_2-TPyP(4). The compounds can be isolated on a preparative scale from Florisil columns eluted with CH_2Cl_2 mixed with a more polar solvent (101). Thus, the mono-4-pyridyl requires 1% to 5% acetone, the *trans* requires 5% to 15% acetone, the *cis* requires 20% to 50% acetone, the tri(4-pyridyl)-mono-phenyl requires 2% MeOH and 10% MeOH for H_2-TPyP(4). The initial aldehyde ratio can be adjusted to produce more or less of a given component.

Sterically hindered tetraaryl porphyrins containing 2,6-dichloro, 2,6-dibromo, or 2,6-dimethylphenyl, and 2,4,6-trimethylphenyl groups are produced in low yield from the Adler-Longo propionic acid procedure, but are often readily synthesized with the Lindsay room temperature method (77). The aldehyde and pyrrole in a 1:1 ratio (each approximately 10^{-2} M) are mixed in CH_2Cl_2 or $CHCl_3$ containing 0.75% EtOH and approximately 10^{-3} M BF_3-OEt_2 (as the acid catalyst) and stirred for several hours at 25°C. The cyclic porphyrinogen formed is then oxidized in the same pot to the porphyrin with DDQ or p-chloranil at reflux, and the impurities are removed by chromatography.

7. N-ALKYLATIONS TO PREPARE CATIONIC PORPHYRINS

It should be noted that all alkylating agents are hazardous, and extreme caution should be taken when working with these substances. Many workers methylate the H_2-TPyP(X) compounds in hot or refluxing chloroform in the presence of excess CH_3I, and the solid iodide salts of H_2-TMPyP(X) precipitate from solution. Since the iodides are not very soluble in water, the product is stirred with the chloride form of an ion-exchange resin either in water, or in water–methanol, and warmed until the solid dissolves. After filtration, the solution is slowly passed through a long column of chloride resin, and the water is removed by lyophilization (90). In some cases, both the tri- and tetra-N-methylated iodides precipitate from chloroform, as indicated by electrophoresis studies on the products (16), and thus full tetra-N-methylation is not always achieved in chloroform with CH_3I. The N-methylation is perhaps best done in DMF with methyl-p-toluenesulfonate (MTS). In a typical procedure, 0.5 g of porphyrin is added to 50 mL of DMF in a 100-mL flask (83). The solution is warmed, and before boiling, 2 mL of MTS is added. The solution is refluxed for 4 hours, and the tosylate salt of the porphyrin is removed from the cooled solution by filtration and ion-exchanged into the chloride form. In some instances, the N-alkylated porphyrins decompose if not isolated soon after the reaction is complete. To ascertain the degree of N-alkylation, a sample from the pot is spotted on a silica gel TLC plate, and the plate is developed with a 1:1:8 (vol/vol/vol) mixture of saturated aqueous KNO_3-water-acetonitrile (10). During the course of the reaction, six bands are observed, with the slowest moving and last remaining the tetra (N-alkylated)-porphyrin. Other workers use 3:3:1:2:1 isopropanol-H_2O-acetone-acetic acid-concentrated NH_3 for the separation of differently charged cationic porphyrins and metalloporphyrins. The sterically hindered H_2-TPyP(2) was also tetra-N-methylated in neat dimethyl sulfate at 110°C. The same N-methylation techniques in DMF are used to prepare the tetrakis[N-methyl-4 (or 3) quinolyl]porphyrins (1), and the popular tetra (4-N,N,N-trimethylanilinium)porphyrin, H_2-TAPP from tetra(4-N,N-dimethylanilinium)porphyrin (65). Evaporating the water from an aqueous solution of M-TAPP in the oven leads to loss of the N-methyl groups.

Several examples of water-soluble "picket-fence" porphyrins have been prepared. The starting material, tetra(2-nitrophenyl)porphyrin, is synthesized by the Adler-Longo technique in acetic acid (26). This compound is dissolved in concentrated HCl and reduced to the tetra(2-aminophenyl)porphyrin with $SnCl_2$ at 70°C. The H_2-T(2-NH_2P)P is a mixture of four atropisomers, with the amino groups above and below the porphyrin plane. A TLC method to separate these isomers is given in section 2.2.1. An 8- × 30-cm column filled with a silica gel-chloroform slurry was used on the preparative scale. The column was loaded with a chloroform solution of the atropisomers, and the three undesired and less polar compounds removed with 1:1 benzene:diethyl ether. The target and most polar *cis*-$\alpha,\alpha,\alpha,\alpha$ isomer was then eluted with 1:1

acetone:diethyl ether. The other three isomers were re-equilibrated at 100°C in $CHCl_3$-toluene, forming more of the desired α,α,α,α species. More efficiently, the isomer mixture is refluxed overnight in benzene in the presence of silica gel. As it forms, the α,α,α,α is preferentially adsorbed on the solid and can be removed with 1:1 acetone: ether (76). The reaction of nicotinic anhydride at room temperature in CH_2Cl_2-pyridine with the amino compound forms the α,α,α,α-tetrakis(*o*-nicotinamidophenyl)porphyrin (85). This species can be gently N-methylated in dry trimethyl phosphate by the addition of methyl trifluoromethylsulfonate, with added 2,6-lutidine to scavenge protons. The ortho-isonicotinamido compound has also been prepared (50,113). The four atropisomers of the water-soluble Cu(II)-TMPyP(2) could be separated on silica gel TLC plates developed with 2-butanone-concentrated NH_3-NH_4PF_6-n-butylamine. The Zn(II) and Ni(II) isomers, but not those of the metal-free H_2-TMPyP(2) or its Mn(III) complex, could also be separated under such conditions (64).

Refluxing the cobalt(II) complex of the meso-tetrakis(pentafluorophenyl) porphyrin overnight in DMF (61) leads to the production of meso-tetrakis-[2,3,5,6,tetrafluoro-4-(dimethy lamino)phenyl]porphyrinato cobalt(II). This complex can be converted into the water-soluble triflate salt (68,69) using methyl trifluoromethanesulfonate in trimethyl phosphate overnight at 60°C under N_2. The metallo triflate salts are stable at room temperature, while the solid chlorides decompose within days.

The electron withdrawing tetrafluorophenyl groups reduce the electron density at the central nitrogen atoms, and a larger effect can be achieved by halogenations at the β-pyrrole positions (31). Thus, Cu(II)-TMPyP(4) dissolved in DMF can be β-octabromonated (96) by addition of Br_2(l), and the metal-free H_2-β-Br_8TMPyP(4) is prepared by removal of the copper with concentrated H_2SO_4. While most water-soluble manganese porphyrins are produced in the 3+ oxidation state, the Mn(II)-β-Br_8-TMPyP(4) is the stable form of this electron deficient porphyrin having a deformed nuclear structure (11). One to four chlorine atoms can be added to the β-pyrroles of H_2-TPyP(2) by refluxing the compound in $CHCl_3$ with N-chlorosuccinimide (60). The products are separated by chromatography, and the H_2-β-Cl_xTEt-PyP(4) are then formed in DMF by the addition of ethyl-p-toluenesulfonate. The sterically hindered 2,6-dichloro-TMPyP(4) has been prepared (57), as well as an octacationic derivative (54). A series of compound containing (N-methyl-4-pyridyl) groups on the β-pyrrole positions have also been synthesized (34). Tetraphenyl type porphyrins with $-CH_2X$ substituents in the para positions, with X = N^+Et_3, N^+Ph_3, NH_2, and PO_3^{2-} are known, and porphyrins have been made water-soluble by the addition of sugar residues (47). Other compounds contain three (N-methyl-4-pyridyl) groups for water solubility, and the fourth phenyl or pyridyl is derivatized with substituents that can interact with nucleic acids (75). Four moles of ethylenediamine (and related diamines) have been added to protoporphyrin IX DME to form acid-soluble compounds (117), and similar species containing two moles of ethylenediamine can be prepared from meso or deuteroporphyrin IX DME. These protoporphyrin derivatives are soluble over a wider pH range if the $-NH-(CH_2)_2{}^-N^+Me_3$ forms are prepared, using techniques similar to those described above. Then an octacationic tetrakis[2,4,6-trimethyl-3,5-bis($-CH_2N^+Me_3$) phenyl]porphyrin is known (5).

8. NEGATIVELY CHARGED PORPHYRINS

8.1. Synthetic Derivatives

The synthetic tetranegatively charged

tetra(4-carboxyphenyl)porphyrin, H_2-TPPC$_4$ is prepared by the Adler-Longo method in propionic acid, and is water-soluble above pH 7.0 due to ionization of the carboxylic acid groups (78). Porphyrins with carboxylic acids in the meta- and ortho-phenyl positions are also known. It is often best to prepare these compounds as their methyl esters, which can be purified by chromatography, and hydrolyze the esters in base at a later stage (28). An enormous amount of work has been done with tetrakis(4-sulfonatophenyl)porphyrin, H_2-TPPS$_4$ and its metal complexes. This porphyrin is soluble in water down to pH approximately 2.0, and, at lower pHs, appears colloidal in solution. To prepare this compound, H_2-TPP is added to concentrated H_2SO_4, and the mixture is heated at 100° to 110°C (66). To monitor the extent of sulfonation, a sample is neutralized (110) and spotted on a reverse phase KC-18 TLC plate (Whatman, Clifton, NJ, USA), and developed with 80/20 MeOH-H_2O (pH approximately 7.4, 0.01 M phosphate buffer). The R_f values are 0.94 for the fully sulfonated H_2-TPPS$_4$, 0.88 for the trisulfonated H_2-TPPS$_3$, 0.74 for *trans* -H_2-TPPS$_2$, 0.59 for *cis*-TPPS$_2$, and 0.12 for H_2-TPPS$_1$. When the reaction is complete, ice is added to the green solution, and the H_2SO_4 is carefully neutralized with concentrated NaOH, adding more ice as needed. The transformation of the porphyrin from the green diacid (H_4-TPPS$_4^{2-}$) into the red free base begins at pH approximately 5.0. When the pH reaches about 9.0, the water is evaporated in the oven, and after pulverizing the resulting solid, it is extracted with methanol in a Soxhlet apparatus. The sodium salt of H_2-TPP$_4$ is soluble in MeOH, and the Na_2SO_4 remains in the cup. For further purification, some groups use dialysis techniques, while others add acetone to a concentrated solution of H_2-TPPS$_4$ in methanol to precipitate a brown solid. A useful procedure (59) is to add monoprotonated *o*-phenanthroline to a pH approximately 4.0 solution of H_2-TPPS$_4$. The insoluble (H-Phen$^+$)$_4$/H_2-TPPS$_4^{4-}$.2 H_2O salt precipitates, and can be washed with water to remove extraneous ions. The solid is then slurried with an ion exchange resin in the Na^+ form until dissolved and passed through a sodium ion-exchange column to remove the protonated *o*-phenanthroline. The partially sulfonated compounds can be isolated using low-pressure liquid chromatography columns packed with LiChroprep RP-18 silica gel and eluted with mixtures of MeOH/phosphate-buffered water (110).

Using neat chlorosulfonic acid at 100°C with the tetra(2,6-dichlorophenyl)-porphyrin, the 3-SO_2Cl species was isolated, and hydrolysis produced the 2,6-dichloro-3-SO_3-phenyl derivative (45). Sulfonation of the tetrakis(pentafluorophenyl) porphyrin (7) with oleum for 10 hours at 140°C leads to four -SO_3^- groups on the β-pyrrole positions, while a 3,5-disulfonated product was found for the octabromonated tetrakis(2,4,6-trimethylphenyl)porphyrin (54). With compounds containing both phenyl and 4-pyridyl groups, only the phenyl rings sulfonate (83).

8.2. Anionic Compounds from Natural Porphyrins

Anionic porphyrins, metalloporphyrins, and their derivatives from natural sources have found a wide variety of usage in modern medicine and biochemistry including the field of photodynamic therapy for various disease states, heme oxygenase inhibition for prevention of jaundice, and inhibition and induction of this enzyme as a tool for biochemical research. Some metalloporphyrins have been used as dioxygen detectors in fluids or air via phosphorescence quenching and as MRI contrast agents (47). Of course the porphyrins along the heme and chlorophyll biosynthetic pathways are employed as standards for intermediates excreted in various disease states and for biomedical research of these diseases.

Isolation of many of the porphyrins and chlorins from natural sources has been described in Chapter 2 by Smith. The most common anionic chlorins one encounters in the laboratory are derived from chlorophyll *a* or *b*. Pheophorbide *a* or *b* each have a single free propionic acid group and as such have very limited water solubility. They can be handled in aqueous solutions containing 50% or more water-miscible organics such as methanol and can be purified by chromatography on C-18 silica adsorbents using sodium phosphate-buffered aqueous–organic eluants. Purity can be checked with TLC on C-18 silica plates (Si-C-18; J.T. Baker, Phillipsburg, NJ, USA), eluting with 85% methanol, 15% 0.01 M sodium phosphate buffer at pH 6.85. The R_f values are approximately 0.44 for pheophorbide *a* and approximately 0.30 for pyropheophorbide *a*.

Chlorin e_6 can be obtained from pheophorbide *a* or pheophytin *a* by basic hydrolysis of the refluxing alcoholic solutions using NaOH or KOH (27,37,40) and purified on C-18 silica packing as the free carboxylic acid form similar to the procedure for pheophorbides but with higher aqueous content of the eluant. The TLC system to check for purity is as above, but with the eluant 75% methanol-25% buffer. The R_f values are approximately 0.76 for chlorin e_6 and 0.66 for chlorin e_4, the meso-acetic acid decarboxylation product of chlorin e_6. In all cases the R_f's for the chlorophyll *b* derivatives are slightly greater than found for the corresponding chlorophyll *a* products. The methyl esters of the above chlorins can be purified by silica or alumina column chromatography using $CHCl_3$ or CH_2Cl_2 containing varying amounts of ethyl acetate. Pheophorbide *a* , however, is unstable in the presence of silica or alumina and chromatography must be carried out rapidly. The chloroform, kerosene, and methanol system in a volume ratio of 200:100:7 on silica plates mentioned earlier is extremely useful for determining the purity of these compounds.

Many of the porphyrins, which occur as porphyrinogens along the biosynthetic pathways can be isolated from the natural sources such as protoporphyrin from hemin, coproporphyrin I from the urine or feces of animals or humans having certain types of porphyria (20,97,118), coproporphyrin III from bacterial sources (70,86), and uroporphyrin I from the urine of cattle (118) or humans (98) having congenital porphyria. Porphyrins excreted from these natural sources can usually be concentrated at a neutral pH by collection on a reverse phase adsorbent such as Sep-pak C-18 cartridges (Waters, Milford, MA, USA) for small quantities or bulk C-18 packing in a Buchner funnel for large volumes. The porphyrins may then be partially purified by careful elution with methanol–buffer or acetonitrile–buffer solutions. A note of caution when working with biological samples that may contain porphyrinogens: One should not make the solutions strongly acidic before oxidation to the porphyrins, which can be accomplished with addition of iodine in ethanol, since even at room temperature, we have noted that a substantial amount of scrambling to the isomer mixtures can occur. The porphyrins can be isolated from the above solutions through removal or the organic solvent by rotary evaporation, then flocculation at pH 4.0 followed by collection by centrifugation and washing with water adjusted to pH 3.0 to 4.0 with acetic acid. Further purification can be achieved by reverse phase chromatography or esterification to the methyl esters and silica or alumina chromatography. The methyl esters can be checked by the Elder TLC system described above, and the free carboxylic acid forms can be evaluated on C-18 plates with 70% to 80% methanol–sodium phosphate buffer system for porphyrins with four or fewer carboxy groups and 50% to 60% methanol/1 or 2 M ammonium acetate for porphyrins with four or more carboxyl groups.

8.3. Isolation of Natural Porphyrins from Bacterial Cultures

Many bacteria, especially photosynthetic bacteria, produce substantial amounts of porphyrins, porphyrinogens, and bacteriochlorophyll, or can be made to do so under certain conditions of stress. In general, the porphyrins or porphyrinogens are mostly excreted into the growth media and can be treated separately from the bacteriochlorophyll in the case of photosynthetic bacteria that remain within the cellular structure of the bacterium.

The cells are separated from the medium by centrifugation at a minimum 2000× *g*. The medium is decanted and stirred or shaken while adding 5 mL of 5% iodine in ethanol per liter of the medium. The solution is allowed to stand for 1 hour protected from light to complete oxidation of any porphyrinogens to the corresponding porphyrins. If porphyrin esters are desired, the medium is passed through a layer of diethylaminoethyl (DEAE) cellulose (about 100 mL of aqueous gravity packed adsorbent per liter of medium) on a Buchner funnel, which binds the anionic porphyrin tightly. The packing is washed with water, dried in the oven, or preferably air-dried, or dried under vacuum. The porphyrins are eluted from the cellulose with 5% wt/vol sulfuric acid in methanol or methanol saturated with HCl until color ceases and allowed to stand protected from light for 24 hours at room temperature. The esterifying mixture is diluted with an equal volume of dichloromethane, and washed first with an equal volume of 1 M sodium acetate solution, then twice more with the same volume of deionized water. The volume is reduced on a rotary evaporator, and the porphyrin mixture is applied to a silica or alumina column to effect separation and purification of the components.

If porphyrin esters are not desired, the porphyrins may be collected from the decanted and filtered growth medium directly onto the bulk C-18 silica reverse phase packing such as that available in 55 to 105 µm size from Waters or Millipore (Bedford, MA, USA) activating first with methanol then washing with water. The packing is then washed with water, and the porphyrins eluted with 90% methanol and water vol/vol. The solvent is removed by rotary evaporation, and the porphyrins taken up in water, filtered, and either collected by flocculation at pH 4.0, or applied to a reverse phase column for further purification. These procedures are applicable to most tetrapyrroles found in any aqueous-based solution whether of mammalian origin, such as urine and extracted feces, or aqueous extracts of plant material. One must be careful of course of treating some tetrapyrroles of biological origin with strong acids such as in the esterification steps described. Such porphyrin may require slightly different handling techniques.

9. PORPHYRINS AND METALLOPORPHYRINS IN SOLUTION

9.1. Behavior in Solution at the Molecular Level

Under certain conditions, porphyrins and metalloporphyrins undergo intermolecular association in solution. In water at pH 7.5 in 0.01 M Tris buffer, plots of absorbance versus concentration for H_2-$TPPS_3$ follow Beers law from approximately 5×10^{-7} M to 1.4×10^{-5} M, and the compound is considered monomeric (90). In the presence of 0.1 M KNO_3 at this pH, however, increasingly negative deviations from Beers law are observed as the concentration of the porphyrin increases, consistent with a monomer–dimer equilibrium, where the absorbance of the dimer is less than that of the monomer. Equations have been developed to determine the dimerization constant, K_D, from such absorbance–concentration data.

$$2H_2\text{-}TPPS_3 \rightleftharpoons [H_2\text{-}TPPS_3]_2 \; K_D \qquad \text{[Eq. 1]}$$

Since dimerization increases with ionic strength, overlay spectra of a given concentration of porphyrin measured at differing salt concentrations also provides evidence for the extent of porphyrin aggregation. Another method is to obtain the spectra of the porphyrin at a given salt concentration in a 0.10-cm cell. The solution is diluted 1/10 and the spectra then taken in a 1.0 cm cell, followed by another 1/10 dilution, run in a 10.0-cm cell. If the compound is monomeric, the overlay spectra should be superimposable. If dimers form, the most dilute solution produces the highest absorbance. Isosbestic points are often noted.

Temperature–jump relaxation methods allow the determination of the rate constants for dimer formation (k_f) and dissociation (k_d), and for H_2-$TPPS_3$, $K_D = 4.8 \times 10^4\ M^{-1}$, $k_f = 2.2 \times 10^8\ M^{-1}\ s^{-1}$, and $k_d = 4.6 \times 10^3\ s^{-1}$. Under the same conditions, both with and without added electrolyte, H_2-TMPyP(4) follows Beers law and shows no kinetic relaxation behavior, and thus behaves as a monomer. The electron withdrawing pyridinium groups remove electron density from the porphyrin ring, disfavoring the van der Waals interactions leading to dimerization. Many water-soluble porphyrins, such as H_2-TMPyP(4), are adsorbed strongly on glassware. A flask that once contained this compound when washed with water appears clean, but 0.1 M HCl added to the flask turns green, as acid converts the absorbed free base into the more weakly adsorbed green diacid H_4-$TMPyP^{6+}$. For low porphyrin concentration work, many workers prefer to use a new plastic cuvette for each measurement on freshly prepared solutions. The amphoteric compound tetrakis[N-(2-sulfoethyl)-4-pyridyl]porphyrin is monomeric in the pH range 2.0 to 12.0, and does not adsorb on glass surfaces (55).

The position of substituents on the porphyrin periphery influence the extent of aggregation. While H_2-$TPPS_4$, sulfonated para-substituted TPP species and H_2-TPPC associate, the ortho-and di-ortho-substituted sulfonated TPP derivatives, as well as tetra(2-carboxyphenyl)porphyrin are monomeric (110,111). The electron-rich natural porphyrins such as meso and protoporphyrin IX examined by fluorescence techniques have high K_D values of $2.7 \times 10^6\ M^{-1}$ and $1.9 \times 10^7\ M^{-1}$, respectively (80), while the octanegative uroporphyrin I shows no evidence of dimerization at moderate ionic strengths above pH 7.0. Dimerization also depends on the nature of the coordinated metal ion. The five or six coordinate Zn(II), VO(IV), Cr(III), Mn(III), and Co(III) complexes of $TPPS_4$ are monomeric under conditions in which the four coordinate Cu(II), Pd(II), and Ag(II)-$TPPS_4$ complexes are dimers (67). In nonaqueous solutions, electron spin resonance studies on paramagnetic metalloporphyrins and concentration-dependent NMR work on metal-free compounds are used to access monomer–dimer behavior (115).

An example of the practical consequences of dimerization and larger aggregate formation is the sometimes anomalous behavior of porphyrins during HPLC analysis. We have noted, for instance, the reverse phase HPLC of H_2-$TPPS_4$ in phosphate buffer systems can show a series of peaks eluting at rather regular intervals in what has been shown to be an essentially pure sample by other chromatographic means. This can be explained as a separation of the monomeric, dimeric, and higher aggregated species, which are not rapidly dissociated under these HPLC conditions. Changing the concentration of the injected sample or the solvent composition of the injection media changes the relative size of the eluting peaks.

In aqueous solution, certain porphyrin systems exhibit supramolecular behavior. The diacid H_4-$TPPS^{2-}$ has a Soret band at 433 nm. The new narrow peak at approximately 489 nm, which appears below pH 2.0 at high ionic strengths, is attributed to the presence of J-aggregates, edge-to-edge ribbon-like zwitterionic electronically coupled species in which the central diacid protons of one porphyrin interact with the sulfonic acid groups of another (87). At higher porphyrin concentrations, another peak at 422 nm appears, due to even larger H-aggregates, which are face-to-face associations of the J-species, and involve hundreds of thousands of interacting porphyrin units (95). Resonance light scattering experiments (92) indicate that the *trans* diphenyl/di(4-sulfonatophenyl) porphyrin, as the free base at pH 6.0 and as the diacid at pH 3.0 show supramolecular behavior, as does the diacid H_4-(β-Br_8-$TPPS)_4^{2-}$ at pH approximately 1.0 and the free base and Cu(II) complex of *trans* diphenyl/di(N-methyl-4-pyridyl)porphyrin. Supramolecular chiral H- and J-aggregates of H_4-$TPPS_4^{2-}$ are formed on poly-L-glutamate at pH 2.9, but only in the presence of the cationic Zn(II), Mn(III), and Au(III)-TMPyP(4) species. The positive porphyrins act as spacers, allowing the anionic porphyrins to approach the polypeptide and gain chirality (94).

Heteronuclear dimers are formed between oppositely charged porphyrins and metalloporphyrins in solution. Thus, 1:1 complexes in acetone–water were found between H_2-$TAPP^{4+}$ (and Zn-$TAPP^{4+}$) with H_2-$TPPS_4^{4-}$, Cu-$TPPS_4^{4-}$, and Zn-$TPPS_4^{4-}$ with K_D values in the 10^5-10^6 M^{-1} range (88). Addition of salts such as $NaClO_4$ lead to dimer dissociation. Job's law mole-ratio spectrophotometric studies indicated that the double decker porphyrin Ce^{III}-$[TMPyP(4)]_2^{7+}$ reacted with two moles of Ni-$TPPS^{4-}$ or VO-$TPPS^{4-}$, presumably with one porphyrin on either face of the cerium dimer (18). Also, two moles of Ce^{IV}-$[TAPP]_2^{8+}$ complexed with one mole of either Ni(II) or VO-$TPPS^{4-}$.

Only 1:1 molecular complexes were formed between indigo di-, tri-, and tetrasulfonates (18) with Cu(II) and Zn(II)-$TAPP^{4+}$ and Zn-$TMPyP(4)^{4+}$. The magnitudes of the association constants for molecular complexes involving H_2-$TPPC_4^{4-}$, H_2-$TAPP^{4+}$, and tetrakis(N-propyl-4-pyridyl)porphyrin with various ligands that bind in a face-to-face fashion can be predicted (102). Molecular complexes between uroporphyrin-I and a variety of large organic cations and neutral heterocyclic molecules such as caffeine, *o*-phenanthroline, methyl viologen, nicotinamide, and adenine have been investigated (82).

Porphyrins show acid base behavior, and the first two acid dissociation constants are defined as follows, where the charges of the peripheral groups are neglected:

$$H_4\text{-}P^{2+} \rightleftharpoons H_3\text{-}P^{+} + H^{+} \quad K_4 \qquad \text{[Eq. 2]}$$

$$H_3\text{-}P^{+} \rightleftharpoons H_2\text{-}P + H^{+} \quad K_3 \qquad \text{[Eq. 3]}$$

The equilibria are measured spectrophotometrically, and it is important to make sure that the porphyrin is monomeric under the pH titration conditions, and that the buffers used do not complex with the porphyrins (56). For example, H_2-$TMPyP(4)^{4+}$ shows $pK_4 = 0.8$, $pK_3 = 1.4$, while for the more basic H_2-$TAPP^{4+}$, $pK_4 = 3.95$ and $pK_3 = 4.11$. The pK_2 and pK_1 values for most porphyrins are above 12, although the H_2-[β-Br_8-$TMPyP(4)]^{4+}$ with eight electron-withdrawing bromines on the β-pyrrole positions (96) gives $pK_2 = 6.5$ and $pK_1 = 10.2$. The magnitudes of these acid dissociation constants depend on the ionic strength and temperature. The pK_3 values are used to rank the relative basicities of water-soluble porphyrins (47), as in the series H_2-TMPyP(2) [-0.9], H_2-TMPyP(4) [1.4], H_2-TMPyP(3) [1.8], H_2-TAPP [4.11], H_2-$TPPS_4$ [4.76], H_2-$TPPC_4$ [5.5], uroporphyrin I [6.0], and hematoporphyrin IX [6.1]. The cationic porphyrins are usually less basic than the anionic compounds.

Detergents have been used to bring 3,8-disubstituted deuteroporphyrin IX DME compounds into aqueous solution (46). With cationic detergents such as cetyltrimethylammonium bromide, the monocation H_3-P^+ is destabilized, and only the diacid–free base equilibria (K_3K_4) can be observed. In 2.5% sodium laurel sulfate, both pK_3 and pK_4 can be obtained, and typical pK_3 values for these esters are 5.8 for mesoporphyrin, 5.5 for deuteroporphyrin, 4.8 for protoporphyrin, and 3.3 for the 3,8 diacetyldeuteroporphyrin (23). Electron withdrawing groups decrease the proton affinity of the central nitrogen atoms.

The reduction potential of the free base porphyrin into its radical anion, $E_{1/2}$ (1), have been measured under the same conditions in DMF for over a hundred water-insoluble porphyrins (119,120), and such constants are also a measure of relative basicities that allows comparisons between meso and β-pyrrole substituted compounds.

$$H_2\text{-P} + e^- \rightleftharpoons H_2\text{-P}^- \qquad E_{1/2}(1) \qquad \text{[Eq. 4]}$$

The more basic the porphyrin, the less tendency it has to add an electron, and the more negative its reduction potential. A very basic porphyrin is OEP with $E_{1/2}(1)$ = -1.85 V, followed by -1.82 for meso DME, -1.77 for deutero DME, -1.70 for proto DME, -1.66 for the unsubstituted porphyrin porphin, -1.56 for H_2-TPP, -1.48 for the 3,8-diacetyl deutero DME and, -1.32 for the less basic 3,8-dicyano deutero DME. Partial potential values could be assigned to substituent groups, such that when added to the potential of the reference compound porphin, allow the calculation of $E_{1/2}(1)$ for new derivatives. For a series of porphyrins, a variety of spectroscopic, kinetic, and equilibrium data can be correlated with either pK_3 or $E_{1/2}$ (1).

9.2. Practical Aspects of Porphyrin Dissolution

As noted in the previous section, most porphyrins exist in an aggregated state in aqueous solution. This can translate into the more fundamental problem of how to successfully dissolve some of the less hydrophilic porphyrins and stabilize solutions long enough to do meaningful biological experiments. This problem is generally associated with mono- and dicarboxylic porphyrins and chlorins and their divalent metallo derivatives, which lack hydrophilic substituents other than the carboxylic acid groups. Protoporphyrin IX is an example of this type of porphyrin in that it is colloidal in aqueous alkali (21). As a general rule, these porphyrins can be dissolved by treating them with dilute base (0.01 to 0.1 M NaOH, KOH, or better yet, ammoniacal bases such as Tris or NH_4OH if usage will allow), then diluting to about 50% with aqueous soluble organic solvents such as ethanol, DMF, dimethylsulfoxide, etc. Reasonably concentrated stock solutions on the order of several millimolar can usually be prepared in this way. The stock solution can then be diluted into a large volume of buffered aqueous solution or media to adjust the pH and minimize contribution from the organic solvent and base. Both solutions are likely to be unstable with time resulting in increasing aggregation and precipitation, and should be prepared freshly and used immediately for each experiment. Although the porphyrins likely exist as very large aggregates in the final buffered solution, they apparently become absorbed by lipophilic membranes and are monomerized in some manner during the course of cellular and in vivo experiments.

10. METALATION METHODS

10.1. Water-Soluble Porphyrins

Herrmann and coworkers have developed a novel heterogeneous procedure to incorporate many metal ions into water-soluble porphyrins (53). Using H_2-$TPPS_4$ as an example, 50 mg of porphyrin are added to 100 mL distilled water containing 0.5 to 1.0 gram of an insoluble metal oxide or acid-etched metal, and the mixture is brought to reflux. The reaction is followed spectrophotometrically, and the four H_2-P bands between 700 and 500 mn are replaced by one or two peaks characteristic of the metalloporphyrin. The reactions typically occur in 1 hour, and the cooled solution is filtered through a 0.22-μm filter, the aqueous metalloporphyrin filtrate evaporated under reduced pressure, and the solid is briefly dried under vacuum at 120°C. Metal and CHN analyses were presented for Cu(II), Zn(II), Co(II), Ni(II), and Cd-$TPPS_4$, and a variety of other ions were shown to incorporate under these conditions. Certain oxides (Cr_2O_3, Mn_2O_3, PdO) and metals (Ti, V, Ru, Ag) were unreactive, and in other cases (HgO, PbO, MnO, CaO) irreversible absorption of the porphyrin on the oxide surface limited the amount of product isolated. Others have noted that the final solid can be contaminated with metal ions bound to the peripheral -SO_3^- groups, and further treatment is required to obtain pure compounds (45). Cationic porphyrins such as H_2-TMPyP(4) can be metalated by this method, but the reactions are often slower than found with anionic derivatives.

The most common incorporation method involves simply refluxing the porphyrin in water with a water-soluble metal salt. The diacid and monocations usually do not incorporate metal ions, so the pH should be kept high enough such that an appreciable amount of H_2-P is present. It is best to run the reaction until all of the porphyrin is metalated, as it is difficult to remove a small amount of H_2-P from M-P at later times. With anionic porphyrins, the solution is filtered and slowly run through an ion exchange column in the Na^+ form to remove uncomplexed metal ions. The eluate is lyophilized, and the metalloporphyrin is purified from the salts by procedures mentioned above for the metal-free derivatives, i.e., recrystallization, passage through Sephadex resins, precipitation with $HPhen^+$, etc. For positive porphyrins, sodium iodide or sodium perchlorate are often added to precipitate the cationic porphyrin salts. (**CAUTION**: Porphyrin perchlorates are potentially explosive, and iodide sometimes reduces a fraction of a trivalent metalloporphyrin to the divalent state). The solids are slurried with a chloride cation exchange resin (heating is often required) and slowly passed through a column of the resin, followed by lyophilization. A safer and more elegant method for small quantities of cationic porphyrins involves the addition of NH_4PF_6 as the precipitating agent, washing the solid with 1:1 2-propanol-ether, and vacuum drying at room temperature (98). The PF_6^- salt is then dissolved in acetone, filtered, and the chloride salt of the porphyrin precipitated with tetrabutyl-ammonium chloride, washed with acetone, and dried in vacuo.

Divalent cadmium (104), lead (52), and magnesium ions are in pH-dependent equilibria with the corresponding metalloporphyrin in aqueous solution, for example:

$$Cd^{2+} + H_2\text{-}P \rightleftharpoons Cd\text{-}P + 2H^+ \quad K_{Cd} \qquad \text{[Eq. 5]}$$

$$Cd\text{-}P + Cu^{2+} \rightarrow Cu\text{-}P + Cd^{2+} \qquad \text{[Eq. 6]}$$

Typical values for K_{Cd} are 7.9×10^{-7} M for TMPyP(2) and 4.2×10^{-11} M for $TPPS_4$. The deformed Cd-P reacts with Cu^{2+} (and Fe^{2+}, Zn^{2+}, Mn^{2+} ions) approximately 10^2 to 10^3 times faster than Cu^{2+} incorporates into H_2-P itself. Such room temperature metal-catalyzed electrophilic substitution reactions have been used to insert metal ions into picket-fence type porphyrins, where refluxing the solution would lead to atropisomerization (85,113). Mercury(II) in acid forms Hg_2-P^{2+} complexes, and similar displacement reactions occur after initial loss of a mercury ion (99). Lithium ions are in equilibrium with Li-P^- complexes of TMPyP(X) (56) and β-Br_8-TMPyP(4) (96) in base. Deformed centrally mono-N-alkylated porphyrins react with metal ions several orders of magnitude faster than do the parent compounds (71). This fact has been used for the rapid preparation of short half-life radiolabeled porphyrins of divalent Cu, Co, and Pd, where the central N-benzyl group is lost upon metalation (72).

In cases where high temperatures in nonaqueous solvents are necessary for metalation with water-insoluble or organometallic reagents, it is often best to first metalate the water-insoluble precursor, which can usually be purified by chromatography. The water-soluble metalloporphyrin is then formed in a subsequent step. For example, H_2-TPyP(4) in trichlorobenzene was reacted with n-BuLi at room temperature to form Li-TPyP(4)$^-$, and after addition of $Ce(acac)_3.H_2O$, the solution was refluxed until metalation was complete (17). The Ce(IV)-$[TPyP(4)]_2$ sandwich complex was purified by chromatography on alumina, and after reaction in DMF with MTS (100°C for 5 days), the water-soluble Ce(III)-$[TMPyP(4)]_2$ was formed. The Ce(IV)-$[TAPP]_2$ was made from the cerium(IV)-N.N-dimethylanilinium precursor by N-methylation in $CHCl_3$/EtOH with CH_3I, and the Ce(IV)-$[TPPC_4]_2$ was produced by basic hydrolysis of the tetramethyl ester. As noted, the oxidation state of the coordinated metal may or may not change during the reactions. A list of the metal ions that have been incorporated into water-soluble porphyrins has been compiled (47).

10.2. Water-Insoluble Porphyrins

Adler's DMF method is often employed for insertion of various metal ions into water-insoluble porphyrins (4). The free base porphyrin and a metal salt (acetate, chloride) are refluxed in DMF until the absorption spectra indicates that metalation is complete. The addition of water to the cooled solution precipitates the metalloporphyrin, which can then be purified by chromatography. An example of this procedure is given below. One or two molecules of dimethylamine are often found bound to trivalent complexes. Buchler has developed techniques of incorporation of high oxidation state metal ions in which the reactions are run in imidazole or phenol melts, and he has reviewed other useful metalation systems (15). These include reactions in acetic acid–sodium acetate, in pyridine and benzonitrile for acid labile complexes, and the uses of metallo acetylacetonates, phenoxides, and organometallic reagents as metal carriers. Buchler's "stability index" S_i (the product of the Pauling electronegativity and cation charge divided by the ionic radius in picometers) is a guide to the tendency of a metalloporphyrin to be demetalated by acids of various concentrations (14), and relationships between the acid-catalyzed demetalation rate constants for a series of M-TAPP complexes and S_i have been explored (2). The loss of the metal ion by acid solvolysis reactions is usually first-order in metalloporphyrin and second-order in (H^+).

The incorporation of many metals requires high temperatures, which can be problematic for most anionic porphyrins derived from natural sources. These porphyrins and chlorins often have peripheral groups that are labile or reactive with the solvents at high temperature. In the case of vinyl or other unsaturated groups, this can be as low as 80°C depending on the solvent, but in most cases, temperatures in excess of 150°C tend to cause the most difficulty. Synthetic procedures involving protection and regeneration of vinyl groups on porphyrins have been described by Smith et al. (107). Metalation of hematoporphyrin even at room temperature generally results in some dehydration of the hydroxyethyl groups to vinyl groups, and if not during the metalation, then certainly during the isolation and drying process.

In general, it is best to do metal incorporations on the ester form of porphyrins with carboxyl groups. This tends to protect these groups from decarboxylation, anhydride formation, and unwanted interactions with solvents or metalating agents. Purification of metalloporphyrin esters is generally easier than the free acid forms using chromatographic and crystallization techniques. The resulting products can be hydrolyzed with strong base, e.g., a stirred mixture of 2 to 4 M NaOH or KOH (24) with the metalloporphyrin ester dissolved in an equal volume of tetrahydrofuran. Complete hydrolysis is usually accomplished in 12 to 24 hours at room temperature and can be ascertained by reverse phase TLC. Hydrolysis is usually marked by precipitation of the product as the Na^+ or K^+ salt or the observed transfer of the compound from the tetrahydrofuran (THF) into the aqueous part of the two phase system. Removal of the THF, which is dissolved in the aqueous layer by rotary evaporation, allows collection of the free acid metalloporphyrin by flocculation at pH 4.0. Methanol and 1% KOH with a trace of water can also be used for hydrolysis provided the ester has some solubility in this mixture (44).

Porphyrins having acetic acid side chains are prone to decarboxylate or undergo other types of degradation if attempts are made to metalate even the ester forms at high temperature. Thus, porphyrins such as uroporphyrin are usually not successfully metalated in refluxing solvents such as phenol, benzonitrile, dichlorobenzene, and imidazole.

Insertion of such metals as Al, the lanthanides, Pt, Sc, VO, TiO, and Zr into these porphyrins is generally not successful. Cobalt incorporation into porphyrins containing free carboxyl groups, even at room temperature, usually results in predominantly insoluble black polymer-like products. This can sometimes be avoided by the addition of large amounts of pyridine to the metalating solution or in some cases by starting with the porphyrin ester and hydrolyzing the purified product. Use of pyridine may result in a product with one or two pyridines coordinated to the central metal ion. These ligands can usually be removed by washing with strong acid, but often this results in the formation of insoluble polymer-like materials, or in certain cases, loss of the coordinated metal through acid hydrolysis reactions. The procedure below is an example of the incorporation of iron into OEP, using the DMF method of Adler (4). With its eight ethyl groups on the β-pyrrole positions, OEP is the most widely used model compound for the natural protoporphyrins, which have eight β-pyrrole substituents.

❖ Procedure 4. Incorporation of Iron into Octaethylporphyrin

1. Under a well ventilated hood and wearing gloves, pour 1.2 L of DMF and approximately 10 mL of acetic acid into a 4-L beaker containing 9.0 g (16.8 mM) of OEP and stirring bar.

The absorption spectra of the metal--free H_2-OEP in this solution has bands (and relative peak heights) at 651.5 nm (1.0), 593.0 nm (1.41), 533.5 nm (1.56), and 518.0 nm (3.45).

2. The beaker is placed on a stirrer–hot plate and slowly heated to approximately 100°C. At this stage, 13.4 g of iron(II) chloride tetrahydrate (64.3 mM) are carefully added in portions to the hot solution, and the temperature is raised until the mixture refluxes.
3. Heating is continued until the spectra of an aliquot in DMF indicates the complete disappearance of the metal-free peaks (especially the 651.5 nm absorbance), with the appearance of new bands due to the Fe(III) porphyrin at 629.0 nm (1.0), 532.5 nm (1.92), and 504.5 nm (1.90).
4. While H_2-OEP is not terribly soluble in hot DMF, the porphyrin goes into solution as the more soluble Fe^{III}-OEP forms. The incorporation usually takes 20 minutes, and small amounts of DMF are occasionally added to keep the volume at approximately 1 L.
5. The solution is then allowed to come to room temperature and Buchner filtered, and then 2 L of 0.1 M HCl are added to essentially quantitatively precipitate the metalloporphyrin, which is collected by filtration.
6. The brown solid is washed with 0.1 M HCl, then water, and dried in an oven at 70°C overnight.
7. The purification of this crude Fe(III)-OEP Cl on an alumina column is described in Procedure 1.

ABBREVIATONS

Br_8-TMPyP(4), TMPyP(4) with 8 bromines on the β-pyrroles; Cl_X-TEPyP-(4), meso-tetrakis(N-ethyl-4-pyridyl)porphyrin with X chlorines on the β-pyrroles; DDQ, 2,3-dichloro-5,6-dicyano-1,4-benzoquinone; DME, dimethylester; DMF, N,N-dimethylformamide; EDTA, ethylenediaminetetraacetic acid; ETIO-I, etioporphyrin-I; $H\text{-}PHEN^+$, monoprotonated 1,10-phenanthroline; HPLC, high-pressure liquid chromatography; MTS, methyl para-toluenesulfonate; OEP, octaethylporphyrin; TAPP, meso-tetrakis(4-N,N,N-trimethylanilinium)porphyrin; TMPyP-(X), meso-tetrakis(N-methyl-X-pyridyl) porphyrin, X = 2, 3, or 4; T(2-NH_2P)P, meso-tetrakis(2-aminophenyl)porphyrin; TPP, meso-tetraphenylporphyrin; $TPPC_4$, meso-tetrakis(4-carboxyphenyl)porphyrin; $TPPS_4$, meso-tetrakis(4-sulfonatophenyl)porphyrin; $TPPS_3$, monophenyl-tri(4-sulfonatophenyl)porphyrin; $TPPS_2$, diphenyl-di(4-sulfonatoaphenyl)porphyrin; $TPPS_1$, triphenyl-mono(4-sulfonatophenyl)porphyrin; TPyP(X) meso-tetrakis(X-pyridyl)porphyrin, X = 2, 3, or 4.

ACKNOWLEDGMENTS

P.H. thanks the Howard University CSTEA project (NASA Contract No. NCC S-184) for financial support. We thank Sabrina L. Bailey and Jeff Yearyean for helpful discussions.

REFERENCES

1. **Adeyemo, A., Shamim, P. Hambright, and R.F.X. Williams.** 1982. *meso*-Tetrakis[N-methyl-4(or 3)quinolyl]porphyrins: metallation rate/basicity correlations. Indian J. Chem. *21*A:763-766.
2. **Adeyemo, A., A. Valiotti, C. Burnham, and P. Hambright.** 1981. Acid solvolysis kinetics of copper and nickel porphyrins: a rate-stability index correlation. Inorg. Chim. Acta Lett. *54*:L63-L65.
3. **Adler, A.D., F.R. Longo, J.D. Finarelli, J. Goldmacher, J. Assour, and L. Korsakoff.** 1967. A simplified synthesis for meso-tetraphenylporphyrin. J. Org. Chem. *32*:476-477.
4. **Adler, A.D., F.R. Longo, F. Kampas, and J. Kim.** 1970. On the preparation of metalloporphyrins. J. Inorg. Nucl. Chem. *32*:2443-2445.

5. **Almarsson, O., H. Adalsteinsson, and T.C. Bruice.** 1995. Synthesis and characterization of an octacationic iron tetraphenylporphyrin, which is soluble in water and non-m-oxo dimer forming. J. Am. Chem. Soc. *117*:4524-4532.

6. **Anton, J.A. and P.A. Loach.** 1975. Synthesis of some mono, di and trisubstituted tetraarylporphyrins. J. Heterocyclic. Chem. *12*:573-576.

7. **Artaud, I., H. Grennberg, and D. Mansuy.** 1992. Selective oxidative cleavage of 1,2-dimethoxyarenes to muconic diesters catalyzed by an iron -sulfonated-tetrakis(pentafluorophenyl)porphyrin. J. Chem. Soc., Chem. Commun. 1036-1038.

8. **Bachmann, R.C. and B.F. Burnham.** 1969. Purification of porphyrin esters by gel-filtration. J. Chromatogr. *41*:394-399.

9. **Barnett, G.H., M.F. Hudson, and K.M. Smith.** 1975. Concerning meso-substituted tetraphenylporphyrin purification. J. Chem. Soc. [Perkins 1] 1401-1403.

10. **Batinic-Haberle, I., L. Benov, I. Spasojevic, and I. Fridovich.** 1998. The ortho effect makes manganese(III) meso-tetrakis(N-methylpyridinium-2-yl)porphyrin a powerful and potentially useful superoxide dismutase mimic. J. Biol. Chem. *273*:24521-24528.

11. **Batinic-Haberle, I., S.I. Liochev, I. Spasojevic, and I. Fridovich.** 1997. A potent superoxide dismutase mimic: manganese-β-octabromo-meso-tetrakis-(N-methylpyridinium-4-yl)porphyrin. Arch. Biochem. Biophys. *343*:225-233.

12. **Batinic-Haberle, I., I. Spasojevic, P. Hambright, L. Benov, A.L. Crumbliss, and I. Fridovich.** 1999. The relationship between redox potentials, proton dissociation constants of pyrroic nitrogens and in vivo superoxide dismutase activities of manganese(III) and iron(III) cationic and anionic porphyrins. Inorg. Chem. *38*:4011-4022.

13. **Brisbin, D.A. and J.O. Asgill.** 1974. The micro-determination of porphyrins. An intergrated laboratory experiment. J. Chem. Ed. *51*:211-213.

14. **Buchler, J.W.** 1975. Static coordination chemistry of metalloporphyrins, p. 157-231. *In* K.M. Smith (Ed.), Porphyrins and Metalloporphyrins, Elsevier, Amsterdam.

15. **Buchler, J.W.** 1979. Synthesis and properties of metalloporphyrins, p. 389-483. *In* D. Dolphin (Ed.), The Porphyrins, Vol. I, Part A. Academic Press, New York.

16. **Buchler, J.W., F.M. Kunzel, U. Mayer, and M. Nawra.** 1994. Metal complexes with tetrapyrrole ligands. Part 64. Electrophoresis of water-soluble porphyrins and their metal complexes. Fresenius J. Anal. Chem. *348*:371-376.

17. **Buchler, J.W. and M. Nawra.** 1994. Metal complexes with tetrapyrrole ligands. 67. Synthesis and spectroscopic properties of water-soluble cerium bisporphyrinate double-decker ions. Inorg. Chem. *33*:2830-2837.

18. **Buchler, R.A. and M. Nawra.** 1996. Metal complexes with tetrapyrrole ligands. 71. Heteroaggregation of cationic metal mono- and bisporphyrinates with anion metal porphyrinates or indigosulfonates. Inorg. Chim. Acta. *251*:227-237.

19. **Byrn, M.P., C.J. Curtis, I. Goldberg, Y. Hsiou, S.J. Khan, P.A. Sawin, S.K. Tendick, and C.E. Strouse.** 1991. Porphyrin sponges: structural systematics of the host lattice. J. Am. Chem. Soc. *113*:3617-3621.

20. **Cardinal, R.A., I. Bossenmaier, Z.J. Petryka, L. Johnson and C.J. Watson.** 1968. Isolation of porphyrins from porphyria urine by preparative thin-layer chromatography. J. Chromatogr. *38*:100-105.

21. **Caughey, W. S., A. Adler, B.F. Burnham, D. Dolphin, S.F. MacDonald, D.C. Mauzerall, and Z.J. Petyka.** 1972. Porphyrins, p. 185-216. *In* Specification and Criteria for Biochemical Compounds, 3rd ed. National Academy of Sciences, Washington, D.C.

22. **Caughey, W.S., R.M. Deal, C. Weiss, and M. Gouterman.** 1965. Electronic spectra of substituted metal deuteroporphyrins. J. Mol. Spectrosc. *16*:451-463.

23. **Caughey, W.S., W.Y. Fujimoto, and B. Johnson.** 1966. Substituted deuteroporphyrins. II. Substituent effects on electronic spectra, nitrogen basicities, and ligand affinities. Biochemistry *5*:3830-3843.

24. **Cavaleiro, J.A.S., G.W. Kenner, and K.M. Smith.** 1974. Pyrroles and related compounds. Part XXXII. Biosynthesis of protoporphyrin IX from coproporphyrininogen III. J. Chem. Soc. [Perkins I]. 1188-1194.

25. **Chu, T.C. and E.J.-H. Chu.** 1966. Thin-layer chromatography of methyl esters of porphyrins, chlorins and related compounds with Eastman "chromagram". J. Chromatogr. *21*:46-51.

26. **Collman, J.P., R.R. Gagne, C.A. Reed, T.P. Halbert, G. Lang, and W.T. Robinson.** 1975. "Picket fence porphyrins." Synthetic models for oxygen binding hemoproteins. J. Am. Chem. Soc. *97*:1427-1439.

27. **Conant, J.B. and W.W. Moyer.** 1930. Studies in the chlorophyll series. III. Products of the phase test. J. Am. Chem. Soc. *52*:3013-3023.

28. **Datta-Gupta, N. and T.J. Bardos.** 1966. Synthetic porphyrins. I. Synthesis and spectra of some *para*-substituted *meso*-tetraphenylporphines. J. Heterocyclic. Chem. *3*:495-502.

29. **DiNello, R.K. and C.K. Chang.** 1978. Isolation and modification of natural porphyrins, p. 289-339. *In* D. Dolphin (Ed.), The Porphyrins, Vol. I, Part A. Academic Press, New York.

30. **Dixon, D.W., R. Schinazi, and L.G. Marzilli.** 1990. Porphyrins as agents against the human immunodeficiency virus. Ann. NY Acad. Sci. *616*:511-513.

30a. **Dolphin, D., (Ed.).** 1978. The Porphyrins, Vol. I-VII. Academic Press, New York.

31. **Dolphin, D., T.G. Traylor, and L.Y. Xie.** 1997. Polyhaloporphyrins: unusual ligands for metals and metal catalyzed oxidations. Acc. Chem. Res. *30*:251-259.

32. **Drabkin, D.L.** 1941. Spectrophotometric studies VIII. The microdetermination of iron in cytochrome c and hemin preparations. J. Biol. Chem. *140*:387-396.

33. **Elder, G.** 1971. Separation of porphyrin methyl esters by two-dimensional thin layer chromatography. J. Chromatogr. *59*:234-236.

34. **Endisch, C., J.-H. Fuhrhop, J. Buschmann, P. Luger, and U. Siggel.** 1996. β-Tetraethyl-β′-tetrapyridin-4-yl porphyrins, their N-methylated tetracations, and heterodimers with ms-tetraphenylsulfonato porphyrins. J. Am. Chem. Soc. *118*:6671-6680.

35. **Falk, J.E.** 1964. Porphyrins and Metalloporphyrins. Elsevier, Amsterdam.

36. **Ferrer-Sueta, G., I. Batinic-Haberle, I. Spasojevic, I. Fridovich, and R. Radi.** 1999. Catalytic scavengening of peroxynitrite by isomeric Mn(III) N-methylpyridylporphyrins in the presence of reductants. Chem. Res. Toxicol. *12*:442-449.

37. **Fischer, H. and R. Baumler.** 1929. Uber phao und phyllerythro-porphyrine. Ann. Chem. *474*:65-120.

38. **Fischer, H. and H. Orth.** 1934. Die Chemie des Pyrroles. Vol 1. Akademische Verlagsgellschaft, Liepzig.

39. **Fischer, H. and H. Orth.** 1937. Die Chemie des Pyrroles. Vol 2, part 1. Akademische Verlagsgellschaft, Liepzig.

40. **Fischer, H. and H. Siebel.** 1932. Uber phaeophorbide a, chorin e and chlorophyll a. Ann. Chem. *499*:84-108.

41. **Fischer, H. and A. Stern.** 1940. Die Chemie des Pyrroles. Vol 2, part 2. Akademische Verlagsgellschaft, Liepzig.

42. **Fleischer, E.B., R. Foust, D. Jeter, and R. Near.** 1973. The solubilities of tetra(monofluorophenyl)porphines. Inorg. Nucl. Chem. Lett. *9*:1219-1220.

43. **Flynn, S.M., S.T. George, L. White, W. Devonish, and G.B. Takle.** 1999. Water-soluble, meso-substituted cationic porphyrins—a family of compounds for cellular delivery of oligonucleotides. BioTechniques *26*:736-740.

44. **Fuhrhop, J.-H. and K.M. Smith.** 1975. *In* K.M. Smith (Ed.), Porphyrins and Metalloporphyrins, Chapter 19. Elsevier, Amsterdam.

45. **Gonsalves, A.M.d'A.R., R.W.A. Johnstone, M.M. Pereira, A.M.P. de SantAna, A.C. Serra, A.J.F.N. Sorbal, and P.A. Stocks.** 1996. New procedures for the synthesis and analysis of 5,10,15,20-tetrakis(sulfophenyl)-porphyrins and derivatives through chlorosulfonation. Heterocycles *43*:829-838.

46. **Hambright, P.** 1975. Dynamic coordination chemistry of metalloporphyrins, p. 233-278. *In* K.M. Smith (Ed.), Porphyrins and Metalloporphyrins, Elsevier, Amsterdam.

47. **Hambright, P.** 2000. The chemistry of water-soluble porphyrins, Ch. 18. *In* K. Kadish, K.M. Smith and R. Guilard (Eds.), The Porphyrin Handbook. Academic Press, New York.

48. **Hambright, P., A. Adeyemo, A. Shamim, and S. Lemelle.** 1985. [[4,4′,4″,4‴-porphyrin-5,10,15,20-tetrayltetrakis(1-methylpyridiniumato](2-)]-indium(III) pentaperchlorate. Inorg. Synth. *23*:55-59.

49. **Hambright, P. and E.B. Fleischer.** 1970. The acid-base equilibria, kinetics of copper ion incorporation and acid-catalyzed zinc ion displacement from the water-soluble α,β,γ,δ-tetra(4-N-methylpyridyl)porphyrin. Inorg. Chem. *9*:1757-1761.

50. **Hambright, P., A. Turner, J.S. Cohen, R.C. Lyon, A. Katz, P. Neta, and A. Adeyemo.** 1987. An iron(III) porphyrin that exhibits minimal dimerization in aqueous solution. Inorg. Chim. Acta *128*:L11-L14.

51. **Han, F.X., R.T. Wheelhouse, and L.H. Hurley.** 1999. Interactions of TMPyP4 and TMPyP2 with quadruplex DNA. Structural basis for differential effects on telomerase inhibition. J. Am. Chem. Soc. *121*:3561-3570.

52. **Haye, S.E. and P. Hambright.** 1984. Equilibrium constants and electrophylic exchange reactions of lead(II) porphyrins. Inorg. Chem. *23*:4777-4779.

53. **Herrmann, O., S.H. Mehdi, and A. Corsini.** 1978. Heterogeneous metal-insertion: a novel reaction with porphyrins. Can. J. Chem. *56*:1084-1089.

54. **Hoffmann, P., G. Labat, A. Robert, and B. Meunier.** 1990. Highly selective bromination of tetramesitylporphyrin: an easy access to robust metallo-porphyrins, M-Br8-TMP and M-Br8-TMPS. Tetrahedron Lett. *31*:1991-1994.

55. **Igarashi, S. and T. Yotsuyanagi.** 1984. New "amphoteric ion type" water-soluble porphines as a spectrophotometric agent. Chem. Lett. 1871-1874.

56. **Islam, M.Q. and P. Hambright.** 1998. Lithium complexes and the kinetics of interactions of zinc ions with tetra(N-methyl-4-pyridyl)porphyrins in basic solution. Trans. Met. Chem. *23*:727-733.

57. **Ito, Y., K. Kunimoto, S. Miyachi, and T. Kato.** 1991. Photocatalytic cleavage of 1,2-diols by a cofacial hindered water-soluble iron(III) porphyrin. Tetrahedron Lett. *32*:4007-4010.

58. **Izbicka, E., R.T. Wheelhouse, E. Raymond, K.K. Davidson, R.A. Lawrence, D. Sun, B.E. Windle, L.H. Hurley, and D.D. Von Hoff.** 1999. Effect of cationic porphyrins as G-quadruplex interactive agents in human tumor cells. Cancer Res. *59*:639-644.

59. **Jimenez, H.R., M. Julve, J.M. Moratal and J. Fuas.** 1987. Stability constants of metalloporphyrins. a study of the protonation, deprotonation and formation of copper(II) and zinc(II) complexes of meso-tetra(p-sufonatophenyl)porphyrin. J. Chem. Soc. Chem. Commun. 910-911.

60. **Kachadourian, R., I. Batinic-Haberle, and I. Fridovich.** 1999. Synthesis and superoxide dismuting activities of partially (1-4) β-chloronated derivatives of manganese(III) meso-tetrakis(N-ethylpyridinium-3-yl)porphyrin. Inorg. Chem. *38*:391-396.

61. **Kadish, K.M., C. Araullo-McAdams, B.C. Han, and M.M. Franzen.** 1990. Synthesis and spectroscopic characterization of (T(p-Me_2N)F4PP)H_2 and (T(p-Me_2N)F_4PP)M where T(p-Me_2N)F4PP is the dianion of meso-tetrakis(o,o,m,m,-tetrafluoro-p-(dimethylamino)phenyl)porphyrin and M = Co(II), Cu(II) or Ni(II). Structures of (meso-tetrakis(pentafluorophenyl)porphyrinato)cobalt(II), (TF_5PP)Co. J. Am. Chem. Soc. *112*:8364-8368.

61a. **Kadish, K., K.M. Smith, and R. Guilard (Eds.).** 2000. The Porphyrin Handbook. Academic Press, New York.

62. **Kalyanasundaram, K.** 1984. Photochemistry of water-soluble porphyrins: comparative study of isomeric tetrapyridyl and tetraksi(N-methylpyridiniumyl)-porphyrins. Inorg. Chem. 23:2453-2459.

63. **Kalyansundaram, K.** 1991. Photochemistry of Polypyridine and Porphyrin Complexes. Academic Press, New York.

64. **Kaufmann, T., B. Shamsai, R.S. Lu, R. Bau, and G.M. Miskelly.** 1995. Separation of the rotational isomers of tetrakis(N-methyl-2-pyridiniumyl)-porphyrin and the crystal structure of α,α,α,β-(tetrakis(N-methyl-2-pyridiniumyl)porphyrin copper hexacyanoferrate. Inorg. Chem. *34*:5073-5079.

65.**Krishnamurthy, M.** 1977. Synthesis and characterization of a new water-soluble porphyrin. Indian J. Chem. *15B*:964-966.

66.**Krishnamurthy, M.** 1977. Kinetics of anation reactions of a water-soluble rhodium(III) porphyrin. Inorg. Chim. Acta. *25*:215-218.

67.**Krishnamurthy, M., J.R. Sutter, and P. Hambright.** 1975. Monomer-dimer equilibration of water-soluble porphyrins as a function of the coordinated metal ion. J. Chem. Soc. Chem. Commun. 13-14.

68.**La, T., R.A. Richards, R.S. Lu, R. Bau, and G.M. Miskelly.** 1995. Solution chemistry and crystal structure of nickel tetrakis(2,3,5,6-tetrafluofo-N,N,N-trimethyl-4-aniliniumyl)porphyrin trifluoromethanesulfonate ($NiTF_4TMAP(CF_3SO_3)4$). Inorg. Chem. *34*:5632-5640.

69.**La, T., R.A.Richards, and G.M. Miskelly.** 1994. Synthesis and characterization of the cationic porphyrin meso-tetrakis(2,3,5,6-tetrafluoro-N,N,N-trimethyl-4-aniliniumyl)porphyrin. Inorg. Chem. *33*:3159-3163.

70.**Lascelles, J.** 1964. Tetrapyrrole biosynthesis and its regulation. Benjamin, New York.

71.**Lavallee, D.K.** 1987. The Chemistry and Biochemistry of N-Substituted Porphyrins. VCH Publishers, New York.

72.**Lavallee, D.K., A. White, A. Diaz, J.-P. Battioni, and D. Mansuy.** 1986. Efficient metalloporphyrin synthesis under mild conditions using N-benzylporphyrins. Tetrahedron Lett. *27*:3521-3524.

73.**Lee, J., J.A. Hunt, and J.A. Groves.** 1998. Manganese porphyrins as redox-coupled peroxynitrite reductases. J. Am. Chem. Soc. *120*:6053-6061.

74.**Lemberg, R. and J.W. Legge.** 1949. Haematin Compounds and Bile Pigments. Interscience, New York.

75.**Li, H. and L. Czuchajowski.** 1994. Ribofuransoides N-substituted with meso-porphyrin as nucleoside-like compounds. Tetrahedron Lett. *35*:1629-1630.

75a.**Lim, C.K. (Ed.).** 2001. Porphyrins. *In* Methods of Chromatography, Vol. 2. World Scientific Publishers, New York.

76.**Lindsey, J.** 1980. Increased yield of a desired isomer by equilibria displacement of binding to silica gel, applied to meso-tetrakis(*o*-aminophenyl)-porphyrin. J. Org. Chem. *45*:5215.

77.**Lindsey, J.S. and R.W. Wagner.** 1989. Investigation of the synthesis of ortho-substituted tetraphenylporphyrins. J. Org. Chem. *54*:828-836.

78.**Longo, F. R., M. G. Finarelli, and J. Kim.** The synthesis and some physical properties of ms-tetra(pentafluorophenyl)porphin and ms-tetra(pentachloro-phenyl) porphin. 1969. J. Heterocycl. Chem. *6*:927-931.

79.**Lyon, R.C., P.J. Faustino, J.S. Cohen, A. Katz, F. Mornex, D. Colcher, C. Baglin, S.H. Koenig and P. Hambright.** 1987. Tissue distribution and stability of metalloporphyrin MRI contrast agents. Magn. Reson. Med. *4*:24-33.

80.**Margalit, R. and M. Rotenberg.** 1984. Thermodynamics of porphyrin dimerization in aqueous solution. Biochem. J. *219*:445-450

81.**Marzilli, L.G.** 1990. Medical aspects of DNA-porphyrin interactions. New J. Chem. *14*:409-420.

82.**Mauzerall, D.** 1965. Spectra of molecular complexes of porphyrins in aqueous solution. Biochemistry. *4*:1801-1810.

83.**Meng, G.G., B.R. James, K.A. Skov, and M. Korbelik.** 1994. Porphyrin chemistry pertaining to the design of anti-cancer drugs; part 2, the synthesis and in vitro test of water-soluble porphyrins containing, in the meso positions, the functional groups: 4-methylpyridinium, or 4-sulfonatophenyl, in combination with phenyl, 4-pyridyl, 4-nitrophenyl, or 4-aminophenyl. Can. J. Chem. *72*:2447-2457.

84.**Milgrom, L.R.** 1997. The Colours of Life: An Introduction to the Chemistry of Porphyrins and Related Compounds. Oxford Press, New York.

85.**Miskelly, G.M., W.S. Webley, C.R. Clark, and D.A. Buckingham.** 1988. Acidity and dimerization of three water-soluble iron(III) porphyrin cations: (*meso*-α,β,α,β-tetrakis(*o*-(Nmethylnicotinamido) phenyl) porphyrinato)iron(III), (meso α,α,α,α-tetrakis(*o*-N-methylnicotinamido) phenyl)porphyrinato)iron(III), and (meso-tetrakis(1-methylpyridinium-4-yl)porphyrinato)iron(III). Inorg. Chem. *27*:3773-3781.

86.**Neilands, J.B. and J.A. Garibaldi.** 1960. Coproporphyrin III tetramethyl ester. Biochem. Prep. *7*:36-38.

87.**Ohno, O., Y. Kaizu, and H. Kobayashi.** 1993. J-aggregate formation of a water-soluble porphyrin in acidic aqueous media. J. Chem. Phys. *99*:4128-4439.

88.**Ojadi, E., R. Selzer, and H. Linschitz.** 1985. Properties of porphyrin dimers, formed by pairing cationic and anionic porphyrins. J. Am. Chem. Soc. *107*:7783-7784.

89.**Pasternack, R.F. and E.J. Gibbs.** 1996. Porphyrin and metalloporphyrin interactions with nucleic acids. *In* A. Sigel and H. Sigel (Eds.), Metal Ions In Biological Systems 33. Marcel Decker, New York.

90.**Pasternack, R.F., P.R. Huber, P. Boyd, G. Engasser, L. Francesconi, E. Gibbs, P. Fasella, G. Cerio Venturo, and L. deC. Hinds.** 1972. On the aggregation of meso-substituted water-soluble porphyrins. J. Am. Chem. Soc. *94*:4511-4517.

91.**Pasternack, R.F., H. Lee, P. Malek, and C. Spenser.** 1977. Solution properties of tetrakis-(4-N-methyl) pyridylporphineiron(III). J. Inorg. Nucl. Chem. *39*: 1865-1870.

92.**Pasternack, R.A., K.F. Schaefer, and P. Hambright.** 1994. Resonance light scattering studies of porphyrin diacid aggregates. Inorg. Chem. *33*:2062-2065.

93.**Petho, G. and L.G. Marzilli.** 1994. A new spectrophotometric method for determination of cationic porphyrinis. Microchem. J. *50*:178-183.

94.**Purrello, R., L. Monsu' Scolaro, E. Bellacchio, S. Gurrieri, and A. Romer**. 1998. Chiral H- and J-type aggregates of meso-tetrakis(4-sulfonatophenyl)porphine on α-helical polyglutamic acid induced by cationic porphyrins. Inorg. Chem.*37*:3647-3648.

95.**Ribo, J.M., J. Crusats, J.-A. Farrera, and M.L. Valero.** 1994. Aggregation in water solutions of tetrasodium diprotonated meso-tetrakis(4-sulfonato-phenyl)porphyrin. J. Chem. Soc. Chem. Commun. 681-682.

96.**Richards, R.A., K. Hammons, M. Joe, and G.M. Miskelly.** 1996. Observation of a stable water-soluble lithium porphyrin. Inorg. Chem 35:1940-1944.

97.**Rimington, C.** 1952. Haems and porphyrins in health and disease. II. Acta Med. Scand. 143:177-196.

98.**Rimington, C. and C.A. Miles.** 1951. A study of the porphyrins excreted in urine by a case of congenital porphyria. Biochem. J. 50:202-206.

99.**Robinson, L.R. and P. Hambright.** 1992. Mercury(II) reactions with water-soluble porphyrins. Inorg. Chem. *31*:652-656.

100.**Rousseau, K. and D. Dolphin.** 1974. A purification of meso-tetraphenyl-porphyrin. Tetrahedron Lett. *48*: 4251-4254.

101.**Sari, M.A., J.P. Battioni, D. Dupre, D. Mansuy, and J.B. Le Pecq.** 1990. Interaction of cationic porphyrins with DNA: importance of the number and position of the charges and minimum structural requirements for interaction. Biochemistry *29*:4205-4215.

102.**Schneider, H.-J. and M. Wang.** 1994. Ligand-porphyrin complexes: quantitative evaluation of stacking and ionic contributions. J. Org. Chem. *59*:7464-7472.

103.**Schwartz, S., M.H. Berg, I. Bossenmaier, and H. Dinsmore.** 1960. Determination of porphyrins in biological material, p. 221-293. *In* D. Glick (Ed.), Methods of Biochemical Analysis, Vol. VIII. Wiley & Sons, New York.

104.**Shamim, A. and P. Hambright.** 1980. An equilibrium and kinetic study of water-soluble cadmium porphyrins. Inorg. Chem. *19*:564-566.

105.**Shamim, A., P. Worthington, and P. Hambright.** 1981. Synthesis and characterization of phenyl/4-pyridyl meso substituted porphyrins. J. Chem. Soc. Pakistan *3*:1-3.

106.**Smith, K.M. (Ed.) 1975.** Porphyrins and Metalloporphyrins. Elsevier, Amsterdam.

107.**Smith, K.M., E.M. Fujinara, K.C. Langrey, D.W. Parish, and H.D. Tabba.** 1983. Manipulation of vinyl groups in protoporphyrin IX: introduction of deuterium and carbon-13 labels for spectroscopic studies. J. Am. Chem. Soc. *105*:6638-6646.

108.**Sternberg, E.D., D. Dolphin, and C. Bruckner.** 1998. Porphyrin based photosensitizers for use in photodynamic therapy. Tetrahedron *54*:4151-4202.

109.**Sugata, S., S. Yamanouchi, and Y. Matsushima.** 1977. Meso-tetrapyridyl-porpyrins and their metal complexes. Synthesis and physico-chemical properties. Chem. Pharm. Bull. *25*:884-889.

110.**Sutter, T.P.G., R. Rahimi, P. Hambright, J. Bommer, M. Kumar, and P. Neta.** 1993. Steric and inductive effects on the basicity of porphyrins and on the site of protonation of porphyrin dianions: radiolytic reduction of porphyrins and metalloporphyrins to chlorins or phlorins. J. Chem. Soc. Faraday Trans. *89*: 495-502.

111.**Turay, J., P. Hambright and N. Datta-Gupta.** 1978. Intermolecular association of natural and synthetic water-soluble porphyrins, J. Inorg. Nucl. Chem. *40*:1687-1688.

112.**Turk, H. and W.T. Ford.** 1991. Epoxidation of styrene with aqueous hypochlorite catalyzed by a manganese(III) porphyrin bound to collodial anion-exchange particles. J. Org. Chem. *56*:1253-1260.

113.**Valiotti, A., A. Adeyemo, R.F.X. Williams, L. Ricks, J. North and P. Hambright.** 1981. A water-soluble "picket fence" porphyrin and its isomers. J. Inorg. Nucl. Chem. *43*:2653-2658.

114.**Varadi, V., F.R. Longo, and A.D. Adler.** Nonchromatographic methods of purification of porphyrins. 1978, p. 581-588. *In* D. Dolphin (Ed.), The Porphyrins, Vol. I, Part A. Academic Press, New York.

115.**White, W.I.** 1978. Aggregation of porphyrins and metalloporphyrins, p. 303-339. *In* D. Dolphin (Ed.), The Porphyrins, Vol. V. Chapter 7. Academic Press, New York.

116.**White, W.I., R.C. Bachman, and B.F. Burnham.** 1978. Chromatography of porphyrins and metalloporphyrins, p 553-580. *In* D. Dolphin (Ed.), The Porphyrins, Vol. I, Part A. Academic Press, New York.

117.**White, W.I. and R.A. Plane.** 1974. A homologous series of water-soluble porphyrins and metalloporphyrins: synthesis, dimerization, protonation and self-complexation. Bioinorg. Chem. *4*:21-35.

118.**With, T.K.** 1958. Preparation of crystalline porphyrin esters from bovine porphyria urine. Biochem. J. *68*:717-720.

119.**Worthington, P., P. Hambright, R.F.X. Williams, M.R. Feldman, K.M. Smith, and K.C. Langry.** 1980. Reduction potentials of beta-substituted free base porphyrins. Inorg. Nucl. Chem. Lett.*16*:441-447.

120.**Worthington, P., P. Hambright, R.F.X. Williams, J. Reid, C, Burnham, A. Shamim, D.M. Bell, R. Kirkland, R.G. Little, N. Datta-Gupta, and U. Eisner.** 1980. Reduction potentials of seventy-five free base porphyrin molecules: reactivity correlations and the production of potentials. J. Inorg. Biochem. *12*:281-291.

4

Enzymatic Preparation of Tetrapyrrole Intermediates

Martin J. Warren[1] and Peter M. Shoolingin-Jordan[2]
[1]School of Biological Sciences, Queen Mary Westfield College, London, England, UK; [2]School of Biological Sciences, University of Southampton, Southampton, England, UK

1. INTRODUCTION

Tetrapyrroles are intensely colored natural products of vital importance in the biosphere for essential processes such as respiration and photosynthesis and are also of key importance as cofactors in a number of other enzyme reactions. Tetrapyrroles may either be linear in nature, as found in the bilins, or cyclic as in the hemes, chlorophylls, and corrins. In the cyclic tetrapyrrole group, the four centrally located pyrrole nitrogen atoms of the macrocyclic ring offer a range of possibilities for metal chelation. Modulation of the properties of the metallotetrapyrrole prosthetic groups by individual proteins give rise to a remarkably versatile family of powerful bio-organic reagents.

The structural complexity of tetrapyrroles is reflected in a highly intricate branched biosynthetic pathway. For organisms such as *Rhodobacter spheroides* and *Pseudomonas aeruginosa*, which can biosynthesize four different classes of modified tetrapyrrole, there are over 40 separate enzymes dedicated to tetrapyrrole synthesis and modification. Despite their prime metabolic significance, tetrapyrroles and their derivatives are biosynthesized in surprisingly small quantities and, prior to the age of genetic engineering, it was difficult to isolate large quantities of pathway intermediates and even more challenging to study the enzymes themselves. As a result, many investigations prior to the 1980s were carried out with isotopic tracers to enable biosynthetic conversions to be followed. The advent of molecular biology has had a dramatic effect in the tetrapyrrole field, allowing milligrams of recombinant enzymes to be prepared that can be used to manufacture substantial amounts of tetrapyrrole products as well as permitting detailed structural investigations of the enzymes. Central to any of these studies is the availability of the encoded gene or cDNA specifying the enzyme of interest and suitable bacterial hosts for their expression. In this chapter, we have confined ourselves to methods for the enzymatic synthesis of intermediates along the porphyrin and siroheme biosynthetic pathways, most

Heme, Chlorophyll, and Bilins: ***Methods and Protocols***
Edited by A.G. Smith and M. Witty

Table 1. List of Strains and Plasmids Described in this Chapter

Strain Plasmid	Properties	Reference
JMA18 = (JM109/pA19)	*R. spheroides hemA* cloned into pUC19. Constitutive expression of ALAS in *E. coli*	4
MS1 = (TB1/pMS1)	*E. coli hemB* cloned into pUC19. Constitutive expression of ALAD in *E. coli.*	20
BM3 = (TB1/pBM3)	*E. coli hemC* cloned into pUC18. Constitutive expression of PBGD.	32
SD2 = BL21(DE3)pLysE/ pET14b-HemD	*E. coli hemD* cloned into pET14b. Inducible expression of His-tagged UROS.	Raux, Davlin, and Warren, unpublished.
ER293 = BL21(DE3)pLysS/ pER291	*B. stearothermophilus cobA* cloned into pET14b. Inducible expression of His-tagged uroporphyrinogen methylase.	Raux and Warren, unpublished.
ER262 = BL21(DE3)pLysS/ pER259	*S. cerevisiae MET8* cloned into pET14b. Inducible expression of His-tagged Met8p.	25
SW500 = BL21(DE3)pLysS/ pET14b-CysG	*E. coli cysG* cloned into pEt14b. Inducible expression of His-tagged CysG.	Woodcock and Warren, unpublished
BL21(DE3)pLysS/pHT#77	Human cDNA for UROD cloned into pAED4. Inducible expression of His-tagged UROD.	24
JM109/pHHCPO	Human cDNA for CPO cloned into a modified pBTac-1 plasmid. Expression of His-tagged CPO.	22
JM109/pMx-PPO	*M. xanthus hemG* cloned into pTF20E, a derivative of pBTac-1. Allows constitutive expression of His-tagged PPO.	6
JM109/pLUG18e2	*B. subtilis hemH* cloned into pUC18. Constitutive expression of ferrochelatase.	11

of which utilize recombinant proteins. For brevity, we have identified one enzyme for each stage of the pathway from a source that we believe is the easiest to obtain and handle. The clones for these various enzymes can be obtained by contacting the relevant authors as referenced in Table 1. More comprehensive information on each enzyme, as well as on the history of the pathway elucidation, may be sourced from a recent review (28).

2. OVERVIEW OF THE TETRAPYRROLE BIOSYNTHESIS PATHWAY

The heme biosynthetic pathway together with the bifurcation points for the synthesis of the other modified tetrapyrroles is outlined in Figure 1. This chapter is structured around the various enzymes highlighted in the diagram, and considers the synthesis of the following compounds:

- 5-aminolevulinic acid
- porphobilinogen
- preuroporphyrinogen
- uroporphyrinogen III, using multiple enzymes
- precorrin-2, sirohydrochlorin and siroheme from uroporphyrinogen III
- coproporphyrinogen III
- protoporphyrinogen IX
- protoporphyrin IX from coproporphyrinogen III, using multiple enzymes
- protoheme

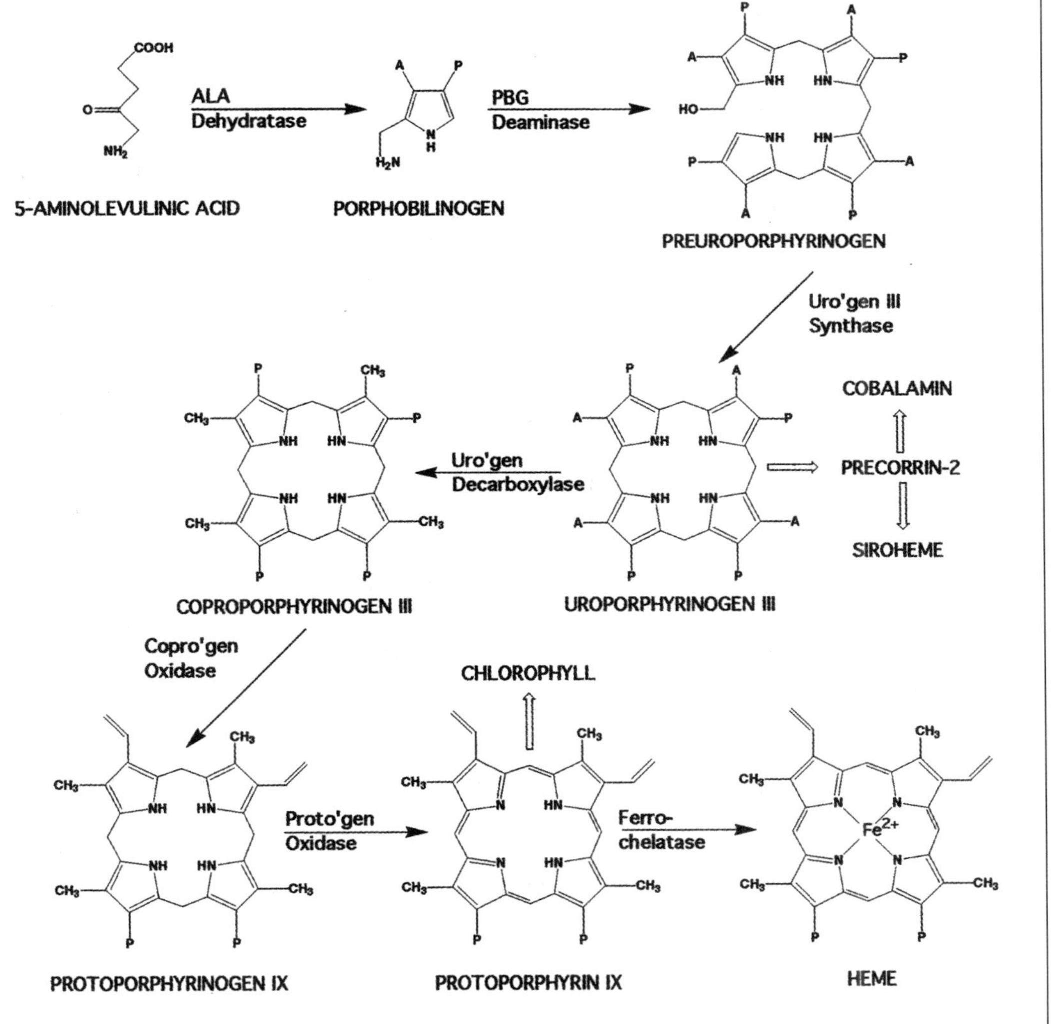

Figure 1. Biosynthesis of heme from ALA. The figure also highlights uroporphyrinogen III as the branchpoint for siroheme and cobalamin synthesis. Abbreviations used: A, acetate side chain; p, propionate side chain.

3. THE ENZYMATIC SYNTHESIS OF 5-AMINOLEVULINIC ACID

5-Aminolevulinic acid (ALA) is formed by two different biosynthetic pathways (Figure 2). One, found in plants, algae, and most bacteria, originates from glutamate, with glutamyl-tRNA and glutamate 1-semialdehyde as intermediates (18), and is traditionally referred to as the C5 pathway. The other pathway, found in mammals, fungi, and some photosynthetic bacteria, involves a single enzymatic step catalyzed by 5-aminolevulinic acid synthase (ALAS) (17). This latter route, often referred to as the Shemin, or C4, pathway, involves condensation between glycine and succinyl-CoA in a reaction in which the carboxyl group of glycine is lost by decarboxylation. ALAS is the rate-determining step in mammalian and fungal heme synthesis, and intracellular levels of the enzyme are tightly regulated. Two enzymes exist in mammalian systems; a ubiquitous enzyme, ALAS1, which is encoded on chromosome 13 and which is subject to tight control in all tissues, and the erythroid enzyme, ALAS2, which is encoded on the X-chromosome and expressed constitutively in developing erythrocytes (9). The photosynthetic bacterium, *R. spheroides*, used for the isolation of the enzyme also has two genes, *hemA* and *hemT* (23).

Aminolevulinic acid can be synthesized using purified ALAS and the procedure can be adapted to prepare isotopically labeled ALA for enzyme synthesis of labeled later pathway intermediates. The ease of using

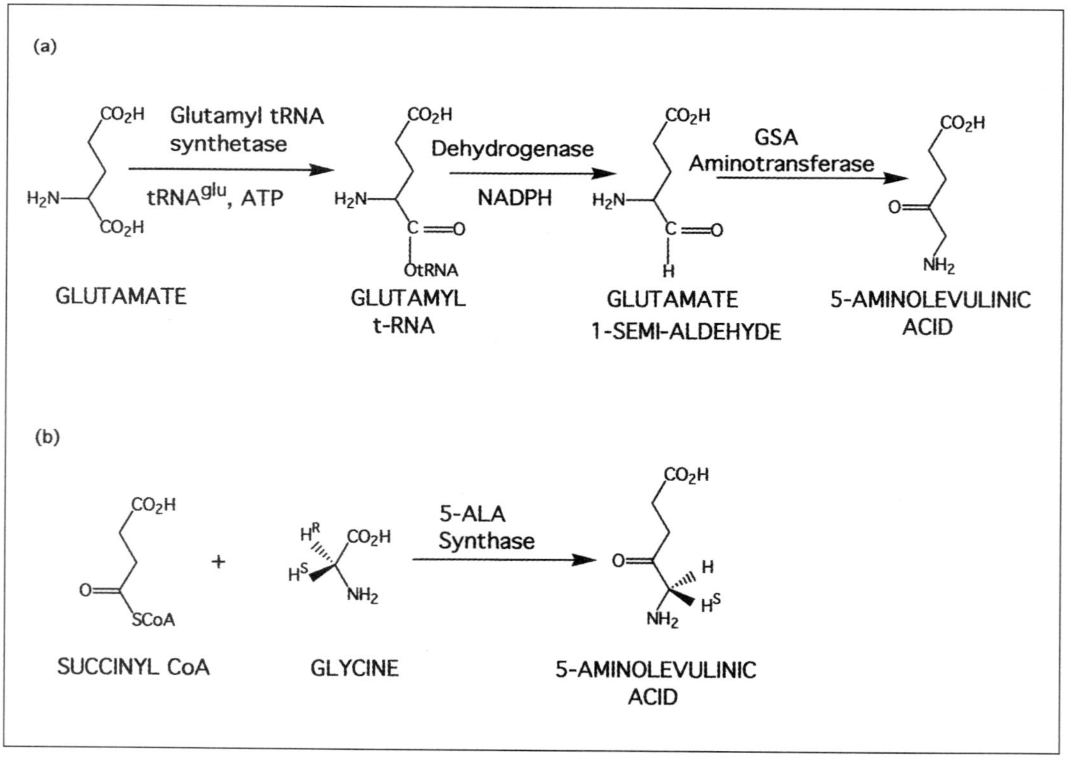

Figure 2. The biosynthesis of ALA. (a) ALA can be synthesized from glutamate by the C-5 pathway or (b) from glycine and succinyl-CoA by the Shemin route. In the case of the latter, it is known that the *proR*-hydrogen of glycine is removed in the overall transformation into ALA.

ALAS has been greatly enhanced by the availability of the recombinant enzyme from *R. spheroides* arising from the cloned and overexpressed *hemA* gene (4).

3.1. Enzyme Purification of *R. spheroides* ALAS Expressed in *Escherichia coli*

ALAS can be purified from wild-type *R. spheroides* (NCIB) according to published methods (27). However, preparing the media for the growth of this organism is tedious, and the yield of purified enzyme is low. To overcome these problems, we have produced a recombinant strain of *Escherichia coli* (JMA19) that overexpresses the *R. spheroides* ALAS (HemA) (Table 1), derived from strain JM109 that had been transformed with the plasmid pA19. The plasmid (pA19) was constructed from a *Hin*dIII/*Eco*RI fragment containing the *hemA* gene from *R. spheroides*, which had been modified at the 5′ end by polymerase chain reaction (PCR) and cloned into pUC19 (Table 1) (4).

❖ Procedure 1. Preparation of *E. coli* Lysate Containing Recombinant *R. spheroides* ALAS

1. Bacterial growth: From an agar plate of recombinant *E. coli* harboring the *R. spheroides hemA* gene (JMA18), a single bacterial colony is removed and used to inoculate a starter culture (5 mL) of Luria-Bertani (LB) medium containing 50 µg/mL ampicillin.
2. The culture is grown for between 5 to 10 hours at 37°C and then used to inoculate a larger (1 L) culture, which is grown overnight at 37°C with rotary shaking (160–180 rpm) for 18 hours.
3. Harvesting and cell lysis: The cells are harvested by centrifugation at 3000× *g* for 20 minutes, and the cell pellet is resuspended in 10 mL of 20 mM sodium phosphate buffer, pH 7.2, containing 0.5 mM pyridoxal 5′ phosphate, 2 mM EDTA, 10% glycerol, 5 mM 2-mercaptoethanol, and 100 µM phenylmethanesulfonyl fluoride (PMSF). All subsequent stages are carried out at 4°C.
4. The suspension is sonicated by placing a large sonicator probe (e.g., a SANYO Soniprep 150 Ultrasonic Disintegrator, Integrated Services TCP, Palisades Park, NJ, USA) about one third of the way into the bacterial suspension and sonicating the solution at medium amplitude (10–12 µm) for 4 1-minute bursts with 2 minutes cooling in between. Cooling is achieved by placing the vessel containing the bacterial solution in an ice-water slurry.
5. After sonication, the extract is centrifuged at 15 000× *g* for 20 minutes to remove cell debris. The clarified straw-colored supernatant contains the active soluble enzyme.

To those unfamiliar with the procedures of protein purification, they are encouraged to read an excellent account of the common procedures employed in protein isolation (26).

❖ Procedure 2. Purification of Homogeneous Recombinant *R. spheroides* ALAS

1. Ammonium sulfate fractionation: Fractionation with solid ammonium sulfate is the first step of the purification process. This procedure is sometimes referred to as salting out and is dependent upon the concentration of the protein solution and the amount of salt that is added. In the case of ALAS, the enzyme is known to precipitate from solution when the solution is saturated with 60% ammonium sulfate. To the clarified bacterial extract, solid ammonium sulfate is added to a saturation of

30% by adding 16.6 g of ammonium sulfate per 100 mL of extract; to ease the speed of solubility, the ammonium sulfate may be finely powdered in a pestle and mortar.
2. After stirring for 10 minutes, the solution is clarified by centrifugation at 10 000× *g* for 15 minutes, and the pellet is discarded. The supernatant is then made 60% with respect to ammonium sulfate by the addition of a further 18.4 g of solid ammonium sulfate per 100 mL of extract.
3. After stirring for a further 10 minutes, the suspension is centrifuged again, but this time the supernatant is discarded, and the protein pellet is retained. The pellet is resuspended in 5 to 10 mL of the above buffer, but without PMSF, and the extract is dialyzed overnight against 5 L of the same buffer.
4. Gel filtration chromatography: The dialyzed extract is further purified by Sepharose® S-200 chromatography (Amersham Pharmacia Biotech, Piscataway, NJ, USA). This is a size exclusion procedure, which separates the protein mixture on the basis of native molecular mass. As a homodimer, ALAS has a native molecular mass of about 90 kDa. Using a column (100 × 5 cm) that had been pre-equilibrated with the same buffer, the dialyzed ammonium sulfate fraction is carefully placed on the top of the column, and the system is developed at a flow rate of about 1 mL/minute.
5. Fractions containing ALAS are determined by the presence of ALAS activity (see below) and are pooled. PMSF is added to give a final concentration of 200 µM, and the extract is diluted 2-fold with distilled water.
6. Anion exchange chromatography: The diluted ALAS solution is next subject to anion exchange chromatography, a procedure that separates proteins according to their negative charge. The ALAS solution is applied to a diethylaminoethyl (DEAE)-Sephacel™ chromatography column (Amersham Pharmacia Biotech) (25 × 2.7 cm) that had been pre-equilibrated in 10 mM sodium phosphate buffer, pH 7.2, containing 0.5 mM pyridoxal 5′ phosphate, 2 mM EDTA, 10% glycerol, 5 mM 2-mercaptoethanol, and 100 µM PMSF. The ALAS is eluted from the column by the application of a linear gradient extending from 0 to 500 mM NaCl in a total volume of 500 mL using the same buffer.
7. Fractions containing ALAS activity, which normally elutes between 30 to 50 mM NaCl, are pooled and dialyzed overnight against the same buffer.
8. Hydroxyapatite chromatography: The dialyzed enzyme preparation is applied to a hydroxyapatite column (25 × 2.7 cm) prepared from hydroxyapatite (HTP) (Bio-Rad Laboratories, Hercules, CA, USA). The column is washed with 100 mL of the same buffer as above, and the enzyme is eluted by the application of a linear gradient extending from 0 to 500 mM NaCl in the same buffer. Fractions eluting at about 25 mM NaCl (total volume 50 mL) containing the pure ALAS are pooled and dialyzed against 20 mM sodium phosphate buffer, pH 7.2, containing 0.5 mM pyridoxal 5′ phosphate, 2 mM EDTA, 10% glycerol, 5 mM 2-mercaptoethanol, and 100 µM PMSF.
9. Storage: The purified enzyme is concentrated to 10 mL under nitrogen using an Amicon® concentration cell fitted with a PM-10 membrane (Millipore, Bedford, MA, USA) and is stored at -20°C, where it is known to remain active for at least 3 months. The purified protein can be visualized after

sodium dodecyl sulfate polyacrylamide gel electrophoresis (SDS-PAGE), where it migrates as a single polypeptide with a molecular mass of about 45 kDa. One liter of culture should produce about 5 mg of purified enzyme.

3.2. Enzyme Assay and Incubation Protocol

ALA may be generated, using ALAS, with the following incubation mixture and substrates.

1. Incubation mixture: Stock reaction buffer (100 μL) consisting of 20 mM potassium phosphate buffer, pH 7.2, containing 0.5 mM pyridoxal 5′ phosphate, and 250 mM glycine is mixed with 5 μL of purified ALAS, and the reaction is initiated by the addition of 25 μL of 10 mM succinyl-CoA. Incubation is carried out at 37°C for up to 30 minutes. The incubation can be scaled up as required for the synthesis of ALA. The yield of ALA should be in excess of 90%, and lower yields are normally associated with an underestimation in the amount of succinyl-CoA.
2. Preparation of succinyl-CoA: Succinyl-CoA can either be purchased commercially (e.g., Sigma, St. Louis, MO, USA) or may be prepared freshly by reacting 8 mg of CoA, 1 mg of freshly powdered succinic anhydride, and 4.5 mg of sodium bicarbonate in 1 mL of distilled water on ice for 30 minutes with stirring. More concentrated succinyl-CoA solutions can be obtained by using less water.

3.2.1. Discontinuous Assay of ALAS

The ALAS may be quantified using the discontinuous chemical assay of Mauzerall and Granick (21). In this case, the above reaction is made to a final volume of 175 μL. The reaction is terminated by the addition of 150 μL of 10% (wt/vol) trichloroacetic acid, and any protein precipitate is removed by centrifugation. A known volume (300 μL) of the supernatant is transferred to a fresh tube containing 300 μL of 1 M sodium acetate buffer (pH 4.6), and 25 μL of acetylacetone is then added. The mixture is heated to 100°C for 10 minutes and, after cooling, an equal volume of modified Ehrlich's reagent (prepared by dissolving 1 g of p-dimethylaminobenzaldehyde in 42 mL of acetic acid and 8 mL of perchloric acid [62% wt/vol]) is added. After allowing 10 minutes for the color (pink) to develop fully, the absorbance of the resultant solution is measured at 553 nm in a spectrophotometer. Enzyme rates are calculated using an extinction coefficient of 6.04×10^4 M^{-1} cm^{-1}.

3.2.2. Continuous Assay of ALAS

A more convenient, though less sensitive, enzyme-linked spectrophotometric assay can also be employed to monitor the activity of ALAS, in which the liberated CoA is coupled to the formation of acetyl-CoA and reduced nicotinamide adenine dinucleotide (NADH) with the enzyme pyruvate dehydrogenase (27).

$$\text{Glycine} + \text{succinyl-CoA} \rightarrow \text{ALA} + \text{CoASH} + CO_2$$

$$\text{CoASH} + \text{pyruvate} + \text{NAD} \rightarrow \text{acetyl-CoA} + CO_2 + \text{NADH} + H^+$$

An alternative enzyme-linked continuous assay using 2-oxoglutarate dehydrogenase involves the regeneration of succinyl-CoA from liberated CoA and 2-oxoglutarate, also forming NADH that can be monitored spectroscopically (12).

3.3. Preparation of Isotopically Labeled ALA

An adaptation of the above assay method can be used to generate isotopically labeled ALA. For instance, either [^{13}C] or [^{14}C]-label at the C5 position of ALA may be introduced from glycine, appropriately labeled at C2. *2RS*-[$^3H_2$]-glycine may be used to label *5S*-[^{3}H] ALA, where the label is stereospecifically located on the aminomethyl methylene carbon atom.

Labeled succinyl-CoA may be used for introducing label at ALA positions C1 through C4. ALA, randomly or stereospecifically tritiated at the C2 and C3 positions, may be generated from 2-oxoglutarate using 2-oxoglutarate dehydrogenase, followed by decarboxylation to succinate, and chemical conversion to succinyl-CoA. However, because of the instability of ALA, it is essential to transform the ALA synthesized with ALAS rapidly into porphobilinogen (PBG), using purified 5-aminolevulinic acid dehydratase (ALAD), in order to stabilize any labeled hydrogen atoms (1). The procedure for coupling ALAS to ALAD is covered in the next section. Labeled PBG prepared from stereospecifically tritiated or deuterated ALA in this way has proved important for mechanistic studies on ALAS and ALAD, as well as on enzymes further along the heme pathway (28).

4. THE ENZYMATIC SYNTHESIS OF PORPHOBILINOGEN

ALAD catalyzes the first of three steps for the transformation of ALA into uroporphyrinogen III, which are found in all living organisms that synthesize tetrapyrroles (13). The enzymes exist as homo-octamers with subunit molecular masses of 35 to 45 kDa, depending on the source organism, and catalyze the condensation of two molecules of ALA into the pyrrole PBG (Figure 3). Comparisons between the amino acid sequences derived from nucleotide sequencing indicate that the enzyme structure is strongly conserved, and this is confirmed by crystallographic studies that show that both prokaryotic and eukaryotic dehydratases have a similar $(\alpha\beta)_8$ barrel subunit structure (7,8). The active site is located at the center of the barrel with two juxtaposed lysines and an aspartic acid playing essential roles in catalysis. One of the lysines, K247 in the *E. coli* enzyme, forms a Schiff base with the substrate molecule at the P-site, so called because it binds the ALA molecule that ultimately becomes the propionic acid side chain of the product PBG. Pairs of

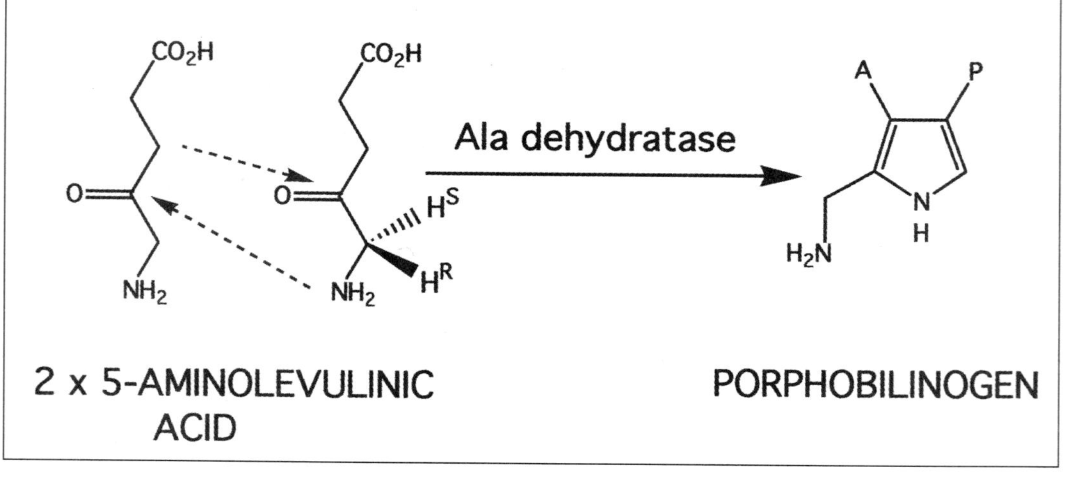

Figure 3. The biosynthesis of porphobilinogen from 2 molecules of ALA. It has been established that the *proR* hydrogen of the ALA molecule occupying the P (propionate) site is stereoselectively removed during the reaction.

subunits are arranged as dimers, held together by long N-terminal arms, with four dimers arranged in D_4 symmetry to form the octamer. The conservation of the quaternary structure through evolution may, in part, be as a consequence of a second and somewhat surprising function of the protein, namely, as the inhibitory complex of the proteasome (10).

ALADs differ in their requirement for divalent metal ions. Those found in animals require only zinc for activity, those found in plants require only magnesium, and others require zinc but are activated by magnesium (14). *E. coli* ALAD, used for the methods below, is of the magnesium-activated zinc type. The metal ion in the zinc-dependent enzymes is chelated to a triple cysteine motif at the active site and appears to be an essential part of the active site that binds the second molecule of ALA at the A-site. The zinc-dependent dehydratases are exceptionally sensitive to low levels of lead, which displaces the zinc ion and inactivates the enzyme. Although ALADs may be purified from a variety of natural sources, the most convenient purification (31) is from a recombinant strain of *E. coli* harboring the *E. coli hemB* gene.

4.1. Purification of ALAD from *E. coli*

Traditionally, ALADs have been isolated from sources that make large quantities of either heme or chlorophyll, such as liver, erythrocytes, and plants. However, more recent cloning strategies have led to the production of large quantities of recombinant forms of the enzyme. In this section, we will detail the purification of a recombinant version of the *E. coli* ALAD. Because of the way the protein is folded, it is not possible to tag the enzyme, for instance, with a polyhistidine epitope to enable affinity purification. Thus, overproduced ALAD has to be purified using conventional chromatographic procedures.

❖ Procedure 3. Purification of Recombinant ALAD from *E. coli*

1. Bacterial growth: *E. coli* strain TB1 containing the plasmid pUC19 harboring the *E. coli hemB* gene in a modified *Eco*RI-*Bam*HI fragment (Table 1) as constructed by Li et al. (20) is grown in 500 mL of LB medium containing ampicillin (50 μg/mL) for 24 hours after inoculation from a starter culture.
2. Harvesting and cell lysis: The cells are harvested by centrifugation at 3000× *g* for 30 minutes and washed to remove excess medium. Approximately 4 g of cell paste are then suspended in 20 mL of 50 mM potassium phosphate buffer, pH 6.0, containing 100 μM $ZnSO_4$, and 20 mM 2-mercaptoethanol. The cells are disrupted by sonication as outlined in section 1.1 and the cell debris is removed by centrifugation at 10 000× *g* for 20 minutes.
3. Ammonium sulfate fractionation: The resulting supernatant is treated with solid ammonium sulfate to bring the saturation to 30% by the addition of 16.6 g of solid ammonium sulfate per 100 mL of extract, and the precipitate is discarded. Addition of a further 5.7 g of ammonium sulfate per 100 mL of extract is added to the supernatant to bring the saturation to 40%, and the precipitate containing the enzyme is collected by centrifugation. The pellet is subsequently resuspended in 3 mL of the above buffer.
4. Gel filtration chromatography: The enzyme is further purified by chromatography using a Sephacryl® S-300 gel filtration column (Amersham Pharmacia Biotech), previously equilibrated in the same buffer. Fractions containing the majority of the ALAD activity (for assay see below) are collected from the column, concentrated to 20 mg/mL, and dialyzed against 50 mM

potassium phosphate buffer, pH 7.0, containing 100 μM $ZnSO_4$, and 20 mM 2-mercaptoethanol.

5. High resolution anion exchange chromatography: Final purification may be achieved by chromatography using a Mono Q™ 5HR FPLC column (Amersham Pharmacia Biotech) equilibrated with the same buffer. The enzyme is eluted in buffer with a linear gradient from 0 to 1 M KCl, and active fractions are collected and pooled.
6. Storage: The pooled active fractions are concentrated to about 2 mg/mL, and the purified enzyme is filter-sterilized for storage at 4°C in 50 mM potassium phosphate buffer, pH 6.0, containing 100 μM $ZnSO_4$, and 20 mM 2-mercaptoethanol. Activity is maintained for 2 weeks. From a 0.5-L culture, between 10 to 20 mg of purified ALAD is obtained.

4.2. Enzyme Assay and Incubation Protocol

Different protocols may be employed for the enzymatic synthesis of PBG, depending on whether a small- or large-scale preparation is required. Indeed, the small-scale synthesis is identical to that used for the assay of the enzyme.

4.2.1. Assay and Small-Scale Enzymatic Synthesis of PBG

Purified *E. coli* ALAD (1–10 μg) is preincubated in a total volume of 500 μL of 50 mM potassium phosphate buffer, pH 8.0, containing 50 μM $ZnSO_4$, and 10 mM 2-mercaptoethanol. The reaction is initiated by the addition of ALA to give a final concentration of 5 mM. Incubation is carried out at 37°C for 3 minutes, after which time an equal volume (500 μL) of 10% trichloroacetic acid containing 0.1 M $HgCl_2$ is added to terminate the reaction and to precipitate the thiol and protein. After centrifugation, an aliquot of the supernatant is mixed with an equal volume of modified Ehrlich's reagent (21), and the absorbance is measured at 555 nm (E_{555} = 6.02×10^4 M^{-1} cm^{-1}). Most ALADs are susceptible to end-product inhibition, a factor that tends to limit the yields of PBG. Typically a yield of 80% is achieved.

4.2.2. Large-Scale Preparation of PBG

ALAD (500 U) is incubated in a stoppered conical flask at 37°C in 1.9 L of 5 mM potassium phosphate buffer, pH 6.8, containing 5 μM $ZnSO_4$, 5 mM 2-mercaptoethanol, and ALA. The 5-aminolevulinic acid hydrochloride (1 g; Sigma) is dissolved in about 95 mL of the same buffer, adjusted carefully to pH 6.8 with 0.1 M NaOH, and made up to 100 mL before adding to the above flask. Incubation is carried out under nitrogen, typically, for 10 hours or until the rate of PBG production has ceased. The reaction is followed by removing 10 μL of the incubation mixture at intervals and adding to 490 μL of 10% trichloroacetic acid containing 0.1 M $HgCl_2$ to precipitate the thiol. After centrifugation, 0.4 mL of the supernatant is mixed with an equal volume of modified Ehrlich's reagent, and the absorbance is measured as above. The PBG is purified from the reaction mixture by adjusting the pH to 7.5 and passing the incubation mixture slowly through a column (2 × 12 cm) of Dowex 1 × 8 acetate (200–400 mesh). The column is first washed with 1 L of distilled water, and the PBG is eluted with 1 M acetic acid and collected by freeze-drying the solution or rapid-flash evaporation below 30°C. The PBG is recrystallized by dissolving it in a minimum volume of 1 M ammonia and adding 1 M acetic acid to bring the pH to the isoelectric point of 5.5. After allowing the crystallization to proceed for 1 hour, the crystals are filtered off and washed with a minimum volume of ice-cold methanol, followed by dry ether, and stored desiccated in vacuo at -20°C. The overall yield of purified PBG is about 50% after recrystallization.

4.2.3. Labeled PBG Synthesis

For the preparation of radioactive PBG from 5-amino[4-^{14}C]levulinic acid, the concentration of potassium phosphate buffer should be reduced to 5 mM and sufficient *E. coli* ALAD units used to ensure quantitative conversion within 20 to 30 minutes. It is essential to adjust the pH of the ALA prior to addition to the enzyme, particularly if it is dissolved in 0.1 M HCl. After synthesis of the PBG, the volume of the solution is reduced, for example using a Speedivac (centrifugation under reduced pressure) or by lyophilization, and the PBG is purified by chromatography using preparative cellulose glass plates developed in *n*-butanol:acetic acid:water (4:1:1 vol/vol). After carefully drying the plates in a cool nitrogen stream, PBG is eluted from the cellulose with water and lyophilized for storage in liquid nitrogen. The PBG can be detected on the plate by spraying the edge of the plate with modified Ehrlich's reagent.

4.2.4. Coupled Enzymatic Synthesis of Labeled PBG Samples from ALA Generated from Glycine and Succinyl-CoA by ALAS

The difficulty of isolating labeled ALA, prepared either from labeled glycine or succinyl-CoA, may be overcome by coupling the reaction to ALAD to convert rapidly any ALA formed into PBG. The latter is more stable and easier to isolate and purify and fulfills the additional requirement that any labeled hydrogen atoms are located in stable positions.

Tritiated or deuterated succinyl-CoA may be used for introducing either randomly or stereospecifically located label at C2 and C3 of ALA. This is accomplished by labeling 2-oxoglutarate with either tritium or deuterium, either nonenzymically or using 2-oxoglutarate dehydrogenase, followed by nonenzymatic decarboxylation to succinate, cyclization to succinic anhydride with dicyclohexylcarbodiimide, and conversion to succinyl-CoA as described above (see also Reference 10). The succinyl-CoA is then transformed into PBG in 5 mM Tris-HCl buffer, pH 6.8, containing 80 mM glycine, 10 μM pyridoxal 5′ phosphate, ALAS (20 U), and ALAD (35 U). The reaction is started by adding labeled succinyl-CoA to give a final concentration of 10 mM and a total volume of 1 mL, and the incubation is continued until the reaction is complete, typically in 30 to 60 minutes.

PBG is then separated from any ALA and glycine by adjusting the mixture to pH 7.2 and application to a Dowex 2 × 8 acetate (400 mesh) column (2 × 10 cm). ALA is removed by washing the column with water (50 mL), and PBG is eluted with 20 mL of 1 M acetic acid and purified by cellulose chromatography as above.

Glycine labeled with ^{13}C, ^{14}C, ^{3}H, or ^{2}H label at C2 may be used to label ALA at the C5 position. Thus, *2RS*-[$^{3}H_2$]-glycine incubated with ALAS generates stereospecifically labeled 5S-[^{3}H] ALA, which can be transformed by ALAD into *11S*-[^{3}H]-PBG. In this case, the reaction mixture is prepared at 0°C in a volume of 1 mL containing 10 mM Tris-HCl buffer, pH 6.8, 3 mM *2RS*-[$^{3}H_2$]-glycine, 5 μM pyridoxal 5′-phosphate, ALAS (20 U), and ALAD (30 U). The reaction is started by the addition of 50 μL of succinyl-CoA (1 μmol) and by raising the temperature to 37°C. Further aliquots of succinyl-CoA may be added at 10-minute intervals. After incubation for 30 minutes, the PBG is purified using a Dowex 2 × 8 acetate column as above.

5. SYNTHESIS OF PREUROPORPHYRINOGEN

Porphobilinogen deaminase (PBGD) also known as hydroxymethylbilane synthase (HMBS) and incorrectly as uroporphyrinogen I synthase, catalyzes the formation of preuroporphyrinogen from 4 molecules of PBG (Figure 1). Preuroporphyrinogen is a highly unstable 1-hydroxymethylbilane that acts as the substrate for uroporphyrinogen III synthase to yield uroporphyrinogen III, the common tetrapyrrole precursor for other tetrapyrroles. PBGDs have been isolated from a number of sources, and their properties have been well established (for a review see Reference 28). All PBGDs exist as monomeric species with molecular mass values between 33 and 45 kDa. The nucleotide sequences of genes/cDNAs specifying the deaminases from bacterial, plant, and animal sources show considerable conservation in the deduced protein sequences, suggesting that all the enzymes are likely to be structurally related to one another. Investigations with the deaminase from *E. coli* have identified a novel prosthetic group, named the dipyrromethane cofactor (16), made up of 2 PBG-derived units linked together and covalently attached to the enzyme. The cofactor acts as a primer for the synthesis of the linear tetrapyrrole (bilane) chain that is built onto the free α-position of the cofactor. This occurs by the sequential condensation of 4 PBG molecules with the holoenzyme through enzyme intermediate complexes, termed ES, ES_2, ES_3, and ES_4. The product, preuroporphyrinogen, is liberated from ES_4 by hydrolysis, regenerating the holoenzyme with the cofactor still covalently attached.

5.1. Enzyme Purification

The deaminase is conveniently isolated from a variety of sources (for reviews see Reference 1). In this case, PBGD is expressed from strain BM3 (Table 1), consisting of *E. coli* TB1 harboring a plasmid (pBM3) constructed by cloning a 1.68 kb *Bam*HI-*Sal*I DNA fragment, containing the *E. coli hemC* gene from pST48, into pUC18 (32).

❖ Procedure 4. Purification of Recombinant *E. coli* PBGD

1. Bacterial growth: Sterilized bacterial medium (4 L) containing 50 mg/mL ampicillin is inoculated from a starter culture and incubated at 37°C overnight in 4 baffled flasks (2 L).
2. Harvesting and cell lysis: The cells are collected by centrifugation at 3000× *g* for 30 minutes and resuspended (3–4 mL/g of cells) in 0.1 M potassium phosphate buffer, pH 8.0, containing 14 mM 2-mercaptoethanol. PMSF, dissolved in ethanol, is added to give a final concentration of 0.1 mM. The bacteria are broken by sonication as described in Procedure 1, and the sonicated extract is clarified by centrifugation at 15 000× *g* for 15 minutes.
3. Heat treatment: PBGDs are thermostable enzymes, and this property is utilized during the purification procedure. The sonicated sample is heat-treated by placing the sample in a water bath for 10 minutes at 60°C, followed immediately by cooling to 0°C in an ice–salt bath. The precipitated protein is removed by centrifugation at 10 000× *g* for 20 minutes at 4°C.
4. Ammonium sulfate fractionation: Solid ammonium sulfate is added slowly to the above extract (protein 30 mg/mL) to give 30% saturation by the addition of 16.6 g of ammonium sulfate per 100 mL of extract. The solution is allowed to equilibrate with stirring for 10 minutes at 4°C, and the supernatant is removed by centrifugation at 10 000× *g*

for 20 minutes at 4°C. Further ammonium sulfate is added to give 60% saturation by the addition of a further 19.8 g per 100 mL of extract. The pellet containing the enzyme is collected by centrifugation and resuspended in 30 mL of 0.1 M potassium phosphate buffer, pH 8.0, containing 14 mM 2-mercaptoethanol. The sample is then dialyzed against 2 L of the same buffer at 4°C for at least 4 hours with stirring.

5. Ion exchange chromatography: Ion exchange chromatography is performed using a DEAE-Sephacel column (2.5 × 20 cm) equilibrated and eluted in an isocratic fashion by the passage of a 2 L of 0.1 M potassium phosphate buffer, pH 8, containing 14 mM 2-mercaptoethanol, through the column. The column fractions containing the deaminase enzyme are located by assay (see below) and by SDS-PAGE.
6. Storage: Active fractions found to be free from any major contaminating proteins are concentrated to 10 to 15 mL by ultrafiltration, using a 100-mL ultrafiltration cell fitted with a PM-10 membrane, and the deaminase is desalted into distilled water using a small gel filtration system such as a PD-10 column. The purified enzyme (specific activity 30–40 U/mg) is then lyophilized to yield a white solid that is stable for several months when stored at -20°C under nitrogen. This protocol generates about 10 mg of purified PBGD per liter of culture.

5.2. Assay of Enzyme

PBGD is assayed using a stopped assay. To 750 µL of 0.1 M Tris-HCl buffer, pH 8.0, is added 40 µL of enzyme containing between 0.1 to 1 µg of purified enzyme. The incubation mixture is equilibrated at 37°C in a water bath, and the reaction is started by the addition of 100 µL of 1 mM PBG. After 10 minutes at 37°C the reaction is terminated by the addition of 200 µL of 5 N HCl. A further 10 µL of benzoquinone (1 mg/mL in methanol) is added, and the mixture allowed to oxidize for a further 20 minutes under bright light. The absorbance is determined at 405 nm (E_{405} = 5.48 × 10^5 M^{-1} L), and reading should fall between 0 to 1 OD units on the spectrophotometer. It may be necessary to dilute the sample 10-fold in 1 N HCl to achieve such readings. One unit is defined as the amount of PBGD enzyme needed to consume 1 µmol of PBG per hour.

5.3. Preparation of Preuroporphyrinogen

Preuroporphyrinogen (0.1 µmol) is generated from 0.5 µmol of PBG in a final volume of 1 mL of degassed Tris-HCl buffer, pH 9.1, using 100 mg of purified PBGD over a period of 1 minute at 37°C. The reaction is performed at a higher pH than the assay to help stabilize the preuroporphyrinogen. The sample is rapidly cooled to 0°C in liquid nitrogen, and the preuroporphyrinogen is separated from the holo-deaminase by ultrafiltration through a PM-10 membrane fitted to a 5-mL concentration cell under nitrogen at 4°C in a cold room. The preuroporphyrinogen is used at once or frozen in liquid nitrogen for up to 1 hour, under nitrogen, until required. The yield of preuroporphyrinogen is in excess of 80%.

6. MULTIENZYME SYNTHESIS OF UROPORPHYRINOGEN III

The enzyme uroporphyrinogen III synthase (UROS) (also known as uroporphyrinogen III cosynthase) catalyzes a remarkable reaction in which preuroporphyrinogen is rearranged and cyclized to yield uroporphyrinogen III (Figure 4). Uroporphyrinogen III is the common precursor for hemes, chlorophylls, vitamin B_{12}, and all other tetrapyrroles (for a review see Reference 28). The UROS substrate, preuroporphyrinogen, is generated by the preceding enzyme of the tetrapyrrole pathway, PBGD (see above section) by a reaction that involves the polymerization of 4 molecules of the monopyrrole precursor PBG. Preuroporphyrinogen has a half-life of less than 5 minutes at neutral pH values (15) cyclizing spontaneously to uroporphyrinogen I, a physiologically unimportant isomer (Figure 4). Uroporphyrinogen III, however, represents an important transitory intermediate in the synthesis of the modified tetrapyrroles, since it represents the first branchpoint in the pathway of cobalamin, siroheme, or coenzyme F_{430}, while decarboxylation of the 4 acetate side chains of uroporphyrinogen III by the enzyme uroporphyrinogen III decarboxylase produces coproporphyrinogen. The ability to produce uroporphyrinogen III in good yields is therefore important for the study of these branchpoint enzymes. Uro-

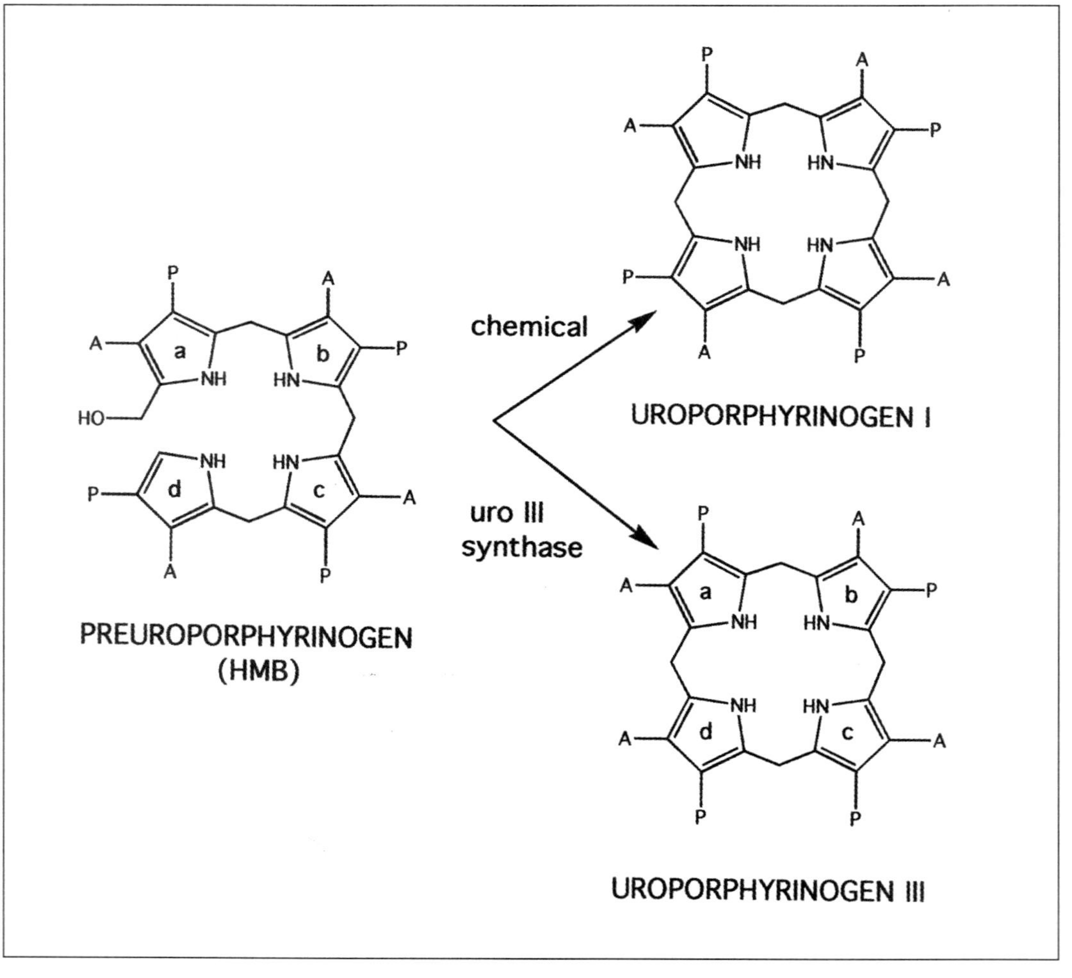

Figure 4. The synthesis of uroporphyrinogen I and III from preuroporphyrinogen. Note the action of UROS, which is able to invert the orientation of ring d during the macrocyclic ring closure process.

porphyrinogens can be generated either non-enzymically, by the reduction of uroporphyrin or, in situ, using a coupled enzyme system. The non-enzymic reduction of uroporphyrin is achieved by the use of sodium amalgam. Although the reaction is easy to perform, problems can arise from the pH of the solution, which becomes very high by the end of the reductive process. For these reasons, we have favored the generation of uroporphyrinogen by enzymatic transformation of PBG, employing the actions of the enzymes PBGD and UROS.

6.1. Purification of UROS

UROS can be purified from a recombinant version of the *E. coli hemD* that has been modified to incorporate a 6-histidine (His) tag at the N terminus of the protein. The *E. coli hemD* was amplified by PCR with appropriately designed primers such that the gene was cloned into the *Nde*I and *Bam*HI sites of pET14b, giving the plasmid pET14b-HemD (Table 1). When transformed into *E. coli* BL21(DE3)pLysE, the strain was found to overproduce the His-tagged version of the protein, which has a molecular mass of 29 kDa. The strain harboring the plasmid is comparatively unstable, and fresh transformants are required when cultures are to be grown. The following protocol can be adapted for purification of all the His-tagged enzymes described in this chapter.

❖ Procedure 5. Purification of His-Tagged UROS from *E. coli*

1. Bacterial growth: The bacteria are grown from a starter culture in 2-L baffled flasks containing 1 L of LB media (with appropriate antibiotics) at 37°C with vigorous shaking until an A_{600} = 0.6 is reached, at which point isopropyl-β-D-thiogalactoside (IPTG) is added to a final concentration of 0.4 mM, and the cells are grown for another 2 hours.
2. Harvesting and cell lysis: The bacteria are collected by centrifugation (10 000× *g* at 4°C). The bacterial pellet is resuspended in 10 mL of binding buffer (5 mM imidazole, 0.5 M NaCl, 20 mM Tris-HCl, pH 7.9). The bacterial suspension is sonicated as described in Procedure 1, and the solution is centrifuged (10 000× *g* at 4°C) to remove the cellular debris.
3. His-bind column: The His-tag sequence of the fusion protein can bind to divalent metal cations such as Co^{2+} and Ni^{2+} immobilized on to His-bind resin (Novagen, Madison, WI, USA; however, many suppliers make different forms of metal chelate resin and readers are encouraged to browse the multitude of catalogues available). After unbound proteins are washed away, the His-tagged protein is eluted with imidazole. The resin (poured into a small column, 1 × 2.5 cm) is initially prepared by rinsing with 15 mL of water, charged with 25 mL of a 50 mM divalent cation solution (normally Ni^{2+}) (charge buffer), and equilibrated with 15 mL binding buffer. The supernatant is loaded onto the charged His-bind column. The column is washed with 10 column volumes of binding buffer, 6 column volumes of wash buffer (100 mM imidazole, 0.5 M NaCl, 20 mM Tris-HCl, pH 7.9), and finally the protein is eluted in 6 column volumes of elution buffer (400 mM imidazole, 0.5 M NaCl, 20 mM Tris-HCl, pH 7.9). The protein eluting from the column can be detected by the use of the Bio-Rad protein assay and SDS-PAGE.
4. Storage: Fractions containing the modified UROS are pooled and desalted by passing through a PD-10 column, previously equilibrated in 50 mM Tris-

HCl, pH 7.8. The protein is lyophilized and is stable in this form for up to 1 year. In comparison to some of the other enzymes described in this chapter, UROS is poorly expressed, and a yield of about 2 mg/L of culture is normally achieved.

6.2. Enzymatic Preparation of Uroporphyrinogen III

Uroporphyrinogen III can be synthesized in vitro using PBG and purified PBGD and UROS. The reaction can be undertaken in a range of buffers between pH 7.5 and 9.0, although the uroporphyrinogen III is generally more stable at the higher pH values. To prevent oxidation of the product, the buffers are normally thoroughly degassed by freeze–thawing under a vacuum of less than 1 mbar. For efficient transformation of PBG into uroporphyrinogen III, the reaction mixture should contain PBGD at 10 μg/mL, UROS at 2 μg/mL, and PBG at 100 μM. The reaction is effectively quantitative, thus producing uroporphyrinogen III at a concentration approaching 25 μM. This can be verified by taking 50 μL of the incubation, mixing with 950 μL of 1 N HCl, and leaving under a bright light for 20 minutes. After centrifugation in an Eppendorf model microfuge at 13 000 rpm for 5 minutes, the absorbance of the solution at 405 nm can be measured, and the concentration of porphyrin can be determined using the extinction coefficient of 5.48×10^5 M^{-1} L.

So long as the enzymatic incubation is kept in an anaerobic environment under reduced light, the uroporphyrinogen III is stable for several hours. The solution should appear colorless, but if it starts to turn pink then this is diagnostic of the solution starting to oxidize. To isolate the uroporphyrinogen III from the incubation (i.e., to remove the enzymes from the reaction mixture) the solution can be filtrated in an ultrafiltration unit fitted with a PM-10 membrane. The filtrate should be kept under argon to help prevent any oxidation. The yield of uroporphyrinogen III from PBG is normally in excess of 95%.

The uroporphyrinogen I isomer can also be synthesized by this method simply by omitting UROS from the incubation.

7. SYNTHESIS OF PRECORRIN-2 (DIHYDROSIROHYDROCHLORIN), SIROHYDROCHLORIN, AND SIROHEME

Enzymatic transformations of uroporphyrinogen III into precorrin-2 are dependent upon the presence of the enzyme uroporphyrinogen III methyltransferase (Figure 5), which requires *S*-adenosyl-L-methionine (SAM) as a methyl donor (3). There are a number of sources of this enzyme including *Pseudomonas denitrificans*, *Bacillus megaterium*, and *Bacillus stearothermophilus*. The CysG enzyme from both *E. coli* and *Salmonella typhimurium* can also be used, although CysG is, in fact, a multifunctional enzyme responsible for the conversion of precorrin-2 into siroheme (30). However, in the presence of only SAM and uroporphyrinogen III, the enzyme will effectively transform uroporphyrinogen III into precorrin-2. The uroporphyrinogen methyltransferases are normally homodimers with a subunit molecular mass of about 30 kDa, while the CysG proteins, which are also homodimers, have a subunit molecular mass of 50 kDa.

7.1. Purification of Uroporphyrinogen Methyltransferases

Although the uroporphyrinogen methyltransferases can be purified from recombinant sources, the preparations are often laborious and in low yields. We have favored

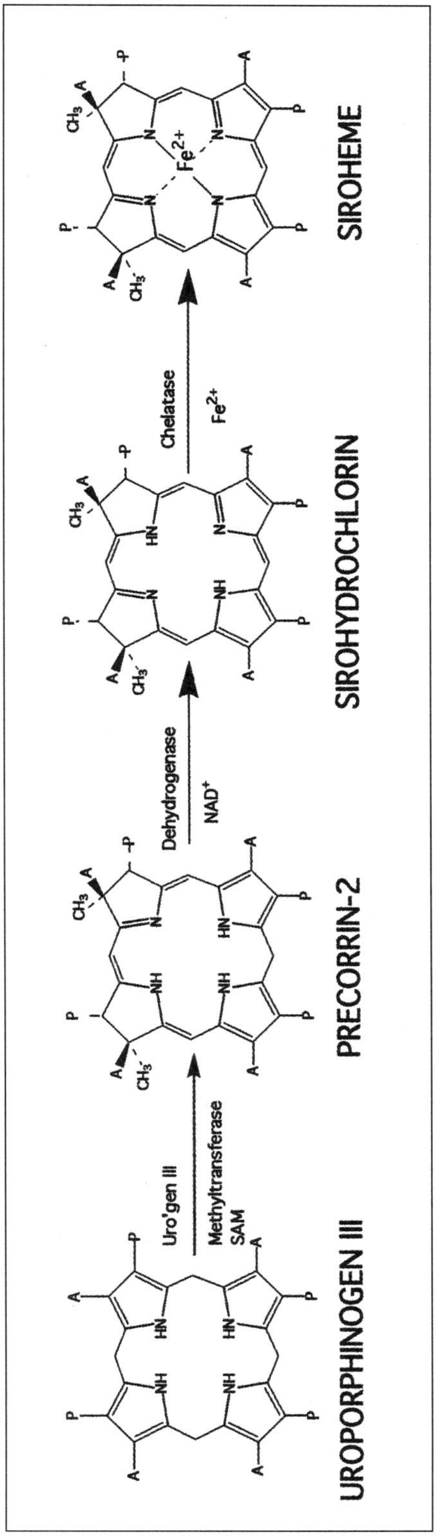

Figure 5. The biosynthesis of siroheme from uroporphyrinogen. Uroporphyrinogen III is methylated at positions 2 and 7 to give precorrin-2 by the enzyme uroporphyrinogen methyltransferase, while dehydrogenation of precorrin-2 gives sirohydrochlorin and finally ferrochelation produces siroheme.

the use of His-tagged enzymes, including the *B. stearothermophilus* CobA and the *E. coli* CysG, since these can be purified easily in a couple of hours by metal chelate chromatography. For instance, the *B. stearothermophilus* CobA can be purified from strain ER262 (Table 1), which is BL21(DE3) pLysS transformed with pER259 (*cobA* cloned into pET14b). As for all His-tagged enzymes, the isolation procedure is very similar to that described in Procedure 5 (Section 6.1). About 15 mg of purified enzyme can be obtained per liter of culture.

7.2. Assay of Uroporphyrinogen Methyltransferase

The enzyme is very difficult to assay. Accurate activity for uroporphyrinogen methyltransferases can be obtained by measuring the incorporation of label from [methyl-^{3}H]SAM into the uroporphyrinogen III framework as previously described (3). The enzyme is incubated in 50 mM Tris-HCl buffer containing 50 μM SAM (10 μCi.μmol^{-1}) and 5 μM uroporphyrinogen III at either 30° or 37°C for up to 1 hour in a final volume of 1 mL. After incubation, the mixture is quickly applied to a small column (e.g., 0.5 mL bed volume) of DEAE Sephacel. After washing the column with 10 column volumes of buffer, the tetrapyrrole compounds were eluted in 3 mL of 1 M HCl. After mixing with an appropriate scintillant, the amount of radioactivity transferred to uroporphyrinogen III can be determined.

7.3. Generation of Product by Incubation of Recombinant Enzyme

Since many of the uroporphyrinogen III methyltransferases display substrate inhibition, uroporphyrinogen III is normally incubated with the enzyme at a final concentration of 5 μM, with SAM at a concentration of 50 μM (3). The high concentration of SAM helps to overcome inhibition with *S*-adenosyl-L-homocysteine. The reaction should be undertaken at pH 8.0 in 50 mM Tris-HCl buffer at either 30° or 37°C. As precorrin-2 is so unstable, we recommend that a high concentration of the uroporphyrinogen methyltransferase be used in the reaction at a concentration of about 50 µg/mL. This ensures a rapid synthesis of precorrin-2, which can be monitored visually since the solution turns a bright yellow color. In fact, precorrin-2 has a broad absorption maximum around 350 to 400 nm. The newly synthesized precorrin-2 can be separated from the other components of the incubation mixture by ion exchange chromatography. After mixing in a few milliliters of ion exchange resin such as DEAE Sephacel, the solution is slowly stirred for about 1 minute. Once the resin has settled, the majority of the supernatant can be decanted, and the resin slurry can be transferred to a small plastic column. The resin is washed with buffer, and buffer containing 250 mM NaCl, to remove the more loosely bound proteins, and the precorrin-2 is eluted in buffer containing 2 M NaCl. Precorrin-2 is highly unstable with a tendency to form mono- and dilactones. The compound is difficult to store and should be used immediately. The uroporphyrinogen methyltransferases are very susceptible to feedback inhibition by S-adenosyl-L-homocysteine, and therefore, to achieve high yields of precorrin-2 (in excess of 90%), a high concentration of enzyme and SAM are required in the incubation mixture.

Sirohydrochlorin can be synthesized from precorrin-2 by the inclusion of either CysG or Met8p together with NAD^+ to the above incubation (25). These enzymes are purified in the same manner as described for the CobA (above) from the appropriate strains shown in Table 1. The

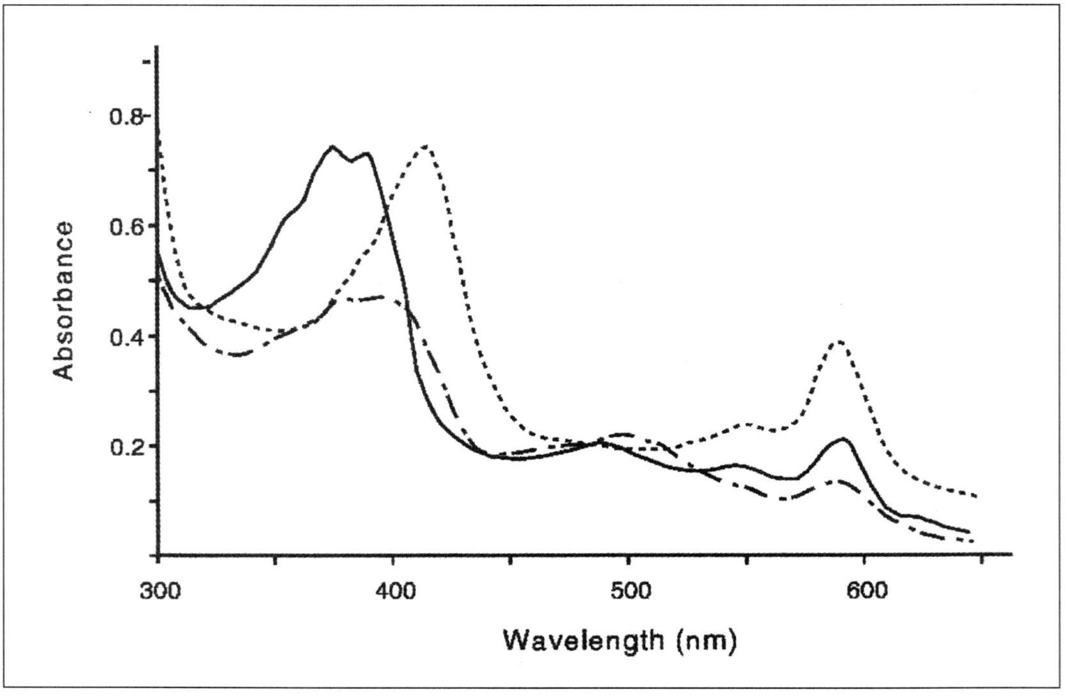

Figure 6. Spectra of precorrin-2, sirohydrochlorin, and cobalt-sirohydrochlorin. The spectrum of precorrin-2 (large dashed line) has a broad absorption maximum around 350 to 400 nm. The spectrum of sirohydrochlorin (filled line) has a more defined absorption maximum at 378 nm, while cobalt sirohydrochlorin (dashed line) has defined maxima at 410 and 595 nm.

CysG or Met8p should be added at a concentration of 50 µg/mL with NAD^+ at 25 µM. Sirohydrochlorin is characterized by the appearance of a new absorption maximum at 378 nm (Figure 6).

Siroheme can be synthesized by the inclusion of ferrous iron with the incubation. However, this reaction is difficult to follow, and a clearer spectral difference can be obtained by the use of cobalt, which produces a spectrum with absorption maxima at 410 and 595 nm (Figure 6). The metal ions should be added to a concentration no higher than 10 µM, otherwise the enzymes become inactivated.

8. SYNTHESIS OF COPROPORPHYRINOGEN

Decarboxylation of the 4 acetate side chains of uroporphyrinogen III leads to the synthesis of coproporphyrinogen III. The enzyme that catalyzes this reaction is uroporphyrinogen III decarboxylase (UROD). The best characterized enzyme is that from human, which can be expressed to high levels in *E. coli* cells as a His-tagged recombinant enzyme. The enzyme does not require any metal ions or cofactors for activity, since it most likely catalyzes the reaction by forming a protonated pyrrole within the porphyrinogen substrate, which acts as an electron sink. The enzyme is prone to acylation of cysteine residues and also to oxidation from bound porphyrinogens. The enzyme would appear to be dimeric with a subunit molecular mass of around 40 kDa. The overproduction of the human enzyme as a recombinant protein has allowed its crystallization, and a detailed 3-dimensional structure is now available (33). In humans, a number of mutations within the UROD gene are known to cause hereditary forms of porphyria, while the enzyme is also prone to inactivation by a number of porphyrinogenic compounds. The dysfunction of UROD is manifested as the most common form of porphyria, porphyria cutanea tarda (19).

8.1. Purification of UROD

A His-tagged recombinant form of UROD has been described recently (24), in which the His-tag does not appear to interfere with the catalytic activity of the enzyme. In this case, the cDNA corresponding to human UROD was cloned into a T7 inducible plasmid with an N-terminal His-tag (Table 1). Expression of the enzyme is achieved by transformation into *E. coli* BL21(DE3)pLysS. The purification of the His-tagged UROD is essentially similar to that described in Procedure 5 (Section 6.1), with yields in excess of 15 mg of purified enzyme per liter of culture.

8.2. Assay of UROD

The simplest way to monitor the activity of UROD is to employ a fluorometric method that relies on the difference in fluorescence between uroporphyrin and coproporphyrin (29). The reaction mixture (3 mL) is stopped by the addition of trichloroacetic acid (to a final concentration of 5%), and the porphyrinogens are then oxidized to their corresponding porphyrins by the addition of 60 µL of H_2O_2 (30%). After 20 minutes, the amount of coproporphyrin can be estimated from its emission fluorescence at 610 nm after excitation at 406 nm. The fluorescence is compared to a standard curve made from commercially obtained coproporphyrin. This technique can only be used as a rough guide to the activity of the enzyme. More accurate assays rely on the exact quantities of porphyrin isomers that are formed during the assay. This is generally achieved after esterification of the reaction products and separation by HPLC (see Chapter 5).

8.3. Synthesis of Coproporphyrinogen

Coproporphyrinogen III can be efficiently generated by the following protocol. An incubation mixture containing 50 mM Tris-HCl buffer, pH 8.0, 2 mM dithiothreitol (DTT), and 5 µM uroporphyrinogen III is prepared. The uroporphyrinogen III is made as described in Section 4. The buffer should be thoroughly degassed by freeze–thawing under reduced pressure. The reaction is started by the addition of purified UROD (5 µg/mL), and the incubation is performed at 37°C under dim light. The coproporphyrinogen III can be removed from the enzyme mixture by ultrafiltration through a PM-10 membrane in an ultrafiltration unit. The solution should appear colorless, and any appearance of reddish coloration should be taken as a sign of oxidation. The coproporphyrinogen should be used immediately, although it may be possible to freeze the solution so long as it is kept under argon. The yield of coproporphyrinogen from uroporphyrinogen is in excess of 95%.

9. SYNTHESIS OF PROTOPORPHYRINOGEN

The synthesis of protoporphyrinogen requires the decarboxylation of the two propionate side chains on rings a and b of the coproporphyrinogen III isomer by the enzyme coproporphyrinogen oxidase (CPO). There are two independent enzyme systems that achieve this transformation, representing aerobic (encoded by *hemF*) and anaerobic processes (encoded by *hemN*). However, the aerobic enzyme is much better characterized, where purified recombinant *hemF*-encoded CPO has been shown to require two molecules of oxygen during the reaction with the release of two molecules of carbon dioxide (22). Although some reports have suggested the enzyme has a requirement for metal ions for activity, the human enzyme appears functional in the absence of any metal or cofactors. Indeed, the simplest source of the enzyme is a His-tagged version of the human enzyme, which is easily overproduced in *E. coli*, yielding in excess of 10 mg/L.

9.1. Purification of CPO

Although the human CPO is thought to be associated with the outer surface of the inner membrane of the mitochondrion, when expressed in *E. coli* it is easily solubilized in the presence of 0.5% *n*-octyl-β-D-glucopyranoside (22). Recombinant expression of the human CPO was achieved by cloning the cDNA into the expression vector pBTac such that the cDNA was cloned in-frame with a 6-histidine tag at the 5′ end (Table 1). The resulting plasmid, termed pHHCPO, was transformed into *E. coli* JM109. Purification of the enzyme is essentially as described in Procedure 5 (Section 6.1), except that the resuspension buffer for the cell pellet (step 2) is 50 mM NaH_2PO_4, 300 mM NaCl, 0.5% *n*-octyl-β-D-glucopyranoside, and 100 mM Tris-HCl, pH 8.0, containing 1 mM PMSF. The Ni-column is washed with resuspension buffer containing 20 mM imidazole, and the CPO is eluted from the column in resuspension buffer plus 250 mM imidazole. After dialysis against resuspension buffer to remove the imidazole, the enzyme can be stored frozen at -20°C for several months. The yield of purified enzyme is in excess of 10 mg/L of culture.

9.2. Assay of CPO and Synthesis of Protoporphyrinogen

The incubation for the synthesis of protoporphyrinogen is undertaken in 50 mM Tris-HCl, pH 8.0, containing 0.2% Tween® 20 and 2.5 mM glutathione. Coproporphyrinogen III is generated by the reduction of coproporphyrin III dihydrochloride (Porphyrin Products, Logan, UT, USA) with 3% sodium amalgam. The reduction itself should be undertaken in 100 Tris-HCl, pH 8.0, and once the solution turns colorless, or nearly colorless, the solution is passed through a small 10-mL column of glass wool. This not only serves to remove the amalgam and mercury, but the glass wool also appears to bind the porphyrin while allowing the porphyrinogen to pass through (Dailey, personal communication). The pH of the solution is then adjusted back to around 7.0 to 8.0 by addition of 2 M morpholinepropanesulfonic acid (MOPS), pH 7.0. The coproporphyrinogen is added to the incubation mixture at a final concentration of 5 μM and incubated with CPO at a concentration of 10 μg/mL. The synthesis of protoporphyrin can be followed by coupling a small portion of the incubation with protoporphyrinogen oxidase (PPO) (see Section 8). Alternatively, the conversion of coproporphyrinogen to protoporphyrinogen can be determined by analysis of the oxidized methyl esters and quantified using an HPLC system (see Chapter 5). The incubation should be performed at 37°C under dim light. The enzyme can be removed from the incubation mixture by ultrafiltration through a PM-10 membrane attached to an ultrafiltration cell. As with the other porphyrinogens, protoporphyrinogen should be used immediately. The yield of protoporphyrinogen from coproporphyrinogen is in excess of 95%.

10. SYNTHESIS OF PROTOPORPHYRIN

The conversion of protoporphyrinogen into protoporphyrin is mediated by the enzyme PPO. The enzyme catalyzes the removal of six electrons and six protons from the porphyrinogen ring, thereby introducing three new double bonds. In aerobic organisms, the enzyme would appear to require the services of a flavin cofactor and passes the electrons onto molecular oxygen. The corresponding anaerobic oxidation of protoporphyrinogen remains poorly understood, but in *E. coli* it would appear to be a multiprotein complex that is coupled to the respiratory chain of the cell. From a commercial standpoint, the enzymatic oxidation of protoporphyrinogen represents an important target for a number of herbicides, diphenyl ether derivatives, which selectively inhibit the enzyme. Defects in the human enzyme are associated with variegate porphyria, the form of porphyria that is particularly common in South Africa (19).

10.1. Purification of PPO

Perhaps the simplest recombinant source of this enzyme is the PPO from *Myxococcus xanthus*, as described by Dailey and Dailey (6). This is a PPO that uses molecular oxygen as the terminal electron acceptor and is a single subunit enzyme. In eukaryotes, the enzyme is found on the cytosolic side of the inner mitochondrial membrane or associated with chloroplast membranes, while in bacteria, it is a peripheral membrane protein. The gene corresponding to the *M. xanthus* PPO was amplified and modified such that the N terminus encodes for a 6-histidine tag. The construct was subsequently cloned into a Tac-driven derivative of pBTac-1, yielding the plasmid pMx-PPO (Table 1).

❖ Procedure 6. Purification of Recombinant PPO from *M. xanthus*

1. Bacterial growth: *E. coli* cells harboring pMx-PPO are grown, and the harvested cells are sonicated as described above for CPO overproduction.
2. Membrane preparation: The lysed cells

are centrifuged at 100 000× *g*, and the supernatant discarded, then this membrane fraction is resuspended in 60 mL of NaH_2PO_4, pH 7.4, 300 mM NaCl, and 0.5% *n*-octyl-β-D-glucopyranoside. The suspension is centrifuged again at 100 000× *g* to separate the solubilized enzyme from the remaining membranes.

3. His-bind column: This is carried out as for CPO, except that PPO is eluted in buffer containing 150 mM imidazole. The recombinant protein can be detected by SDS-PAGE, migrating with a molecular mass of about 50 000 Da. The purified protein is yellow in color due to the presence of the flavin cofactor and has a characteristic flavoprotein UV/VIS spectrum.
4. Storage: The protein can be stored frozen at -20°C for several months. Purified PPO is obtained in excess of 10 mg/L of culture.

10.2. Synthesis of Protoporphyrin and Assay

The synthesis of protoporphyrin can be achieved either by use of a coupled enzyme system from coproporphyrinogen III or by chemical reduction of protoporphyrin by sodium amalgam. The use of a coupled enzyme system is perhaps more attractive and will be discussed here. The incubation is set up as described for coproporphyrinogen synthesis above. The incubation for the synthesis of protoporphyrinogen is undertaken in 50 mM Tris-HCl, pH 8.0, containing 0.2% Tween 20, and 2.5 mM glutathione. Coproporphyrinogen is added to the incubation mixture at a final concentration of 5 µM and incubated with CPO at a concentration of 10 µg/mL and PPO at 20 µg/mL. The synthesis of protoporphyrin can be followed fluorometrically by making a 1:10 dilution of the incubation mixture with buffer and determining the fluorescence at 635 nm after excitation at 405 nm (29). Coproporphyrin emits at 610 nm, so it is important to make sure that the auto-oxidation of coproporphyrinogen is not being observed. The increase in protoporphyrin fluorescence is measured over a 10-minute period, and the enzyme activity can be deduced with reference to a calibration curve for the fluorescence of a standard solution of protoporphyrin. The yield of protoporphyrin from protoporphyrinogen is in excess of 95%.

11. SYNTHESIS OF PROTOHEME

The final step in the synthesis of protoheme is the insertion of ferrous iron in a reaction that is catalyzed by ferrochelatase. In eukaryotes, this enzyme is normally peripherally associated with the inner membrane of the mitochondrion. Quite surprisingly, the human enzyme contains an iron sulphur center, although no immediate role has been forwarded for its presence. Defects in the human enzyme are associated with erythropoietic protoporphyria, a relatively severe form of porphyria that can cause severe liver damage (19). In *B. subtilis*, ferrochelatase exists as a soluble protein and represents one of the simplest sources of the enzyme (11). The increased solubility of the *Bacillus* enzyme was a major expedient in the crystallization of the enzyme (2).

11.1. Enzyme Purification

The *B. subtilis hemH* is cloned into pUC18 under control of the *lac* promoter to give plasmid pLUG18e2. When transformed into *E. coli* JM109 cells, the plasmid causes the bacteria to constitutively overproduce the enzyme to a level of about 10 mg/L of culture (Table 1). The strain harboring pLUG18e2 is somewhat unstable, and fresh transformants need to be used for new cultures.

❖ Procedure 7. Purification of Recombinant Ferrochelatase from *B. subtilis*

1. Bacterial growth: The strain is grown in LB media in 2-L flasks containing 1 L of media supplemented with ampicillin at 100 µg/mL at 37°C with vigorous shaking.
2. Harvesting: The cells are collected by centrifugation (10 000× *g* for 10 min), and the cell pellet is suspended in 25 mL of 30 mM Tris-HCl, pH 8.0, containing 20% (wt/vol) sucrose, lysozyme (0.25 mg/mL), and EDTA (15 mM). After incubation at 25°C for 30 minutes, the resulting spheroplasts are harvested by centrifugation at 7000× *g* for 15 minutes. The pellet is resuspended in 12 mL of 50 mM Tris-HCl, pH 7.4, containing 5 mM $MgSO_4$.
3. Sonication: The spheroplast suspension is sonicated as described in Procedure 1, and the lysate is centrifuged at 48 000× *g* for 30 minutes at 4°C. The pellet is discarded, and the supernatant is retained.
4. Ammonium sulfate fractionation: The supernatant is made 70% with respect to ammonium sulfate by the addition of 44.2 g/100 mL of extract. After 45 minutes at 0°C, the solution is centrifuged at 10 000× *g* for 10 minutes at 4°C. The pellet is discarded, and the supernatant is made 90% with respect to ammonium sulfate by the addition of a further 13.6 g/100 mL of extract. After 45 minutes at 0°C, the solution is centrifuged at 10 000× *g* for 10 minutes at 4°C, and the 70%–90% pellet is kept and resuspended in 4 mL of 20 mM Tris-HCl, pH 7.4.
5. Anion ·exchange chromatography: After dialysis against 5 L of the same buffer, the enzyme fraction is applied to a column of DEAE Sephacel (40-mL bed volume) and the column is washed with one bed volume of buffer. The ferrochelatase is eluted from the column by application of a linear gradient of 0 to 0.6 M NaCl in 20 mM Tris-HCl. The enzyme elutes at approximately 0.3 M NaCl.
6. Gel filtration chromatography: Fractions containing the enzyme are pooled and concentrated to approximately 5 mL in an ultrafiltration unit fitted with a PM-10 membrane. The concentrated sample is then applied to a column of Sephacryl S-100 HR (2.6 × 100 cm). The ferrochelatase elutes from the column as a single peak in a homogeneous form.
7. Storage: The enzyme can be concentrated and stored at -20°C for several months without loss of activity. The purified ferrochelatase is obtained in a yield of about 0.5 mg/L of culture.

11.2. Incubation Protocol and Assay

Ferrochelatase activity is best monitored by recording the disappearance of protoporphyrin (5,11). This can be monitored by a decrease in fluorescence as a divalent metal ion (normally zinc in assays) is chelated into the porphyrin macrocycle. The reaction is normally undertaken in a 3-mL cuvette with a 2.5-mL standard reaction mixture consisting of: 100 mM Tris-HCl, pH 7.2, 0.3 mg/mL Tween 80, 100 µM $ZnCl_2$, and 1 to 5 µg of purified ferrochelatase. The reaction is normally started by the addition of 1.5 µM protoporphyrin, prepared as described below, to the incubation, and the reaction is monitored for up to 10 minutes. The excitation wavelength is 407 nm, and the emission of fluorescence at 635 nm is recorded.

Protoheme can be synthesized from protoporphyrin IX and ferrous iron using the following procedure. The incubation mixture contains 100 mM Tris-HCl, pH 7.2, protoporphyrin IX at 2 μM, 0.3 mg/mL Tween 80, 20 μM Fe^{2+}, 6 mM DTT, 5 mM sodium dithionite, and 2 μg/mL ferrochelatase. Fe^{2+} is prepared daily as a stock solution of 50 mM $(NH_4)_2Fe(SO_4)_2$ in 0.3 M DTT. Protoporphyrin IX is prepared as a stock of 100 μM disodium protoporphyrin dissolved in water containing 15 mg/mL Tween 80. The insertion of ferrous iron can also be followed spectrofluorometrically by measuring the rate of protoporphyrin disappearance, as described above. The yield of protoheme is in excess of 90%.

ABBREVIATIONS

ALA, 5-aminolevulinic acid; ALAS, 5-aminolevulinic acid synthase; PBG, porphobilinogen; PBGD, porphobilinogen deaminase; CPO, coproporphyrinogen oxidase; Da, Dalton molecular mass unit; LB medium, Luria-Bertani medium; PMSF, phenylmethanesulfonyl fluoride; PPO, protoporphyrinogen oxidase; SAM, *S*-adenosyl-L-methionine; SDS-PAGE, sodium dodecyl sulfate polyacrylamide gel electrophoresis.

REFERENCES

1. **Akhtar, M. and C. Jones.** 1986. Preparation of stereospecifically labelled porphobilinogens. Methods Enzymol. *123*:375-383.
2. **Al-Karadaghi, S., M. Hansson, S. Nikonov, B. Jonsson, and L. Hederstedt.** 1997. Crystal structure of ferrochelatase: the terminal enzyme in heme biosynthesis. Structure *5*:1501-1510.
3. **Blanche, F., L. Debussche, D. Thibaut, J. Crouzet, and B. Cameron.** 1989. Purification and characterization of S-adenosyl-L-methionine:uroporphyrinogen methyltransferase from *Pseudomonas denitrificans*. J. Bacteriol. *171*:4222-4231.
4. **Bolt, E.L., L. Kryszak, J. Zeilstra-Ryalls, P.M. Shoolingin-Jordan, and M.J. Warren.** 1999. Characterisation of the *R. sphaeroides* 5-aminolevulinic acid synthase isoenzymes, HemA and HemT, isolated from recombinant *Escherichia coli*. Eur. J. Biochem. *265*:1-11.
5. **Dailey, H.A.** 1977. Purification and characterisation of the membrane bound ferrochelatase from *Spirillum itersonii*. J. Bacteriol. *132*:302-307.
6. **Dailey, H.A. and T.A. Dailey.** 1996. Protoporphyrinogen oxidase of *Myxococcus xanthus*. J. Biol. Chem. *271*:8714-8718.
7. **Erskine, P.T., N. Senior, S. Awan, R. Lambert, G. Lewis, I.J. Tickle, M. Sarwar, P. Spencer, P. Thomas, M.J. Warren et al.** 1997. X-ray structure of 5-aminolaevulinic acid dehydratase, a hybrid aldolase. Nat. Struct. Biol. *4*:1025-1031.
8. **Erskine, P.T., E. Norton, J.B. Cooper, R. Lambert, A. Coker, G. Lewis, P. Spencer, M. Sarwar, S.P. Wood, M.J. Warren, and P.M. Shoolingin-Jordan.** 1999. X-Ray structure of 5-aminolevulinic acid dehydratase from *Escherichia coli* complexed with the inhibitor levulinic acid at 2.0 A resolution. Biochemistry *38*:4266-4276.
9. **Ferreira, G.C. and J. Gong.** 1995. 5-Aminolaevulinate synthase and the first step of heme biosynthesis. J. Bioenerg. Biomembr. *27*:151-159.
10. **Guo, G.G., M. Gu, and J.D. Etlinger.** 1994. 240-kDa proteasome inhibitor CF-2. is identical to delta-aminolevulinic acid dehydratase. J. Biol. Chem. *269*:12399-12402.
11. **Hansson, M. and L. Hederstedt.** 1994. Purification and characterisation of a water-soluble ferrochelatase from *Bacillus subtilis*. Eur. J. Biochem. *220*:201-208.
12. **Hunter, G.A. and G.C. Ferreira.** 1995. A continuous spectrophotometric assay for 5-aminolevulinate synthase that utilizes substrate cycling. Anal. Biochem. *226*:221-224.
13. **Jaffe, E.K.** 1995. Porphobilinogen synthase, the first source of heme's asymmetry. J. Bioenerg. Biomembr. *27*:169-179.
14. **Jaffe, E.K.** 2000. The porphobilinogen synthase family of metalloenzymes. Acta Crystallogr. D *56*:115-128.
15. **Jordan, P.M., G. Burton, H. Nordlöv, M.M. Schneider, L. Pryde, and A.I. Scott.** 1979. J. Chem. Soc., Chem. Commun. 204-205.
16. **Jordan, P.M. and M.J. Warren.** 1987. Evidence for a dipyrromethane cofactor at the catalytic site of *E. coli* porphobilinogen deaminase. FEBS Lett. *225*:87-92.
17. **Jordan, PM.** 1991. The biosynthesis of 5-aminolaevulinic acid and its transformation into uroporphyrinogen III, p. 1-66. *In* A. Neuberger and L.L.M. van Deenen (Eds.), and P.M. Jordan (Vol. Ed.), New Comprehensive Biochemistry, Vol. 19, Biosynthesis of Tetrapyrroles. Elsevier, Amsterdam.
18. **Kannangara, C.G., R.V. Andersen, B. Pontoppidan, R. Willows, and D. von Wettstein.** 1994. Enzymic and mechanistic studies on the conversion of glutamate to 5-aminolaevulinate, p. 3-25. *In* D.J. Chadwick, and K. Ackrill (Eds.), The Biosynthesis of Tetrapyrrole Pigments, Ciba Foundation Symposium 180. John Wiley & Sons, New York.
19. **Kappas, A., S. Sassa, R.A. Galbraith, and Y. Nordmann.** 1995. The porphyrias, p. 2103-2160. *In* C.R. Scriver, A.L. Beaudet, W.S. Sly, and D. Valle (Eds.), The Metabolic and Molecular Basis of Inherited Disease, 7th ed. McGraw Hill, New York.
20. **Li, J.M., C.S. Russell, and S.D. Cosloy.** 1989. The structure of the *E. coli hemB* gene. Gene *75*:177-184.
21. **Mauzerall, D. and S. Granick.** 1956. The occurrence and determination of δ-aminolevulinic acid and porphobilinogen in urine. J. Biol. Chem. *219*:435-446.

22. **Medlock, A.E. and H.A. Dailey.** 1996. Human protoporphyrinogen oxidase is not a metalloprotein. J. Biol. Chem. *271*:32507-32510.
23. **Neidle, E.L. and S. Kaplan.** 1993. Expression of *Rhodobacter sphaeroides hemA* and *hemT* genes encoding two 5-aminolaevulinic acid synthase isoenzymes. J. Bacteriol. *175*:2292-2303.
24. **Phillips, J., F.G. Whitby, J.P. Kushner, and C.P. Hill.** 1997. Characterisation and crystallization of human uroporphyrinogen decarboxylase. Prot. Sci. *6*:1343-1346.
25. **Raux, E., T. McVeigh, S.E. Peters, T. Leustek, and M.J. Warren.** 1999. The role of *Saccharomyces cerevisiae* Met1p and Met8p in siroheme and cobalamin biosynthesis. Biochem. J. *338*:701-708.
26. **Scopes, R.K.** 1987. Protein Purification, Principles and Practice, 2nd ed. Springer Verlag, Basel.
27. **Shoolingin-Jordan, P.M., J.E. LeLean, and A.J. Lloyd.** 1997. Continuous coupled assay for 5-aminolevulinate synthase. Methods Enzymol. *281*:309-316.
28. **Shoolingin-Jordan, P.M. and K.-M. Cheung.** 1999. Biosynthesis of heme, p. 61-107. *In* D.H.R. Barton, K. Nakanishi, and O. Meth-Cohn (Eds.), and J.W. Kelly (Vol. Ed.), Comprehensive Natural Products Chemistry, Vol. 4, Amino Acids, Peptides, Porphyrins and Alkaloids. Elsevier, Amsterdam.
29. **Smith, A.G. and W.T. Griffiths.** 1993. Enzymes of chlorophyll and heme biosynthesis. Methods Plant Biochem. *9*:299-343.
30. **Spencer, J.B., N.J. Stolowich, C.A. Roessner, and A.I. Scott.** 1993. The *Escherichia coli cysG* gene encodes the multifunctional protein, siroheme synthase. FEBS Lett. *335*:57-60.
31. **Spencer, P. and P.M. Jordan.** 1993. Purification and characterisation of 5-aminolaevulinic acid dehydratase from *E. coli* and a study of reactive thiols at the metal binding domain. Biochem. J. *290*:279-287.
32. **Thomas, S.D. and P.M. Jordan.** 1986. Nucleotide sequence of the *hemC* locus encoding porphobilinogen deaminase of *Escherichia coli* K12. Nucleic Acids Res. *14*:6215-6226.
33. **Whitby, F.G., J.D. Phillips, J.P. Kushner, and C.P. Hill.** 1998. Crystal structure of human uroporphyrinogen decarboxylase. EMBO J. *17*:2463-2471.

5 Analysis of Biosynthetic Intermediates, 5-Aminolevulinic Acid to Heme

Chang Kee Lim
MRC Bioanalytical Science Group, School of Biological and Chemical Sciences, Birkbeck College, University of London, London, England, UK

1. INTRODUCTION

Chromatographic techniques are widely used for the analysis of heme and its precursors. Recent and continuing improvements in column packing materials for high-performance liquid chromatography (HPLC) have led to much better column efficiency and resolution. There have also been great advances in the direct coupling of liquid chromatography (LC), including capillary electrophoresis (CE), to mass spectrometry (MS) to provide highly sensitive and specific methods of analysis.

The separation and detection of the biosynthetic intermediates from 5-aminolevulinic acid (ALA) to heme are described in detail in this chapter. The emphasis is in HPLC and CE, and the well-established thin-layer chromatography will not be included.

2. 5-AMINOLEVULINIC ACID AND PORPHOBILINOGEN

ALA and porphobilinogen (PBG) are usually separated by ion exchange chromatography, converted into the p-dimethylamino-benzaldehyde derivatives, and then determined spectrophotometrically at 553 nm (16). The procedures, widely described in textbooks, are also available, with technical instructions, from ion exchange resins suppliers, e.g., Bio-Rad Laboratories (Hercules, CA, USA). The method is recommended for the routine qualitative and quantitative measurement of ALA and PBG.

ALA and PBG have been separated by HPLC (11) and micellar electrokinetic capillary chromatography (13). They were detected with a UV detector at 220 to 240 nm.

A simple CE method has been developed for the separation of PBG. The compound was effectively separated on a 75-cm fused-silica capillary (75 µm inner diameter) with 50 mM ammonium acetate buffer (pH 5.16 adjusted with acetic acid) as the running buffer and 20 kV and 30°C as the running voltage and temperature, respectively. PBG was detected at 220 nm with a detection limit of 1 µg/mL. Under the CE conditions, the charged PBG molecule could also be detected at 400 to 420 nm, although the detection was less sensitive than at 220 nm.

Heme, Chlorophyll, and Bilins: ***Methods and Protocols***
Edited by A.G. Smith and M. Witty

The above CE method has been modified by inclusion of 10% (vol/vol) acetonitrile in the running buffer (50 mM ammonium acetate, pH 5.20) and coupled on-line to electrospray ionization mass spectrometry (ESIMS) to provide an extremely sensitive and specific analytical method for ALA and PBG (12). The detection limits were estimated to be 100 and 10 amol of ALA and PBG on column, respectively. The sensitivity could be further improved by the use of selected ion recording (SIR) scans or nanospray ionization, or both.

Figure 1 shows the separation and detection of ALA and PBG by CE-ESIMS. The protonated ion of ALA is at m/z 132 and that of PBG is at m/z 227. However, the protonated PBG was found to lose ammonia (NH_3) easily in the electrospray source to give an intense ion at m/z 210, corresponding to a methylenepyrrolenine ion. PBG was, therefore, detected at m/z 210 for the methylenepyrrolenine ion and multiple reaction monitoring (MRM) acquisitions could be used for PBG by monitoring the transition from m/z 227 to m/z 210. This method is recommended for applications where high sensitivity and specificity are required.

3. ANALYSIS OF PORPHYRINS

The naturally occurring porphyrins exist in complex mixtures including isomeric forms. Effective analysis, therefore, requires high resolution coupled with sensitive detection. To date, the best technique for the separation of porphyrins and their isomers is HPLC. The resolution achieved by HPLC has not been reproduced by other separation methods.

3.1. Extraction of Porphyrins from Biological Materials for HPLC Analysis

Sample preparation is an important and integrated part of the successful application of HPLC to the analysis of porphyrins in biological materials. A good sample preparation procedure minimizes quantitative errors and places less demand on the chromatography, allowing faster and better analysis.

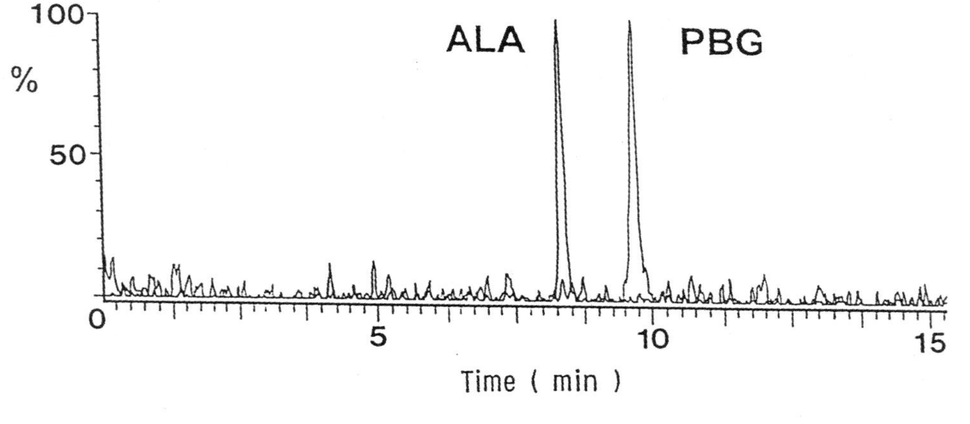

Figure 1. CE-ESIMS of ALA and PBG. Capillary, 70 cm × 75 µm i.d.; running buffer, 50 mM ammonium acetate, pH 5.2:acetonitrile (90:10, vol/vol); running voltage, 20 kV; ESI voltage, 3.5 kV.

It is recommended that, whenever possible, porphyrins should be extracted and analyzed as the free acids. The separation of porphyrin free acids is superior to that of the corresponding methyl esters (7). Methyl esterification of porphyrins may also cause structural modification of the parent compounds. The deconjugation and transmethylation of protoporphyrin glycoconjugates following esterification and extraction of porphyrins from rat Harderian gland is a typical example. The procedure led to the incorrect identification of protoporphyrin glycoconjugates as the unconjugated protoporphyrin (9).

3.1.1. Preparation of Urine Samples

Fresh urine (200 to 500 µL) may be injected after centrifugation into the HPLC for analysis. Sediments or precipitates are often seen in stored urine, and these may adsorb porphyrins. The urine (1 mL) should be thoroughly mixed with concentrated HCl (40 µL) to dissolve the precipitated material before HPLC separation.

3.1.2. Extraction of Porphyrins from Feces

The following procedure (19), which provides a relatively clean extract, is recommended.

❖ Procedure 1. Extraction of Porphyrins from Feces

1. Weigh about 50 mg of feces into a 15-mL graduated centrifuge tube.
2. Add 1 mL of concentrated HCl and vortex mix for 1 minute or until the particles disintegrate.
3. Add 3 mL of diethyl ether (peroxide-free) and vortex mix for 1 minute.
4. Add 3 mL of water and vortex mix for 1 minute.
5. Centrifuge at 2000× *g* for 10 minutes to give an upper ether layer, a pad of insoluble material at the interface, and a lower layer of aqueous acid.
6. Discard the ether layer that contains the unwanted carotenoid pigments and chlorophyll derivatives.
7. Record the volume (usually 4.5 mL) of the aqueous acid layer, which contains the extracted porphyrins, and transfer about 2 mL into a clean tube using a Pasteur pipet.
8. Filter the solution, e.g., through a syringe filter assembly, to remove any particulate material. The solution may be used for HPLC or spectrophotometric analysis.

The above procedure should be carried out in subdued light, e.g., red safety light, in order to minimize undue alteration to light sensitive porphyrins, especially protoporphyrin.

3.1.3. Extraction of Plasma and Red Cell Porphyrins

For the extraction of porphyrins in plasma (19), the sample (0.5 mL) is vortex mixed with 5 mL of ether:acetic acid (4:1, vol/vol) followed by centrifugation to remove the precipitated protein. The supernatant is then vortex mixed with 3 mL of 2.7 M HCl. The lower aqueous acid layer is used for HPLC analysis.

Plasma porphyrins may also be extracted by vortex mixing 100 µL of sample with 200 µL of acetonitrile:dimethyl sulfoxide (DMSO) (4:1, vol/vol). The supernatant after centrifugation is used for HPLC separation. This method, also suitable for the extraction of red cell porphyrins, is recommended for rapid porphyrin profile analysis by HPLC.

3.1.4. Extraction of Porphyrins from Tissues

Porphyrins in tissues can be effectively extracted by homogenizing the sample in acetonitrile-DMSO (4:1, vol/vol), using 1 mL of homogenizing medium per 100 mg of tissues. Repeated extraction may be necessary for complete recovery. The supernatant after centrifugation is thoroughly mixed with 2 volumes of water or HPLC aqueous phase buffer before separation. Injection of the organic extract without suitable adjustment of the aqueous content resulted in peak distortion.

3.2. Separation of Porphyrin Isomers

The separation of isomers, particularly the type I and type III isomers, is important for the differential diagnosis of certain porphyrias. For example, the coproporphyrin excreted in the urine and feces of patients with congenital erythropoietic porphyria (CEP) is type I, while in hereditary coproporphyria it is type III.

3.2.1. Uroporphyrin I, II, III, and IV Isomers

Uroporphyrin I and III isomers can be rapidly and effectively separated by isocratic reversed-phase (RP)-HPLC on octadecylsilyl (ODS) C_{18} columns with 13% (vol/vol) acetonitrile in 1 M ammonium acetate buffer, pH 5.16 (adjusted with acetic acid), as eluent (Figure 2). Optimization studies have shown that the molar concentration and pH of ammonium acetate buffer significantly affected the retention and resolution of uroporphyrin isomers (18). The optimum buffer concentration was 1 M, and the best pH range for chromatography on a conventional ODS column was between 5.10 and 5.20.

For separation on a base-deactivated (BDS) C_{18} column, the optimum pH was 5.55 (Figure 2), although 5.16 was also suitable. BDS C_{18} columns are columns with

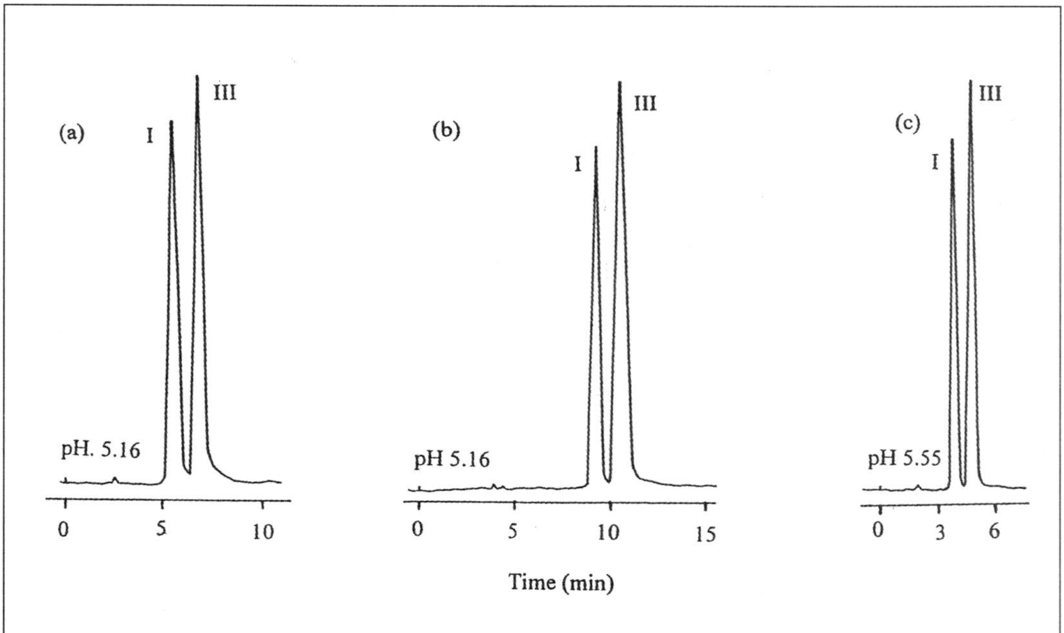

Figure 2. Separation of uroporphyrin I and III isomers. (a) On Hypersil-BDS C_{18} with acetonitrile:1 M ammonium acetate, pH 5.16 (9:91, vol/vol), as eluent; (b) on Hypersil-ODS with acetontrile:1 M ammonium acetate, pH 5.16 (13:87, vol/vol), as eluent; and (c) on Hypersil-BDS C_{18} with acetonitrile:1 M ammonium acetate, pH 5.55 (9:91, vol/vol), as eluent. Flow-rate, 1 mL/minute.

fewer residual silanol groups through exhaustive end-capping or made by different bonding procedures to those normally used for conventional ODS columns. Residual silanol groups on silica-based C_{18} columns interact adversely with basic compounds, causing peak tailing or broadening. In general, BDS C_{18} columns give better resolution and faster separation for porphyrins than conventional C_{18} columns.

Methanol should not be used as the organic modifier for the separation of uroporphyrins, especially when isocratic elution is used. It causes severe peak tailing and excessive retention with loss of resolution. Methanol is a hydrogen-bonding organic modifier. A layer of methanol adsorbed onto the hydrophobic hydrocarbonaceous C_{18} stationary phase surface can form extensive hydrogen bonds with the 8 carboxylic acid groups of uroporphyrin. The result is long retention and peak tailing. This phenomenon is not observed for porphyrins with one or more methyl groups, since the interaction is dominated by hydrophobic interaction between the hydrophobic methyl group(s) and the stationary phase surface.

A small proportion (e.g., 10%) of acetonitrile can be added to methanol, and the mixture (10% acetonitrile and 90% methanol) can be used as the organic modifier. The more hydrophobic acetonitrile, which is also a nonhydrogen bonding organic modifier, will be preferentially adsorbed onto the stationary phase surface, thus preventing hydrogen bond formation.

The complete separation of uroporphyrin I, II, III, and IV isomers has not been achieved. They were resolved into three peaks in the elution order of I, III + IV, and II (7,18). The resolution was not improved by employing a BDS C_{18} column.

3.2.2. Type I and Type III Heptacarboxylic Acid Porphyrins

The four type III isomers of heptacarboxylic acid porphyrin could not be completely separated by RP-HPLC, although the type I isomer easily resolved from the type III isomers either with 15% (vol/vol) acetonitrile in 1 M ammonium acetate buffer, pH 5.16, as eluent on a conventional C_{18} column or with 28% acetonitrile: methanol (10:90) in 1 M ammonium acetate buffer, pH 5.55, as eluent on a BDS C_{18} column. The four type III isomers were resolved into three peaks in the elution order of 7c, 7d, and 7a + 7b (7), with 28% acetonitrile:methanol (10:90) in 1 M ammonium acetate buffer, pH 5.16, as eluent. The letters a, b, c, and d are used throughout this chapter to denote the positions of methyl groups on rings A, B, C, and D, respectively.

3.2.3. Type I and Type III Hexacarboxylic Acid Porphyrins

There are two type I and six type III hexacarboxylic acid porphyrin isomers. The two type I isomers (6Iab and 6Iac) have been separated from the most abundant type III isomer (6IIIad) by isocratic RP-HPLC with 16% (vol/vol) acetonitrile in 1 M ammonium acetate buffer, pH 5.16, as eluent on a Hypersil-ODS column (ThermoQuest, Bellafonte, PA, USA). The complete separation of all 8 isomers has not been achieved. Using the above system, 6IIIac coeluted with 6IIIbd, and 6IIIab coeluted with 6IIIbc (7).

3.2.4. Type I and Type III Pentacarboxylic Acid Porphyrins

There are four type III and one type I pentacarboxylic acid porphyrin isomers. These 5 isomers have been separated by RP-HPLC on a Hypersil-ODS column with 45% (vol/vol) acetonitrile:methanol (10:90) in 1 M ammonium acetate buffer, pH 5.16, as eluent (Figure 3a). The elution order was 5I, 5bcd, 5abc, 5acd, and 5abd. A reversal of elution order between 5I and 5abd was observed when 19% acetonitrile in 1 M ammonium acetate, pH 5.16, was used as the mobile phase (Figure 3b). The presence of methanol in the mobile phase resulted in an overall improvement in resolution.

3.2.5. Coproporphyrin I, II, III, and IV Isomers

Coproporphyrin isomers are easily separated by RP-HPLC (7). The separations of the 4 isomers on a Hypersil-ODS and a Hypersil-BDS C_{18} column are shown in Figure 4 (a and b). Better resolution with faster elution times was achieved on the Hypersil-BDS C_{18} column. The Hypersil-BDS C_{18} column also required less acetonitrile (23%) for elution than the Hypersil-ODS column (30%), which is an obvious advantage.

3.2.6. Protoporphyrin, Heme, and Related Compounds

Dicarboxylic acid porphyrins, heme, and related compounds are much more hydrophobic than the other porphyrins described above. They require a much higher proportion of organic modifier for elution. Since acetonitrile is immiscible with 1 M ammonium acetate above the proportion of about 35%, it cannot be used as the sole organic modifier for the separation of this group of compounds. Either a mixture of acetonitrile:methanol (10:90) or methanol alone can be used instead. Methanol is completely miscible with 1 M ammonium acetate.

The separation of dicarboxylic porphyrins and metalloporphyrins by RP-HPLC has been described (10). A typical separation of protoporphyrin and heme-related compounds on a Hypersil-ODS

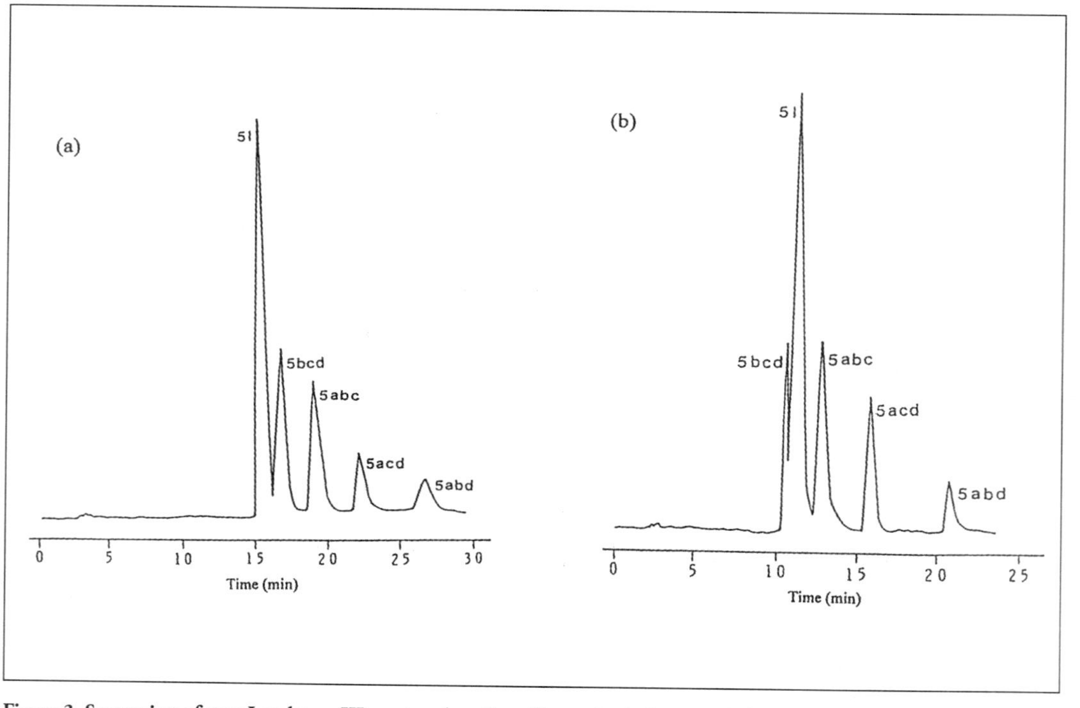

Figure 3. Separation of type I and type III pentacarboxylic acid porphyrin isomers. Column, Hypersil-ODS; eluent (a), 45% acetonitrile:methanol (10:90, vol/vol) in 1 M ammonium acetate, pH 5.16; eluent (b), 19% (vol/vol) acetonitrile in 1 M ammonium acetate, pH 5.16. The letters a, b, c, and d denote the positions of methyl groups on rings A, B, C, and D, respectively.

column with 86% (vol/vol) methanol in 1 M ammonium acetate buffer, pH 5.16, as eluent is shown in Figure 5.

3.2.7. Separation of Porphyrin Mixtures from Uroporphyrin to Protoporphyrin

From uroporphyrin to protoporphyrin, the compounds differ widely in hydrophobicity. Gradient elution is therefore essential for the simultaneous separation of these porphyrins, including their isomers.

A 15-minute linear gradient elution from 13% to 30% acetonitrile in 1 M ammonium acetate, pH 5.16, has been described for the separation of type I and type III isomers of 8-, 7-, 6-, 5-, and 4-carboxylated porphyrins (7). The system is applicable to analysis where the separation of dicarboxylated porphyrins is not required, e.g., urinary porphyrins. The elution of dicarboxylated porphyrins requires an acetonitrile content higher than its miscibility with 1 M ammonium acetate. It should be emphasized that using this gradient system the acetonitrile content should not be allowed to exceed 35%, and the column must not be washed with acetonitrile at the end of the separation due to

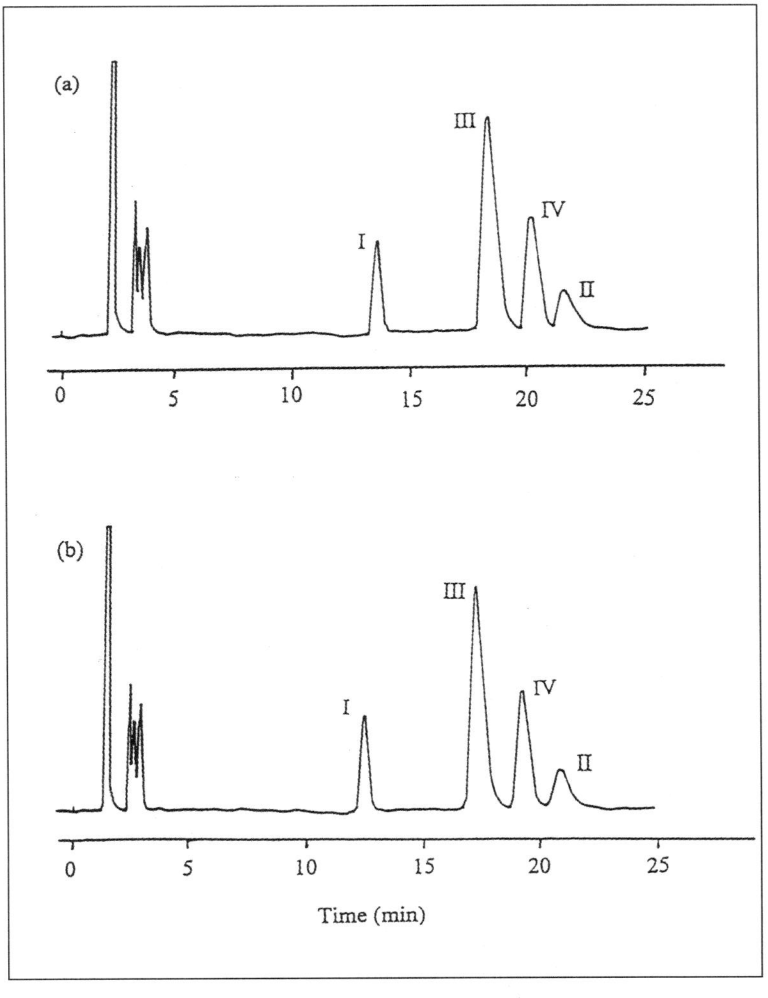

Figure 4. Separation of coproporphyrin I, II, III, and IV isomers. (a) On Hypersil-ODS with 30% (vol/vol) acetonitrile in 1 M ammonium acetate, pH 5.16, as eluent; (b) on Hypersil-BDS C_{18} with 23% (vol/vol) acetonitrile in 1 M ammonium acetate, pH 5.16, as eluent. Flow-rate, 1 mL/minute.

the immiscibility problem. It is also important to remember that whenever acetonitrile and 1 M ammonium acetate is used for elution, as in the separation of the individual group of porphyrin isomers by isocratic elution, the column should not be washed with acetonitrile before removal of ammonium acetate with a solvent in which it is completely miscible. The column may be washed with 90% methanol or acetonitrile:methanol in water.

Porphyrin mixtures including protoporphyrin are best separated by gradient elution RP-HPLC with (1,7,17) or without ion-pairing agents (7). Columns of silica gel chemically bonded with different hydrocarbon chain lengths, from C_1 to C_{18}, have all been successfully used for the RP-HPLC separation of porphyrin mixtures in biological materials (7,8). With the increasing use of on-line HPLC-MS in analysis, including the tetrapyrroles, gradient elution solvent mixtures that are fully compatible with MS are necessary. This rules out systems that use involatile inorganic phosphate in separation.

A simple RP-HPLC system with 0.1% trifluroroacetic acid (solvent A) and acetonitrile (solvent B) as the gradient elution solvent mixture has been used for the separation of porphyrins. The system, fully compatible with MS, is able to resolve the type I and III isomers of 6-, 5-, and 4-carboxylated porphyrins. Separation of uro- and heptacarboxylic acid porphyrin isomers, however, was not achieved (Figure 6). The system is best used for the separation of porphyrins with fewer numbers of carboxylic acid groups, including the dicarboxylic acid porphyrins.

The ammonium acetate buffer system that is fully compatible with MS and provides high efficiency separation of porphyrins is the buffer of choice. It is recommended that the following gradient mixtures are used for elution: solvent A, 10% (vol/vol) acetonitrile in 1 M ammonium acetate buffer, pH 5.16; solvent B, 10% (vol/vol) acetonitrile in methanol. Various elution programs can be used, depending on the applications, with these two solvent mixtures for porphyrin separation (15). The pH of the buffer, 5.16, is optimal for the separation of porphyrin mixtures.

Figure 7 shows the separation of porphyrins in the feces and urine of a patient with porphyria cutanea tarda (PCT) on a C_{18}-bonded RP column (Hypersil-ODS). It clearly demonstrates the flexibility and applicability of the system.

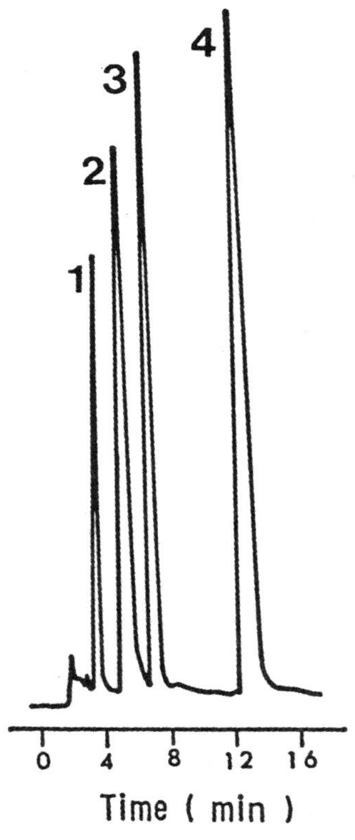

Figure 5. Separation of protoporphyrin and metalloporphyrins. Column, Hypersil-ODS; eluent, methanol:1 M ammonium acetate, pH 5.16 (86:14, vol/vol); flow rate, 1 mL/minute. Peaks: 1 = Zn-deuteroporphyrin; 2 = heme; 3 = Zn-protoporphyrin; 4 = protoporphyrin.

4. RETENTION MECHANISM OF PORPHYRINS AND METALLOPORPHYRINS IN RP-HPLC

Understanding the retention behavior and mechanism is useful in the prediction and elucidation of the possible nature of substituent groups present in unknown porphyrins.

The most dominant mechanism of retention in RP-HPLC is hydrophobic interaction. In porphyrins, this is between the side-chain substituents and the hydrophobic hydrocarbonaceous (ODS) stationary phase surface. The hydrophobicity of the porphyrin side-chain substituents increases in the order:

$CH_2COOH<CH_2CH_2COOH<CH_3<CH_2CH_3<CH{=}CH_2$

The number of the most hydrophobic substituents available for interaction, therefore, determines the relative retention of the porphyrin. This is dominated by the number of alkyl (especially methyl) groups. Thus, the elution of porphyrins in the order of uroporphyrin (8COOH), heptacarboxylic acid porphyrin (7COOH, $1CH_3$), hexacarboxylic acid porphyrin (6COOH, $2CH_3$), pentacarboxylic acid porphyrin (5COOH, $3CH_3$), coproporphyrin (4COOH, $4CH_3$), mesoporphyrin (2COOH, $4CH_3$, $2CH_2CH_3$), and protoporphyrin (2COOH, $4CH_3$, $2CH{=}CH_2$) was observed.

The hydrophobic interaction mechanism also explains the elution order of porphyrin isomers. The elution in the order of I, III, IV, and II, for coproporphyrin isomers is an example. Coproporphyrin II has two pairs of adjacent CH_3 groups at positions 1 and 8 and 4 and 5, respectively, using Fischer's numbering system. These provide the largest hydrophobic surface areas available for interaction. It is therefore the longest retaining isomer. The symmetrical coproporphyrin I has no adjacent CH_3 groups and, thus, has the smallest hydrophobic surface areas available for interaction. It is, therefore, the fastest eluting isomer. Coproporphyrin III and IV isomers each have a pair of adjacent CH_3 groups, at positions 1 and 8 and 2 and 3, respectively. In this situation, the relative distance between the adjacent CH_3 pair and the remaining two nonadjacent CH_3 groups becomes an important factor in determin-

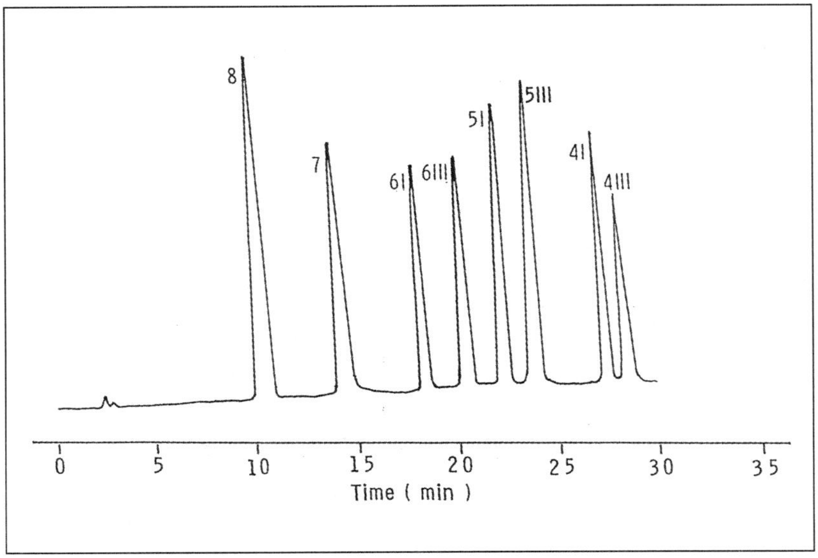

Figure 6. Separation of a standard mixture of porphyrins. Column, Hypersil-ODS (25 cm × 4.6 mm i.d., 5 µm particle size); solvents, 0.1% trifluoroacetic acid (solvent A) and acetonitrile (solvent B); elution, linear gradient from 25% solvent B (75% solvent A) to 50% solvent B in 30 minutes; flow rate, 1 mL/minute; detection, absorbance 404 nm. Peaks: 4, 5, 6, 7, and 8 refer, respectively, to tetra(copro)-, penta-, hexa-, hepta-, and octa(uro)carboxylic acid porphyrin; I and III denote type I and type III isomers.

ing the relative hydrophobicity. In coproporphyrin III, these are 5 (from position 1 to 3) and 6 (from position 8 to 5) bond lengths apart, respectively. In coproporphyrin IV, each of the adjacent CH_3 groups is 5 bond lengths away (from positions 2 to 8 and 3 to 5) from their nearest nonadjacent CH_3 group. The slightly shorter distance (one bond length) between one of the adjacent pair CH_3 groups and its nearest nonadjacent CH_3 group (5 instead of 6 bonds apart) is sufficient to make the type IV isomer slightly more hydrophobic and, therefore, be retained longer than the type III isomers. Similar arguments apply to the separation of the penta- and hexacarboxylic acid porphyrin isomers.

The type III heptacarboxylic acid porphyrin isomers each have only a single CH_3 group that dominated the hydrophobic interaction. These isomers are, therefore, very similar in hydrophobicity and difficult to separate.

In metalloporphyrins, hydrophobic interaction between the side-chain substituents and stationary phase surface is also an important retention mechanism. However, the insertion of a metal ion, which completely occupies the center of the porphyrin hole, significantly altered the electronic environment around the central N atoms. The retention of metalloporphyrins is also dependent on the ability, and therefore the species, of the inserted

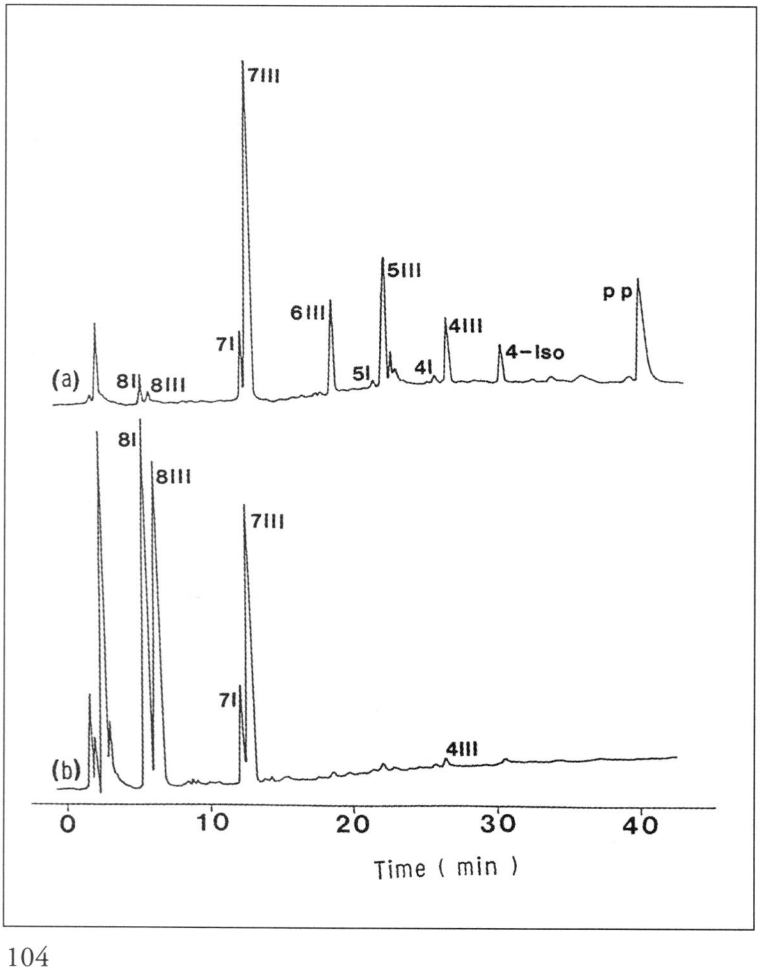

Figure 7. Separation of porphyrins in (a) the fecal extract and (b) the urine of a patient with PCT. Column, Hypersil-ODS (250 × 4.6 mm, 5 µm particle size); solvent A, 10% acetonitrile in 1 M ammonium acetate buffer (pH 5.16); solvent B, 10% acetonitrile in methanol; elution, linear gradient at 1 mL/minute from 10% to 90% solvent B in 30 minutes, followed by isocratic elution at 90% solvent B for a further 10 minutes; detection, 404 nm. Peaks; 4, 5, 6, 7, and 8 refer, respectively, to tetra(copro), penta-, hexa-, hepta-, and octa (uro)carboxylic acid porphyrin; 4-Iso = isocoproporphyrin; pp = protoporphyrin.

metal ion to accept axial ligands from the mobile phase. Addition of polar axial ligands leads to a decrease in the overall hydrophobicity of a molecule and, consequently, its retention (10).

5. SEPARATION OF PORPHYRIN METHYL ESTERS

In some applications, it is more convenient to separate porphyrins as their methyl ester derivatives. The majority of methods reported for the separation of porphyrin methyl esters are by normal phase (adsorption) chromatography on silica gel columns (7). RP-HPLC, however, provides better resolution and, although unable to separate the type I and III isomers of uro- and heptacarboxylic acid porphyrin methyl esters, is able to simultaneously separate the type I and III isomers of hexacarboxylic acid, pentacarboxylic acid, and coproporphyrin methyl esters in a single gradient elution run. Furthermore, the polar hydroxylated porphyrins, e.g., hydroxyuroporphyrin, which are difficult to elute from a silica gel column, are easily eluted from a RP column.

The separation of a mixture of porphyrin methyl esters on an ODS column by gradient elution from 70% (vol/vol) acetonitrile in water to 100% acetonitrile in 30 minutes is shown in Figure 8. The system is also fully compatible with MS.

6. SEPARATION OF PORPHYRINOGENS

The porphyrinogens are the true intermediates in the biosynthesis of heme. Porphyrinogens are unstable to oxidation by air, especially under acidic conditions and in the presence of light, to the corresponding porphyrins, and are therefore rarely analyzed. However, studies have shown that for isomer separation, the porphyrinogens are usually better resolved than the porphyrins (3–6). Isomerically pure porphyrinogens are needed as substrates for enzymic experiments, and their separation may also have application in situations where the complete resolution of isomers is essential. For example, investigation of the preferred order of uroporphyrinogen decarboxylation requires the complete separation of all four type III heptacarboxylic porphyrinogen isomers (14).

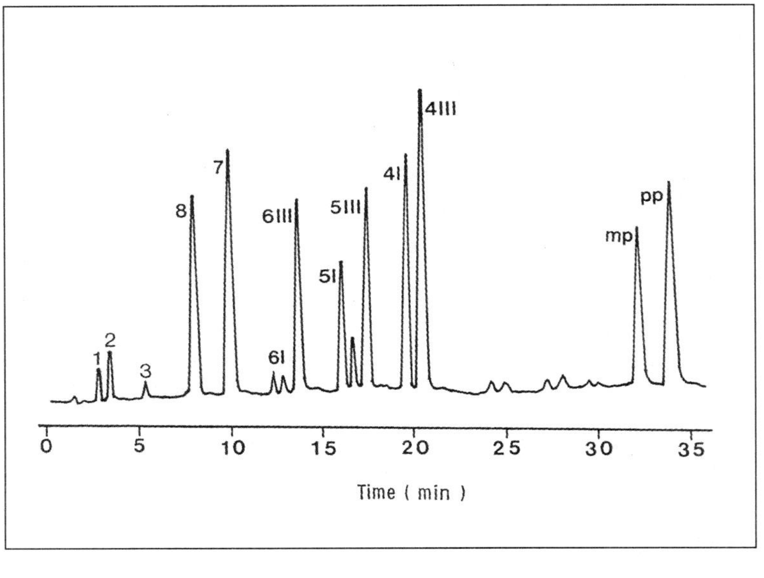

Figure 8. RP-HPLC of porphyrin methyl esters. Column, Hypersil-ODS (250 × 4.6 mm, 5 μm particle size); eluent, linear gradient elution from 70% acetonitrile in water to 100% acetonitrile in 30 minutes; flow-rate, 1 mL/minute. Peaks; 1, 2, and 3 = hydroxylated porphyrins; 4, 5, 6, 7, and 8 refer, respectively, to tetra(copro)-, penta-, hexa-, hepta-, and octa-(uro)carboxyl porphyrin; I and III denote type I and type III isomers; mp = mesoporphyrin; pp = protoporphyrin.

Porphyrinogens are easily prepared by reduction of porphyrins, usually with 3% (wt/wt) sodium amalgam, as described below.

❖ Procedure 2. Preparation of Porphyrinogens by Reduction of Porphyrins

1. Dissolve the porphyrin in 0.01 M KOH in a tube. Cover the tube with foil to prevent exposure to light.
2. Flush the tube and solution with nitrogen and add excess 3% sodium amalgam.
3. Shake or vortex mix the stoppered tube vigorously until no red fluorescence is detected under UV light.
4. Transfer the porphyrinogen solution immediately into a clean tube containing a small volume of 0.1 M EDTA, flush with nitrogen, and keep the stoppered tube on ice in the dark until analysis. The porphyrinogens are stable for at least 1 hour under these conditions.

Uroporphyrinogen I and III isomers have been separated on an ODS column with 6% (vol/vol) acetonitrile in 1 M ammonium acetate buffer, pH 5.16, as eluent. The elution order was III before I, which is the reversed of that observed for the corresponding porphyrins. The porphyrinogens are also more hydrophilic than the porphyrins, requiring less organic modifier for elution.

The four type III heptacarboxylic acid porphyrin isomers, which could not be separated by RP-HPLC, were completely resolved as the porphyrinogens with acetonitrile:methanol:1 M ammonium acetate buffer, pH 5.16 (7:3:90, by vol), as eluent (Figure 9). The elution order was 7a, 7c, 7b, and 7d, which is also different from that of the porphyrins (6,14).

The six type III hexacarboxylic acid porphyrinogen isomers have been separated by RP-HPLC with acetonitrile:methanol:1 M ammonium acetate buffer, pH 5.16 (8:12:80, by vol) as mobile phase (4,7). The superior resolution of porphyrinogen isomers is again demonstrated.

The complete separation of the five type I and III pentacarboxylic acid porphyrinogen isomers was achieved on a Hypersil-ODS column with methanol:1 M ammonium acetate buffer, pH 5.16 (40:60, vol/vol), as eluent (3,7).

The type I, II, III, and IV isomers of coproporphyrinogen could be rapidly separated on an ODS column with acetonitrile:1 M ammonium acetate, pH 5.16 (25:75, vol/vol) as mobile phase. With this system, the four coproporphyrinogen isomers could also be simultaneously separated from the four coproporphyrin isomers (5).

Protoporphyrinogen was easily separated from protoporphyrin by RP-HPLC with methanol:1 M ammonium acetate buffer, pH 5.16 (90:10, vol/vol), as eluent. The separation of these two compounds may be useful for studying their interconvertion either chemically or enzymatically.

7. RETENTION BEHAVIOR OF PORPHYRINOGENS IN RP-HPLC

Although hydrophobic interaction is undoubtedly still the mechanism governing the retention of porphyrinogens in RP-HPLC, the reversal in the elution order observed for many of the porphyrinogen isomers in comparison to porphyrins requires explanation. For example, while coproporphyrin isomers were eluted in the order I, III, IV, and II based on the hydrophobic interaction hypothesis, the corresponding coproporphyrinogen isomers were eluted in the order of I, II, III, and IV, an apparent contradiction to the hydrophobic interaction hypothesis. However, reduction of the methine bridges of the rigid porphyrin macrocycle resulted in a relatively flexible porphyrinogen molecule, which is able to adopt various conformations. In the flexible coproporphyrinogen molecules, the small CH_3 group in each isomer may be subjected to varying degrees of steric hindrance or shielding by the larger propionic acid groups, depending on the adopted conformation. This alters the expected hydrophobic surface areas available for interaction, which results in a change in elution order. Since the conformations of the various groups of porphyrinogens under RP-HPLC conditions are unknown, it makes prediction of the elution order difficult.

8. HPLC DETECTORS FOR PORPHYRINS, METALLOPORPHYRINS, AND PORPHYRINOGENS

Porphyrins and metalloporphyrins have an intense absorption band at the 400 nm region (Soret band). Detection at the Soret band region with an UV-visible detector set at 400 to 405 nm allows the simultaneous detection of all porphyrins and metalloporphyrins.

Porphyrins also have intense red fluorescence, and a fluorescence detector set at excitation and emission wavelengths of 400 to 415 nm and 600 to 620 nm, respectively, provides a highly sensitive and specific method of detection. Heme is a nonfluorescence compound and cannot be detected fluorimetrically.

The porphyrinogens do not fluoresce and have only relatively weak UV absorption at the 220 to 240 nm region. They can be detected with a UV detector set at 220 nm, but are best detected electrochemically with an amperometric or coulometric detector because of their ease of oxidation. Porphyrins and metalloporphyrins are also electro-active and can be detected with an electrochemical detector.

In terms of sensitivity, specificity, and ability to positively identify and characterize compounds, the best "detector" currently available is the mass spectrometer. On-line HPLC-ESIMS and tandem mass spectrometry (MS/MS) provide unsurpassed specificity for the analysis of tetrapyrroles.

The methyl esters of porphyrins are usually used in MS analysis because they ionize better in the ion source than free acid porphyrins. Porphyrins with a higher number of carboxylic acid groups, e.g., uro- and heptacarboxylic acids porphyrins, are more difficult to ionize than those with a lesser number of carboxylic acid groups, such as copro- and protoporphyrins. With the recent devel-

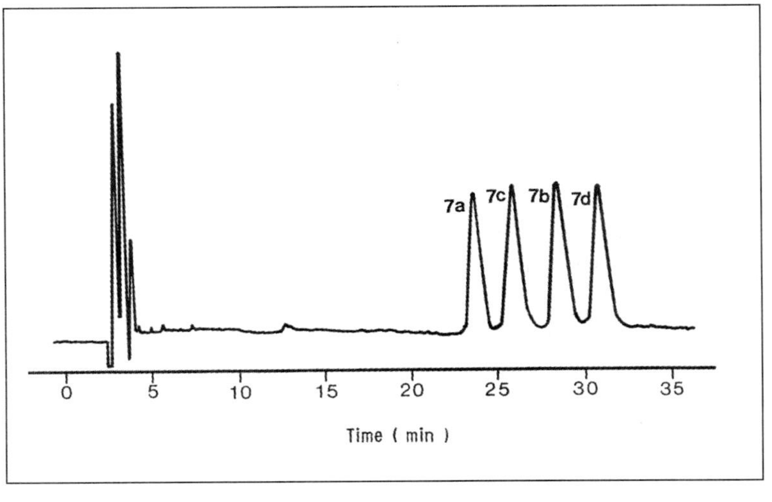

Figure 9. Separation of heptacarboxyl porphyrinogen isomers. Column, Hypersil-ODS (250 × 4.6 mm, 5 µm particle size); eluent, acetonitrile:methanol:1 M ammonium acetate, pH 5.16 (7:3:90, by vol); flow-rate, 1 mL/minute; detection, amperometric at +0.70 V.

opment and introduction of high sensitivity high resolution hybrid electrospray ionization orthogonal quadrupole-time of flight mass spectrometer (e.g., Q-Tof II; Micromass, Altrincham, Cheshire, England, UK), however, all free acid porphyrins can now be analyzed with great sensitivity without the need for derivatization.

9. SEPARATION OF PORPHYRINS BY CAPILLARY ELECTROPHORESIS

CE has been used for the separation of porphyrins and the majority of the methods used the micellar electrokinetic capillary chromatographic (MECC) mode in conjunction with visible absorbance or fluorescence detection (2,20). The separation of porphyrins in the urine of a patient with CEP is shown in Figure 10. The running buffer was 100 mM sodium dodecyl sulfate (SDS) and 20 mM 3-(cyclohexylamino)-1-propane sulphonic acid (CAPS) at pH 11.0 and the running voltage and temperature were 25 kV and 45°C, respectively.

Isomers of porphyrins are generally difficult to separate by CE, and this procedure has not achieved the same resolution as HPLC.

Detection in CE is also less sensitive than in HPLC. Although CE is highly sensitive in terms of mass detection, samples at low concentration, e.g., urinary porphyrins, can be difficult to analyze because of the small sample volumes (approximately 10 nL) injected. The requirement for surfactants (e.g., SDS) in the running buffer for porphyrin separation makes it incompatible with MS analysis.

10. FUTURE DEVELOPMENTS

Further development in MS can be expected with the introduction of user-

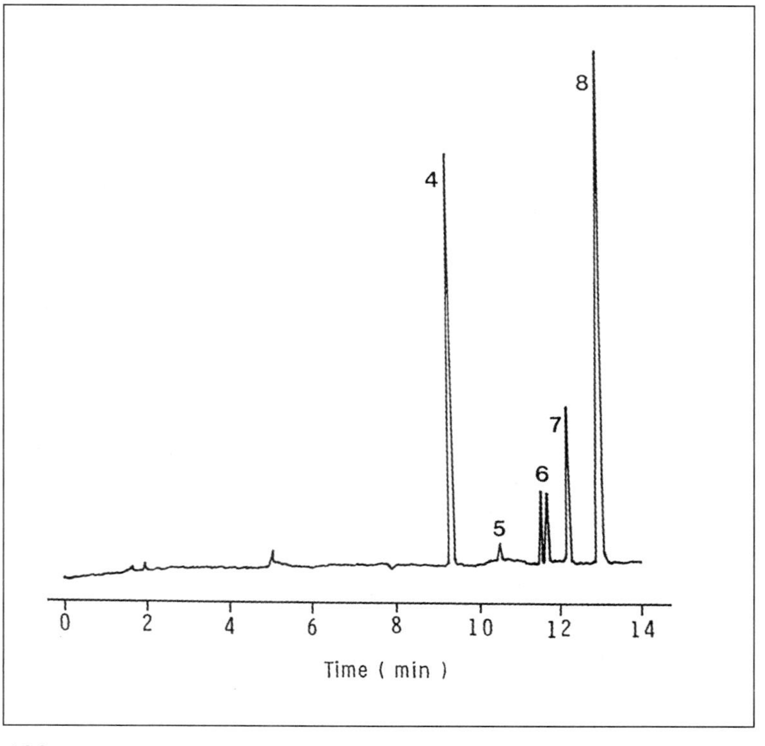

Figure 10. Separation of porphyrins in the urine of a patient with CEP by CE. Running buffer, 100 mM SDS, 20 mM CAPS (pH 11.0); voltage, 25 kV; temperature, 450°C; detection, absorbance 404 nm. Peaks: 4, 5, 6, 7, and 8 refer, respectively, to tetra(copro)-, penta, hexa-, hepta-, and octa(uro)carboxylic acid porphyrin.

friendly instruments of very high sensitivity and resolution. This will allow the porphyrins to be analyzed with even greater sensitivity and specificity, especially in HPLC-MS/MS analysis. The use of microcolumn HPLC-nanospray MS will further enhance the sensitivity of detection and should be explored for possible application in the analysis of the biosynthetic intermediates from ALA to heme.

ABBREVIATIONS

ALA, 5-aminolevulinic acid; CAPS, 3-(cyclohexylamino)-1-propane sulfonic acid; CE, capillary electrophoresis; CEP, congenital erythropoietic porphyria; DMSO, dimethyl sulfoxide; ESIMS, electrospray ionization mass spectrometry; HPLC, high-performance liquid chromatography; LC, liquid chromatography; MECC, micellar electrokinetic capillary chromatography; MRM, multiple reaction monitoring; MS, mass spectrometry; MS/MS, tandem mass spectrometry; ODS, octadecylsilyl; PBG, porphobilinogen; RP, reversed-phase; SDS, sodium dodecylsulfate; SIR, selected ion recording.

REFERENCES

1. **Bonnett, R., A.A. Charalambides, K. Jones, I.A. Magnus, and R.J. Ridge.** 1978. The direct determination of porphyrin carboxylic acids. High-pressure liquid chromatography with solvent systems containing phase-transfer agents. Biochem. J. *173*:693-695.
2. **Chiang, S.C.C. and S.F.Y. Li.** 1997. Separation of porphyrins by capillary electrophoresis in fused-silica and ethylene vinyl acetate copolymer capillaries with visible absorbance detection. Biomed. Chromatogr. *11*:366-370.
3. **Li, F., C.K. Lim, and T.J. Peters.** 1987. Separation and characterization of pentacarboxylic porphyrinogen isomers by high-performance liquid chromatography with electrochemical detection. Biochem. J. *243*:421-423.
4. **Li, F., C.K. Lim, and T.J. Peters.** 1989. Preparation, high-performance liquid chromatographic separation and characterization of hexacarboxylic porphyrinogens. J. Chromatogr. *461*:353-359.
5. **Lim, C.K., F. Li, and T.J. Peters.** 1986. High-performance liquid chromatography of uroporphyrinogen and coproporphyrinogen isomers with amperometric detection. Biochem. J. *234*:629-633.
6. **Lim, C.K., F. Li, and T.J. Peters.** 1987. High-performance liquid chromatography of type-III heptacarboxylic porphyrinogen isomers. Biochem. J. *247*:229-232.
7. **Lim, C.K., F. Li, and T.J. Peters.** 1988. High-performance liquid chromatography of porphyrins. A review. J. Chromatogr. Biomed. Appl. *429*:123-153.
8. **Lim, C.K. and T.J. Peters.** 1984. Urine and faecal porphyrin profiles by reversed-phase high-performance liquid chromatography in the porphyrias. Clin. Chim. Acta *139*:55-63.
9. **Lim, C.K., M.A. Razzaque, J. Luo, and P.B. Farmer.** 2000. Isolation and characterization of protoporphyrin glycoconjugates from rat Harderian gland by HPLC, capillary electrophoresis and HPLC/electrospray ionization MS. Biochem. J. *347*:757-761.
10. **Lim, C.K., J.M. Rideout, and T.J. Peters.** 1984. High-performance liquid chromatography of dicarboxylic porphyrins and metalloporphyrins: retention behaviour and biomedical applications. J. Chromatogr. *317*:333-341.
11. **Lim, C.K., J.M. Rideout, and D.M. Samson.** 1979. Determination of 5-aminolaevulinic acid and porphobilinogen by high-performance liquid chromatography. J. Chromatogr. *185*:605-611.
12. **Lord, G.A., J.L. Luo, and C.K. Lim.** 1999. Capillary zone electrophoresis/mass spectrometry of 5-aminolaevulinic acid and porphobilinogen. Rapid. Comm. Mass Spec. *14*:314-316.
13. **Luo, J.L., J. Deka, and C.K. Lim.** 1996. Determination of 5-aminolaevulinic acid dehydratase activity in erythrocytes and porphobilinogen in urine by micellar electrokinetic capillary chromatography. J. Chromatogr. *722*:353-357.
14. **Luo, J. and C.K. Lim.** 1993. Order of uroporphyrinogen III decarboxylation on incubation of porphobilinogen and uroporphyrinogen III with erythrocyte uroporphyrinogen decarboxylase. Biochem. J. *289*:529-532.
15. **Luo, J. and C.K. Lim.** 1995. Isolation and characterization of new porphyrin metabolites in human porphyria cutanea tarda and in rats treated with hexachlorobenzene by HPTLC, HPLC and liquid secondary ion mass spectrometry. Biomed. Chromatogr. *9*:113-122.
16. **Mauzerall, D. and S. Granick.** 1956. The occurrence and determination of δ-aminolevulinic acid and porphobilinogen in urine. J. Biol. Chem. *219*:435-446.
17. **Meyer, H.D., W. Vogt, and K. Jacob.** 1984. Improved separation and detection of free porphyrins by high-performance liquid chromatography. J. Chromatogr. *290*:207-213.
18. **Rideout, J.M., D.J. Wright, and C.K. Lim.** 1983. High performance liquid chromatography of uroporphyrin isomers. J. Liq. Chromatogr. *6*:383-394.
19. **Rossi, E. and D.H. Curnow.** 1986. Porphyrins, Ch. 10, p. 261-303. *In* C.K. Lim (Ed.), HPLC of Small Molecules. IRL Press, Oxford.
20. **Weinberger, R., E. Sapp, and S. Moring.** 1990. Capillary electrophoresis of urinary porphyrins with absorbance and fluorescence detection. J. Chromatogr. *516*:217-285.

6 Analysis of Intermediates and End Products of the Chlorophyll Biosynthetic Pathway

Constantin A. Rebeiz
University of Illinois, Urbana, IL, USA

1. INTRODUCTION

1.1. Multibranched Chlorophyll Biosynthetic Pathway

Since the 1963 seminal review of Smith and French (62), our understanding of the chlorophyll (chl) biosynthetic pathway has changed dramatically. Several factors have contributed to this phenomenon, among which are: *(i)* development of systems capable of Chl and thylakoid membrane biosynthesis in organello and in vitro (17,28,42,43,46); *(ii)* development of powerful analytical techniques that allowed the qualitative and quantitative determination of various intermediates of the pathway (this chapter); and *(iii)* recognition of the probability that the structural and functional complexity of thylakoid includes a multibranched heterogeneous Chl biosynthetic pathway (44).

Chlorophyll biosynthetic heterogeneity (41,47,49) refers to the biosynthesis of Chl via multiple biosynthetic routes. The multibranched Chl *a*/*b* biosynthetic pathway depicted in Figure 1 may be a template of a Chl-protein biosynthesis center, where the assembly of photosystem I, photosystem II, and light harvesting Chl-protein complexes (LHC) takes place (44). The multiple Chl biosynthetic pathways are considered, individually or in groups of two or three adjacent pathways, to constitute Chl-apoprotein biosynthesis subcenters earmarked for the coordinated assembly of individual pigment–protein complexes. Apoproteins destined for some of the biosynthesis subcenters may possess signals for specific Chl biosynthetic enzymes peculiar to that subcenter, such as 4-vinyl reductases, formyl synthetases, or Chl *a* and Chl *b* synthetases. Once an apoprotein formed in the cytosol or in the plastid reaches its biosynthesis subcenter destination and its signal is removed, it may bind nascent Chl formed via one or more biosynthetic pathways. During Chl binding, it may fold properly and act, at that location, as a template for other thylakoid apoproteins having appropriate signals (44).

Figure 1 depicts various biosynthetic intermediates and end products of the 15

Heme, Chlorophyll, and Bilins: *Methods and Protocols*
Edited by A.G. Smith and M. Witty

biosynthetic pathways that constitute the multibranched Chl *a*/*b* biosynthetic pathway. This chapter will describe various analytical techniques that permit the qualitative and quantitative analysis of the intermediates of this complex pathway. Chapter 10 in this volume describes in more general terms the isolation, quantification, and characterization of total Chl *a*/*b*.

2. MATERIALS AND METHODS

All tetrapyrrole intermediates and end products of the Chl biosynthetic pathway are loosely or tightly bound to plastid membrane lipoproteins (4,31,32,38). Before qualitative and quantitative analysis, these tetrapyrroles need to be extracted from their native environment with organic solvents. On the basis of their polarity, plant tetrapyrroles fall into two broad categories: *(i)* very apolar fully esterified tetrapyrroles which are soluble in apolar solvents such as petroleum ethers and hexane; and *(ii)* less apolar tetrapyrroles soluble in more polar solvents such as acetone. Once tetrapyrroles have been partitioned between polar and apolar solvents, they can be subjected to a variety of purification and analytical techniques. Before discussing details of purification and analysis, tetrapyrrole extraction and partitioning into apolar and polar fractions will be described.

2.1. Tetrapyrrole Extraction

Tetrapyrrole extraction from green tissues can be carried out under laboratory light (about 5 μmol/m^{2}s). However, pigment extraction from etiolated tissues should be performed under a green safelight that does not photoconvert protochlorophyllide *a* (Pchlide *a*) and Pchlide *a* ester (Pchlide *a* E) to chlorophyllide *a* (Chlide *a*) and Chl *a* (66). In our laboratory, we routinely use a low irradiance green light with a transmission maximum at 503 nm, a bandwidth of 40 nm, and a photon density of about 0.01 μmol/m^{2}s.

❖ Procedure 1. Preparation of an Ammoniacal Extract

1. Since homogenization of plant tissues disrupts the plant vacuoles and releases organic acids, NH_4OH is added to the acetone to neutralize the medium and prevent loss of the acid-labile central Mg atom from the Mg-tetrapyrroles (19). For many of the compounds which are labile, such as protoporphyrin IX (Proto), the extraction should be carried out under subdued lighting and at ice bucket temperatures.
2. Cut the plant tissue into small pieces about 1 cm in length.
3. Homogenized in acetone: 0.1 N NH_4OH (9:1 vol/vol) at 0° to 4°C for 60 seconds, using a homogenizer (PT 10/35 probe; Brinkman Instruments, Westbury, NY, USA), at a ratio of 7 mL of solvent per gram of tissue.
4. Centrifugation at 39 000× *g* for 12 minutes at 1°C.
5. Decant the ammoniacal acetone extract and store at -80°C until required.

Isolated plastids and subplastidic preparations can be used, for example, to study precursor product relationships in organello or in vitro. Plastids or subplastidic preparations (see Chapter 12), which are usually suspended in buffers with a pH range of about 7.0 to 8.0, are extracted with acetone: 0.1 N NH_4OH (9:1 vol/vol) at a rate of 10 mL per 1 mL of preparation and then treated as above.

2.2. Tetrapyrrole Partitioning Between Apolar and Polar Solvents

Once an ammoniacle extract has been made, partitioning tetrapyrroles between apolar and polar solvents is an easy way of achieving partial purification prior to

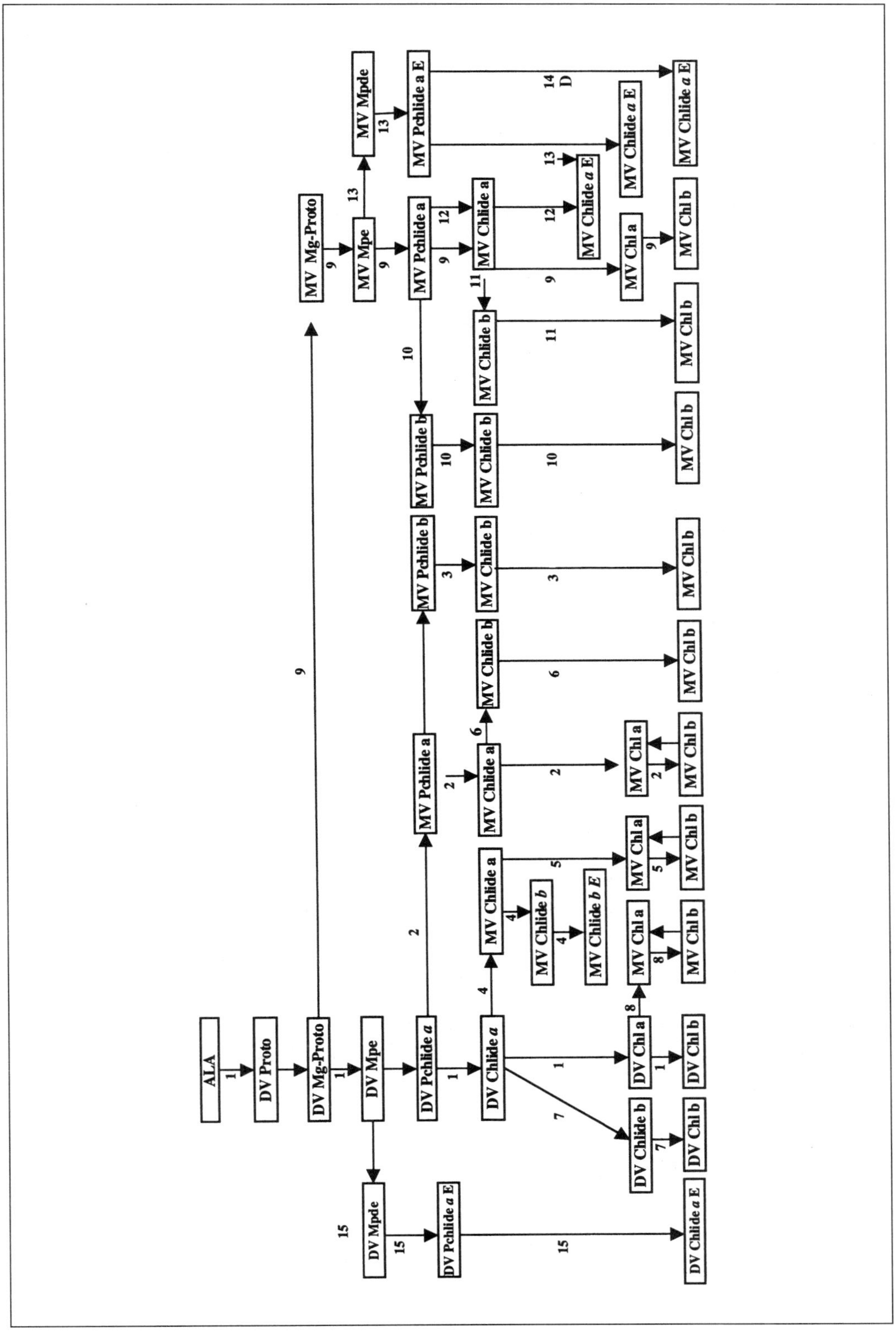

Figure 1. Integrated Chl *a/b* biosynthetic pathway adapted from Reference 44. To facilitate understanding of the text, various biosynthetic pathways are designated by the numbers 1 to 14.

quantitative and qualitative analysis.

❖ Procedure 2. Preparation of a Hexane-Extracted Acetone Residue

1. Prepare an ammoniacal acetone extract as above.
2. Using a separation funnel, transfer chlorophylls and other fully esterified tetrapyrroles to hexane by extraction with an equal volume of hexane.
3. Using conical centrifuge tubes, carry out a second extraction with one-third volume of hexane. Emulsions are resolved using a benchtop centrifuge at room temperature.
4. After centrifugation, remove the lower phase (hypophase) that consists of hexane-extracted acetone (HEAR) using a Pasteur pipet.
5. The HEAR hypophase contains monocarboxylic tetrapyrroles such as Pchlide, Chlide, and Mg-protoporphyrin monomethyl ester (Mpe), and dicarboxylic tetrapyrroles such as Proto and Mg-Proto.
6. The HEAR epiphase contains fully esterified tetrapyrroles such as chlorophylls and protochlorophyllide esters. The epiphase is usually contaminated by trace amounts of hypophase pigments and in some situations it is desirable to purify further the epiphase pigments prior to quantitative analysis.

2.3. Spectrofluorometry

The use of fluorescence emission and excitation spectrofluorometry has been instrumental in the detection and spectral analysis of various tetrapyrroles. It is therefore, appropriate to precede the description of the detection and quantitative determination of various tetrapyrroles by a brief introduction to spectrofluorometry.

When light of a particular wavelength (excitation light) is absorbed by a fluorescent compound, the absorbed light is reemitted at a longer wavelength. The reemitted light is called fluorescence. In contrast to absorbance spectroscopy, fluorescence is a 2-window technique as it allows the monitoring of emission or excitation spectra. For example, by excitation at an appropriate fixed wavelength, the emitted light can be scanned, thus generating an emission spectrum. On the other hand, by monitoring the emitted light at a fixed emission wavelength, the absorbed light can be scanned, thus generating an excitation spectrum.

In addition to its high sensitivity (subpicomole quantities of tetrapyrroles are readily detectable), fluorescence measurements are highly selective. For example, let us consider a mixture of several fluorescent compounds with approximately an equal propensity to fluoresce, i.e., having approximately similar relative quantum yields, and different absorption and emission properties. It is possible to elicit the fluorescence of any compound in the mixture by selective excitation at a specific wavelength that minimizes the detection of the other fluorescent species. Of course, the other compounds in the mixture may also exhibit some fluorescence that needs to be corrected for during quantitative measurements. Conversely, the excitation spectrum of any one of the compounds in the mixture may also be observed by recording an excitation spectrum (also referred to as an action spectrum) at a selected emission wavelength that minimizes the detection of the other compounds in the mixture. Furthermore, if the fluorescence measurements are made at 77 K, significant narrowing and sharpening of the emission and excitation bands are observed. This in turn results in considerable improvement in band resolution.

The correct use of spectrofluorometry is slightly more complicated than the use of absorbance spectroscopy. The excitation source, which in most cases is a Xenon arc, has a limited life span. As the arc lamp ages, so does its output. As a consequence, the sensitivity of the instrument decreases as a function of the age of the excitation lamp. This can be corrected for by adjusting the sensitivity of the instrument on a daily basis against a stable fluorescent standard, such as an acrylic block of rhodamine b, by adjustment of the photomultiplier voltage. The stability of the excitation source also varies continuously as the voltage in the laboratory fluctuates as a function of power demands. This caveat can simply be corrected for by recording spectra in the ratio mode. In that mode, the excitation beam is split into two beams. One beam passes through the sample (sample beam), and the other beam, the reference beam, passes through a quantum counter such as a cuvette filled with a concentrated rhodamine b solution (about 10 mg/mL polyethylene glycol 600). Since rhodamine b exhibits a constant quantum yield between 380 and 600 nm by monitoring the ratio of the sample and reference beams instead of monitoring only the sample beam, the effect of any fluctuation in the electrical current on signal intensity is eliminated. The recording of fluorescence spectra in the ratio mode is routine in most commercial scanning spectrofluorometers.

Finally, the photomultiplier response of the instrument as a function of wavelength is not constant and usually decreases dramatically in the red region of the spectrum. In computer-interfaced instruments, this is corrected for on-line by multiplying the signal ratios at every wavelength by a correction factor. This feature is standard on most sophisticated scanning spectrofluorometers.

2.4. Purification of Tetrapyrroles

Prior to some analyses, it is desirable to purify individual tetrapyrroles from crude mixtures. In the following, a brief description of common chromatographic procedures used in tetrapyrrole separation and purification are described.

2.4.1. Paper Chromatography

Paper chromatography is one of the oldest techniques used for the separation of various tetrapyrroles including Chls. Although more powerful techniques have since been developed, it is still a viable and cheap alternative for the separation of Proto by ascending chromatography, from more polar tetrapyrroles such as uroporphyrins (Uro) and coproporphyrins (Copro) (37).

2.4.2. Thin Layer Chromatography

For the past 35 years, thin layer chromatography (TLC) has been the workhorse of tetrapyrrole purification. However, a comprehensive treatment of TLC is beyond the scope of this chapter. In our laboratory, we prepare our own TLC plates using a variety of adsorbents including cellulose, silica gel H, and polyethylene.

Silica gel H plates are prepared by mixing 42 g of silica gel H (EM Science, Gibbstown, NJ, USA) and 135 mL of water. Cellulose (Cellulosepulver, MN 300; Macherey-Nagel, Duren, Germany) plates are prepared by mixing 22 g of cellulose and 135 mL of water. Polyethylene plates are prepared by mixing 40 g of chromatographic grade polyethylene powder (Polysciences, Warrington, PA, USA) and 120 mL of pure or 90% aqueous acetone. Mixing is carried out for 60 seconds in a Waring blender. The plates (0.25 or 0.5 mm in thickness) are prepared by using a Desaga spreader. After air-drying for about 30 minutes, the plates are activated by heating at 105°C overnight. The activated plates are stored dry in a dessicator until further use. Unless otherwise indicated, we favor development at 4°C in darkness. More detail about specific TLC systems will be described as needed.

2.4.3. High-Pressure Liquid Chromatography

In some cases, high-pressure liquid chromatography (HPLC), also referred to as high-performance liquid chromatography, is very useful for qualitative and quantitative tetrapyrrole analysis. Prior to quantitative analysis, it is advisable to check the purity of the segregating tetrapyrrole peaks by high-resolution spectroscopic techniques such as spectrofluorometry. It has been our experience that separated HPLC tetrapyrrole peaks are often contaminated by very small amounts of other tetrapyrroles.

Unless otherwise indicated, we currently use an Applied Biosystems HPLC system (Foster City, CA, USA) that consists of a quaternary LC-620 pump, variable wavelength absorption (LC-95) and fluorescence (LC-240) detectors, and a cooled autosampler (ISS 200). Ten- to fifty-microliter aliquots of the acetone:ammonium hydroxide extract are injected. Separations are performed on a an Applied Biosystems Pecosphere 3 × 3C, C-18 reversed phase, 4 × 0.5 cm column or other columns as indicated for specific analyses. A PEEK-C18 guard column (Applied Biosystems) is used to protect the separation column, and elution is usually achieved with isocratic or gradient solvent systems. The elutants are monitored by on-line spectrofluorometry where fluorescence can be stimulated by excitation at 400, 420, or 440 nm, and signal is monitored with a total emission accessory. The amounts of various tetrapyrroles are determined from a calibration curve using Proto, Mg-Proto, Pchlide, Chl, and pheophytin standards. When needed, the HPLC pump is stopped at each emission peak, and an emission spectrum of the eluting band is recorded between 580 and 700 nm.

3. ANALYSIS OF DIVINYL PROTOPORPHYRIN IX

Proto is a divinyl (DV) dicarboxylic tetrapyrrole with vinyl groups at position 2 and 4 of the macrocycle (Figure 2). In extracting Proto from plant tissues or isolated organelles, it is essential to eliminate loss of Mg from Mg-Proto and Mpe, since Mg loss would generate Proto and Proto monomethyl ester that would contaminate the endogenous Proto pool. Elimination of Mg loss is achieved by extraction with ammoniacal acetone as described in section 2.1.

3.1. Quantitative Determination of Proto by Room Temperature Spectrofluorometry

Upon tetrapyrrole extraction from plant tissues and partitioning of the pigments between hexane and acetone, Proto passes into the HEAR phase (section 2.2) along with other dicarboxylic tetrapyrroles such as DV and monovinyl (MV) Mg-Proto (Figure 3) and monocarboxylic tetrapyrroles such as DV and MV Mpe (Figure 3), DV and MV Pchlide *a* (Figure 4), DV and MV Chlide *a* and *b* (Figure 5), and pheophorbide (Pheobide) *a* and *b* (i.e., demetalated Chlide *a* and *b*). The fluorescence emission properties of Proto are such that it is possible to determine the amount of Proto in the presence of other di- and monocarboxylic tetrapyrroles without further purification. For example, in HEAR at room temperature, Proto exhibits respective excitation and emission maxim at 404 and 633 nm, while Mg-Proto and Mpe exhibit respective excitation and emission maxima at 418 and 595 nm, Pchlides *a* at 438 and 638 nm, Chlide *b* at 461 and 660 nm, Chlide *a* at 432 and 675 nm, pheophytin and Pheobide *a* [Pheo(bide) *a*] at 413 and 674 nm and Pheo(bide) *b* at 439 and 660 nm. By excitation of the HEAR fraction at 400 nm, i.e., close to the Soret excitation maximum of Proto, and by recording an emission spectrum between 580 and 700 nm, it is possible to maximize the fluorescence emission of Proto at 633 nm. The fluorescence emission of Mg-Proto and Mpe are far removed from the emission maximum of Proto and do not interfere with its determination. Only fluorescence emissions from Copro at 622 nm, from Pchlide *a* at 638 nm and to a minor extent from Chlide *a* and *b*, are likely to interfere. The possible contribution of these fluorescences to Proto fluorescence is eliminated by spectral deconvolution. This is achieved by exciting the tetrapyrrole mixture at 440 nm, i.e., close to the Soret excitation maximum of Pchlide *a* and Chlide *a* and *b*, and by recording a second emission spectrum of the mixture between 580 to 700 nm. The net fluorescence amplitude due only to Proto emission is deconvoluted from the Copro, Pchlide *a*, and Chlide *a* and *b* emissions with the use of Equation 1 (45).

Equation 1. Deconvolution of Fluorescence Data

Proto $(F_{633}\ E_{400}) = 1.055\ F\ (F_{633}\ E_{400}) - 0.1190\ F\ (F_{622}\ E_{400}) - 0.442\ F\ (F_{638}\ E_{440}) - 0.0141\ F\ (F_{675}\ E_{440})$

Where:

Proto $(F_{633}\ E_{400})$ = deconvoluted net fluorescence emission amplitude of Proto at 633 nm upon excitation of the tetrapyrrole mixture at 400 nm.

$F\ (F_{633}\ E_{400})$ = undeconvoluted total fluorescence emission amplitude observed at 633 nm upon excitation of the tetrapyrrole mixture at 400 nm. This fluorescence amplitude is usually contaminated by contributions from Copro, Pchlide *a,* and Chlide *a* and *b* emissions.

$F\ (F_{622}\ E_{400})$ = undeconvoluted total fluorescence emission amplitude observed at 622 nm upon excitation of the tetrapyrrole mixture at 400 nm. This factor corrects for Copro fluorescence emission.

$F\ (F_{638}\ E_{440})$ = undeconvoluted total fluorescence emission amplitude observed at 638 nm upon excitation of the tetrapyrrole mixture at 440 nm. This factor corrects for Pchlide *a* fluorescence emission.

$F\ (F_{675}\ E_{440})$ = undeconvoluted total fluorescence emission amplitude observed at 675 nm upon excitation of the tetrapyrrole mixture at 440 nm. This factor corrects for Chlide *a* and *b* fluorescence emission.

See Table 2 for further equations.

The various fluorescence amplitudes to be substituted into Equation 1 are read from the recorded emission spectra. Next, the net fluorescence emission amplitude of Proto, as calculated from Equation 1, is converted to Proto concentration by reference to a standard calibration curve. The latter is prepared from standard Proto solutions of known concentrations and from their fluorescence emission amplitudes. The latter are recorded under the same instrumental conditions by excitation at 400 nm. In this manner, Proto can be determined with a precision of about 4% (45). Minimum detection levels are about 0.2 pmol/mL. Equation 1 can be used with any spectrofluorometer capable of recording corrected fluorescence emission and excitation spectra. The user has to construct, however, the required calibration curve that relates net fluorescence emission amplitudes to Proto concentrations.

In the Laboratory of Plant Biochemistry and Photobiology, Proto calculations as well as all other spectrofluorometric calculations are automatically carried out by the PC interfaced with the spectrofluorometer, using Porphyrin Analytical Tools® (PAT) software (48).

The washed diethyl ether extract can be used for further manipulations. If the original HEAR fraction contains both Proto and Mg-Proto, and one wishes to separate these tetrapyrroles, additional chromatographic steps are required (section 3.3).

3.2. Extraction of Proto from HEAR

Prior to structural analysis, it is often necessary to extract Proto from the HEAR fraction into less polar solvents such as diethyl ether. Since Proto is a dicarboxylic tetrapyrrole (Figure 3), it is quite soluble in the aqueous HEAR fraction. As a consequence, its extraction into less polar solvents requires a procedure adapted for dicarboxylic tetrapyrroles (26). Essentially, the procedure involves first the transfer of the less polar monocarboxylic tetrapyrroles to diethyl ether. This is followed by extraction of dicarboxylic tetrapyrroles such as Proto and Mg-Proto into diethyl ether after adjustment of the HEAR pH.

❖ Procedure 3. Extraction of Monocarboxylic and Dicarboxylic Tetrapyrroles with Diethyl Ether

1. Prepare a HEAR fraction hypophase as above and transfer to a conical centrifuge tube.
2. Transfer monocarboxylic tetrapyrroles to diethyl ether by adding:
   1/5 volume of diethyl ether,
   1/17 volume of saturated NaCl,
   1/70 volume of 0.37 MKH_2PO_4 at pH 7.7.
3. Mix thoroughly and resolve the phases by centrifugation at 1500× *g* for 30 seconds at room temperature.
4. Remove the ether phase using a Pasteur pipet.
5. Re-extract the HEAR fraction 4 to 5 times with small volumes of diethyl ether using centrifugation to resolve the emulsion.
6. Combine the ether extracts for further

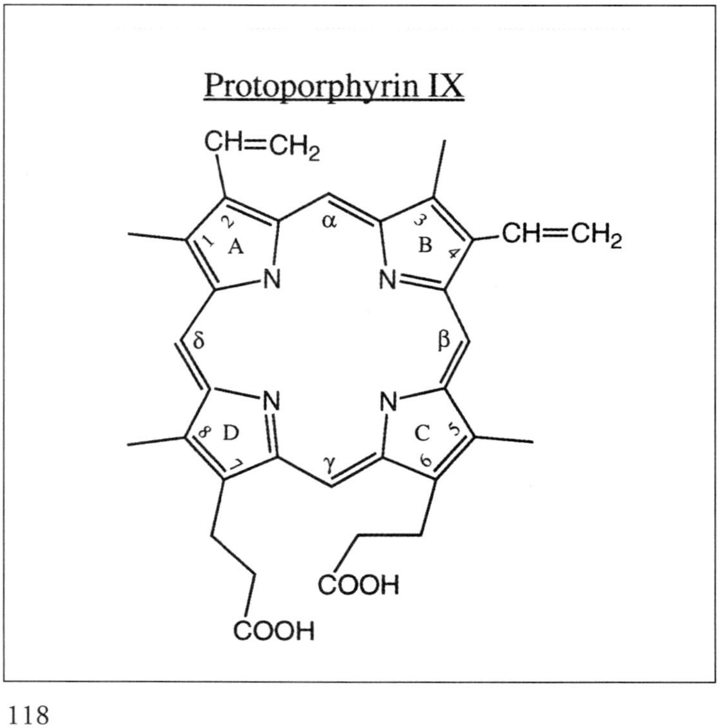

Figure 2. Structure of protoporphyrin IX, which is the last common intermediate of chlorophyll and heme.

manipulation of monocarboxylic tetrapyrroles.

7. For extraction of dicarboxylic tetrapyrroles from the remaining ether-extracted HEAR fraction, adjust the pH to 4.0 to protonate the free carboxylic groups.
8. Extract 5 times with small volumes of diethyl ether using centrifugation to breakdown the emulsion.
9. Wash the combined ether extracts containing dicarboxylic tetrapyrroles by passing through a 0.5 M solution of KH_2PO_4 adjusted to pH 4.8. This washing step is extremely important for the recording of sharp 77 K spectra in diethyl ether. It is best achieved by removing the diethyl ether extract with a Pasteur pipet and slowly releasing it at the bottom of a 15-mL test tube filled with the phosphate buffer. As the diethyl ether bubbles rise to the top, the aqueous contaminants pass into the buffer. The washed diethyl ether extract can then be removed using a Pasteur pipet.

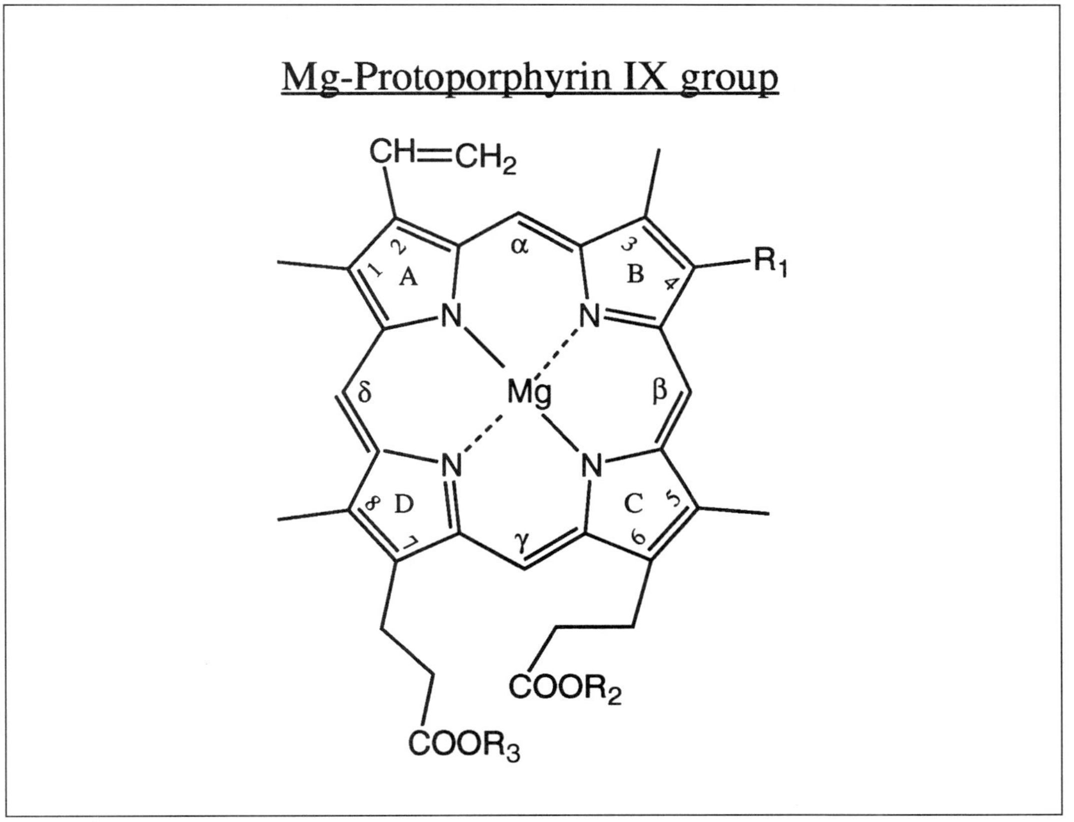

Figure 3. Mg-protoporphryin IX and its derivatives.

Compound	Abbreviation	R1	R2	R3
Divinyl-Mg-protoporphyrin IX	DV Mg-Proto	vinyl	H	H
Monovinyl-Mg-protoporphyrin IX	MV Mg-Proto	ethyl	H	H
Divinyl-Mg-protoporphyrin IX monomethylester	DV Mpe	vinyl	methyl	H
Monovinyl-Mg-protoporphyrin IX monomethylester	MV Mpe	ethyl	methyl	H
Divinyl-Mg-protoporphyrin IX diester	DV Mpde	vinyl	methyl	LCFA
Monovinyl-Mg-protoporphyrin IX diester	MV Mpde	ethyl	methyl	LCFA

3.3. Chromatographic Separation of Proto from other Tetrapyrroles

3.3.1. Separation by Paper Chromatography

Although more powerful techniques have since been developed, it is still a viable and cheap alternative for the separation of Proto from more polar tetrapyrroles such as Uro and Copro by ascending chromatography (37). Essentially, the tetrapyrrole mixture, as in an aliquot of HEAR or acetone extract, is spotted at the origin of a 5 × 15 cm piece of No. 1 or No. 3 chromatography paper (Whatman, Clifton, NJ, USA), allowed to dry, and then the paper is inserted into a cut 1000-mL glass cylinder containing about 10 mL of 2,4- or 2,6-lutidine: 0.05 N NH_4OH (5:3.5 vol/vol). The cylinder is capped with a piece of aluminum foil. Development is carried out under subdued light at room temperature in a ventilated hood. When the solvent front has migrated about 10 cm, the chromatogram is viewed under a 360 nm UV

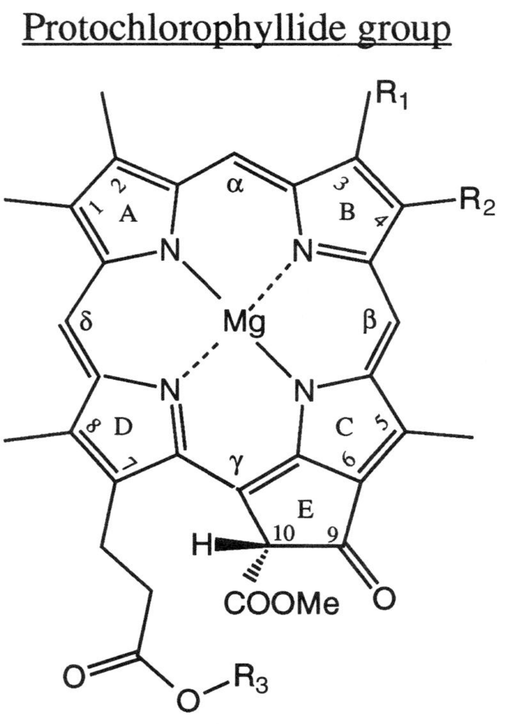

Figure 4. Protochlorophyllide group. Tetrapyrroles which incorporate the fifth isocyclic ring, ring E. LCFA = long-chain fatty acid.

Compound	Abbreviation	R1	R2	R3
Divinyl-protochlorophyllide *a*	DV Pchlide *a*	methyl	vinyl	H
Monovinyl-protochlorophyllide *a*	MV Pchlide *a*	methyl	ethyl	H
Divinyl-protochlorophyllide *a* ester	DV Pchlide *a* E	methyl	vinyl	LCFA
Monovinyl-protochlorophyllide *a* ester	MV Pchlide *a* E	methyl	ethyl	LCFA
Divinyl-protochlorophyllide *b*	DV Pchlide *b*	formyl	vinyl	H
Monovinyl-protochlorophyllide *b*	MV Pchlide *b*	formyl	ethyl	H
Divinyl-protochlorophyllide *b* ester	DV Pchlide *b* E	formyl	vinyl	LCFA
Monovinyl-protochlorophyllide *b* ester	MV Pchlide *b* E	formyl	ethyl	LCFA

light. The separated tetrapyrroles are detected by their red fluorescence. In this system, Proto migrates with a retention factor (R_f) of about 0.8, while Copro and Uro migrate with R_fs of about 0.6 and 0.3, respectively (37). Although paper chromatography is useful for separating Proto form other tetrapyrroles, it is not convenient for preparative purposes because of the difficulty of eluting the separated tetrapyrroles.

3.3.2. Chromatographic Separation and Determination by HPLC

If HPLC instrumentation is available, it can be efficiently used for the quantitative separation of Proto from other tetrapyrroles. Using large columns, it can also be conveniently used for preparative purposes.

We have successfully used the solvent system of Ho (23) for separating Proto from Uro, Copro, and other multicaboxylic tetrapyrroles in HEAR fractions prepared from cancer cell cultures (50). The same system can be used with HEAR preparation of plant sources. With this solvent system, there is no need to convert tetrapyrroles to their methyl esters prior to chromatography. If feasible, it is always advantageous to avoid derivatization prior

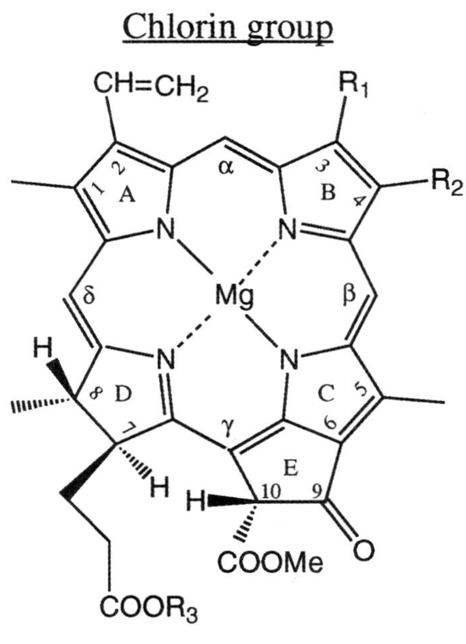

Figure 5. Chlorin group. Tetrapyrroles reduced to the oxidation state of chlorophylls, i.e., a single bond between carbons 7 and 8.

Compound	Abbreviation	R1	R2	R3
Divinyl-chlorophyllide *a*	DV Chlide *a*	methyl	vinyl	H
Monovinyl-chlorophyllide *a*	MV Chlide *a*	methyl	ethyl	H
Divinyl-chlorophyllide *b*	DV Chlide *b*	formyl	vinyl	H
Monovinyl-chlorophyllide *b*	MV Chlide *b*	formyl	ethyl	H
Divinyl-chlorophyll *a*	DV Chl *a*	methyl	vinyl	phytol
Monovinyl-chlorophyll *a*	MV Chl *a*	methyl	ethyl	phytol
Other divinyl-chlorophyll *a*	Other DV Chl *a*	methyl	vinyl	LCFA

to chromatography, since derivatization may generate artifacts and often results in lower yields that necessitate correction for percent recoveries during quantitative determinations.

Separation of Proto from tri-, tetra-, penta-, hexa-, and octacarboxylic tetrapyrroles can be achieved on a Bondapak reverse phase C18-bonded column (Waters, Milford, MA, USA). The column is equilibrated with the first binary mobile phase (0.1 M sodium phosphate in acetonitrile, 29:16 vol/vol, pH 3.0) for 10 minutes. After sample injection, the mobile phase is changed to 0.1 M sodium phosphate in acetonitrile (18:129 vol/vol, pH 3.0) for 15 minutes. The flow rate is usually about 1.10 mL/minute. The eluant is monitored by spectrofluorometry. Fluorescence is elicited by excitation at 405 nm and is monitored at an emission wavelength of 630 nm. The amounts of various tetrapyrroles can be determined from a calibration curve using tetrapyrrole chromatographic standards (Porphyrin Products). In this system, Proto exhibits a retention time of about 12.68 minutes. Most HPLC manuals describe the use of internal and external standards for quantitative determinations. The procedures usually vary slightly from manufacturer to manufacturer depending on the used software. The method can be scaled up for preparative purposes. In that case, the recovered fraction can be treated as described in section 3.2 to transfer the purified Proto to diethyl ether.

3.4. Quantitative Determination of Purified Proto

3.4.1. Quantitative Determination by Spectrofluorometry

Purified Proto can be determined in HEAR as described in section 3.1 for tetrapyrrole mixtures. It can also be determined in any solvent from emission spectra elicited by excitation close to its Soret absorbance maximum at around 400 to 405 nm. In each case, a calibration curve using known amounts of Proto should be constructed in order to convert fluorescence amplitudes to Proto concentrations. At each concentration, an emission spectrum elicited by excitation at the chosen Soret excitation wavelength should be recorded, and a calibration curve relating fluorescence emission amplitudes of pure Proto solutions to Proto concentration is constructed.

3.4.2. Quantitative Determination by Spectrophotometry

The concentration of purified Proto solutions can be determined in a variety of organic solvents by absorbance spectroscopy. The procedure is only limited by the availability of molar extinction coefficient values for the solvent under consideration and is usually used for the preparation of Proto stock solutions of known concentrations.

At room temperature, Proto exhibits a stepladder (etio) absorption spectrum in various organic solvents with a very strong Soret absorption band between 340 and 440 nm and 4 absorption bands of decreasing intensity between 480 and 650 nm (19). The most suitable wavelength for the quantitative determination of Proto is at its Soret absorbance maximum, which varies from 400 to 406 nm depending on the solvent. At the Soret absorbance maximum, Proto has a large molar extinction coefficient that allows the quantitative determinations of dilute solutions. Since many other pigments, such as carotenoids, Chls, and Pchlides, absorb light in this spectral region, it is important to use highly purified Proto. It is also possible to use Proto absorbance in the red region of the spectrum, at 622 nm, for quantitative determinations. At this wavelength, the Proto molar extinction coefficient is very low, and such determinations require the use of more concentrated solutions.

The molar extinction coefficient is related to Proto absorbance and concentration by Equation 2 (15).

Equation 2. Beer's Law

$A = \varepsilon c l$

Where: A = Absorbance.
ε = Molar absorption coefficient.
c = Concentration of Proto in moles per liter.
l = Optical length in cm of the used cuvette, which in most cases is 1 cm.

4. ANALYSIS OF OTHER TETRAPYRROLES

The quantitative determination of the other intermediates of the Chl biosynthetic pathway (Figure 1) can be carried out using essentially the same procedure as described for Proto. Table 1 reports the absorption coefficients for the compounds in various solvents, while Table 2 depicts the equations used to deconvolute the spectra for quantitative determinations. For each compound, the analysis involves the following steps:

1. Extraction in ammoniacal acetone from tissues or organelles.
2. Quantitative determination by spectrofluorometry in crude HEAR extracts.
3. Transfer of the tetrapyrrole from acetone to hexane and then from HEAR to diethyl ether.
4. Chromatographic purification on thin layer plates or HPLC.
5. Quantitative determination of ether-extracted tetrapyrroles or purified tetrapyrroles by spectrofluorometry and/or spectrophotometry.

In the following sections, each group of compounds are considered in turn, and similarities and differences in analytical procedures are highlighted as appropriate.

5. ANALYSIS OF MG-PROTOPORPHYRIN IX

The Mg-Proto pool is a mixed DV-MV dicarboxylic tetrapyrrole pool (Figure 3). Under most conditions, the proportion of DV Mg-Proto is much larger than that of MV Mg-Proto.

5.1. Quantitative Determination of Mg-Proto in Crude Extracts by Spectrofluorometry

Upon tetrapyrrole extraction from plant tissues and partitioning of the pigments between hexane and acetone (sections 2.1 and 2.2), the mixed DV and MV Mg-Proto pool passes into the HEAR phase along with other dicarboxylic and monocarboxylic tetrapyrroles. Since in HEAR, DV and MV Mg-Proto exhibit similar emission maxima at 594 to 596 nm at room temperature, it is possible to determine the concentration of the mixed DM-MV Mg-Proto pool in the presence of other di- and monocarboxylic tetrapyrroles without further purification. Since Mg-Proto and Mpe exhibit identical spectrophotometric and fluorescence properties, it is necessary to extract the monocarboxylic Mpe pool from the HEAR fraction prior to Mg-Proto determination.

Extraction of monocarboxylic tetrapyrroles from the HEAR fraction into diethyl ether is achieved exactly as described in section 3.2. Once this is done, the ether-extracted HEAR fraction is excited at 420 nm, i.e., close to the Soret excitation maximum of the mixed Mg-Proto pool, and an emission spectrum is recorded between 580 and 700 nm. The Mg-Proto pool exhibits a well-defined emission band

Table 1. Molar Extinction Coefficients (Ext Coefficient) of Metabolic Intermediates of the Chl *a* Biosynthetic Pathway in Different Solvents at Room Temperature

Tetrapyrrole	Solvent	Abs (nm)	Ext Coefficient	Reference
DV Proto DME	Diethyl ether	404	158000	19
DV Proto DME	Dioxan	406	164000	19
DV Proto DME	$CHCl_3$	407	171000	19
DV Proto DME	Pyridine	409	163000	19
Disodium DV Proto	80% Acetone	402	108244	This work
Disodium DV Proto	80% Acetone	629	3883	This work
DV Mg-Proto DME	Diethyl ether	419	308000	19
DV Mg-Proto DME	Diethyl ether	551	18200	19
DV Mg-Proto DME	Diethyl ether	589	18200	19
Disodium DV Mg-Proto	80% Acetone	417	165900	This work
Disodium DV Mg-Proto	80% Acetone	550	13817	This work
Disodium DV Mg-Proto	80% Acetone	589	12537	This work
MV Pchl(ide) *a*	Diethyl ether	432	187517	30
MV Pchl(ide) *a*	Diethyl ether	623	22620	30
MV Pchl(ide) *a*	Acetone	432	165817	30
MV Pchl(ide) *a*	Acetone	623	21394	30
MV Pchl(ide) *a*	Methanol	434	108685	30
MV Pchl(ide) *a*	Methanol	629	16674	30
DV Pchl(ide) *a*	Diethyl ether	437	205000	23
DV Pchl(ide) *a*	Diethyl ether	622	22100	23
MV Pchlide *b* phytyl ester	Diethyl ether	442	111519	56
MV Pchlide *b* phytyl ester	Diethyl ether	630	24357	56
MV Pchlide *b* phytyl ester	80% Acetone	446	88585	56
MV Pchlide *b* phytyl ester	Diethyl ether	632	18171	56
MV Chl *a*	Diethyl ether	430	104090	34
MV Chl *a*	Diethyl ether	660	83450	34

Table 1, continued.

Tetrapyrrole	Solvent	Abs (nm)	Ext Coefficient	Reference
MV Chl *a*	80% Acetone	430	81110	34
MV Chl *a*	80% Acetone	663	73300	34
DV Chl *a*	Diethyl ether	435	110820	57
DV Chl *a*	Diethyl ether	660	78340	57
DV Chl *a*	80% Acetone	438	95120	57
DV Chl *a*	80% Acetone	664	69290	57
MV Chl *b*	Diethyl ether	455	129090	34
MV Chl *b*	Diethyl ether	643	48630	34
MV Chl *b*	80% Acetone	460	118210	34
MV Chl *b*	80% Acetone	645	41370	34
DV Chl *b*	Diethyl ether	462	118900	57
DV Chl *b*	Diethyl ether	643	43910	57
DV Chl *b*	80% Acetone	468	87830	57
DV Chl *b*	80% Acetone	651	35780	57
MV Pheophytin *a*	Diethyl ether	665	50530	70
MV Pheophytin *a*	Diethyl ether	410	109770	70
MV Pheophytin *a*	80% Acetone	666	49310	66
MV Pheophytin *a*	80% Acetone	409	114040	66
DV Pheophytin *a*	Diethyl ether	667	45680	57
DV Pheophytin *a*	Diethyl ether	417	107900	57
DV Pheophytin *a*	80% Acetone	665	16520	57
DV Pheophytin *a*	80% Acetone	419	41300	57
MV Pheophytin *b*	Diethyl ether	654	32750	70
MV Pheophytin *b*	Diethyl ether	433	173500	70
MV Pheophytin *b*	80% Acetone	655	31600	66
MV Pheophytin *b*	80% Acetone	436	160220	66
DV Pheophytin *b*	Diethyl ether	657	29580	57
DV Pheophytin *b*	Diethyl ether	440	117300	57
DV Pheophytin *b*	80% Acetone	657	21060	57
DV Pheophytin *b*	80% Acetone	443	70960	57

Molar extinction coefficients of Pchlide *a* were calculated from the specific absorption coefficients reported in Reference 30 and a molecular weight of 613 for Pchlide *a*. In Reference 30 the cited specific absorption coefficients were mistakenly assigned to Pchlide *a* E by the authors.

with a maximum at 594 to 596 nm. In this spectral region, interference by other natural tetrapyrroles is not encountered, and correction for fluorescence band overlap is unnecessary. The fluorescence amplitude at 595 to 596 nm or the integrated area under the fluorescence emission band is next converted to Mg-Proto concentration by reference to calibration curves prepared under the same instrumental and analytical conditions. The calibration curves should relate authentic Mg-Proto concentrations to fluorescence emission amplitude at 595 to 596 nm or to integrated fluorescence areas under the Mg-Proto emission band. In this manner, Mg-Proto can be determined with a precision of about 7% to 8%. Minimum detection levels are about 0.2 pmol/mL.

5.2. Quantitative Determination of DV and MV Mg-Proto in Crude Extracts by Spectrofluorometry

Before structural studies or determination of DV and MV Mg-Proto, it is necessary to extract Mg-Proto from the Mpe-free ether-extracted HEAR fraction into less polar solvents such as diethyl ether. Extraction of DV and MV Mg-Proto from the ether-extracted Mpe-free HEAR fraction is achieved exactly as described in section 3.2 for Proto.

The use of low temperature (77 K) is essential to differentiate between DV and MV Mg-Proto and all other DV and MV Mg-tetrapyrroles. At 77 K, due to considerable fluorescence excitation and emission band narrowing, the DV and MV Mg-Proto pools exhibit respective excitation and emission maxima at 424 and 591 nm (DV Mg-Proto) and 417 and 589 nm (MV Mg-Proto) (12). It is not possible, however, to directly determine the amounts of DV and MV Mg-Proto from 77 K spectra, since for a given amount of DV and MV Mg-Proto, the overall fluorescence amplitudes vary with the freezing pattern of the sample. For a given spectrum, however, the ratio of fluorescence amplitudes at any two or more wavelengths is independent of the freezing pattern.

As a consequence, determination of the amounts of DV and MV Mg-Proto is a 2-step process. First, the total amount of DV plus MV Mg-Proto in the ether-extracted HEAR fraction is determined at room temperature, exactly as described in section 3.2. Next, the DV/MV ratio of the Mg-Proto pool is determined from 77 K spectra recorded in diethyl ether. The amount of DV and MV Mg-Proto is then calculated from the total amount of Mg-Proto, as determined from the ether-extracted HEAR fraction at room temperature and from the DV/MV Mg-Proto ratio as determined at 77 K in diethyl ether (63).

To calculate the DV/MV Mg-Proto ratio, two sharp 77 K excitation spectra in diethyl ether need to be recorded. For the recording of sharp DV and MV Mp-Proto and all other 77 K DV and MV Mg-tetrapyrroles spectra, it is essential to thoroughly wash the diethyl ether fraction with a 0.5 M solution of KH_2PO_4 adjusted to pH 4.8 (or to pH 7.0 for monocarboxylic Mg-tetrapyrroles) as described in section 3.2. At a pH lower than 4.8, Mg loss becomes a problem, and at a pH higher than 4.8, the dicarboxylic Mg-Proto pool will be lost by passing into the aqueous phase.

The first 77 K excitation spectrum is recorded from 380 to 500 nm by positioning the emission monochromator at the emission maximum of DV Mg-Proto at 591 nm. In this spectrum, DV Mg-Proto will exhibit a sharp and narrow excitation band with a maximum at 424 nm (63). The second 77 K excitation spectrum is recorded from 380 to 500 nm by positioning the emission monochromator at 587 nm, just below the 589 nm emission maximum of MV Mg-Proto. In this spectrum,

MV Mg-Proto will exhibit a sharp and narrow excitation band with a maximum at 417 nm (63). In most cases, if both DV and MV Mg-Proto are present, two sharp maxima are observed, one at 424 nm for DV Mg-Proto and one at 417 nm for MV Mg-Proto. The net fluorescence excitation amplitudes of DV and MV Mg-Proto are then deconvoluted and calculated with Equations 3 and 4 (Table 2) (63).

The various undeconvoluted fluorescence excitation amplitudes to be substituted into Equation 3 and 4 are read from the recorded excitation spectra. Next, the ratio of the deconvoluted DV Mg-Proto excitation amplitude to the deconvoluted MV Mg-Proto excitation amplitude is calculated. This ratio is referred to as the apparent DV/MV Mg-Proto excitation ratio. That ratio is converted to an authentic DV/MV Mg-Proto concentration ratio by reference to a calibration curve. The latter relates apparent ratios of DV/MV Mg-Proto excitation amplitudes to known ratios of DV/MV Mg-Proto concentrations. The calibration curve is constructed as follows: first, known concentrations of DV and MV Mg-Proto dissolved in diethyl ether are mixed in different proportions. Second, fluorescence excitation spectra are recorded on every mixture at 77 K as described above. Third, the apparent fluorescence excitation ratios at 424/417 nm are plotted against the authentic concentration ratio for every mixture. In this manner, DV and MV Mg-Proto can be determined with a precision of about 5% to 6% (63).

5.3. Chromatographic Separation of Mg-Proto

5.3.1. Chromatographic Separation on Thin Layers of Silica Gel H

The mixed DV-MV Mg-Proto pool can be separated from monocarboxylic Mg-porphyrins on thin layers of silica gel H (12). Essentially, the tetrapyrrole mixture, as in an aliquot of HEAR or acetone extract, is spotted at the origin of a 5 × 20 cm TLC plate of silica gel H. While the plate is still wet, it is inserted into a cut 1000-mL glass cylinder containing about 10 mL of toluene:ethyl acetate:ethanol (8:2:2 vol/vol/vol). The cylinder is capped with a piece of aluminum foil. Development is carried out at 4°C in darkness. After the solvent front has migrated about 10 cm, the plate is viewed under 366 nm UV. The separated tetrapyrroles are detected by their red fluorescence. In this system, Proto remains at the origin, and Mg-Proto moves with an R_f of about 0.09. Monocarboxylic Mg-porphyrins such as Mpe, Pchlides, and Chlides move toward the center of the plate. Mg-Proto is eluted in methanol:acetone (4:1 vol/vol). Elution is achieved by scraping the wet silica gel H band into a small beaker containing a few milliliters of the methanol:acetone mixture. This is followed by transfer of the slurry to a conical centrifuge tube with a Pasteur pipet and centrifugation at 1800× *g* for 1 minute on a table top centrifuge. The supernatant containing Mg-Proto, with or without Proto, is removed using a Pasteur pipet. The proportions of DV/MV Mg-Proto components can be determined by spectrofluorometry in diethyl ether at 77 K as described in section 5.2. The Mg-Proto pool is transferred to ether, as described in section 3.2, after dilution with a small volume of HEAR. The ether extract should be washed with a 0.5 M solution of KH_2PO_4 adjusted to pH 4.8 prior to 77 K spectroscopy.

5.3.2. Chromatographic Separation of DV Mg-Proto from MV Mg-Proto on Thin Layers of Polyethylene

Separation of the purified Mg-Proto pool into DV and MV components can be achieved on thin layers of polyethylene developed in 90% aqueous acetone (12).

Table 2. Spectrofluorometric and Spectrophotometric Equations Used to Deconvolute Spectral Band Overlaps

Compound	Solvent	DL	Eq	Equation
DV Mg-Proto (E424 F591)	Diethyl ether	0.1	3	1.014 F (E424 F591) - 0.106. F (E417 F587)
MV Mg-Proto (E417 F587)	Diethyl ether	0.1	4	1.013 F (E417 F587) - 0.131. F (E424 F591)
Pchlide *a* (F638 E440)	HEAR	0.1	5	1.0080 F (F638 E440) - 0.0141 F (F675 E440) - 0.0197 F (F633 E400) + 0.0028 F (F622 E440)
DV Pchlide *a* (E453 F625)	Diethyl ether	0.1	6	1.061 F (E451 F625) - 0.068 F (E437 F625) 1.3307
MV Pchlide *a* (E437 F625)	Diethyl ether	0.1	7	1.060 F (E437 F625) - 0.964 F (E451 F625) 0.8056
Pchlide *a* (623)	Diethyl ether	–	8	41.10 (Abs 623) - 4.93 (Abs 663) - 4.93 (Abs 644)
Pchlide *a* (626)	80% Acetone	–	9	33.20 (Abs 626) - 4.48 (Abs 663) - 7.58 (Abs 645)
MV Pchlide *b* (E463 F643)	Diethyl ether	0.1	10	1.0022 F (E463 F643) - 0.0507 F (E455 F660)
MV Chlide *b* (E455 F660)	Diethyl ether	0.1	11	1.0022 F (E455 F660) - 0.0438 F (E463 F643)
MV Pchlide *b* (F643 E443)	Diethyl ether	0.1	12	1.04 F (F643 E463) - 0.54 F (F635 E440)
Pchlide *a* (F635 E440)	Diethyl ether	0.1	13	1.04 F (F635 E440) - 0.08 F (F643 E463)
MV Chlide *a* (E433 F674)	HEAR	0.2	14	[1.2600 F (E433 F674) - 0.1661 F (E412 F674) - 0.0894 F (E460 F660) - 0.2796 F (E438 F 660)] - 0.1 Chlide *a* (E433 F674)
DV Chlide *a* (E458 F674)	Diethyl ether	0.1	15	1.0205 F (E458 F674) - 0.0557 F (E447 F674)
MV Chlide *a* (E447 F674)	Diethyl ether	0.1	16	1.0205 F (E447F674) - 0.3749 F (E458 F674)
Chlide *a* (663)	Diethyl ether	–	17	12.162 (Abs 663) - 1.08 (Abs 644) - 0.32 (Abs 624)
Chlide *a* (663)	80% Acetone	–	18	14.1803 (Abs 663) - 2.9099 (Abs 645) - 0.2238 (Abs 626)
MV Chlide *b* (E460 F660)	HEAR	0.2	19	[1.0520 F (E460 F660) - 0.1450 F (E438 F660) - 0.0551 F (E433 F674) - 0.0030 F (E412 F674)] + 0.1810 Chlide *b* (E460 F660)

Table 2, continued.

Compound	Solvent	DL	Eq	Equation
DV Chlide *b* (E498 F666)	Diethyl ether	0.1	20	1.0184 F (E498 F666) - 0.0425 F (E475 F660)
MV Chlide *b* (E475 F660)	Diethyl ether	0.1	21	1.0185 F (E475 F660) - 0.4421 F (E498F666)
Chlide *b* (644)	Diethyl ether	–	22	20.96 (Abs 644) - 3.31 (Abs 663) - 0.59 (Abs 624)
Chlide *b* (645)	80% Acetone	–	23	26.0064 (Abs 645) - 4.6613 (Abs 6663) - 0.3636 (Abs 626)
MV Pheo(bide) *a* (E412 F674)	HEAR	0.2	24	[1.132 F (E412 F674) - 0.9713 F (E433 F674) - 0.0682 F (E460 F660) + 0.06496 F(E438 F 660)] + 0.0682 Pheo(bide) *a* (E412 F674)
MV Pheo(bide) *b* (E438 F660)	HEAR	0.2	25	[1.179 F (E438 F660) - 0.4033 F (E460 F660) - 0.541 F (E433 F674) + 0.04873 F (E412 F 674)] + 0.095 Pheo(bide) *b* (E438 F660)

E = excitation wavelength; F = emission wavelength; Abs = absorbance wavelength. The (E…F…) terminology is used to refer to equations that use excitation spectra, the (F….E…) terminology is used to refer to equations that use emission spectra, while the Abs terminology is used in equations that use absorbance spectra. For example: (E424 F591) = Soret excitation amplitude at 424 nm recorded at an emission wavelength of 591 nm; (F638E440) = fluorescence emission amplitude at 638 nm elicited by excitation at 440 nm; Abs 663 = absorbance at 663 nm. DL = detection limit (pmol/mL). Eq = equation number. See section 3.1, Equation 1 for a fully described example using figures for protoporphyrin IX.

In this solvent, DV and MV Mg-Proto migrate with approximate R_fs of 0.58 and 0.66, respectively. The separated bands can be eluted by scraping into a small beaker containing diethyl ether and centrifugation of the slurry to recover the diethyl ether supernatant. The diethyl ether fractions should be washed with a 0.5 M solution of KH_2PO_4 adjusted to pH 4.8 prior to 77 K spectroscopy as in section 3.2.

5.4. Quantitative Determination of Purified DV and MV Mg-Proto

Purified DV and MV Mg-Proto can be determined in any solvent from room temperature emission spectra elicited by excitation close to their Soret absorbance maxima between 417 and 424 nm. As described in section 3.4.1 for purified Proto, a calibration curve using known amounts of DV or MV Mg-Proto should be constructed in order to convert fluorescence data to Mg-Proto concentrations.

The concentration of purified DV Mg-Proto solutions can be determined in a variety of organic solvents by room temperature absorbance spectroscopy. The procedure is only limited by the availability of molar extinction coefficients for the solvent under consideration. Such determinations are usually used for the preparation of DV Mg-Proto stock solutions of known concentrations.

At room temperature, DV-Mg Proto exhibits a 3-banded absorption spectrum in various organic solvents with a very strong Soret absorption band between 360 and 440 nm and 2 absorption bands (the α and β bands) of about equal intensity between 530 and 610 nm (22). The most suitable wavelength for the quantitative determination of DV Mg-Proto is at its Soret absorbance maximum, which varies from 417 to 420 nm depending on the solvent. At the Soret absorbance maximum, DV Mg-Proto has a large molar extinction coefficient that allows the quantitative determinations of dilute solutions. Since many other pigments such as carotenoids, Chls, and Pchlides absorb light in this spectral region, it is important to use a highly purified DV Mg-Proto fraction. It is also possible to use DV Mg-Proto absorbance in the red region of the spectrum, at about 550 or 589 nm, for quantitative determinations. At these wavelengths, the molar extinction coefficient is very low, and quantitative determinations require more concentrated solutions. The molar extinction coefficients are related to DV Mg-Proto absorbance and concentration by Beer's law (Equation 2). Table 1 reports absorption maxima and molar extinction coefficient values for DV Mg-Proto in various solvents.

At room temperature, MV-Mg Proto exhibits a 3-banded absorption spectrum similar to that of DV Mg-Proto. The absorption maxima are blue-shifted however in comparison to DV Mg-Proto. In diethyl ether at room temperature, the Soret absorption maximum is observed at 412 nm, and the α and β absorption bands exhibit maxima at 546 and 586 nm (12). To our knowledge, precise molar extinction coefficients for MV Mg-Proto have not been determined. As an approximation, the molar extinction coefficients of DV Mpe may be used at the MV Mg-Proto maxima.

6. ANALYSIS OF MG-PROTOPORPHYRIN IX MONOMETHYL ESTER

The Mg-Proto monomethyl ester (Mpe) pool is a mixed DV-MV monocarboxylic tetrapyrrole pool (Figure 4). The same molar extinction coefficients (Table 1) and Equations (Table 2) used for Mg-Proto are also used for Mpe determination. What differs is the extraction of the two pools, since Mpe is a less polar monocarboxylic tetrapyrrole, while Mg-Proto is a more polar dicarboxylic tetrapyrrole.

6.1. Quantitative Determination of Mpe in Crude Extracts by Spectrofluorometry

Upon tetrapyrrole extraction from plant tissues and partitioning of the pigments between hexane and acetone (sections 2.1 and 2.2), the mixed DV-MV Mpe pool passes into the HEAR along with other dicarboxylic and monocarboxylic tetrapyrroles. Since Mg-Proto and Mpe exhibit similar spectrophotometric and fluorescence properties, it is necessary to correct for the presence of Mg-Proto. In a first step, the total Mg-Proto plus Mpe content is determined on an aliquot of HEAR as described in section 5.1 for Mg-Proto. Next, the Mpe pool is extracted into diethyl ether as described in section 3.2 for monocarboxylic tetrapyrroles. The Mg-Proto content is then determined on the ether-extracted HEAR fraction as described in section 5.2. The amount of Mpe is calculated by subtracting the amount of Mg-Proto from the total amount of Mg-Proto plus Mpe.

Similarly, DV and MV Mpe are determined, as for Mg-Proto (section 5.2), after transfer of the Mpe pool to diethyl ether.

6.2. Chromatographic Separation of Mpe from other Tetrapyrroles

The mixed DV-MV Mpe pool can be separated from other tetrapyrroles as described section 5.3.1 for Mg-Proto. Mpe moves with an R_f of about 0.31, just ahead of Pchlide *a* and Chlide (4), while Mg-Proto moves with an R_f of about 0.09. Mpe is eluted in methanol:acetone (4:1 vol/vol) as described for Mg-Proto. The proportions of DV/MV Mpe components are determined by 77 K spectrofluorometry in washed diethyl ether as described in section 5.2.

Separation of the crude (i.e., prior to silica gel H purification) or silica gel H-purified Mpe pool into DV and MV components can be achieved on thin layers of polyethylene developed in 90% aqueous acetone in darkness as described in section 5.3.2. In this solvent, DV and MV Mpe migrate with approximate R_fs of 0.54 and 0.68, respectively.

6.3. Quantitative Determination of Purified DV and MV Mpe

As in the case of DV and MV Mg-Proto, the concentration of purified DV and MV Mpe can be determined in a variety of organic solvents either by fluorescence or absorbance spectroscopy, using the same Equations and molar extinction coefficients.

7. ANALYSIS OF MG-PROTOPRPHYRIN IX DIESTER

The Mg-Protoporphyrin IX diester (Mpde) pool (35) is a mixed DV-MV fully esterified tetrapyrrole pool (Figure 3). The proportion of DV and MV Mpde appears to depend upon the plant species (12). The same equations used for the determination of Mg-Proto and Mpe are also used for Mpde determination. What differs are the extraction procedures, since Mpde is an apolar fully esterified tetrapyrrole.

7.1. Quantitative Determination of Mpde in Crude Extracts by Spectrofluorometry

Upon tetrapyrrole extraction from plant tissues and partitioning of the pigments between hexane and acetone, the mixed DV-MV Mpde pool passes into the apolar hexane phase along with other fully esterified tetrapyrroles. It is possible to determine the concentration of the mixed DM-

MV Mpde pool in the presence of other fully esterified tetrapyrroles. This is best achieved in etiolated tissues devoid of Chl or in greening tissues containing only small amounts of Chl. If large amounts of Chl are present, prior separation of the Mpde pool from Chl is required (see below).

To determine the amount of Mpde, an aliquot of the hexane extract is dried under a stream of N_2 gas at ice bucket temperature. The residue is dissolved in 80% acetone and monitored by room temperature emission spectrofluorometry as described for Mg-Proto in section 5.1. Alternatively, if large amounts of Mpde are present, a very small aliquot of the hexane extract can be diluted with 80% acetone prior to spectrofluorometric determination.

Similarly, the amounts of DV and MV Mpde are determined from the total amount of Mpde measured at room temperature and the DV/MV Mpde ratio determined at 77 K (section 5.2).

7.2. Separation of Mpde on Thin Layers of Silica Gel H

The mixed DV-MV Mpde pool can be separated from other fully esterified Mg-porphyrins on thin layers of silica gel H (35) as described in section 5.3.1 for Mg-Proto. The mixed DV-MV Mpde and Pchlide *a* E pools move with respective R_fs of about 0.75 and 0.82. If present, Chl moves in between Mpde and Pchlide *a* E. In this system, dicarboxylic tetrapyrroles stay at the origin while monocarboxylic tetrapyrroles, such as Mpe, move with an R_f of about 0.28 to 0.31 (35). Mpde can be eluted with diethyl ether.

7.3. Separation of Various Mpdes by HPLC

The Mg-Proto diester pool appears to be esterified by a number of long chain fatty alcohols probably at position 7 of the macrocycle. Gas chromatographic mass spectroscopic analysis of these long chain fatty alcohols did not result in an elucidation of their chemical structure (35). It is possible, however, to separate the various Mpde components by HPLC. Successful separation was achieved on a 20-cm Spherisorb ODS (Thermo Separation Products, San Jose, CA, USA) (5 µm) column eluted isocratically with water: methanol:acetone (5:85:10 vol/vol/vol) at a flow rate of 1 mL/minute. In this system, standard Mg-Proto dimethyl ester exhibits a retention time of about 3 minutes, while the Mpde pool separates into 3 major components with retention times of about 9.6, 10.9, and 12.6 minutes (35).

7.4. Chromatographic Separation of DV Mpde from MV Mpde

Complete separation of the DV and MV Mpde components has so far met with limited success (Rebeiz, unpublished).

8. ANALYSIS OF PROTOCHLOROPHYLLIDE *a*

The Pchlide *a* pool is a mixed highly dynamic DV-MV tetrapyrrole pool (Figure 4). The DV and/or MV content of this pool vary widely depending on the greening group affiliation of the plant species (24) and the phase of the photoperiod (1,16). Prior to 1979, the possible existence of a DV Pchlide *a* component in the Pchlide *a* pool of algae and higher plants was not recognized. As a consequence, pre-1980 studies of the Pchlide *a* pool may have overlooked the presence of substantial amounts of DV Pchlide *a*.

Two different pools of MV Pchlide *a*, formed via different Chl *a* biosynthetic pathways, coexist in etiolated and green plants (44). One pool is formed from DV Pchlide *a* (Figure 1, pathway 2) by reduction of the vinyl group at position 4 to ethyl, a reaction catalyzed by [4-vinyl]-Pchlide *a* reductase (65). The other is formed by conversion of MV Mpe to MV Pchlide *a* (64), a reaction catalyzed by a putative MV cyclopentanone ring synthetase (Figure 1, pathway 9). MV Pchlide *a* formation from DV Pchlide *a* via pathway 2 takes place in both dark divinyl (DDV)-light-dark divinyl (LDDV), i.e., DDV-LDDV, plant species such as cucumber and in dark monovinyl (DMV)-light-dark monovinyl (LDMV), i.e., DMV-LDMV, plant species such as corn wheat and barley. Conversely, MV Pchlide *a* formation from MV Mpe via pathway 9 appears to predominate in DMV-LDMV plant species (1).

DV Pchlide *a* occupies a central position in the DV carboxylic Chl *a* biosynthetic pathway as a precursor of MV Pchlide *a* and DV Chlide *a* (Figure 1, pathways 1 and 2). In contrast to MV Pchlide *a*, only one pool of DV Pchlide *a*, which is part of biosynthetic pathway 1, exists in etiolated and green plants (44).

8.1. Quantitative Determination of Pchlide *a* in Crude Extracts by Spectrofluorometry

Upon tetrapyrrole extraction from plant tissues and partitioning of the pigments between hexane and acetone (sections 2.1 and 2.2), the mixed DV-MV Pchlide *a* pool passes into the HEAR fraction along with other dicarboxylic and monocarboxylic tetrapyrroles. In HEAR at room temperature, DV and MV Pchlide *a* exhibit similar emission maxima at 637 to 639 nm. In this region of the spectrum, the main interfering tetrapyrrole is DV Proto, which emits at 632 to 633 nm. Lesser interference by the long wavelength emission tail of Copro, and the short wavelength emission tail of Chlide *a* may also be encountered. The concentration of the mixed DV-MV Pchlide *a* in the presence of other di- and monocarboxylic tetrapyrroles can be determined without further purification, after correction for spectral band overlap.

This is achieved by exciting the tetrapyrroles in the HEAR fraction at 440 nm, i.e., close to the Soret excitation maximum of Pchlide *a* and Chlides *a*, and by recording an emission spectrum between 580 to 700 nm. A second emission spectrum is elicited by excitation at 400 nm, i.e., close to the Soret excitation maxima of Proto and Copro. The net fluorescence amplitude due only to Pchlide *a* emission is deconvoluted from the Copro, Proto, and Chlide *a* emissions with the use of Equation 5 (Table 2) (45). The Pchlide *a* net fluorescence amplitude is converted to Pchlide *a* concentration by reference to a standard calibration curve. The latter is prepared as described in section 3.1 from standard Pchlide *a* solutions of known concentrations and from their fluorescence emission amplitudes. Pchlide *a* can be determined with a precision of about 6% (45). Minimum detection levels are about 0.1 pmol/mL.

8.2. Quantitative Determination of MV and DV Pchlide *a* in Crude Extracts by Spectrofluorometry

It is necessary to extract the Pchlide *a* pool from the HEAR fraction into less polar solvents such as diethyl ether prior to structural studies or determination of MV and DV Pchlide *a* (section 3.2).

As in the case of other MV and DV Mg-tetrapyrroles, the use of low temperature (77 K) is essential to differentiate between MV and DV Pchlide *a*. At this temperature, due to considerable fluorescence excitation and emission band narrowing, MV Pchlide *a* exhibits a split Soret excitation band with maxima at 437 and 443 nm and an emission maximum at 625 nm. Likewise, DV Pchlide *a* exhibits split Soret excitation maxima at 443 and 451 nm and an emission maximum at 625 nm (63). In this case too, determination of the amounts of DV and MV Pchlide *a* is a 2-step process. First, the total amount of DV plus MV Pchlide *a* in the HEAR fraction is determined at room temperature exactly as described in section 8.4. Next, the DV/MV ratio of the Pchlide *a* pool is determined at 77 K in diethyl ether. The amount of DV and MV Pchlide *a* is then calculated from the total amount of Pchlide *a* and from the DV/MV Pchlide *a* ratio (63).

To calculate the DV/MV Pchlide *a* ratio, one sharp excitation spectrum needs to be recorded at 77 K in diethyl ether (see section 5.2). The 77 K excitation spectrum is recorded from 380 to 500 nm by positioning the emission monochromator at the emission maximum of MV and DV Pchlide *a* at 625 nm. In this spectrum, the short wavelength MV Pchlide *a* Soret excitation maximum at 437 nm will appear as a sharp distinct peak. Likewise, the long wavelength DV Pchlide *a* Soret excitation maximum at 451 nm will also appear as a sharp distinct peak (63). The MV Pchlide *a* long wavelength excitation maximum and the DV Pchlide *a* short wavelength excitation maximum will overlap completely and appear as a single excitation maximum at 443 nm. For calculation of the DV/MV Pchlide *a* ratio, only the Soret excitation maxima at 437 and 451 nm are used. Because of complete overlap of the DV and MV Soret excitation maxima at 443 nm, this wavelength cannot be used in the calculations. The net fluorescence excitation amplitudes of DV and MV Pchlide *a* are deconvoluted and calculated with the use of Equations 6 and 7 (63). The ratio of the deconvoluted DV Pchlide *a* excitation amplitude to the deconvoluted MV Pchlide *a* excitation amplitude, referred to as the apparent excitation amplitude ratio, is next calculated. That ratio is converted to an authentic DV/MV Pchlide *a* concentration ratio by reference to a calibration curve (see section 5.2). The latter relates apparent ratios of DV/MV Pchlide *a* excitation amplitudes to known ratios of DV/MV Pchlide *a* concentrations. In this manner, DV and MV Pchlide *a* can be determined with a precision of 4% to 7% (63).

8.3. Chromatographic Separation of Pchlide *a*

8.3.1. Chromatographic Separation on Thin Layers of Silica Gel H

The mixed DV-MV Pchlide *a* pool can be separated from other monocarboxylic Mg-porphyrins on thin layers of silica gel H as described in section 5.3.1 (12). Pchlide *a* moves with an R_f of about 0.2 just behind Mpe (R_f of 0.31) (4). Pchlide *a* is eluted in methanol:acetone (4:1 vol/vol) as described for Mg-Proto. The proportions of DV/MV Pchlide *a* components are determined by spectrofluorometry in washed diethyl ether extracts at 77 K as described in section 8.4.

8.3.2. Chromatographic Separation of DV Pchlide a *from MV Pchlide* a

Separation of the of the crude (i.e., prior to silica gel H purification) or silica gel H-purified Pchlide *a* pool into DV and MV components can be achieved on thin layers of polyethylene developed in 90% aqueous acetone in darkness as described in section 5.3.2 (11). In this solvent, DV and MV Pchlide *a* migrate with approximate R_fs of 0.55 and 0.75, respectively. The separated bands are eluted in diethyl ether.

Partial separation of DV and MV Pchlide *a* by HPLC on a polyvinyl alcohol polymer column has been described by Shioi and coworkers (61). The separation was performed on a 250 × 4.6 mm ODP-50 Asahipak™ column (Showa Denko, Tokyo, Japan) packed with octadecyl silica polyvinyl alcohol polymer (ODP). Spherical particle size was 5 µm, pore diameter 250 Å, and carbon loading 17%. After injection of a 20-µL aliquot, elution was at 27°C with a linear mobile-phase gradient from methanol:1 M ammonium acetate (8.2:2 vol/vol) to acetonitrile:acetone (7:3 vol/vol). The duration of the gradient was 28 minutes, after which the final composition of the mobile phase was maintained isocratic for an additional 5 to 10 minutes. Resolution of MV from DV Pchlide *a* is, however, only partial with considerable overlap of the two Pchlides *a*.

8.4. Quantitative Determination of Purified DV and MV Pchlide *a*

Purified DV and MV Pchlide *a* can be determined in organic solvents from room temperature emission spectra elicited by excitation close to their Soret absorbance maxima between 430 and 440 nm. In each case, a calibration curve using known amounts of DV or MV Pchlide *a* should be constructed in order to convert fluorescence data to Pchlide *a* concentrations. At each concentration, an emission spectrum elicited by excitation at the chosen Soret excitation wavelength should be recorded. A calibration curve relating DV or MV Pchlide *a* concentration to fluorescence emission amplitudes of the pure DV and MV Pchlide *a* solutions at their emission maxima should be constructed as described in section 3.4.1. for purified Proto.

Alternatively, the concentration of purified MV Pchlide *a* solutions can be determined in a variety of organic solvents by room temperature absorbance spectroscopy. Such determinations are usually used for the preparation of MV Pchlide *a* stock solutions of known concentrations. At room temperature, MV Pchlide *a* exhibits a 5-banded absorption spectrum in various organic solvents. The wavelengths of absorption maxima depend upon the solvent. A strong Soret absorbance band with an absorption maximum at 432 to 434 nm and a less intense red absorbance band with a maximum at 622 to 629 nm are observed (Table 1). The Soret and red absorbance maxima can both be used for quantitative determinations. If the Soret absorbance is used, a highly purified sample free of carotenoids and other tetrapyrroles is required. Contamination by carotenoids does not interfere however with quantitative measurements when the red absorbance maximum is used. The molar extinction coefficients are related to DV Pchlide *a* absorbance and concentration by Beer's law (Equation 2). Table 1 reports absorption maxima and molar extinction coefficient values for MV Pchlide *a* in various solvents.

At room temperature, DV Pchlide *a* exhibits an absorbance spectrum similar to that of MV Pchlide *a*. Although the red absorbance maximum of MV and DV Pchlide *a* are identical, the Soret absorbance maximum of DV Pchlide *a* is red-shifted by about 5 nm with respect to that of MV Pchlide *a*. Quantitative determination of DV Pchlide *a* by absorbance spectroscopy are performed as described for MV Pchlide *a*. DV Pchlide *a* molar extinction coefficient values are reported in Table 1.

8.5. Quantitative Determination of Pchlide *a* in the Presence of Chlorophyll(ides) by Spectrophotometry

Under certain conditions, as during the early phases of greening of etiolated tissues, it is possible to determine the amount of the mixed MV/DV Pchlide *a* pool in the presence of Chl *a* and *b* and/or Chl(ide) *a* and *b* by absorbance spectroscopy. The technique uses simultaneous equations to correct for Pchlide *a* and Chl(ide) *a* and *b* absorbance band overlap, although these cannot be used when the ratio of Chl(ide) *a*/Pchlide *a* exceeds 6.0.

In diethyl ether, the amount of Pchlide *a* in the presence of Chl(ide) *a* and *b* can be determined from Equation 8 (Table 2).

Equation 8.

Pchlide *a* (623) = 41.10 (Abs_{623}) - 4.93 (Abs_{663}) - 4.93 (Abs_{644}).

Where: Pchlide *a* (623) = Amount of Pchlide *a* in nmoles/mL determined from the absorbance of the pigment mixture at 623 to 624nm.

Abs_{663} = Absorbance of the pigment mixture at 663 nm. This factor corrects for the Chl(ide) *a* overlap.

Abs_{644} = Absorbance of the pigment mixture at 644 nm. This factor corrects for the Chl(ide) *b* overlap.

Equation 8 was adapted from that reported by Koski and Smith (29). In deriving Equation 8, a molar extinction coefficient of 24 457.104 for Pchlide *a* in diethyl ether, at 623 nm, was used. This molar extinction coefficient was calculated from the molecular weight of MV Pchlide *a* (i.e., 613) and the specific absorption coefficient of MV Pchlide *a* (39.9), which was erroneously reported as that of MV Pchlide *a* phytyl ester by Koski and Smith (30). In calculating the Pchlide *a* molar extinction coefficient, the following relationship was used:

> Molar extinction coefficient (ε) = Specific absorption coefficient (α) (molecular weight)

In 80% acetone, the amount of Pchlide *a* in the presence of Chl(ide) *a* and *b* can be determined from Equation 9 (Table 2) in a manner similar to that described for Equation 8, if the ratio of Chl *a*/Pchlide *a* is less than 6.

9. ANALYSIS OF MV PCHLIDE *b*

The Pchlide *b* pool is a MV tetrapyrrole pool (Figure 4) (56). In green cucumber cotyledons, its concentration varies from 400 to 800 pmol per 100 mg of tissue protein. MV Pchlide *b* is not formed and does not accumulate in etiolated tissues (25). It has been proposed that of the 8 possible biosynthetic pathways capable of MV Chl *b* formation only 2 pathways (biosynthetic pathways 3 and 10) proceed via MV Pchlide *b* (44). The photoconversion of MV Pchlide *b* to Chlide *b* has been recently demonstrated in vitro in barley etioplast membranes (54). In reconstituted model systems, Pchlide oxidoreductase A (POR A) appears to bind preferentially MV Pchlide *b* (51). This may indicate that in DDV-LDDV green tissues capable of Pchlide *b* biosynthesis, POR A may be involved in the conversion of Pchlide *b* to Chlide *b*.

9.1. Quantitative Determination of Pchlide *b* in Crude Extracts by Spectrofluorometry

Upon tetrapyrrole extraction from plant tissues and partitioning of the pigments between hexane and acetone (sections 2.1 and 2.2), the MV Pchlide *b* pool passes into the HEAR fraction along with Pchlide *a* and other dicarboxylic and monocarboxylic tetrapyrroles. In 80% acetone or in HEAR at room temperature, MV Pchlide *b* exhibits a broad emission maximum at around 641 nm and a Soret excitation maximum at 447 nm (56). The emission maximum, at 641 nm, overlaps with the 77 K emission maximum of hexacoordinated Pchlide *a* (640 nm) (13) and should not be confused with it. At 77 K in diethyl ether, it exhibits a Soret excitation maximum at 463 nm. Upon excitation at 463 nm, it exhibits a pronounced red emission maximum at 643 nm. The detailed spectral properties of MV Pchlide *b* have been previously described (56).

9.1.1. Determination of Pchlide b *in Crude Extracts Containing Chl(ide)* a *and* b

Under natural conditions, MV Pchlide *b* can only be observed in green tissues or during the advanced stages of greening of etiolated tissues. Spectrofluorometric determination of MV Pchlide *b* in the presence of Pchlide *a* and Chlide *a* and *b* is best achieved in diethyl ether at 77 K. However, as mentioned for other MV and DV Mg-porphyrins, 77 K spectroscopic determinations cannot be used alone for quantitative measurements. A 2-step technique using room temperature and 77 K spectrofluorometry has therefore been developed. In a first step, the amount of MV Chlide *b* in HEAR is determined by room temperature spectrofluorometry as described in section 13.1. In a second step, the MV Chlide *b*/MV Pchlide *b* ratio is determined at 77 K in diethyl ether. The amount of MV Pchlide *b* is then calculated from the total amount of Chlide *b* as determined at room temperature and from the MV Chlide *b*/MV Pchlide *b* ratio as determined at 77 K in diethyl ether (25).

To calculate the MV Chlide *b*/MV Pchlide ratio, two sharp 77 K excitation spectra in diethyl ether need to be recorded. First, the HEAR fraction is extracted with diethyl ether (section 3.2). To optimize the detection of MV Pchlide *b*, the first 77 K excitation spectrum is recorded from 380 to 500 nm in the ether extract by positioning the emission monochromator at 643 nm, the emission maximum of MV Pchlide *b* in diethyl ether at 77 K. In this spectrum, the MV Pchlide *b* Soret excitation maximum at 463 nm will appear as a sharp distinct peak. To optimize the detection of MV Chlide *b*, a second 77 K excitation spectrum is recorded from 380 to 500 nm by positioning the emission monochromator at 660 nm, the emission maximum of MV Chlide *b* in diethyl ether at 77 K. In this spectrum, the MV Chlide *b* Soret excitation maximum at 475 nm will also appear as a distinct peak. For deconvolution and calculation purposes, the best discrimination between MV Pchlide *b* and MV Chlide *b* is accomplished by deconvoluting the net Soret excitation fluorescence amplitudes of the diethyl ether extract at 455 and 463 nm, respectively. The net fluorescence excitation amplitudes of MV Pchlide *b* and MV Chlide *b* at 463 and 455 nm are deconvoluted and calculated with the use of Equations 10 and 11 (25). The apparent MV Pchlide *b*/MV Chlide *b* excitation ratio is converted to an authentic MV Pchlide *b*/MV Chlide *b* concentration ratio by reference to a calibration curve that relates apparent ratios of MV Pchlide *b*/MV Chlide *b* excitation amplitudes to known ratios of MV Pchlide *b*/MV Chlide *b* concentrations (see section 5.2). Using this method MV Pchlide *b* can be determined with a precision of about 10% (25).

9.1.2. Determination of Pchlide b *in Crude Extracts Containing Pchlide* a

Experimental conditions may arise where MV Pchlide *b* needs to be determined in the presence of MV Pchlide *a* and in the absence of MV Chlide *b*. This situation may be encountered in model systems prepared from etiolated tissues and synthetic exogenous MV Pchlide *b* (51,54). Under such conditions, quantitative determination of MV Pchlide by spectrofluorometry can be achieved on an aliquot of HEAR and on a diethyl ether extract of HEAR. In this manner, losses incurred during purification of small amounts of MV Pchlide *b* are avoided. Again, a 2-step technique using room temperature and 77 K spectrofluorometric determinations are used. In a first step, the amount of Pchlide *a* in the HEAR fraction is determined by room temperature spectrofluorometry as described in section 8.4. In a second, step the Pchlide *a*/MV Pchlide *b* ratio is determined at 77 K in diethyl ether. The amount of MV Pchlide *b* is then calculated from the total amount of Pchlide *a* as determined at room temperature and from the Pchlide *a*/MV Pchlide *b* ratio as determined at 77 K in diethyl ether (25).

To calculate the Pchlide *a*/MV Pchlide *b* ratio, two sharp 77 K emission spectra in diethyl ether need to be recorded. First, the Pchlide *a* and *b* pools are extracted from the HEAR fraction with diethyl ether as described in section 3.2 for monocarboxylic tetrapyrroles. To optimize the detection of MV Pchlide *b*, the first 77 K emission spectrum is recorded from 580 to 700 nm by excitation at the Soret excitation maximum of MV Pchlide *b* at 463 nm. In this spectrum, MV Pchlide *b* emission would appear as a sharp distinct peak at 643 nm. To optimize the detection of MV Pchlide *a*, a second 77 K emission spectrum is recorded from 580 to 700 nm by excitation at 440 nm. In this spectrum, MV Pchlide *a* would exhibit a sharp emission maximum at 625 nm and a broader emission band between 632 and 650 nm. The best discrimination between Pchlide *a* and MV Pchlide *b* is accomplished by deconvoluting the net emission fluorescence amplitudes of the crude diethyl ether extract at 635 and 643 nm (25).

The net fluorescence emission amplitudes of MV Pchlide *b* and Pchlide *a* at 643 and 625 nm, respectively, are deconvoluted and calculated with the use of Equations 12 and 13 (25). Next, the calculated apparent Pchlide *a*/MV Pchlide *b* emission ratio is converted to an authentic Pchlide *a*/MV Pchlide *b* concentration ratio by reference to a calibration curve that relates apparent ratios of Pchlide *a*/MV Pchlide *b* emission amplitudes to known ratios of Pchlide *a*/MV Pchlide *b* concentrations (see section 5.2). In this manner, MV Pchlide *b* can be determined with a precision of about 7% (25).

9.2. Chromatographic Separation of Pchlide *b* from other Tetrapyrroles

A promising HPLC technique has been described by Scheumann et al. for synthetic MV Pchlide *b* (54). Separation of MV Pchlide *b* from MV Pchlide *a* is achieved on a C-18 reverse phase silica gel column (Hypersil ODS, 5 µm particle size; Shandon Lipshaw, Pittsburgh, PA, USA). Elution is at a flow rate of 1.0 mL/minute, with a step gradient starting with 34% 25 mM aqueous NH_4OAC, 15% acetone, and 51% methanol, and increasing to 16% water, 60% acetone, and 24% methanol within 20 minutes. Finally, the gradient is held at 100% acetone for 14 minutes. In this system, MV Pchlide *b* and *a* are eluted with respective retention times of about 9 and 15 minutes.

9.3. Quantitative Determination of Purified Pchlide *b*

Purified MV Pchlide *b* can be determined in any appropriate solvent from room temperature emission spectra elicited by excitation close to its Soret absorbance maximum. In each case, a calibration curve using known amounts of MV Pchlide *b* should be constructed in order to relate MV Pchlide *b* concentration to fluorescence emission amplitudes of the pure MV Pchlide *b* solutions at its emission maximum. A similar procedure using Soret excitation spectra instead of emission spectra can also be used.

Alternatively, absorbance spectroscopy may be used. At room temperature, MV Pchlide *b* exhibits a 6-banded absorption spectrum in various organic solvents. The wavelengths of absorption maxima depend on the solvent. In diethyl ether, a strong Soret absorbance band with an absorption maximum at 442 nm and a less intense red absorbance band with a maximum at 630 nm are observed (Table 1). Table 1 reports absorption maxima and molar extinction coefficient values for MV Pchlide *b* phytyl ester in various solvents. MV Pchlide *b* and its esterified analog exhibit identical spectral and molar extinction properties. The Soret and red absorption maxima can both be used for quantitative determinations. If the Soret absorbance is used, a highly purified sample free of carotenoids and other tetrapyrroles is required. Carotenoids do not interfere with quantitative determinations using the red absorbance maximum. The molar extinction coefficients are related to MV Pchlide *b* absorbance and concentration by Beer's law (Equation 2).

10. ANALYSIS OF PCHLIDE ESTER *a*

The protochlorophyllide ester (Pchlide E) *a* pool is a highly heterogeneous fully esterified tetrapyrrole pool. It is present in all etiolated and green tissues so far surveyed (Rebeiz, unpublished). It consists of MV and DV Pchlide *a* E (Figure 4) (10).

The long chain fatty alcohols esterified at position 7 of the macrocycle are variable and have been reported to consist of geranylgeraniol (GG), dihydroGG (DHGG), tetrahydroGG (THGG), and hexahydroGG (i.e., phytol) in *Scenedesmus obliquus* (27), in etiolated and greening cucumber cotyledons (58), and in etiolated leaves of kidney bean (59).

In the integrated Chl *a*/*b* pathway, the MV Pchlide *a* E pool is considered to originate from MV Mpde via biosynthetic pathway 13 (Figure 1). Because of experimental difficulties, a precursor product relationship between MV Mpde and MV Pchlide *a* E has not yet been demonstrated. It has been demonstrated, however, that the bulk of the Pchlide *a* E pool of cucumber cotyledons, which consists mainly of MV Pchlide *a* E, is not formed from MV Pchlide *a* by esterification. Instead, MV Pchlide *a* and MV Pchlide *a* E appear to be formed in parallel from a common precursor at the level of Proto and/or Mg-Proto or Mpe (36). The MV Pchlide *a* GG, dihydroGG, and tetrahydroGG may be considered intermediates on the way to the formation of MV Pchlide *a* P.

The DV Pchlide *a* pool is considered to originate from DV Mpde (Figure 1, pathway 15) via a series of reactions similar to those of MV Pchlide *a* E. Because of experimental difficulties, a precursor product relationship between DV Mpde and DV Pchlide *a* ester has not yet been demonstrated. The DV Pchlide *a* GG, DHGG, and THGG may be considered intermediates on the way to the formation of DV Pchlide *a* E.

10.1. Quantitative Determination of Pchlide *a* E in Crude Extracts by Spectrofluorometry

Upon tetrapyrrole extraction from plant tissues and partitioning of the pigments between hexane and acetone (sections 2.1 and 2.2), the mixed Pchlide *a* E pool passes into hexane along with other fully esterified tetrapyrroles. Pchlide *a* and its ester exhibit identical spectrofluorometric and spectrophotometric properties, so quantitative determination of Pchlide *a* E is exactly as described in section 8.1. In most cases, Pchlide *a* E in the hexane extract will be contaminated by trace amounts of Pchlide *a* carried over from acetone to hexane. This minor Pchlide *a* contamination can be eliminated by washing the hexane with ammoniacal acetone or by purification of Pchlide *a* E on thin layers of silica gel H as described below.

10.2. Chromatographic Separation of Pchlide *a* E from other Mg-Porphyrins

The mixed DV-MV Pchlide *a* E pool can be separated from other fully esterified Mg-porphyrins on thin layers of silica gel H as described in section 5.3.1 (10). The Pchlide *a* E, Chl *a* and *b*, and Pchlide *a* pools move with respective R_fs of about 0.9, 0.8, and 0.5. Pchlide *a* E is eluted with diethyl ether.

After separation on thin layers of silica gel H, the Pchlide *a* E can be purified further on polyethylene plates. The ether extract is washed with 0.5 M KH_2PO_4 buffer, pH 7.0, to remove traces of chromatographic solvents and dried under a stream of N_2 gas. The dried residue is dissolved in diethyl ether and applied to thin layers of polyethylene. The plates are allowed to dry completely by holding at room temperature for 10 minutes under subdued light before developing in the dark, at room temperature, in 2-propanol:acetone (1:1 vol/vol) (10). MV and DV Pchlide *a* E move with respective R_fs of 0.5 and 0.3. The bands fluoresce weak red under UV light of 366 nm and are eluted in diethyl ether. The eluted bands are dried under a stream of N_2 gas before redissolving in diethyl ether for spectroscopic analysis. Because of the low concentration of Pchlide *a* E in most plant tissues, and because of low recoveries, the red fluorescence of the separated MV and DV bands may be barely visible. Nevertheless, the fluorescence of the eluted bands is readily detected by high-resolution spectrofluorometry.

After separation of MV from DV Pchlide *a* E, HPLC is the method of choice for separating various MV or DV Pchlide *a* esters. For example, after separation of DV from MV Pchlide *a* E on Fractogel TSK DEAE-650 (Merk, Darmstadt, Germany), the different MV and DV Pchlide *a* E of dark-grown wheat roots and *S. obliquus* were separated on a 4 × 200 mm Nucleosil C18-reversed phase column, 5 μm particle size (Macherey and Nagel). Elution was with a linear gradient of 20% to 80% ethyl acetate in 80% aqueous methanol for 30 minutes at a flow rate of 1 mL/minute. Sharp separation of Pchlide *a* GG from Pchlide *a* DHGG, THGG, and phytol was achieved (27,60).

Most etiolated tissue of higher plants contains mainly MV and/or DV Pchlide *a*, lesser amounts of MV Pchlide *a* E, and only trace amounts of DV Pchlide *a* E. In most cases, it is possible to separate the various Pchlide *a* esters with minimal manipulation, using an isocratic HPLC solvent system. Fifty microliters of the crude acetone:ammonium hydroxide extract (see section 2.1) of etiolated tissues is injected onto a Pecosphere 3 × 3C, C-18 reverse phase, 4 × 0.5 cm column, and eluted with an isocratic solvent system that consists of water:acetone:methanol (5:10:75 vol/vol/vol), at a rate of 1 mL/minute. The elutants are monitored by on-line spectrofluorometry. The amounts of eluting Pchlide *a* and various Pchlide *a* E can be determined from a calibration curve using Pchlide *a* as a standard and appropriate commercial software. With this system, 4 Pchlide *a* E with retention times of about 8.2, 9.9, 12.2, and 15.1 minutes are quantitatively resolved (Rebeiz, unpublished).

10.3. Quantitative Determination of Purified DV and MV Pchlide *a* E

As was described for purified DV and MV Pchlide *a* in section 8.4, the concentration of purified MV and/or DV Pchlide *a* E can be determined either by spectrofluorometry or absorbance spectroscopy.

The red absorbance maxima of MV and DV Pchlide *a* E are identical, but the Soret absorbance maximum of DV Pchlide *a* E is red-shifted by about 5 nm with respect to that of MV Pchlide *a* E. Quantitative determination of purified DV Pchlide *a* E by absorbance spectroscopy are performed as described for MV Pchlide *a* E. Molar extinction coefficient for DV Pchlide *a* E are reported in Table 1.

10.4. Determination of Pchlide *a* E in the Presence of Chlorophyll(ides) *a* and *b*

The same absorbance equations used for the determination of Pchlide *a* in the presence of Chl(ide) *a* and *b* in diethyl ether and 80% acetone (section 8.5) can be used for Pchlide *a* E determination in the presence of Chl(ide) *a* and *b*.

11. ANALYSIS OF MV PCHLIDE *b* ESTER

The Pchlide *b* E pool is a MV tetrapyrrole pool (Figure 4). Its occurrence in green plants was first reported by Shedbalkar et al., but is less ubiquitous than that of MV Pchlide *b* (57).

Upon tetrapyrrole extraction from plant tissues and partitioning of the pigments between hexane and acetone, MV Pchlide *b* E passes into hexane along with Pchlide *a* E and Chl *a* and *b*. Since Pchlide *b* E exhibits identical electronic spectroscopic properties as MV Pchlide *b*, the same spectrofluorometric equations used for MV Pchlide *b* determination can be used for Pchlide *b* E analysis (section 9.1).

Pchlide *b* E can be separated from other tetrapyrroles by chromatography on thin layers of silica gel H in darkness at 4°C in toluene:ethyl acetate:ethanol (8:2:2 vol/vol/vol). It migrates with a R_f of about 0.56. It is eluted in diethyl ether and rechromatographed in darkness at room temperature on thin layers of cellulose developed in ligroin (60–90):*n*-propanol (99:1 vol/vol) (57).

12. ANALYSIS OF CHLOROPHYLLIDE *a*

The Chlide *a* pool is a mixed highly dynamic DV-MV tetrapyrrole pool (Figure 5). The DV and/or MV content of this pool vary widely depending on the greening group affiliation of the plant species (24), the phase of the photoperiod (9), and light pretreatment of the plant tissue (18). DV Chlide *a* (Figure 5) occurs as a single pool that occupies a central position in the integrated Chl *a*/*b* pathway as a precursor of MV Chlide *a* (Figure 1, pathway 4), of DV Chlide *b* (Figure 1, pathway 7), and DV Chl *a* (Figure 1 pathway 1).

In contrast, 3 different pools of MV Chlide *a*, formed via different Chl *a* biosynthetic pathways, coexist in etiolated and green plants (44). One pool is formed by photoreduction of MV Pchlide *a* via pathway 2 (Figure 1). The second is formed from DV Chlide *a* via pathway 4 by conversion of the vinyl group at position 4 to ethyl, a reaction catalyzed by (4-vinyl)-Chlide *a* reductase (4VCR) (Figure 1) (13). The third is formed by conversion of MV Pchlide *a* via pathways 9 and 12 (Figure 1).

12.1. Spectrofluorometric Determination of Chlide *a* in Crude Extracts Containing Mainly MV Chlide *a*

Upon tetrapyrrole extraction from plant tissues and partitioning of the pigments between hexane and acetone (sections 2.1 and 2.2), the mixed DV-MV Chlide *a* pool passes into HEAR along with other dicarboxylic and monocarboxylic tetrapyrroles. In most higher and lower plants, except in prochlorophyte picoplankton and the *Nec1* corn mutant where the DV forms abound, the Chlide *a* pool consists of MV Chlide *a* (673 nm emission) and trace amounts of DV Chlide *a* (also 673 nm emission). It is usually accompanied by MV Chlide *b* (657 nm emission), MV pheophorbide a, i.e., demetalated, de-esterified Chl *a* (673 nm emission) and smaller amounts of MV Pheophorbide *b* (660 nm emission). It is also accompanied by Pchlide *a* (638 nm emission), and in some cases by small amounts of Mg-Proto and Mpe (595 nm emission).

Under these circumstances, it is possible to determine the amounts of MV Chlide *a* without prior purification by room temperature excitation spectrofluorometry. Since the emission maxima of MV Chlide *a* and *b* and MV pheophorbide *a* and *b* fall in the red region of the spectrum, far away from the emission of other Mg-tetrapyrroles, it is possible to record Soret excitation spectra with minimum Soret excitation overlap from other Mg-tetrapyrroles. The concentration of Chlide *a* can then be determined from the excitation spectra after correction for excitation band overlap. This is achieved by recording 2 Soret excitation spectra in HEAR at room temperature. One spectrum is recorded at an emission wavelength near the emission maximum of MV Chlide *a*, at 674 nm. The other spectrum is recorded at an emission wavelength near the emission maximum of MV Chl *b* at 660 nm. The net Soret fluorescence excitation amplitudes due only to MV Chlide *a* is deconvoluted with the use of Equation 14 (7). The net fluorescence excitation amplitude of MV Chlide *a*, as calculated from Equation 14, are converted to MV Chlide *a* concentration by reference to a standard calibration curve relating MV Chlide *a* concentrations to fluorescence excitation amplitudes. In this manner, MV Chlide *a* can be determined with a precision of about 1% for pure MV Chlide *a* to 24% for extracts containing only 6% Chlide *a* (7). Minimum detection levels are about 0.2 pmol/mL.

12.2. Spectrofluorometric Determination of Chlide *a* in Crude Extracts Containing Mainly DV Chlide *a*

In some cases, the Chlide *a* pool consists mainly of DV Chlide *a* with or without smaller amounts of MV Chlide *a*. This situation prevails in the *Nec1* corn mutant (6), in the primitive prochlorophyte picoplankton of the subtropical waters of the North Atlantic as well as in the picoplankton of the euphonic zone of the world tropical and temperate oceans, and the Mediterranean Sea, and in DV-LDDV plant species subjected to alternating light/dark treatments (18). Under these circumstances, total Chlide *a* can be evaluated using Equation 14 to calculate the net Soret fluorescence excitation amplitude at 433 nm, i.e., 5 nm below the DV Chlide *a* excitation maximum at 438 nm. Next, the net fluorescence excitation amplitude of the DV Chlide *a* pool, at 433 nm, is converted to DV Chlide *a* concentration by reference to a standard calibration curve. The latter is prepared from standard DV Chlide *a* solutions of known concentrations and from their fluorescence excitation amplitudes at 433 nm. The latter are recorded at an emission maximum of 674 nm.

12.3. Spectrofluorometric Determination of MV and DV Chlide *a* in Crude Extracts

Prior to structural studies or determination of MV and DV Chlide *a*, it is necessary to extract the Chlide *a* pool from the HEAR fraction into diethyl ether as described in section 3.2 for monocarboxylic tetrapyrroles.

As in the case of other MV and DV Mg-tetrapyrroles, the use of low temperature (77 K) is essential to differentiate between MV and DV Chlide *a*. At 77 K, MV Chlide *a* exhibits a Soret excitation band with a maximum at 447 and an emission maximum at 674 nm. DV Chlide *a* exhibits a red-shifted Soret excitation maximum at 458 nm and an emission maximum at 674 nm (8). In this case too, determination of the amounts of DV and MV

Chlide *a* is a 2-step process. First, the total amount of DV plus MV Chlide *a* in HEAR is determined at room temperature exactly as described in section 12.1. Next, the DV/MV ratio of the Chlide *a* pool is determined at 77 K in diethyl ether. The amount of DV and MV Chlide *a* is then calculated from the total amount of Chlide *a* as determined at room temperature and from the DV/MV Chlide *a* ratio as determined at 77 K in diethyl ether (68).

To calculate the DV/MV Chlide *a* ratio, one sharp 77 K excitation spectrum in diethyl ether need to be recorded (see section 5.2). The 77 K excitation spectrum is recorded from 380 to 500 nm by positioning the emission monochromator at the emission maximum of MV and DV Chlide *a* at 674 nm. In this spectrum, the DV Chlide *a* Soret excitation maximum at 458 nm will appear as a sharp distinct peak. The MV Chlide *a* Soret excitation maximum at 447 nm will also appear as a sharp distinct peak (67). The net fluorescence excitation amplitudes of DV and MV Chlide *a* are deconvoluted and calculated with the use of Equations 15 and 16 (Table 2) (68). The ratio of the deconvoluted DV Chlide *a* excitation amplitude to the deconvoluted MV Chlide *a* excitation amplitude is converted to an authentic DV/MV Chlide *a* concentration ratio by reference to a calibration curve that relates the apparent ratios of DV/MV Chlide *a* excitation amplitudes to known ratios of DV/MV Chlide *a* concentrations (see section 5.2). In this manner, DV and MV Chlide *a* can be determined with a precision of 2% to 6% (68).

12.4. Chromatographic Separation of Chlide *a* from other Tetrapyrroles

The mixed DV-MV Chlide *a* pool can be separated from other monocarboxylic Mg-porphyrins on thin layers of silica gel H as described in section 5.3.1 (8). Chlide *a* and Pchlide *a* move with an R_f of about 0.2 just behind Mpe (8). The mixed Chlide *a*-Pchlide *a* band is eluted in methanol:acetone (4:1 vol/vol). Further separation of Chlide *a* from Pchlide *a* can be achieved after methylation with freshly prepared diazomethane. The methylated pigments are chromatographed on thin layers of silica gel H in toluene:ethyl acetate:ethanol exactly as described above. In this solvent, Pchlide *a* E and Chl run with respective R_f values of about 0.90 and 0.86. Methyl Chlide *a* and methyl Pchlide *a* run with respective values of about 0.68 and 0.75. The methylated Chlide *a* is eluted in diethyl ether. It is usually contaminated with small amounts of methylated Pchlide *a*. The recovery of a highly pure methylated Chlide *a* preparation requires a second purification on silica gel H as described above (8).

12.5. Separation and Analysis of DV and MV Chlide *a*

Separation of silica gel H-purified methyl Chlide *a* into DV and MV components can be achieved on thin layers of polyethylene developed in 2-propanol:acetone (1:1 vol/vol) as described for DV and MV Pchlide *a* E (section 10.2) (10).

An efficient separation of cucumber DV and MV Chlide *a* from DV and MV Pchlide *a* has been described (21). Separation is achieved on a 201 TP 250 × 4.6 mm I.D., polymeric octadecylsilica, 5 μm particle size, and 300 Å pore size column (Vydac/The Separation Group, Hesperia, CA, USA). Elution at a rate of 1.2 mL/minute is with a linear acetone gradient from 30% aqueous acetone to 55% aqueous acetone in 8 minutes, then to 100% acetone in 4 minutes. The mobile phase is kept at 100% acetone for an additional 4 minutes. In this system, MV and DV Chlide *a* eluted separately between 3 and 4 minutes, while MV and DV Chlide *a* eluted separately between 6 and 8 minutes.

12.6. Quantitative Determination of Purified DV and MV Chlide *a*

Purified DV and MV Chlide *a* can be determined in any solvent from room temperature emission spectra elicited by excitation close to their Soret absorbance maxima between 430 and 440 nm. In each case, a calibration curve relating DV or MV Chlide *a* concentration to fluorescence emission amplitudes of the pure DV and MV Chlide *a* solutions, at their emission maxima, should be constructed.

The concentration of purified MV Chlide *a* solutions can be determined in a variety of organic solvents by room temperature absorbance spectroscopy. At room temperature, MV Chlide *a* exhibits an 8-banded absorption spectrum. A strong Soret absorbance band with an absorption maximum at 430 nm and a less intense red absorbance band with a maximum at 660 to 663 nm are observed (Table 1). The Soret and red absorption maxima can both be used for quantitative determinations. If the Soret absorbance is used, a highly purified sample free of carotenoids and other tetrapyrroles is essential, but this is not necessary when the red absorbance maximum is used. The molar extinction coefficients are related to DV Chlide *a* absorbance and concentration by Beer's law (Equation 2). Table 1 reports absorption maxima and molar extinction coefficient values for MV Chlide *a* in diethyl ether and 80% aqueous acetone.

At room temperature, DV Chlide *a* exhibits an absorbance spectrum similar to that of MV Chlide *a*, although the Soret absorbance maximum of DV Chlide *a* is red shifted by about 5 nm with respect to that of MV Chlide *a*.

12.7. Quantitative Determination of Chlide *a* in the Presence of Chlide *b* and Pchlide *a* by Spectrophotometry

Under certain conditions, as during the early phases of greening of etiolated tissues, it is possible to determine the amount of the Chlide *a* pool in the presence of Chlide *b* and Pchlide *a* by absorbance spectroscopy. These simultaneous equations correct for Pchlide *a* and Chlide *a* and *b* absorbance band overlap but do not correct for the slight discrepancy between the molar extinction coefficients of the MV and DV components.

In diethyl ether, the amount of Chlide *a* in the presence of Chlide *b* and Pchlide *a* can be determined from Equation 17 (Table 2), which was adapted from that reported by Koski and Smith (30), using a molar extinction coefficient of 83 450 for Chl(ide) *a* in diethyl ether at 660 nm. This molar extinction coefficient was calculated from the molecular weight of MV Chl *a* (i.e., 893.5) and the specific absorption coefficient of MV Chl *a* (93.4) in diethyl ether (34). Similarly, in 80% acetone, the amount of Chlide *a* in the presence of Chlide *b* and Pchlide *a* can be determined from Equation 18.

13. ANALYSIS OF CHLIDE *b*

In higher plants, the Chlide *b* pool consists exclusively of MV Chlide *b*. In the corn Nec1 mutant (6), it consists of DV Chlide *b* (Figure 5) (Rebeiz, unpublished). In the primitive prochlorophyte picoplankton of the subtropical waters of the North Atlantic as well as in the picoplankton of the euphotic zone of the world tropical and temperate oceans and the Mediterranean sea, the Chlide *b* pool is most probably a mixed DV-MV tetrapyrrole pool. Figure 1 shows the 5 different subpools of MV Chlide *b* formed via different pathways.

13.1. Spectrofluorometric Determination of Chlide *b* in Crude Extracts

Upon extraction from plant tissues and

partitioning between hexane and acetone (sections 2.1 and 2.2), the MV Chlide *b* pool passes into HEAR along with other dicarboxylic and monocarboxylic tetrapyrroles such as MV Chlide *a* (673–674 nm emission), MV pheophorbide *a* (673 nm emission), and smaller amounts of MV Pheophorbide *b* (660 nm emission). It is also accompanied by Pchlide *a* (638 nm emission) and in some cases by small amounts of Mg-Proto and Mpe (595 nm emission).

It is possible to determine the amounts of MV Chlide *b* without prior purification by room temperature excitation spectrofluorometry, since the emission maxima of MV Chlide *a* and *b* and MV pheophorbide *a* and *b* fall in the red region of the spectrum, far away from the emission of other Mg-tetrapyrroles. The concentration of Chlide *b* can be determined from the excitation spectra after correction for excitation band overlap. This is achieved by recording 2 Soret excitation spectra in HEAR at room temperature. One spectrum is recorded close to the emission maximum of MV Chlide *a* at 674 nm. The other spectrum is recorded close to the emission maximum of MV Chlide *b* at 660 nm. The net fluorescence excitation amplitude of MV Chlide b, as calculated from Equation 19 (7), is converted to MV Chlide *b* concentration by reference to a standard calibration curve that relates standard MV Chlide *b* solutions of known concentrations to their fluorescence excitation amplitudes. In this manner, MV Chlide *b* can be determined with a precision of about 1% for pure MV Chlide *b* to 39% for extracts containing only 6% MV Chlide b.

In those cases where the Chlide *b* pool consists mainly or exclusively of DV Chlide *b* with or without smaller amounts of DV Chlide *a* (as described above), total Chlide *b* can be evaluated using Equation 19 to calculate the net Soret fluorescence excitation amplitude at 460 nm, i.e., 8 nm below the DV Chlide *b* excitation maximum. The net fluorescence excitation amplitude of the Chlide *b* pool at 460 nm is next converted to DV Chlide *b* concentration by reference to a standard calibration curve that relates standard DV Chlide *b* excitation amplitudes at 460 nm to concentrations.

13.2. Quantitative Determination of MV and DV Chlide *b* in Crude Extracts by Spectrofluorometry

Before structural studies or determination of MV and DV Chlide *b,* it is necessary to extract Chlide *b* from HEAR into diethyl ether (section 3.2).

As in the case of other MV and DV Mg-tetrapyrroles, the use of low temperature (77 K) is essential to differentiate between MV and DV Chlide b. At 77 K, MV Chlide *b* exhibits a Soret excitation band with an excitation maximum at 475 nm, an excitation shoulder at 485 nm, and an emission maximum at 659 to 660 nm. DV Chlide *b* exhibits a red-shifted Soret excitation maximum at 490 nm, a Soret excitation shoulder at 498 nm, and an emission maximum at 666 nm (67). In this case too, determination of the amounts of DV and MV Chlide *b* is a 2-step process. First, the total amount of DV plus MV Chlide *b* in HEAR is determined at room temperature (section 13.1). Next, the DV/MV ratio of the Chlide *b* pool is determined at 77 K in diethyl ether. The amount of DV and MV Chlide *b* is then calculated from the total amount of Chlide *b* as determined from the HEAR fraction at room temperature and from the DV/MV Chlide *b* ratio as determined at 77 K in diethyl ether (67).

To calculate the DV/MV Chlide *b* ratio, 2 sharp 77 K excitation spectra in diethyl ether need to be recorded. The first 77 K excitation spectrum is recorded from 380 to 500 nm by positioning the emission monochromator at the emission maximum of MV Chlide *b* at 660 nm. In this spectrum, the MV Chlide *b* Soret excitation maximum at 475 nm will appear as a distinct broad peak. The second 77 K excitation spectrum is recorded from 380 to 500 nm by positioning the emission monochromator at the emission maximum of DV Chlide *b* at 666 nm (69). In this spectrum, the DV Chlide *b* Soret excitation maximum at 490 nm will appear as a broad distinct peak. The net fluorescence excitation amplitudes of DV and MV Chlide *b* are deconvoluted and calculated with the use of Equations 20 and 21 (Table 2) (67). Next, the ratio of the deconvoluted DV Chlide *b* excitation amplitude to the deconvoluted MV Chlide *b* excitation amplitude is converted to an authentic DV/MV Chlide *b* concentration ratio by reference to a calibration curve that relates the apparent ratios of DV/MV Chlide *b* excitation amplitudes to known ratios of DV/MV Chlide *b* concentrations (see section 5.2). In this manner, DV and MV Chlide *b* can be determined with a precision of 1% to 6% (68).

13.3. Chromatographic Separation of Chlide *b* from other Tetrapyrroles

In green tissues, the Chlide *b* pool is accompanied by Chlide *a*. To our knowledge, purification of MV and DV Chlide *b* has not been thoroughly investigated, but it might be possible to use the method of separation of methylated derivatives, as described for Chl *a* and *b* in section 14.

13.4. Quantitative Determination of Purified DV and MV Chlide *b*

Purified DV and MV Chlide *b* and their methyl esters can be determined in organic solvents from room temperature emission spectra elicited by excitation close to their Soret absorbance maxima between 455 and 468 nm. In each case, a calibration curve relating known amounts of standard DV or MV Chlide *b* or their methyl esters to their fluorescence emission amplitudes should be constructed.

Alternatively, absorbance spectroscopy may be used. At room temperature, MV Chlide *b* exhibits a 7-banded absorption spectrum with a strong Soret absorbance band, a solvent-dependent absorption maximum at 455 to 460 nm, and a less intense red absorbance band with a maximum at 643 to 645 nm (Table 1). If the Soret absorbance is used for quantitation, a highly purified sample free of carotenoids and other tetrapyrroles is required, but this is not necessary when the red absorbance maximum is used. The molar extinction coefficients are related to DV Chlide *b* absorbance and concentration by Beer's law (Equation 2).

Quantitative determination of DV Chlide *b* by absorbance spectroscopy can be performed as described for MV Chlide *a*, using the values given in Table 1.

13.5. Quantitative Determination of Chlide *b* in Crude Extracts in the Presence of Pchlide *a*

Under certain conditions, as during the early phases of greening of etiolated tissues, it is possible to determine the amount of Chlide *b* in the presence of Chlide *a* and Pchlide *a* by absorbance spectroscopy. As for Chlide *a*, Equations 22 and 23 (Table 2) are used to correct for Pchlide *a* and Chlide *a* absorbance band overlap, but do not correct for the slight discrepancy between the molar extinction coefficients of MV and DV components.

14. ANALYSIS OF CHLOROPHYLL *a*

The Chl *a* pool is a mixed highly dynamic DV-MV tetrapyrrole pool (Figure 5). The DV and/or MV content of this pool vary widely depending on the greening group affiliation of the plant species (24), the phase of the photoperiod (9), and light pretreatment of the plant tissue (2). The chemical structure of MV Chl *a* phytol (Chl *a* P) (Figure 5) was determined by Fischer and Stern (20). In the integrated Chl *a/b* biosynthetic pathway, the MV Chl *a* P pool is considered to consist of 4 different subpools of MV Chl *a* P formed via different Chl *a* biosynthetic pathways (44). Two subpools are formed from MV Pchlide *a* and MV Chlide *a* via pathways 2 and 9 (Figure 1). The third subpool is formed from DV Chlide *a* and MV Chlide *a* via pathways 4 and 5. The fourth subpool is formed from DV Chl *a* by direct conversion of the vinyl group at position 4 to ethyl via pathway 8 (2). It is conjectured that the 4 MV Chl *a* P subpools are part of different Chl-protein complexes having different roles in photosynthesis (44).

During the early phases of greening of etiolated tissues, in addition to MV Chl *a* P, each MV Chl *a* subpool (Figure 5) may also contain Chl *a* GG, Chl *a* DHGG, and Chl *a* THGG. These Chls may be considered as transient intermediates on the way to the formation of Chl *a* P (52,53).

In the integrated Chl *a/b* biosynthetic pathway, 3 additional subpools of MV Chl *a*, with long chain fatty alcohols at position 7 of the macrocycle other than phytol, are depicted. These Chls occur in trace amounts and may be either intermediates, on the way to the formation of MV Chl *a* P, or end products of the Chl biosynthetic pathway. Because of experimental difficulties, the esterifying long chain fatty alcohol(s) at position 7 of the macrocycle have not yet been characterized. The 3 subpools have been detected in DDV-LDMV plant species such as corn and barley during etiolation and/or during the early phases of greening. One subpool is formed from MV Chlide *a* via pathway 12 during the early phases of greening. The second subpool is formed in etiolated plants via pathway 13 by photoconversion of MV Pchlide *a* E (9). The third subpool appears to be formed via pathway 14 by dark reduction of MV Pchlide *a* E during seed germination in total darkness (Rebeiz, unpublished).

In the integrated Chl *a/b* pathway, DV Chl *a* is depicted as a precursor of one of the MV Chl *a* subpools (Figure 1, pathways 1 and 8). Its specific formation from DV Chlide *a* and its conversion to MV Chl *a* were recently documented (2). The nature of the fatty alcohol(s) at position 7 has not been investigated. A second DV Chl *a* pool, with a long chain esterifying fatty alcohol at position 7 of the macrocycle other than phytol, may also occur in some plant tissues enriched in DV Pchlide *a* E such as cucurbit seed coats. Such a pool may be formed by (photo)conversion of DV Pchlide *a* E via pathway 8 (Figure 1).

14.1. Quantitative Determination of Chl *a* in Crude Extracts

Upon tetrapyrrole extraction from green(ing) plant tissues and partitioning of the pigments between hexane and acetone (sections 2.1 and 2.2), Chl *a* passes into hexane along with Pchlide *a* E. In the process, Chl *a* may become contaminated by very small amounts of Chlide *a*, which may be removed by washing the hexane extract with an equal volume of acetone: water:0.1 N NH_4OH (8:1:1 vol/vol/vol).

In most green higher and lower plants, the Chl *a* pool in the hexane fraction consists mainly of MV Chl *a* (673 nm emission). During the early stages of greening of DDV-LDDV plant species such as cucumber, the Chl *a* pool may also contain small amounts of DV Chl *a* (673 nm emission). In green tissues, the Chl *a* pool is usually accompanied by MV Chl *b* (657 nm emission), MV pheophytin *a*, i.e., demetalated Chl *a* (673 nm emission), and smaller amounts of MV Pheophytin *b* (660 nm emission). It is also accompanied by small amounts of Pchlide *a* E (638 nm emission).

Since Chl *a* and Chlide *a* exhibit identical spectral properties, it is possible to determine the amounts of Chl *a* from room temperature excitation spectra without prior purification by using Equation 14 (section 12.1). For green tissue, a very small aliquot of the hexane extract may be diluted in 80% acetone prior to spectrofluorometric determinations. For etiolated tissues subjected to a brief light treatment, Chl *a* is concentrated by drying an aliquot of the hexane extract under a stream of N_2 gas at ice bucket temperature. The residue is dissolved in a small volume of 80% acetone prior to spectrofluorometric analysis.

In some cases, such as in prochlorophyte picoplanktons and the *Nec1* corn mutant, the Chl *a* pool consists mainly of DV Chl a, with or without smaller amounts of MV Chl a. Under these circumstances, total Chl *a* can be evaluated as for Chlide *a* (section 12.2).

14.2. Quantitative Determination of MV and DV Chl *a* in Crude Extracts by Spectrofluorometry

Before structural studies or determination of the amounts of MV and DV Chl *a* in a mixture of the 2 tetrapyrroles, it is necessary to transfer the Chl *a* pool from hexane to diethyl ether. This is achieved either by dilution of a very small aliquot of the washed hexane extract in diethyl ether or by drying a small aliquot of the washed hexane extract under a stream of N_2 gas and dissolving the residue in diethyl ether. The amounts of DV and MV Chl *a* are determined from room temperature and 77 K excitation spectra by using Equations 15 and 16 (Table 2) exactly as described for DV and MV Chlide *a*.

14.3. Chromatographic Separation of Chl *a* from other Tetrapyrroles by Thin Layer Chromatography

To facilitate purification of the Chl *a* and *b* pools, it is possible to partially separate Chl *a* from Chl *b* by solvent–solvent extraction. This can be achieved by addition to the hexane extract enough 84% aqueous methanol at ice bucket temperatures to form two phases: a deep green epiphase enriched in Chl *a* and β-carotene and a pale green hypophase enriched in Chl *b* and xanthophylls.

The mixed DV-MV Chl *a* pool can be separated from other dicarboxylic and monocarboxylic Mg-porphyrins on thin layers of silica gel H as described in section 5.3.1. Chl *a* and *b* move together just behind Pchlide *a* E and away from other Mg-porphyrins with an R_f of about 0.8 (10). The Chl *a* plus *b* pools are eluted in diethyl ether.

Once the Chl *a* plus *b* pools have been separated from other Mg-porphyrins and xanthophylls, DV and MV Chl *a* can be readily separated from DV and MV Chl *b* by chromatography on thin layers of cellulose. Essentially an aliquot of the diethyl ether eluate containing Chl *a* plus *b* is spotted at the origin of a 5 × 20 cm thin layer plate of cellulose. Once the Chl sample has been spotted, the lower edge of the plate is dipped into a beaker containing acetone. As the acetone moves up the plate through the cellulose, it concentrates the diffuse Chl spot into a sharp thin line at the chromatographic origin. This in turn improves considerably the resolution of the separated bands. The plate is next inserted into a cut 1000-mL glass cylinder containing about 10 mL of ligroin (b.p: 60°–80°C):*n*-propanol (99:1 vol/vol). The cylinder is capped with a piece of aluminum foil. Development is carried out at room temperature in darkness. After the solvent front has migrated about 10 to 15 cm, the plate is viewed under 366 nm UV light. The separated Chls are detected by their red fluorescence. MV Chl *a* and *b* migrate with respective R_fs of about 0.5 and 0.21, while DV Chl *a* and *b* migrate with respective R_fs of about 0.56 and 0.16 (68). The Chl bands are eluted in diethyl ether.

14.4. Chromatographic Separation of Chl *a* by HPLC

HPLC has become a very popular analytical tool for the quantitative separation of complex plant extracts containing Chl *a* and *b*. A few HPLC applications will be described below.

14.4.1. Separation of Chl a *from Chl* b *and Pheophytins*

In addition to Chl *a* and *b*, the extracts of green plants often contain demetallated Chl known as pheophytins. These tetrapyrroles can be separated and their concentration can be determined by HPLC. Reversed phase C-18 bonded columns are used. Five- to 50-µL aliquots of the ammoniacal acetone extract are injected. Elution at room temperature and at flow rate of 1 mL/minute is achieved with an isocratic solvent system consisting of water:methanol:acetone (5:75:20 vol/vol/vol). On an ODS C-18 Spherisorb column (Thermo Separation Products) Chl *b*, Chl *a*, pheophytin *b*, and pheophytin a elute with the following retention times: 2.04, 2.88, 6.83, and 7.99 minutes, respectively (39).

14.4.2. Separation of DV Chl a *from MV Chl* a

Several HPLC protocols have been developed for the separation of DV and MV Chl *a* from other plant pigments (5,21,33,61). For example, the protocol developed by Barlow et al. uses a 3-µm Hypersil MOS2, end-capped, C-8 (6.2%–6.8% carbon), 120 Å pore size, 100 × 4.6-mm column (Shandon Lipshaw) at 30°C. Pigments are separated at a flow rate of 1 mL/minute by a linear gradient expressed by minute; % solvent A; % solvent B and programmed as follows: (0; 75; 25), (1; 50; 50), (20; 30; 70); (25; 0; 100), and (32; 0; 100). The column is reconditioned to original conditions over a 7-minute period. In this system, solvent A consists of methanol:1 M ammonium acetate (70:30 vol/vol). Solvent B consists of 100% methanol. DV and MV Chl *a*, elute much slower (26–27 min) than DV and MV Chl *b* (20–21 min).

14.4.3. Separation and Determination of Chl a *Esterified with Alcohols other than Phytol*

Chl *a* pools esterified with alcohols other than phytol have been observed during dark germination and during the very early stages of greening of etiolated tissues.

In dark-grown barley and corn seedlings, we have consistently observed the presence of trace amounts of Chl *a* esterified with long chain fatty alcohols (LCFA) other than phytol (Figure 1, pathway 14). Experimental difficulties caused by the low concentration of these tetrapyrroles have so far hindered determination of the nature of the esterifying LCFA at position 7 of the macrocycle. Because of interference by the sizable Pchlide *a* E pool, the dark-formed Chlide *a* E pool constituents can only be observed after separation of the fully esterified tetrapyrroles by HPLC. In the HPLC system described in section 10.2, the dark-formed Chlide *a* Es exhibit retention times of about 10.4 to 11.1 minutes (Rebeiz, unpublished).

The HPLC system described in section 10.2 can also be used to monitor the formation of Chlide *a* Es in etiolated tissues during the first 500 milliseconds of greening. As evidenced by on-line emission spectra of eluting peaks, the formation of 4 different Chlide *a* E can be detected during the first 500 milliseconds, following a 2.5-millisecond light treatment of etiolated barley and corn (Figure 1, pathway 13). The various Chlide *a* Es exhibit retention times of 4.5, 5.2, 6.3, and 7.7 minutes. The chemical nature of the esterifying LCFAs at position 7 of the macrocycle is still undetermined. It is not clear either whether the various Chlide *a* E are formed by photoconversion of Pchlide *a* Es or by rapid esterification of newly formed Chlide *a* (Rebeiz, unpublished).

14.4.4. Determination of Chl a *Esterified with Geranylgeraniol Derivatives*

During the early stages of greening of etiolated tissues, newly formed Chlide *a* is converted to Chl *a* P via Chl *a* GG, Chl *a* DHGG, and Chl *a* THGG (53). Separation of these various Chl *a* has been achieved after pheophytinization (i.e., demetalation) on a LiChrosorb RP-8, 5 to 10 µm particle size column (Knauer, Oberusel, Germany). Isocratic elution is at a flow rate of 1.5 mL/minute, in methanol:water (95:5 vol/vol) at room temperature (55). After extraction with ethyl acetate, pheophytinization is achieved by treatment with HCL (14). In this system, the pheophytinized GG Chl *a* derivatives exhibit the following retention times: pheophytin a GG (Pheo *a* GG) = 10 to 11 minutes, Pheo *a* DHGG = 11.8 to 12 minutes, Pheo *a* THGG = 14 to 14.2 minutes, and Pheo *a* P = 16 to 16.4 minutes (55).

14.5. Quantitative Determination of Purified DV and MV Chl *a*

Since DV and MV Chl *a* exhibit identical absorbance properties to DV and MV Chlide *a*, the concentration of purified Chl *a* solutions is determined exactly as described for Chlide *a* (sections 12.6 and 12.7).

15. ANALYSIS OF CHL *b*

In most higher plants, the Chl *b* pool consists exclusively of MV Chl *b* (Figure 5). In the *Nec1* corn mutant, it consists exclusively of DV Chl *b* (Figure 5) (6,68). In primitive prochlorophyte picoplanktons, the Chlide *b* pool is a mixed DV-MV tetrapyrrole pool.

In the integrated Chl *a*/*b* pathway, the MV Chl *b* pool consists of 8 different subpools formed via different Chl *b* biosynthetic pathways (Figure 1). Four subpools are formed from DV Pchlide *a* via pathways 2, 5, 6, and 8. Two subpools are formed from MV Pchlide *b* via pathways 3 and 10. The last 2 subpools are formed from MV Chlide *a* via pathways 9 and 11. MV Chl *b* formation via pathways 5 and 8 is confined to DDV-LDDV plant species such as cucumber, while its formation via pathways 9, 10, and 11 is confined to DDV-LDMV plant species such as corn barley and wheat and to DMV-LDMV plant species such as Johnson grass (Figure 1). Biosynthetic pathways 2, 3, and 6 occur in all three greening groups.

As far as we know, the esterifying alcohol at position 7 of the MV Chl *b* macrocycle is phytol. Because of the extreme biosynthetic heterogeneity of the Chl *b* pool, it is possible that a more thorough investigation of the esterifying alcohol at position 7 of the Chl *b* macrocycle may reveal the presence of minor amounts of LCFAs other than phytol.

In the integrated Chl *a*/*b* pathway, the DV Chl *b* pool consists of 2 different subpools formed via 2 different Chl *b* biosynthetic pathways (Figure 1). One subpool is formed from DV Chlide *b* via pathway 7. The second subpool is formed from DV Chl *a* via pathway 1. DV Chl *b* formation from DV Chlide *a* via ancestral pathways 1 and 7, takes place only in DDV-LDDV plant species. The accumulation of massive amounts of DV Chl *b* is observed in the *Nec1* corn mutant and in primitive prochlorophyte picoplanktons.

The isolation and purification of Chl *b* has been described in sections 14.3 and 14.4. Similarly, because the spectral properties of Chl *b* are identical to Chlide b, the former may be quantified using the procedures described in section 13.

16. ANALYSIS OF PHEOPHYTINS AND PHEOPHORBIDES *a*

The first degradation products of Chl *a* are pheophytin *a* and/or pheophorbides *a* [Pheo(bides) *a*] (40). Pheophytins are demetalated Chls while pheophorbides are demetalated Chlides. During senescence or after treatment with certain chemicals, green plants tend to accumulate substantial amounts of pheophytins and pheophorbides (3). Spectrofluorometric techniques have been developed to determine the concentration of pheophorbides and pheophytins in crude extracts containing Chlides and Chls. Pheophorbides are determined on aliquots of HEAR, while pheophytins are determined on aliquots of the hexane extract after drying in a stream of N_2 gas at ice bucket temperature and dissolution in 80% aqueous acetone.

16.1. Analysis of Pheo(bide) *a* in Crude Extracts by Spectrofluorometry

Upon tetrapyrrole extraction from plant tissues and partitioning of the pigments between hexane and acetone (sections 2.1 and 2.2), pheophorbide *a* passes into the HEAR fraction along with other dicarboxylic and monocarboxylic tetrapyrroles, while pheophytin *a* passes into hexane along with Chls. In most higher and lower plants, except in prochlorophyte picoplanktons and the *Nec1* corn mutant, the pheophytin and pheophorbide *a* pools are MV in nature.

It is possible to determine the amounts of pheophytin *a* and pheophorbide *a* without prior purification by room temperature excitation spectrofluorometry. Since the emission maximum of MV Pheo(bide) *a* falls in the red region of the spectrum far away from the emission of other tetrapyrroles, it is possible to record Soret excitation spectra with minimum Soret excitation overlap. The concentration of Pheo(bide) *a* can then be determined from the excitation spectra after correction for excitation band overlap. This is achieved by recording 2 Soret excitation spectra in HEAR or 80% acetone at room temperature. One spectrum is recorded close to the emission maxima of MV Chl(ide) *a* and Pheo(bide) *a* at 674 nm. The other spectrum is recorded close to the emission maxima of MV Chl(ide) *b* and Pheo(bide) *b* at 660 nm. The net Soret excitation amplitude due only to MV Pheo(bide) *a* is deconvoluted with the use of Equation 24 (7). The net fluorescence excitation amplitude of MV Pheo(bide) *a*, as calculated from Equation 24, is converted to MV Pheo(bide) *a* concentration by reference to a standard calibration curve that relates known concentrations of standard Pheo(bide) *a* to their fluorescence excitation amplitudes. In this manner, MV Pheo(bide) *a* can be determined with a percent error of about 1% for pure MV Pheo(bide) *a* to 13% for extracts containing only 5% MV Pheo(bide) *a* (7). Minimum detection levels are about 0.2 pmol/mL.

16.2. Separation of Pheophytin *a* from other Tetrapyrroles

On silica gel H, pheophytin *a* moves with Chl *a* and is eluted in diethyl ether. Pheophytin *a* can then be readily separated from Chl *a* and *b* by chromatography on thin layers of cellulose. In ligroin (b.p: 60°–80°C):acetone:*n*-propanol (99:10:0.45 vol/vol/vol), pheophytin *a* migrates close to the solvent front with an R_f of about 0.9, while Chl *a* and *b* move with respective R_fs of about 0.46 and 0.23 (7). The pheophytin *a* band is eluted in diethyl ether.

Separation of pheophytin *a* and its derivatives by HPLC has been described in section 14.4.

16.3. Quantitative Determination of Purified Pheo(bide) *a*

Purified MV Pheo(bide) *a* can be determined in any appropriate solvent from room temperature emission spectra elicited by excitation close to its Soret absorbance maximum between 410 and 414 nm. In each case, a calibration curve using known amounts of MV Pheo(bide) *a* should be constructed in order to convert fluorescence data to Pheobide *a* concentrations.

The concentration of purified MV and DV Pheo(bide) *a* solutions can be determined in a variety of organic solvents by room temperature spectrophotometry. At room temperature, between 380 and 700 nm, MV and DV Pheobide *a* exhibit a 7-banded absorption spectrum in various organic solvents. The wavelengths of absorption maxima depend upon the solvent. A strong Soret absorbance band with an absorption maximum at 409 to 410 nm [MV Pheo(bide) *a*] or 417 to 419 nm [DV Pheo(bide) *a*] and a less intense red absorbance band with a maximum at 665 to 666 nm [MV Pheo(bide) *a*] or 665 to 667 nm [DV Pheo(bide) *a*] are observed (Table 1). The Soret and red absorbance maxima can both be used for quantitative determinations. If the Soret absorbance is used, a highly purified sample free of carotenoids and other tetrapyrroles should be used. Contamination by carotenoids does not interfere however with quantitative determinations when the red absorbance maximum is used. The molar extinction coefficients are related to MV and DV Pheobide absorbance and concentration by Beer's law (Equation 2). Table 1 reports absorption maxima and molar extinction coefficient values for MV and DV Pheo(bide) *a* in diethyl ether and 80% aqueous acetone.

17. ANALYSIS OF PHEO(BIDE) *b*

Upon tetrapyrrole extraction from plant tissues and partitioning of the pigments between hexane and acetone (sections 2.1 and 2.2), pheophorbide *b* passes into HEAR along with other dicarboxylic and monocarboxylic tetrapyrroles, while pheophytin *b* passes into hexane along with Chls. In most higher and lower plants, except in prochlorophyte picoplanktons and the *Nec1* corn mutant, the pheophytin and pheophorbide *b* pools are MV in nature.

Quantification of Pheo(bide) *b* in crude extracts is carried out as described in section 16.1, but using Equation 25 (7). The net fluorescence excitation amplitude of MV Pheo(bide) *b*, as calculated from Equation 25, is converted to MV Pheo(bide) *b* concentration by reference to a standard calibration curve prepared from standard MV Pheo(bide) *b* solutions of known concentrations and from their fluorescence excitation amplitudes. In this manner, MV Pheo(bide) *b* can be determined with a percent error of about 1% for pure MV Pheo(bide) *b* to 14% for extracts containing 14% MV Pheo(bide) *a* (7). At concentrations of 6% Pheobide *b* or less, the analytical error becomes substantial (133.5%). Minimum detection levels are about 0.2 pmol/mL.

Similarly, the purification and quantitation of MV Pheo(bide) *b* is as described for MV Pheo(bide) *a*, using the appropriate molar extinction coefficients given in Table 1.

ABBREVIATIONS

Unless preceded by MV or DV, abbreviations for tetrapyrroles are used to designate metabolic pools that may consist of MV and DV components. ALA, δ-aminolevulinic acid; Chl(ide), Chlide and/or Chlide ester; Chl, chlorophyll; Chlide, chlorophyllide; Copro, coproporphyrin; D, dark; DHGG, dihydrogeranylgeraniol; DME, dimethyl ester; DV, divinyl; GG, geranylgeraniol; HEAR, hexane-extracted acetone residue; HPLC, high-pressure liquid chromatography; L, light; LCFA, long chain fatty alcohol; LHC, light harvesting Chl; Mpde, Mg-Proto diester; Mpe, Mg-Proto monomethyl ester; MV, monovinyl; P, phytol; Pchl(ide), Pchlide and/or Pchlide ester; Pchlide E, Pchlide ester; Pchlide, protochlorophyllide; Pheo(bide), pheophorbide and/or pheophytin; POR, Pchlide oxidoreductase; Proto, protoporphyrin IX; THGG, tetrahydrogeranylgeraniol; Uro, uroporphyrin.

ACKNOWLEDGMENTS

This work was supported by funds from the Illinois Agricultural Experiment Station, by the John P. Trebellas Photobiotechnology Research Endowment, and by the C.A. and C.C. Rebeiz Endowment for Fundamental Research.

REFERENCES

1. **Adb El Mageed, H.A., K.F. El Sahhar, K.R. Robertson, R. Parham, and C.A. Rebeiz**. 1997. Chloroplast biogenesis 77. Two novel monvinyl and divinyl light-dark greening groups of plants and their relationship to the chlorophyll *a* biosynthetic heterogeneity of green plants. Photochem. Photobiol. *66*:89-96.
2. **Adra, A.N. and C.A. Rebeiz**. 1998. Chloroplast biogenesis 81. Transient formation of divinyl chlorophyll *a* following a 2.5 ms light flash treatment of etiolated cucumber cotyledons. Photochem. Photobiol. *68*:852-856.
3. **Amindari, S.** 1992. Structure-Function Photodynamic Herbicidal Studies of Phenanthroline, Dipyridyl and Pyridine Analogs, p. 287. NRES, Urbana.
4. **Amindari, S.M., W.E. Splittstoesser, and C.A. Rebeiz.** 1995. Photodynamic effects of several metabolic tetrapyrroles on isolated chloroplasts, p. 217-246. *In* J.R. Heitz and K.R. Downum (Eds.), Light-Activated Pest Control. Am. Chem. Soc. Washington, DC.
5. **Barlow, R.G., D.G. Cummings, and S.W. Gibb**. 1997. Improved resolution of mono- and divinyl chlorophylls *a* and *b* and zeaxanthin and lutein in phytoplankton extracts using reverse phase C-8 HPLC. Mar. Ecol. Prog. Ser. *161*:303-307.
6. **Bazzaz, M.B.** 1981. New chlorophyll chromophores isolated from a chlorophyll deficient mutant of maize. Photobiochem. Photobiophys. *2*:199-207.
7. **Bazzaz, M.B. and C.A. Rebeiz**. 1979. Chloroplast culture V. Spectrofluorometric determination of chlorophyll(ide) *a* and *b* and pheophytin (or pheophorbide) *a* and *b* in unsegregated pigment mixtures. Photochem. Photobiol. *30*:709-721.
8. **Belanger, F.C., J.X. Dugan, and C.A. Rebeiz**. 1982. Chloroplast biogenesis: identification of chlorophyllide *a* (E458F674) as a divinyl chlorophyllide *a*. J. Biol. Chem. *257*:4849-4858.
9. **Belanger, F.C. and C.A. Rebeiz**. 1980. Chloroplast biogenesis 30. Chlorophyll(ide) (E459F675) and chlorophyll(ide) (E449F675) the first detectable products of divinyl and monovinyl protochlorophyll photoreduction. Plant Sci. Lett. *18*:343-350.
10. **Belanger, F.C. and C.A. Rebeiz**. 1980. Chloroplast biogenesis. Detection of divinyl protochlorophyllide ester in higher plants. Biochemistry *19*:4875-4883.
11. **Belanger, F.C. and C.A. Rebeiz**. 1980. Chloroplast biogenesis. Detection of divinyl protochlorophyllide in higher plants. J. Biol. Chem. *255*:1266-1272.
12. **Belanger, F.C. and C.A. Rebeiz**. 1982. Chloroplast biogenesis. Detection of monovinyl magnesium protoporphyrin monoester and other monovinyl magnesium porphyrins in higher plants. J. Biol. Chem. *257*:1360-1371.
13. **Belanger, F.C. and C.A. Rebeiz**. 1984. Chloroplast biogenesis 47: spectroscopic study of net spectral shifts induced by ligand coordination in metalated tetrapyrroles. Spectrochim. Acta *40A*:807-827.
14. **Benz, J., C. Wolf, and W. Rudiger**. 1980. Chlorophyll biosynthesis: hydrogenation of geranylgeraniol. Plant Sci. Lett. *19*:225-230.
15. **Bladon, P. and G. Eglinton**. 1964. Ultraviolet, visible and infrared spectroscopy, p. 22-34. *In* J.C.P. Schwarz (Ed.), Physical Methods in Organic Chemistry. Oliver & Boyd, London.
16. **Carey, E.E. and C.A. Rebeiz**. 1985. Chloroplast biogenesis 49. Difference among angiosperms in the biosynthesis and accumulation of monovinyl and divinyl protochlorophyllide during photoperiodic greening. Plant Physiol. *79*:1-6.
17. **Daniell, H. and C.A. Rebeiz.** 1982. Chloroplast culture IX. Chlorophyll(ide) *a* biosynthesis in vitro at rates higher than in vivo. Biochem. Biophys. Res. Commun. *106*:466-470.
18. **Duggan, J.X. and C.A. Rebeiz**. 1982. Chloroplast biogenesis 37: induction of chlorophyllide *a* (E459F675) accumulation in higher plants. Plant Sci. Lett. *24*: 27-37.

19.**Falk, J.E.** 1964. Porphyrins and Metalloporphyrins, p. 250-253. Elsevier, Amsterdam.

20.**Fischer, H. and A. Stern**. 1940. Die Chimie des Pyrroles, p. 321-324. Kademische, Verlagsgesellschaft.

21.**Garrido, J.L. and M. Zapata**. 1997. Reversed-phase high performance liquid chromatographic separation of mono- and divinyl chlorophyll forms using pyridine-containing mobile phases and a polymeric octadecylsilica column. Chromatographia *44*:43-49.

22.**Granick, S.** 1948. Protoporphyrin 9 as a precursor of chlorophyll. J. Biol. Chem. *172*:717-727.

23.**Houssier, C. and K. Sauer** 1969. Optical properties of the protochlorophyll pigments II. Electronic absorption, fluorescence and circular dichroism spectra. Biochim. Biophys. Acta *172*:261-266.

24.**Ioannides, I.M., D.M. Fasoula, R.K. Robertson, and C.A. Rebeiz**. 1994. An evolutionary study of chlorophyll biosynthetic heterogeneity in green plants. Biochem. Sys. Ecol. *22*:211-220.

25.**Ioannides, I.M., V.P. Shedbalkar, and C.A. Rebeiz**. 1997. Quantitative determination of 2-monovinyl protochlorophyll(ide) *b* by spectrofluorometry. Anal. Biochem. *249*:241-244.

26.**Kim, J.S. and C.A. Rebeiz**. 1996. Origin of the chlorophyll *a* biosynthetic heterogeneity in higher plants. J. Biochem. Mol. Biol. *29*:327-334.

27.**Knaust, R. and H. Senger**. 1994. Monovinyl and divinyl protochlorophyll in different stages of esterification isolated from mutant C-2A' of the unicellular green alga *Scenedesmu obliquus*. Physiol. Plant. *90*:490-496.

28.**Kolossov, V., I.M. Ioannides, S. Kulur, and C.A. Rebeiz**. 1999. Chloroplast biogenesis 82: development of a cell-free system capable of the net synthesis of chlorophyll(ide) *b*. Photosynthetica *36*:253-258.

29.**Koski, V.M.** 1950. Chlorophyll formation in seedlings of *Zea mays* L. Arch. Biochem. *29*:339-343.

30.**Koski, V.M. and J.H.C. Smith**. 1948. The isolation and spectral absorption properties of protochlorophyll from etiolated barley seedlings. J. Am. Chem. Soc. *70*:3558-3562.

31.**Lee, H.J., M. Ball, and C.A. Rebeiz**. 1991. Intraplastidic localization of the enzymes that convert δ-aminolevulinic acid to protoporphyrin IX in etiolated cucumber cotyledons. Plant Physiol. *96*:910-915.

32.**Lee, H.J., M.D. Ball, R. Parham, and C.A. Rebeiz**. 1992. Chloroplast biogenesis 65: enzymic conversion of protoporphyrin IX to Mg-protoporphyrin IX in a subplastidic membrane fraction of cucumber etiochloroplasts. Plant Physiol. *99*:1134-1140.

33.**Lenning, K.V., J.L. Garrido, J. Aristegui, and M. Zapata**. 1995. Temperature-programmed high performance liquid chromatography of mono- and divinyl chlorophyll forms from marine phytoplankton. Chromatographia *41*:539-543.

34.**Mackinney, G.** 1941. Absorption of light by chlorophyll solutions. J. Biol. Chem. *140*:315-322.

35.**McCarthy, S.A., F.C. Belanger, and C.A. Rebeiz**. 1981. Chloroplast biogenesis: detection of a magnesium protoporphyrin diester pool in plants. Biochemistry *20*:5080-5087.

36.**McCarthy, S.A., J.R. Mattheis, and C.A. Rebeiz**. 1982. Chloroplast biogenesis: biosynthesis of protochlorophyll(ide) *via* the acidic and fully esterified biosynthetic branches in higher plants. Biochemistry *21*:242-247.

37.**Nicolas, R.E.H. and C. Rimington**. 1949. Qualitative analysis of the porphyrins by partition chromatography. Scand. J. Cli. Lab. Invest. *1*:12-18.

38.**Parham, R. and C.A. Rebeiz**. 1995. Chloroplast biogenesis 72: a (4-vinyl) chlorophyllide *a* reductase assay using divinyl chlorophyllide *a* as an exogenous substrate. Anal. Biochem. *231*:164-169.

39.**Rebeiz, C.A.** 1978. The separation of chlorophyll and pheophytins by reversed phase HPLC. Chromatography Rev. *4*:8-9.

40.**Rebeiz, C.A. and M.B. Bazzaz**. 1979. Cell-free agriculture: the concept and its initial implementation, p. 453-471. *In* C.D. Scott (Ed.), Biotechnology in Energy Production and Conservation. John Wiley & Sons, New York.

41.**Rebeiz, C.A., F.C. Belanger, S.A. McCarty, G. Freyssinet, J.X. Duggan, S.M. Wu, and J.R. Mattheis**. 1981. Biosynthesis and accumulation of novel chlorophyll *a* and *b* chromophoric species in green plants, p. 197-212. *In* G. Akoyunoglou (Ed.), Photosynthesis V. Chloroplast Development. Balaban International Services, Philadelphia.

42.**Rebeiz, C.A., and P. Castelfranco**. 1971. Chlorophyll biosynthesis in a cell-free system from higher plants. Plant Physiol. *47*:33-37.

43.**Rebeiz, C.A. and P. Castelfranco**. 1971. Protochlorophyll biosynthesis in a cell-free system from higher plants. Plant Physiol. *47*:24-32.

44.**Rebeiz, C.A., I.M. Ioannides, V. Kolossov, and K.J. Kopetz**. 1999. Chloroplast biogenesis 80. Proposal of a unified multibranched chlorophyll *a/b* biosynthetic pathway. Photosynthetica *36*:117-128.

45.**Rebeiz, C.A., J.R. Mattheis, B.B. Smith, C.C. Rebeiz, and D.F. Dayton**. 1975. Chloroplast biogenesis. Biosynthesis and accumulation of protochlorophyll by isolated etioplasts and developing chloroplasts. Arch. Biochem. Biophys. *171*:549-567.

46.**Rebeiz, C.A., A. Montazer-Zouhoor, and H. Daniell**. 1984. Chloroplast culture X: thylakoid assembly in vitro. Isr. J. Bot. *33*:225-235.

47.**Rebeiz, C.A., R. Parham, D.A. Fasoula, and I.M. Ioannides**. 1994. Chlorophyll biosynthetic heterogeneity, p. 177-193. *In* D.J. Chadwick and K. Ackrill (Eds.), The Biosynthesis of the Tetrapyrrole Pigments. John Wiley & Sons, New York.

48.**Rebeiz, C.A. and D.G. Saab**. 1995. Porphyrin Analytical Tools. Software protected by copyright.

49.**Rebeiz, C.A., S.M. Wu, M. Kuhadje, H. Daniell, and E.J. Perkins**. 1983. Chlorophyll *a* biosynthetic routes and chlorophyll *a* chemical heterogeneity. Mol. Cell. Biochem. *58*:97-125.

50.**Rebeiz, N., S. Arkins, K.W. Kelley, and C.A. Rebeiz**. 1996. Enhancement of coproporphyrinogen III transport into isolated leucocyte mitochondria by ATP. Arch. Biochem. Biophys. *333*:475-481.

51.**Reinbothe, C., N. Lebedev, and S. Reinbothe**. 1999. A protochlorophyllide light-harvesting complex involved in de-etiolation of higher plants. Nature *397*:80-84.

52.**Rudiger, W.** 1993. Esterification of chlorophyllide and its implication for thylakoids development, p. 219-240.

In C. Sundqvist and M. Ryberg (Eds.), Pigment-Protein Complexes in Plastids: Synthesis and Assembly. Academic Press, New York.

53.**Rudiger, W. and S. Schoch**. 1991. The last steps of chlorophyll biosynthesis, p. 451-464. *In* H. Scheer (Ed.), Chlorophylls. Academic Press, New York.

54.**Scheumann, V., H. Klement, M. Helfrish, U. Oster, S. Schoch, and W. Rudiger**. 1999. Protochlorophyllide *b* does not occur in barley etioplasts. FEBS Lett. *445*:445-448.

55.**Schoch, S.** 1978. The esterification of chlorophyllide *a* in greening bean leaves. Z. Naturforsch *33c*:712-714.

56.**Shedbalkar, V.P., I.M. Ioannides, and C.A. Rebeiz**. 1991. Chloroplast biogenesis. Detection of monovinyl protochlorophyll(ide) *b* in plants. J. Biol. Chem. *266*:17151-17157.

57.**Shedbalkar, V.P. and C.A. Rebeiz.** 1992. Chloroplast biogenesis: determination of the molar extinction coefficients of divinyl chlorophyll *a* and *b* and their pheophytins. Anal. Biochem. *207*:261-266.

58.**Shioi, Y. and T. Sasa**. 1883. Formation and degradation of protochlorophylls in etiolated and greening cotyledons of cucumber. Plant Cell Physiol. *24*:835-840.

59.**Shioi, Y. and T. Sasa**. 1983. Compositional heterogeneity of protochlorophyllide ester in etiolated leaves of higher plants. Arch. Biochem. Biophys. *220*:286-292.

60.**Shioi, Y. and T. Sasa.** 1982. Separation of protochlorophylls esterified with different alcohols from inner seed coat of three cucurbitaceae. Plant Cell Physiol. *23*:1315-1321.

61.**Shioi, Y., K. Watanabe, K.-i. Takmiya, J.L. Garrido, and M. Zapata**. 1995. Separation of mono-and divinyl chlorophyll species by high-performance liquid chromatography using an octadecyl polyvinyl alcohol polymer column. Anal. Biochem. *231*:225-229.

62.**Smith, J.H.C. and C.S. French**. 1963. The major accessory pigment in photosynthesis. Ann. Rev. Plant Physiol. *14*:181-224.

63.**Tripathy, B.C. and C.A. Rebeiz**. 1985. Chloroplast biogenesis. Quantitative determination of monovinyl and divinyl Mg-protoporphyrins and protochlorophyll(ides) by spectrofluorometry. Anal. Biochem. *149*:43-36.

64.**Tripathy, B.C. and C.A. Rebeiz**. 1986. Chloroplast biogenesis. Demonstration of the monovinyl and divinyl monocarboxylic routes of chlorophyll biosynthesis in higher plants. J. Biol. Chem. *261*:13556-13564.

65.**Tripathy, B.C. and C.A. Rebeiz.** 1988. Chloroplast biogenesis 60. Conversion of divinyl protochlorophyllide to monovinyl protochlorophyllide in green(ing) barley, a dark monovinyl/light divinyl plant species. Plant Physiol. *87*:89-94.

66.**Vernon, L.P.** 1960. Spectrophotometric determination of chlorophylls and pheophytins in plant extracts. Anal. Biochem. *32*:1144-1150.

67.**Withrow, R.B. and L. Price.** 1957. A darkroom safelight for research in plant physiology. Plant Physiol. *32*:244-248.

68.**Wu, S.M., J.M. Mayasich, and C.A. Rebeiz**. 1989. Chloroplast biogenesis: quantitative determination of monovinyl and divinyl chlorophyll(ide) *a* and *b* by spectrofluorometry. Anal. Biochem. *178*:294-300.

69.**Wu, S.M. and C.A. Rebeiz** 1985. Chloroplast biogenesis. Molecular structure of chlorophyll *b* (E489 F666). J. Biol. Chem. *260*:3632-3634.

70.**Zscheile, F.P. and C.P. Comar**. 1941. Influence of preparative procedure on the purity of chlorophyll components as shown by absorption spectra. Bot. Gazette *102*:463-481.

7 Analysis of Heme and Hemoproteins

Angela Wilks
University of Maryland, Baltimore, MD, USA

1. INTRODUCTION

Heme is perhaps the most ubiquitous cofactor found in nature and the most functionally diverse. Hemoproteins are involved in cell respiration (cytochromes), oxygen-binding and transport (hemoglobin and myoglobin), oxidative biotransformations (cytochrome P-450 and peroxidases), and most recently, as sensors in 2-component regulatory systems (guanylate cyclase, FixL, and CooA). The ability of hemoproteins to carry out extremely diverse reactions arises largely from the protein environment in which the heme molecule resides and specifically the nature of the heme–ligands. Other factors that contribute to the reactivity of the heme are intrinsic to the heme itself, including the substituents on the heme periphery and, in some cases, the covalent attachment of the heme to the protein. The structures of the most common heme–ligands and examples of the hemoproteins in which they occur are found in Table 1.

The advent of modern cloning and molecular biological techniques has opened up the field of hemoprotein research in regard to the elucidation of structure and function. It is, however, pertinent to keep in mind that functional hemoproteins require the binding and coordination of heme, which must be synthesized and incorporated into the protein during protein synthesis. In addition, the expression of a given hemoprotein is dependent on the nature of that particular protein. For example, the membrane-bound cytochrome P-450 enzymes have markedly different requirements than a soluble protein such as myoglobin. Therefore, the expression and purification of the proteins will be discussed in the context of both their function and properties. In light of the excellent reviews on the chemistry of heme and porphyrins in Chapters 1 and 2, this review will focus solely on the expression, purification, and analysis of heme in the context of hemoproteins.

2. HEMOPROTEIN EXPRESSION AND PURIFICATION

Recombinant expression systems for both prokaryotic and eukaryotic hemoproteins are now routinely utilized to obtain

Heme, Chlorophyll, and Bilins: ***Methods and Protocols***
Edited by A.G. Smith and M. Witty

Table 1. Heme Ligand Structure and Function

Heme coordination	Protein class	Protein function
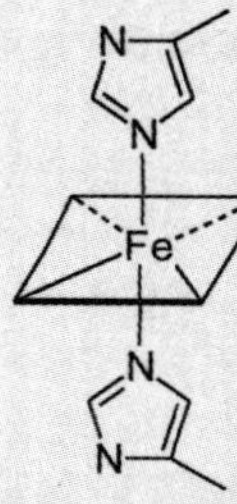 Hexacoordinate	Cytochrome b_5 Cytochrome c (Class II and IV) Cytochrome c oxidase (contains heme a)	Electron transfer
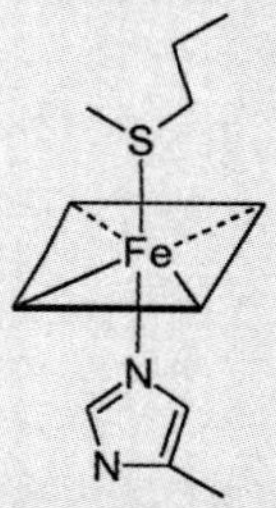 Hexacoordinate	Soluble cytochrome b_{562} Cytochrome c (Class I, IIb and IV)	Electron transfer
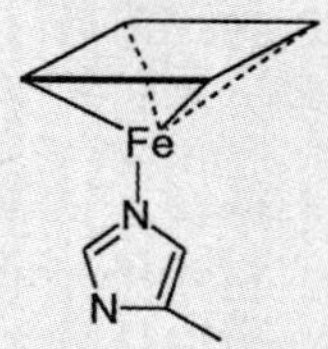 Hexa/Pentacoordinate	Hemoglobin, myoglobin Fix L/CooA Soluble guanylate cyclase Peroxidases Heme oxygenase	O_2-binding O_2/CO-binding NO-binding Oxidoreductase Oxidoreductase
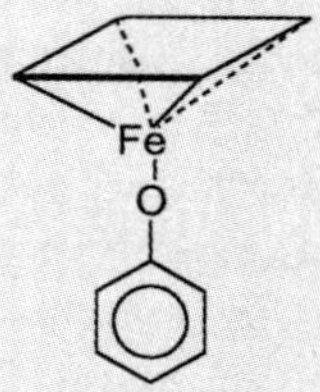 Hexa/Pentacoordinate	Catalase	Oxidoreductase
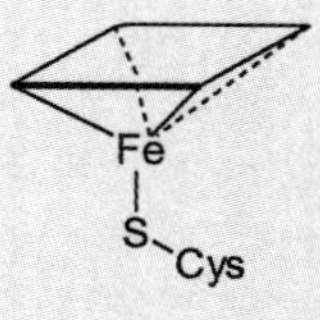 Hexa/Pentacoordinate	Cytochrome P450 Nitric oxide synthases Chloroperoxidase	Oxidoreductases

large quantities of protein for biochemical and biophysical characterization. The advance in molecular biological techniques has not only furthered our understanding of hemoprotein function but also contributed to the rapidly expanding role of hemoproteins in areas such as cell signaling and regulation of gene expression and function.

2.1. Hemoprotein Expression Systems

2.1.1. Oxygen-Binding Proteins

The successful expression and purification of recombinant myoglobins from a number of species have been reported (103,110,122,127). Initial studies on the expression in *Escherichia coli* of human myoglobin utilized a fusion protein consisting of the first 31 amino acids of the phage lambda cII gene and the tetrapeptide Ile-Glu-Gly-Arg, followed by the myoglobin gene sequence (122). The fusion product was then isolated, reconstituted with heme, cleaved with trypsin, and purified to generate the active protein in gram quantities. Subsequent studies focused on the generation of synthetic genes for sperm whale myoglobin, which allowed for optimization of the preferred *E. coli* codon usage and the incorporation of unique restriction sites throughout the gene (110,127). The introduction of multiple restriction sites enabled the design of cassette based primers for site-directed mutagenesis (110,127). The synthetic gene when cloned into pUC18 (Life Technologies, Rockville, MD, USA) and expressed in *E. coli* DH5α™ (Life Technologies), produced active holoprotein indistinguishable from that of the native protein. High expression yields of myoglobin apoprotein as inclusion bodies has been obtained in the T7 promoter based vector pET17b (Novagen, Madison, WI, USA) (49). The authors were able to refold and reconstitute the protein in quantities of 200 mg/L of cells when grown on minimal media alone.

The individual globin chains of hemoglobin can be expressed as fusion proteins, in recombinant *E. coli* systems, or as is more common, with the β-globin chain, the apoprotein itself (27,41,72). It is also possible in *E. coli* cells to co-express the α and β-globins together and obtain active holoprotein (41,42,44,47,100). In all of these systems, however, some heterogeneity is observed. A functionally homogenous protein is best obtained by removal of the heme, separation of the α and β-globin chains, and subsequent reassembly in the presence of cyanohemin to reform the active tetrameric protein (41,42,47).

More recently, a new class of ligand binding hemoproteins that act as biological sensors have been identified in a number of organisms. This class of proteins includes the eukaryotic soluble guanylate cyclase, FixL of *Rhizobia*, and CooA of *Rhodospirillum rubrum*, which sense NO, O_2, and CO, respectively (91). While mammalian soluble guanylate cyclases have been intractable to expression in *E. coli*, some progress has been made in insect cell lines (135). The oxygen sensor FixL is a modular protein consisting of an N-terminal heme domain and a C-terminal kinase transmitter domain. Expression of both the full-length protein and the heme domain alone have been successfully carried out in *E. coli* TG1 under control of the T7 promoter in a pUC8 derived vector (32). The CooA transcription factor has been subcloned into pKK223-3 (Amersham Pharmacia Biotech, Piscataway, NJ, USA) under control of the T7 promoter and expressed in *E. coli* JM109 cells (2,99).

2.1.2. Oxidases

It is only recently that the membrane-bound eukaryotic cytochrome P-450 enzymes have been routinely expressed in recombinant *E. coli* systems. There are a number of factors critical for successful recombinant expression of this family of enzymes. Many of the specific requirements for the successful expression of the cytochrome P-450 enzymes involve the ability of the cell to adjust their synthesis of heme and lipids to a level that allows for correct folding and membrane association. If the rate of expression exceeds the ability of the cell to synthesize these factors, then the protein irreversibly accumulates as inclusion bodies. The requirements for functional cytochrome P-450 expression in *E. coli* were determined to be factors contained in the expression vector, i.e., the promoter and *lac* repressor gene, the structure and/or sequence of the mRNA around the initiation codon, as well as variables such as the *E. coli* strain and culture conditions (3). The powerful T7 phage promoter is capable of synthesizing large amounts of the recombinant protein, but the majority was found to accumulate in inclusion bodies. The most successful vectors for the expression of cytochrome P-450 are based on the *lac* promoter and its derivatives. The two *E. coli* expression vectors that have been successfully used for cytochrome P-450 expression are pCW Ori+, a derivative of pHSe5 (28,71), and pSP19g10L, a derivative of pSPORT-1 (Life Technologies).

The fusion of a promoter and ribosomal binding site of a prokaryotic plasmid and the nucleotide sequence from a eukaryotic gene can have an inhibitory effect on translation due to the formation of mRNA secondary structure. In cytochrome P-450, 17α-hydroxylase (CYP17) changes in the amino terminal codons promoted high level expression, whereas the unmodified DNA failed to produce any detectable protein (3). Utilization of the pCW Ori+ plasmid vector together with the first 8 codons of the modified CYP17 has resulted in the high-level expression of a number of cytochrome P-450 enzymes (35,124).

The eukaryotic cytochrome P-450 enzymes have also been expressed in yeast, baculovirus, and mammalian expression systems (33). The yeast and mammalian expression systems in general produce much less total protein, but have been extremely valuable in the area of metabolite and drug research.

The recent emergence of nitric oxide in physiological functions, such as signal transduction in the cardiovascular and nervous systems and cytostatic functions of the immune system, has led to an extensive body of work on nitric oxide synthase (NOS) (66,77). A number of isoforms of nitric oxide synthase have been expressed in baculovirus, including the human endothelial, inducible and neuronal enzymes, as well as the rat neuronal nitric oxide synthase (11,12,73,90). Large-scale expression of nitric oxide synthase in baculovirus has been achieved, but a critical factor in obtaining soluble active protein was the addition of exogenous heme required for correct folding and binding of the cofactor tetrahydrobiopterin (62,64,98).

The successful expression of nitric oxide synthase isoforms in *E. coli* has developed more recently and was critical to the rapid progress in elucidating many of the structural and mechanistic features of NOS (66, 77). As with the other members of the cytochrome P-450 family, the NOS enzymes were expressed successfully in the pCW Ori+ vectors (30,31,93,131). Factors reported to be critical for maximizing the expression of the rat neuronal nitric oxide synthase were the co-expression of groEL and groES chaperonins in the protease deficient BL21 (DE3) pLysS *E. coli* strain (93). However, Gerber and Ortiz de Montellano have reported successful expression of the rat neuronal nitric oxide synthase in the absence of chaperonins (31). The inducible NOS, unlike the neuronal isoform, requires calmodulin, and co-expression of calmodulin with iNOS is essential for active holoprotein (30,131). All of the isoforms require tetrahydrobiopterin, a cofactor that is not synthesized in vivo by prokaryotes, therefore, reconstitution with this cofactor is required to obtain active protein.

Yeast expression systems have not been as extensively used as recombinant *E. coli* in large part because the yield of active holoprotein is in the 0.5 to 1.0 mg/L range, compared with 5 to 10 mg/L in the recombinant *E. coli* systems. However, both the neuronal and macrophage NOS enzymes have been successfully expressed in *Saccharomyces cerevisiae* (6,50,94).

The prototypical peroxidases, on which most of the mechanistic and structural studies of peroxidase chemistry have been determined, are the plant horseradish peroxidase (HRP) and the yeast cytochrome c peroxidase. The heterologous expression of HRP has been primarily carried out in the baculovirus system, which yields active glycosylated holoprotein (39), and in *E. coli*, where the protein is expressed as inclusion bodies and is subsequently refolded in the presence of heme and calcium (104).

Cytochrome c peroxidase has been expressed primarily in yeast where active holoprotein is obtained (84). Yeast cytochrome c peroxidase has been expressed heterologously in *E. coli* by replacing the *fol* gene in a vector used to overexpress dihydrofolate reductase with the cytochrome c peroxidase gene (26). The expression levels were 15 mg/L of cell culture, of which 10% was holoprotein and 90% apoprotein.

Limited success has been achieved with the mammalian lactoperoxidase (17), myeloperoxidase (113), and prostaglandin synthase (29), which have been expressed exclusively in baculovirus Sf9 cells. In the case of lactoperoxidase, the heme was only partially covalently bound (17). The authors reported that the noncovalently bound heme could be covalently attached to the protein on treatment with hydrogen peroxide and proposed that such an autocatalytic mechanism may be responsible for the covalent attachment of heme in vivo. Myeloperoxidase, which has significant homology to lactoperoxidase and a covalently linked heme, may also have a similar mechanism for covalent attachment of the heme to the protein.

When discussing the expression and purification of the oxidases, it is important to include heme oxygenase in this category, although it is not strictly speaking a hemoprotein. Heme oxygenase catalyzes the rate-limiting step in heme degradation, and the products biliverdin and carbon monoxide have been implicated in antioxidant activity and signal transduction pathways, respectively (20,63,78). Heme oxygenase binds heme as a cofactor and substrate and, in the absence of any reductant, has spectra similar to the oxygen-binding proteins hemoglobin and myoglobin (48,128,132,133). The rat and human heme oxygenase have been expressed in recombinant *E. coli* systems as both the full-length protein and a soluble catalytically active form lacking the hydrophobic C-terminal anchor domain (48,126,128). The full-length rat heme oxygenase (32 kDa) was expressed using the *lac* promoter of pKK233-2 (Amersham Pharmacia Biotech) in *E. coli* XL-1 blue. The protein, although catalytically active, was not homogenous as judged by a proteolytic fragment at 28 kDa on sodium dodecyl sulfate-polyacrylamide gel electrophoresis (SDS-PAGE) (48). A more recent approach, in which the protein was expressed without the C-terminal hydrophobic domain as a soluble fully active protein, resulted in high-level expression of a homogenous 30 kDa protein. The protein was expressed under the control of the *phoA* promoter at high level in *E. coli* DH5α cells (128). The protein, although initially expressed in low phosphate media, was in later experiments expressed in LB media when it was observed that the expression levels in either media were similar. The protein was expressed in LB media for a number of reasons, including the simpler protocol and cost effectiveness. Recently, the soluble bacterial heme oxygenases of *Synechocystis* sp. PCC 6803 and the pathogen *Corynebacterium diphtheriae* have been expressed in *E. coli* as catalytically active proteins (14,129). The *C. diphtheriae* heme oxygenase (HmuO) was expressed under the control of the T7 promoter with and without a 6-histidine tag at the C terminus, and no significant differences in protein yields or activity were noted (129).

2.1.3. Electron Transfer Proteins

As previously described, two of the critical factors in the expression of many mammalian hemoproteins has been the removal of a hydrophobic membrane anchor domain or redesigning the codons encoding the first 5 amino acids of the N terminus. Sligar and coworkers (5) in early studies completely synthesized the genes for both the complete rat hepatic cytochrome b_5 with the membrane anchor and the protease-treated soluble form. In addition, they incorporated an optimal ribosomal binding site and spacer region for expression of the proteins in *E. coli* (5). The soluble protein, when expressed in pUC, accounted for 8% of the total protein, with the membrane-bound protein being somewhat lower and fractionating with the cell membrane. In later studies, the use of the high expression T7 promoter was utilized in the expression of both the rat and human cytochrome b_5. In these later studies, polymerase chain reaction (PCR) was used to engineer a sequence encoding a 4-histidine tag at the N terminus allowing rapid purification by nickel–chelate affinity chromatography (46). Optimization of expression for structural nuclear magnetic resonance (NMR) studies was carried out by Guiles and coworkers in which they expressed the rat liver cytochrome b_5 in the T7 derived pET3C vectors (Novagen) to high levels utilizing *E. coli* BL-21 (pLysS) strain (95). The yields of protein from this vector ranged from 100 mg/L in rich medium to 40 mg/L in minimal media.

It is only recently that *S. cerevisiae* cytochrome c has been expressed heterologously in *E. coli* (81). The successful expression of the holoprotein required the co-expression of cytochrome c heme–lyase for covalent attachment of heme a to the apoprotein. The successful expression of the proteins was achieved by cloning the genes encoding the cytochrome (CYC1) and the heme–lyase (CYC3) in parallel under the control of the Lac and Trc promoters in the vector pUC18. The expression system yielded 15 mg/L of active iso-1-cytochrome c holoprotein.

The expression systems described above encompass only a small fraction of hemoproteins, and while it is hard to generalize on the expression of a given class or a particular hemoprotein, some simple generalizations can be made. First, hemoproteins with covalently attached hemes (cytochrome c) and/or posttranslational modifications, such as glycosylation (myeloperoxidase, lactoperoxidase), have been more successful in eukaryotic expression systems. Second, a critical factor in recombinant *E. coli* expression of membrane bound hemoproteins, such as the cytochrome P-450 enzymes, is the ability of the cells to synthesize heme and other cofactors required for correct folding and activity of the holoprotein.

2.2. Protein Purification Methods

2.2.1. Oxygen-Binding Proteins

Purification of many hemoproteins has been simplified with the development of recombinant expression systems and the recent advances in affinity chromatography. Recombinant sperm whale myoglobin has been purified to homogeneity primarily utilizing ion exchange chromatography (110,127). Procedure 1 outlines how this was achieved.

A significant amount of data on the structure and function of hemoglobin has accumulated to date, however, there is less information available on the great number of naturally occurring hemoglobin mutants, of which many have significant clinical value. The problem of obtaining sufficient quantities of these naturally occurring mutants has been circumvented by the recent development of molecular biological techniques. A similar approach to that taken with myoglobin, in which a synthetic gene utilizing the preferred *E. coli* codon usage and the incorporation of a number of restriction sites for rapid and efficient cassette mutagenesis, has been described (41). Initial studies utilized the co-expression of synthetic α and β-globins from the same operon downstream of the *lac* promoter (41). The two subunits combined intracellularly with endogenously produced heme to give tetrameric hemoglobin in a yield of 5% to 10% of the total protein. This is described in Procedure 2, which has been adapted from Reference 42.

❖ Procedure 1. Purification of Recombinant Sperm Whale Myoglobin

1. Fresh overnight cultures of pMb221 (pUC18 containing the synthetically constructed myoglobin gene) (5 mL) were used to inoculate 4 × 2 L of LB media containing 100 µg/mL ampicillin. The cells were grown for 18 hours and harvested by centrifugation.
2. The cells were resuspended in 50 mM Tris (pH 8.0), 1 mM ethylene diamine tetraacetic acid (EDTA), 1 mM phenylmethylsulfonylfluoride (PMSF), 50 U DNase I/mL, 5 U RNase/mL, and 50 µg/mL lysozyme and stirred for 1 hour at 4°C. The cells were sonicated (5 × 30 s) prior to centrifugation (140 000× *g* for 30 min).
3. The resultant supernatant was stirred at 4°C, and ammonium sulfate was added (final concentration 50%), and the solution was stirred for a further 1 hour. The precipitates were collected by centrifugation, and the supernatant was taken to 95% saturation with ammonium sulfate and stirred for 2 hours. The precipitates were again collected by centrifugation, washed with and then resuspended in 50 mM sodium phosphate (pH 6.0).
4. The protein was applied to a Bio-Gel® P100 column (2.5 × 100 cm; Bio-Rad Laboratories, Hercules, CA, USA) equilibrated with 50 mM Tris (pH 6.0). The fractions containing myoglobin (as judged by the absorbance at 408 nm) were pooled and concentrated.
5. The protein was further purified by fast protein liquid chromatography (FPLC) on a Mono S™ HR-10 column (Amersham Pharmacia Biotech) using a linear gradient of 50 mM sodium phosphate (pH 6.0) to 75 mM sodium phosphate (pH 8.0) over 45 minutes. The purified myoglobin was stored as is at -70°C until further use.

❖ Procedure 2. Purification of Human Hemoglobin

1. Frozen cells (150 g) were resuspended in 300 mL 15 mM Tris-HCl (pH 8.0), 0.1 mM EDTA, 1 mM dithiothreitol (DTT), 1 mM $MgCl_2$, 0.1 mM $MnCl_2$, 40 mg RNase A, 5 mg RNase B, 10 000 U of DNase. The cells were lysed by sonication and centrifuged (20 min at 17 300× *g*). The supernatant was further clarified at 50 000 rpm for 30 minutes.
2. The supernatant was clarified through a coarse Sephadex® G-25 column (4 × 70

cm; Amersham Pharmacia Biotech) equilibrated with 50 mM Tris-HCl (pH 8.0) containing 0.1 mM EDTA and 1 mM DTT (start buffer). The eluate was loaded onto a DE52 column (4 × 8 cm; Whatman, Clifton, NJ, USA) equilibrated with start buffer and washed with a further 500 mL of the same buffer. The hemoglobin fraction was eluted with a gradient of 500 mL each of start buffer and 15 mM Bis-Tris (pH 6.2) containing 0.1 mM EDTA and 1 mM DTT. All of the above buffers were equilibrated with carbon monoxide. The hemoglobin fraction was concentrated through PM30 Amicon® membranes (Millipore, Bedford, MA, USA).

3. The concentrated hemoglobin fraction was passed through a coarse Sephadex G-25 column (as above) equilibrated with 15 mM Tris-HCl (pH 8.0), and the eluate was high-performance liquid chromatography (HPLC)-purified on a DEAE 5PW TSK column (2 × 15 cm; TosoHaas, Montgomeryville, PA, USA) equilibrated with the same buffer. The protein was eluted with a gradient from 15 mM Tris-HCl (pH 8.0) to 15 mM Bis-Tris-HCl (pH 7.0) at a flow rate of 5 mL over 85 minutes.
4. The hemoglobin fraction was then passed over a Sephadex G-25 column equilibrated with 20 mM BisTris-HCl (pH 7.0) and further purified by HPLC on a SP 5PW TSK column (2 × 15 cm; TosoHaas) equilibrated with the same buffer. The protein was eluted with a gradient of 20 mM BisTris-HCl (pH 7.0) to 25 mM Tris-HCl (pH 8.0) at 5mL/minute over 85 minutes.

The resulting protein showed significant heterogeneity within the subunit composition, which was only resolved by disassembling the hemoglobin and reconstituting the individual components in the presence of heme (42,47) as described in Procedure 3.

❖ Procedure 3. Denaturation and Reassembly of Recombinant Human Hemoglobin

1. Aliquots of purified 0.5 mM hemoglobin (5 mL) were added to 200 mL of acid–acetone (2.5 mL HCl/L acetone) at -20°C followed by centrifugation at 7000 rpm in a Sorvall SS34 rotor (Kendro Laboratory Products, Newtown, CT, USA) for 10 minutes at -20°C. The resultant hemoglobin pellet was resuspended in 8 M urea, 5 mM sodium phosphate, and 50 mM mercaptoethanol (pH 6.7) (urea buffer). The sample was then dialyzed overnight in the same buffer. Following dialysis any precipitation was removed by centrifugation at 50 000 rpm in a 55.1 Ti rotor (Beckman Coulter, Fullerton, CA, USA) for 30 minutes at 15°C.
2. The denatured protein was applied to a CM Sepharose® CL6B column (2 × 7 cm; Amersham Pharmacia Biotech) equilibrated in urea buffer. The column was washed with approximately 80 mL of the urea buffer until all of the non-binding protein was eluted. The β-globin was then eluted with a linear gradient of 180 mL each of urea buffer (pH 6.7) and the same buffer containing 0.1 M NaCl. The α-globin was eluted with a second gradient of 0.1 to 0.2 M NaCl in the urea buffer. The eluted globin fractions were concentrated over PM10 membranes (Millipore).
3. The starting concentrations in the urea buffer of the α and β-globins for reconstitution were between 4 and 5 mg/mL, based on the Molar absorption coefficients of 1.0×10^4 and 1.54×10^4 for the α and β subunits, respectively. Cyanohemin was added to a 1.2 molar excess of the combined globin concentration, and the solution was diluted to a final concentration of 0.3

mg/mL in cold deionized water equilibrated with carbon monoxide (CO) by bubbling with CO, and the pH was adjusted to 8.0 with a few crystals of Tris. The solution was left undisturbed overnight at 5°C and was then concentrated through PM30 membranes. The concentrated reassembled hemoglobin was then reduced anaerobically with sodium dithionite in the presence of CO and passed through a Sephadex G-25 column equilibrated with 15 mM Tris-HCl (pH 8.4) to remove the excess hemin and dithionite.

4. Separation of the excess α and β-globin chains was accomplished on a DE52 column (4 × 5 cm; Whatman) in 15 mM Tris-HCl (pH 8.4). The α-globin and intact hemoglobin were then eluted from the column with a gradient of 750 mL each of start buffer and 80 mM Tris-HCl (pH 8.0). The α-globin was eluted first followed by the hemoglobin as the major band with the excess β-globin remaining bound to the column.

2.2.2. Oxidases

The purification of the cytochrome P-450 enzymes has been greatly simplified with the advent of recombinant *E. coli* expression systems and the use of metal-chelate affinity chromatography techniques. For the purpose of this review, we will focus on the purification of cytochrome P-450 enzymes from bacterial systems in light of the considerable advantages in high level of expression, ease of manipulation, and the relatively low cost. A number of laboratories have utilized the pCW Ori+ vector (see Section 2.1.2) in the development of expression and purification systems for many cytochrome P-450 enzymes (30,31,33,35,38,65,131).

The critical step in the purification of cytochrome P-450 enzymes is the preparation and solubilization of the membranes. The selection of a suitable detergent and its concentration in which to solubilize a given cytochrome P-450 enzyme is largely a matter of trial and error. Solubilization of the protein from bacterial cell pellets is usually carried out in 100 mM Tris buffer in the pH range 7.5 to 8.0, containing either 500 mM sucrose or 20% glycerol (18,36). The outer membrane is then solubilized in the presence of lysozyme and protease inhibitors (1 mM PMSF, 2 μM leupeptin, 10 μM bestatin, and 0.04 U/mL aprotinin) with gentle stirring for 1 hour at 4°C. The spheroplasts are collected by centrifugation at 100 000× *g* for 1 hour. The pellets can then be resuspended in the start buffer for solubilization. Solubilization of the cytochrome P-450 is carried out at 4°C for 60 minutes, and the solubilized enzyme is centrifuged at 100 000× *g* for 60 minutes. In the purification of CYP4All, 1% Emulgen 911 appeared the most effective in solubilizing the protein from bacterial membranes (18). In the case of CYP1A1, CYP2C10, and CYP3A4, a combination of 0.625% cholate and 0.62% Triton® N101 was found to yield the greatest level of soluble active holoprotein (36).

Although there is no overall scheme for the purification of recombinant cytochrome P-450 enzymes, two general approaches have been taken. First, the purification of the native protein has largely been carried out by a series of ion exchange chromatography steps on DEAE Sephacel™ (Amersham Pharmacia Biotech) and CM Sepharose (36). More recently, a number of cytochrome P-450 enzymes have been purified utilizing nickel chelate chromatography (18,69). In these proteins, a 6-histidine tail was engineered at the C terminus allowing binding to a nickel–nitriloacetic acid (Ni-NTA) agarose column. The protein can then be eluted with imidazole in a 1-step purification. The nickel–chelate affinity method has also been extensively utilized in the purification of the nitric oxide synthase enzymes (30,92,131).

2.2.3. Electron Transfer Proteins

The purification of the engineered soluble recombinant cytochrome b_5 (5) has been the basis for a number of subsequent purification schemes.

❖ Procedure 4. Purification of Cytochrome b_5

1. *E. coli* TB-1 cells were harvested and lysed in 50 mM Tris-HCl (pH 8.0) containing 0.1 mM DTT, 1 mM EDTA, RNase A (1 U/mL), DNase (16 U/mL), and lysozyme (3 mg/mL). Cell debris was removed by centrifugation, and the supernatant was applied to a DEAE Sephacel column (2.6 × 9.4 cm) equilibrated in 50 mM Tris-HCl (pH 8.0), 0.1 mM DTT, and 1 mM EDTA. The protein was eluted with a linear gradient.
2. The fractions containing cytochrome b_5 were pooled, concentrated by ultrafilitration, and applied to a Bio-Gel P-30 gel filtration column (1.9 × 73 cm; Bio-Rad) equilibrated in 50 mM Tris-HCl (pH 8.0). The fractions containing cytochrome b_5 were pooled and reapplied to a DEAE Sephacel column as described above. The resulting purified protein was then stored in liquid nitrogen.

Minor modifications of the above procedure have been utilized in the purification of the soluble microsomal cytochrome b_5, in which the second ion exchange chromatography step is replaced with a hydroxyapatite column (95). The purification of the outer mitochondrial membrane cytochrome b_5 (89) has also been reported, and the protocol is similar to that described previously (5).

The purification of *S. cerevisiae* iso-1-cytochrome c from *E. coli* cells has recently been reported (81). The preparation of the soluble fraction is essentially the same as previously described for the other soluble hemoproteins such as cytochrome b_5 and myoglobin. Purification of the protein was carried out by successive cation exchange chromatography steps on CM Sepharose CL6B and on Mono S HR 10/10 by FPLC.

2.2.4. Affinity Chromatography of Heme-Binding Proteins

In recent years, the advances in affinity chromatography have expanded dramatically with the development of molecular biological tools for the expression of fusion proteins, such as the polyhistidine-tag system (described above) and the cellulose binding domain system (Novagen), which allow 1-step affinity purification of the target protein. Historically however, affinity chromatography has utilized immobilized coenzymes such as nucleotides or flavins. One cofactor that has largely been overlooked in this aspect is heme, although hemin–agarose is commercially available (Sigma, St. Louis, MO, USA) (120). Hemin–agarose has been used to copurify hemopexin and HBP.93 from rabbit serum (121). However, one drawback in the use of hemin–agarose is the stringent conditions required for elution of the proteins. In the case of the hemopexin and HBP.93 purification, the protein was eluted in 0.1 M acetic acid (121). One unusual characteristic of the resin is the high affinity for heme itself, which eliminates the possibility of hemin being used to specifically elute the bound protein from the column. However, this property can be used to specifically remove heme or concentrate heme from a large sample volume such as deproteinized body fluids.

One area of research that hemin–agarose has been particularly useful is in the identification of heme-binding proteins from pathogenic bacteria. Pathogenic bacteria require iron for their survival, and this requirement is in part linked to their virulence (8). The major source of iron within the host is found in heme and hemoproteins, and many pathogens have developed sophisticated uptake systems for the acquisition of heme from their environment. A number of heme receptors and periplasmic transport proteins have been isolated utilizing hemin–agarose, including outer membrane lipoprotein heme receptors from *Haemophilus influenzae* (60) and *Haemophilus ducreyi* (22), as well as a secreted hemoglobin–protease in pathogenic *E.coli* strain EB1 (80).

3. DETECTION AND QUANTITATION OF HEMOPROTEINS

A unique and useful property of hemoproteins is the ability to determine the nature of the heme as well as quantitate both the heme and protein based on the absorption characteristics of the chromophore. In addition, the absorption properties provide valuable information as to the redox and spin state of individual hemoproteins in multicomponent systems. These techniques are invaluable in the study of complex systems where individual spectral characteristics can be accurately determined.

3.1. Detection of Heme in Microsomes and Whole Cells

A number of methods for heme detection and quantitation in tissue extracts have been described. All of the methods rely on monitoring the distinctive Soret absorbance either directly or following extraction of the chromophore.

3.1.1. Pyridine Hemochrome

Assignment of heme type in isolated hemoproteins as well as quantitation can be carried out using the pyridine hemochrome assay (24,25,114).

❖ Procedure 5. Pyridine Hemochrome Method

1. The isolated hemoprotein in a 4.0-mL sample volume is converted to a pyridine hemochrome by the addition of 0.5 mL of pyridine and 0.5 mL 0.5 N NaOH solution.
2. The reduced spectrum (+ dithionite) minus the oxidized spectrum (no dithionite) is calculated at 550 (protoheme) or 557 nm (heme c). The millimolar extinction coefficients ($\Delta\varepsilon_{ox\text{-}red}$) at each of the wavelengths are 30.0 and 19.1 at 557 and 550 nm, respectively (87).

The method has also been applied to more complicated samples such as mitochondria, yeast, and photosynthetic bacteria (1,4,54,56,87,101). In such cases, it was necessary to remove lipids and contaminating photosynthetic pigments by solvent extraction. Protoheme and heme c can then be separated by differential extraction in acidic organic solvent (25). However, the pretreatment of heme samples must be approached with some caution. Degradation of hematins in alkali solution has been observed, and the pyridine ferrohemochrome of protoheme is especially labile. Differential extraction of heme in HCl/acetone results in the partial extraction of heme c from tissue samples. Improved methods of detection and quantitation of heme have been developed (as described above). These methods are based on the redox difference absorbance measured at 2 wavelengths of the pyridine hemochrome. These methods can be utilized without prior separation of the hemes and organic solvent treatment (52). The method described below has been successfully used to quantitate the hemes in the photosynthetic bacteria *R. sphaeroides* (52).

❖ Procedure 6. Detection of Hemes in Photosynthetically Grown Cells of *R. sphaeroides*

1. Cells grown at 30°C are disrupted with a sonic oscillator.
2. Pyridine (0.5 mL) and 0.5 mL of 0.5 N NaOH solution are added to 4.0 mL of the disrupted cells.
3. The difference absorbance is measured between the oxidized minus reduced samples at 2 wavelength sets $\Delta(\Delta red - ox)_{555.5\text{-}535}$ and $\Delta(\Delta red - ox)_{549.5\text{-}535}$.
4. The amount of protoheme and heme c in the samples can then be calculated from the millimolar absorption coefficients 24.0 for $\Delta(\Delta red - ox)_{555.5\text{-}535}$ and 10.9 $\Delta(\Delta red - ox)_{549.5\text{-}535}$ for protoheme and 6.46 $\Delta(\Delta red - ox)_{555.5\text{-}535}$ and 22.0 $\Delta(\Delta red - ox)_{549.5\text{-}535}$ for heme c.

It is critical the measurements be carried out within a few hours of cell disruption, and that the samples are maintained at 4°C. The redox difference spectra should be measured immediately following the addition of dithionite, since the ferrohemochrome is labile.

3.1.2. Difference Spectrophotometric Analysis

The application of difference spectrophotometry to pigments, as developed by Chance and coworkers, has become a powerful method to study electron-transport reactions in turbid solutions and to monitor changes in oxidation state of membrane bound respiratory hemoproteins (10). In microsomes of liver, kidney, and the adrenal cortex, the main hemoprotein components are cytochromes b_5 and P-450, which can be measured by difference spectrophotometry. The cytochrome P-450s are best characterized by the absorbance band at 450 nm for the reduced carbon monoxide adduct. However, other hemoproteins, such as hemoglobin, also react with carbon monoxide and may interfere with accurate determination of cytochrome P-450 content. In tissue extracts or microsomes contaminated with other membrane fractions, this can be a significant problem, which requires modification of the procedure (23). The amount of cytochrome P-450 is calculated from the difference spectra of the reduced CO sample versus the reduced contents of the reference sample with any contribution from cytochrome b_5 being cancelled out. The change in absorbance between 450 nm relative to 490 nm can then be converted to a concentration based on the millimolar difference absorption coefficient ($\varepsilon_{mM} = 91$). In a separate sample at the same concentration (usually 1.5 mg/mL protein concentration), the reduced minus oxidized spectrum can be measured to determine the cytochrome b_5 content. By again applying the millimolar difference absorption coefficient of 21 for the absorbance change at 556 minus 575 nm, or of 185 from the change at 426 minus 409 nm, the cytochrome b_5 concentration can be calculated. One point of note is that NADH should be used in preference to NADPH for the reduction of cytochrome b_5, as NADPH contributes to endogenous substrate turnover by cytochrome P-450, and therefore cytochrome b_5 may only be reduced 75% to 85% in samples not completely purged of oxygen.

The identification and quantitation of cytochromes by low temperature difference spectral analysis in liver microsomes, yeast, and bacteria have previously been described (61,68,70,82). The major advantage of low temperature spectral analysis is the sharpening of bands which may otherwise be overlapping at room temperature.

In the yeast *S. cerevisiae*, the characteristic α bands in the 550 nm region of the reduced cytochromes b, b_1, c, and c_1 are ideal for measuring transitions in cells undergoing adaptation from an anaerobic to an aerobic environment. In *E. coli*, low temperature difference spectrum (α and β regions) analysis of the succinate-reduced minus oxidized cells has identified absorption peaks at 528 and 556 nm representing the β and α bands, respectively, of b and c type cytochromes. Also identified in the α region, although at much lower intensity, are bands due to cytochrome a_1 (590 nm) and cytochrome d (628 nm) (82). These bands revealed considerable heterogeneity and were further resolved by fourth-order finite difference analysis into components at 542, 545.5, 553.5, and 563 nm and peaks at 520.5, 528, 535, and 537 nm for the α and β bands, respectively.

3.1.3. Substrate Binding and Spin Transitions

While the pyridine hemochrome method and difference spectrophotometric analysis are effective in analyzing the composition and quantitative aspects of heme content in tissues, measurement of hemoprotein spin-transitions is somewhat more problematic. However, the measurement of spin-transitions in cytochrome P-450 systems is particularly relevant given the direct relationship of functional properties of the monoxygenase system with spin-state. A number of systems for cytochrome P-450 have been developed which rely on the difference in Soret absorption spectra between the high-spin (λ_{max} 388–396 nm) and low-spin (λ_{max} 416–418 nm) (34,51,88,112). However, microsomal cytochrome P-450s are problematic in that they undergo spontaneous inactivation to form the P-420 (inactive form) protein. The overlap of the 450 and 420 wavelengths, together with the turbidity of the sample and the overlap of the P-450 absorbance bands with other hemoproteins such as cytochrome b_5, makes direct difference spectral analysis difficult. A functional model for the analysis of microsomal mono-oxygenase systems has been developed based on principal component analysis (PCA) or singular value decomposition (SVD) analysis (45). This method is widely used to interpret fluorescence and absorbance spectral changes in complex biochemical systems (37,45, 85,102). By deducing a set of standard absorbance spectra for the high-spin, low-spin, and P420 states of pure ferric microsomal cytochrome P-450, and using PCA analysis, the thermodynamic parameters of substrate binding and spin transitions in yeast microsomes containing human cytochrome P-450 3A4 have been evaluated (86).

3.1.4. Multicomponent Spectral Analysis of Hemoproteins in Biological Materials

Spectral scans across a range of visible wavelengths (390–640 nm) contain a great deal of information about the oxidation state of individual hemoproteins. Information can be obtained for tissue homogenates or subcellular fractions through comparison to reference spectra by multicomponent analysis. A hemoprotein spectral analysis program (HSAP), a spreadsheet program written with Microsoft® Excel® 4.0, has been described (7). The program measures absorbances at many wavelengths in the range of 500 to 640 nm and obtains molar ratio values for the individual hemoproteins. To correct for light-scattering errors in turbid samples, a turbidity calculation has been introduced to the HSAP (13). This program is effective in measuring reduced hemoproteins, while the unambiguous identification of oxidized hemoproteins was not so straightforward. The assignment of oxidized hemoproteins was accounted for by the addition of spectral data in the strongly absorbing Soret region, where the peak maxima are sufficiently different to allow discrimination between the reduced and oxidized proteins. Subsequent development of a difference spectral analysis program (DSAP) has further aided the separation and identification of individual cytochromes of organelles such as mitochondria (74). The combination of HSAP and DSAP provide a method for interpreting complex spectral information of biological samples and may prove useful with other measurements of tissue oxidation in assessing clinical problems associated with free radical reactions.

3.1.5. Detection of Hemoprotein Peroxidase Activity on SDS-PAGE

Electrophoresis of proteins is commonly used in determination of protein purity, molecular weight, and in the case of isoelectrofocusing (IEF), the pI of a given protein. This technique can be useful in identifying hemoproteins of unknown molecular weight in an impure protein fraction or isolate by taking advantage of the heme-associated peroxidase activity as a visualization tool (96,116). In this procedure, a substrate specific for heme, 3,3′,5,5′-tetramethylbenzidine (TMBZ) is incubated with the SDS-PAGE gel and visualized by the addition of hydrogen peroxide (H_2O_2). One major disadvantage of the use of this staining technique in SDS-PAGE analysis is that the denaturing conditions often lead to loss of noncovalently bound heme from the protein during the electrophoresis run. This problem has been circumvented with the use of lithium dodecyl sulfate (LDS) in place of the SDS. The gels are run as previously described (59), except SDS is replaced by LDS in all buffers. The sample buffer is prepared with a lower concentration of DTT (5 mM), as this can interfere with the TMBZ staining of the gel following electrophoresis. The samples are immediately loaded onto the gel following addition of LDS-PAGE sample buffer and are not boiled so as to maintain the heme-associated peroxidase activity. The gels (12 cm in length and 1.5 mm thickness) are run overnight at 4°C with a constant current (8 mA/gel) (21).

The following procedure has been modified slightly by Dutta and Henry (21) for use in electroblotting experiments on polyvinylene difluoride (PVDF) membrane. Following transfer of the protein to PVDF membranes, the gels are incubated in the TMBZ solution for 3 hours at 4°C (in the dark). H_2O_2 is added to a final concentration of 30 mM, and the staining is visible after 1 hour (storage in the staining solution for periods greater than 3 hours results in irreversible background staining). The membranes can be quickly destained by washing twice in 100% methanol, followed by destaining in 70 mM sodium sulfite for 5 minutes. The membranes can then be stained with Coomassie® Brilliant Blue R-250, 50% methanol, and 10% acetic acid and destained in 90% methanol.

❖ Procedure 7. TMBZ Staining of Polyacrylamide Gels

1. A solution of 6.3 mM TMBZ in methanol is prepared freshly prior to use. Immediately before use, 3 volumes of the TMBZ methanol solution is mixed with 7 parts of 0.25 M sodium acetate, pH 5.0. The gels are then immersed in this solution for a period of 1 to 2 hours (in the dark) and mixed every 15 minutes.
2. Following incubation, the gels are stained with the addition of H_2O_2 to a final concentration of 30 mM. The staining should be visible within 3 to 5 minutes with increasing intensity over a 30-minute period.
3. The gels are then placed in isopropanol:0.25 M sodium acetate (pH 5.0) (3:7). The solution is changed 3 times to remove any precipitated TMBZ and clear the gel background.

4. The gels can be scanned at 690 nm or photographed at this time. Stained gels can be stored for a period of 2 months in the dark at room temperature.
5. The gels can be destained in 70 mM sodium sulfite and washed (3 × 20 min) with 30% isopropanol and stained in Coomassie brilliant blue R-250 (21).

3.2. Detection and Quantitation of Heme-Containing Proteins by Chemiluminescence

The techniques described above for the detection of peroxidase activity in hemoproteins, while useful, has a number of drawbacks including prolonged incubation periods, sensitivity to light, and carcinogenicity. The benzidine staining procedure is also a useful marker for the presence of blood in clinical and forensic samples. To this end, a number of commercially available products have been developed (9,19) which provide both qualitative and semiquantitative evidence of blood in clinical samples and do not utilize the use of TMBZ.

The development of luminol-based chemiluminescence assays for detecting HRP-labeled antibody have been reported to have increased sensitivity to the standard chromogenic techniques (9). With the increased sensitivity of chemiluminescence, the ability to detect peroxidase activity in other hemoproteins has been analyzed. Comparison of the luminescence assay with the standard chromogenic techniques on SDS-PAGE and electroblot for several hemoproteins indicated increased sensitivity for qualitative detection of hemoproteins (19). Utilizing a liquid scintillation procedure, linear and quantitative detection of peroxidase activity could be measured for biological samples containing heme in the pmol range (see methods below).

❖ Procedure 8. Detection of Heme and Hemoproteins by Chemiluminescence

1. The detection of chemiluminescence is based on the peroxidase-dependent breakdown of luminol, and the reactions were carried out by modification of a standard procedure from the ECL® blotting system commercially available from Amersham Pharmacia Biotech.
2. Gel electrophoresis: SDS-PAGE gels, run with or without reducing agents, electroblots, or dotblots, are washed in Dulbecco's phosphate-buffered saline (dPBS) and incubated for 1 minute in a 1:1 mixture of CL reagents 1 and 2 (ECL Blotting Kit). The gels or blots are then placed between transparent acetate sheets and exposed to X-Omat™ film (Eastman Kodak, Rochester, NY, USA) for a period of time (10 s to 18 h).
3. Liquid scintillation methods: Measurement of chemiluminescence was carried out by mixing 10 μL of each reagent (CL1 and CL2) in a microfuge tube followed by 1 μL of sample. The solution was mixed and immediately placed in a plastic scintillation vial, and the emission of light was measured for 60 seconds. The samples were mixed and placed in the instrument individually to keep constant the time between mixing and scintillation spectroscopy. The values obtained were in counts per minute (cpm).

The chemiluminescence assay with HRP and cytochrome c, when compared to the heme staining procedure, shows increased sensitivity. While both methods could detect protein on SDS-PAGE and electroblot to 10 and 100 ng, respectively, the chemiluminescence assay can detect 1 ng of HRP and 10 ng of cytochrome c when exposed for 18 hours (19).

Chemiluminescence methods developed for scintillation detection assays has been utilized for the detection of blood in urine (hematuria). The detection of blood in urine showed a linear relationship over a wide range of concentrations with the detection level as low as 0.1 to 1 pg/μL (19).

The use of chemiluminescence in the detection of hemoproteins has been modified to specifically detect oxidative metabolites of hemoglobin and myoglobin (125). Hemoglobin is a major target for reactive metabolites of drugs as well as environmental toxins, and the relatively long half-life of the red blood cell has led to its use as a marker of exposure to such toxic metabolites (79). Hydrogen peroxide at low concentrations reacts with the heme group of myoglobin and hemoglobin to form a covalent heme–protein adduct. The modified hemoprotein is then capable of generating hydrogen peroxide in an essentially autocatalytic process. Thus, the covalent modification of the protein is an important marker for oxidative damage and, more specifically, in the case of myoglobin, a sensitive marker for ischemic reperfusion injury.

Utilizing the chemiluminescence procedure outlined above, Vuletich and Osawa have modified the reaction conditions to specifically select for protein bound heme adducts on SDS-PAGE gels (125). The use of DTT is largely excluded as a reducing agent as it degrades the heme and, hence, reduces the peroxidase activity of the heme adduct. Replacement of DTT with tris(2-carboxyethyl) phosphine (TCEP), which does not degrade the heme, allows detection of the peroxidase activity of the heme adduct with no contamination from noncovalent heme-containing proteins, as the noncovalent heme is lost on denaturation and electrophoresis.

4. HEMOPROTEIN ACTIVE-SITE PROBES

The lack of any available crystal structure for the membrane-bound P-450 isozymes has led to the use of indirect methods for determining active-site structures, such as homology modeling (69,83), mechanism-based inactivation (57,58), and photoaffinity labeling (67,75,134). Although such probes have not yet been routinely used with hemoproteins, the future development of photoaffinity probes with such a broad specificity opens up the possibility of determining topological information on the active sites of hemoproteins whose structures are unknown.

4.1. Photoaffinity Labeling with Synthesizing Trifluoromethyl-diazirinylphenyl Diazenes

Recently, Tschirret-Guth and Ortiz de Montellano (119) have advanced the active-site labeling methodology developed in their laboratory. In previous studies, it was noted that reaction of arylhydrazines and aryldiazenes with hemoproteins gave an iron–aryl complex (76). Early work utilized the fact that, upon oxidation, the aryl complex shifted to form a mixture of *N*-arylprotoporphyrin-IX regioisomers. The ratio of the 4 regioisomers represented a low-resolution topology of the active-site, but gave no information on the specific amino acid residues within the active-site. The recent synthesis of azidoaryldiazenes as photoaffinity probes for the active-site of hemoproteins has addressed this question (119). The formation of the iron–aryl complex with the heme and subsequent formation of the activated nitrene by oxidation limits the interaction to amino acid residues within the active site of the hemoprotein (Figure 1a). The reaction of azidophenyldiazene with myoglobin was specific for the distal histidine (His-64), suggesting the probe preferentially selected for nucleophilic residues (117). To develop photoaffinity probes that were capable of labeling multiple residues, the authors replaced the azido function with the more reactive trifluoromethyl group by synthesizing 3- and 4-(trifluoromethyldiazirinyl)phenyl diazene (118) (Figure 1b). Irradiation of the sample following formation of the iron–diazirinylphenyl complex at 350 nm generates a highly reactive carbene capable of inserting into unactivated C-H bonds. The reaction of sperm whale myoglobin with the activated carbene and subsequent analysis of the tryptic peptides by mass spectrophotometric techniques identified at least 4 residues within the heme-binding site of the protein, Leu-29, Val-68, His-64, and Ile-107. With the exception of His-64, which was previously identified in the earlier azidophenyldiazene work (117), the 3 remaining residues were nonpolar and less than 3 to 4 Å from the heme.

5. SPECTROPHOTOMETRIC ANALYSIS OF HEMOPROTEINS

The unique chromophore of hemoproteins allows the use of a number of spectrophotometric techniques to yield detailed information on the nature of the heme ligand. Although individual spectroscopy techniques can yield valuable information, it is the combined use of techniques such as UV/visible, magnetic circular dichroism (MCD), resonance Raman, and NMR that can provide powerful insight as to the nature of the coordination, ligation, and oxidation state of the heme.

It will not be the focus of this review to give a detailed account of each technique, but present a brief synopsis of the information that can be gained as it relates to hemoprotein structure and function. On studies of the human heme oxygenase (HO-1), we have previously used a combination of spectrophotometric techniques such as UV/visible, resonance Raman, MCD, and NMR spectrophotometry to determine the ligation, coordination, and electronic nature of the heme. In addition to providing valuable information on the ligation and coordination state of the heme, these techniques have aided greatly in elucidating and confirming the mechanism of action of the enzyme.

HO-1 catalyzes the rate limiting step in the conversion of heme to biliverdin (Figure 2), and although not a hemoprotein on reconstitution with heme, the enzyme forms a 1:1 ferric heme:HO-1 (heme:HO-1) complex with electronic spectra similar to that of myoglobin (128). At pH 6.0, the heme:HO-1 complex has a Soret at 404 nm and a charge transfer band at 632 nm, characteristic of a high spin species (Figure 3b). On increasing the pH from 6.0 to 10.0, the heme Soret shifts from 404 to 409 nm, and the charge transfer band at 630 nm decreases with the appearance of α/β bands at 539 and 574 nm, a process attributed to a pH-dependent high-spin to low-spin conversion of a water ligand to a hydroxide (Figure 3b) (40,111). The UV/visible spectra of the reduced complex in the presence of CO has a Soret at 420 nm and α/β bands at 567 and 537 nm, consistent with a histidine proximal ligand.

Although the UV/visible spectrum can provide some limited information on the proximal ligand of a hemoprotein of unknown structure, it is not conclusive. MCD is a difference technique that will have both positive and negative components yielding a more detailed spectrum than the corresponding UV/visible spectrum. The increased sensitivity of MCD has been shown to be an effective "fingerprinting tool" for the determination of the ligand coordination, oxidation state, and spin state of hemoproteins (15,123). Assignment of the proximal ligand in a hemoprotein of unknown structure by MCD, like UV/visible spectroscopy and resonance Raman (see below), is dependent on comparison to a protein whose proximal ligand is known, such as myoglobin (histidine), catalase (tyrosine), and cytochrome P-450 (cysteine). The sample to be analyzed is prepared with a known sixth ligand such as cyanide, which together with the 4 pyrroles can then be compared to the ferric cyano complex of hemoproteins, whose proximal ligands are known. The electronic absorption and MCD spectra of the cyanoferric heme:HO-1 complex is very similar to that of cyanoferric myoglobin, thus confirming the proximal histidine ligation in the heme:HO-1 complex (Figure 3a) (40). The derivative bands of the heme:HO-1 complex are blue-shifted when compared to myoglobin, but are still very similar in structure (Figure 3b) (40).

Raman spectroscopy is a vibrational spectroscopy technique that can detect transitions between different ligation states of porphyrins as the spin state is changed (107,109). The change in spin state alters the size of the iron and its displacement

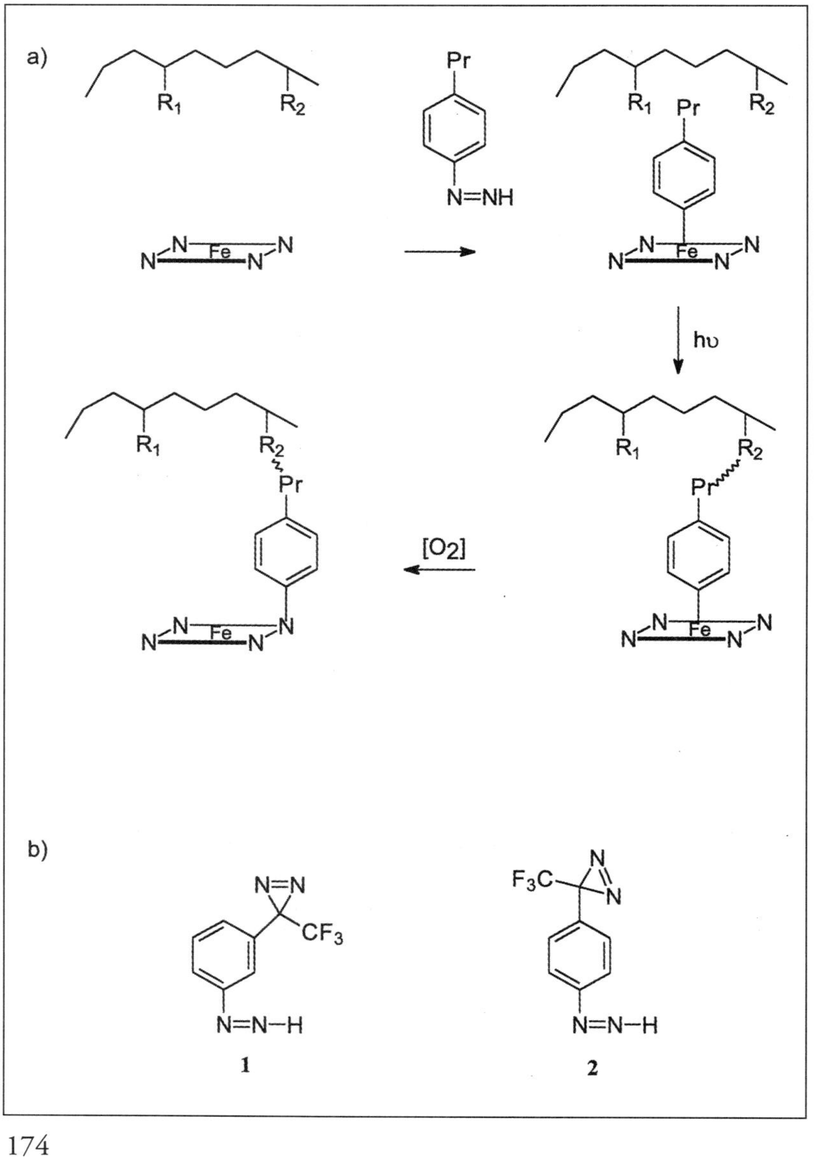

Figure 1. (a) Labeling of active-site residues of hemoproteins with photoactivable probes. (b) Structure of the photoactivatable probes of trifluoromethyldiazirinylphenyldiazene as synthesized by Tschirret-Guth et al. (118).

from the heme plane which directly effects the vibrational frequencies of the porphyrin ring bond stretches. This technique can be used to characterize the spin state and coordination state in both the ferric and ferrous oxidation states of a given hemoprotein (108). As in the MCD experiments, the resonance Raman spectra of the heme:HO-1 complex show a marked change in the heme axial coordination and spin state on changing pH (111). At pH 6.0, the spectrum shows the heme to be primarily 6-coordinate high-spin (6CHS) similar to myoglobin, in which the heme is coordinated through a proximal histidine with water bound as the sixth ligand. Increasing the pH to 8.0 significantly alters the Raman spectrum, with the bands at 1483 (υ_3) and 1565 cm^{-1} (υ_2) of the ferric 6CHS species decreasing, and the bands corresponding to the ferric 6-coordinate low-spin species (6CLS) at 1503 (υ_3), 1582 (υ_2), and 1638 (υ_{10}) increasing (Figure 4). This again supports the ionization of a coordinating water molecule to a hydroxide above pH 8.0.

Resonance Raman studies of the heme: HO-1 complex provided direct confirmation of histidine as the proximal ligand (111). The iron–imidazole or iron–thiolate ligand is not generally detectable due to the weak coupling of the π electronic system. However, in the case of the iron–imidazole (Fe-His) ligation, the reduced ferrous 5-coordinate species shows significant enhancement due to the iron being out of the plane of the porphyrin. This transition provides a characteristic marker for the Fe-His stretching mode in the low frequency ferrous spectra. Interestingly, the position

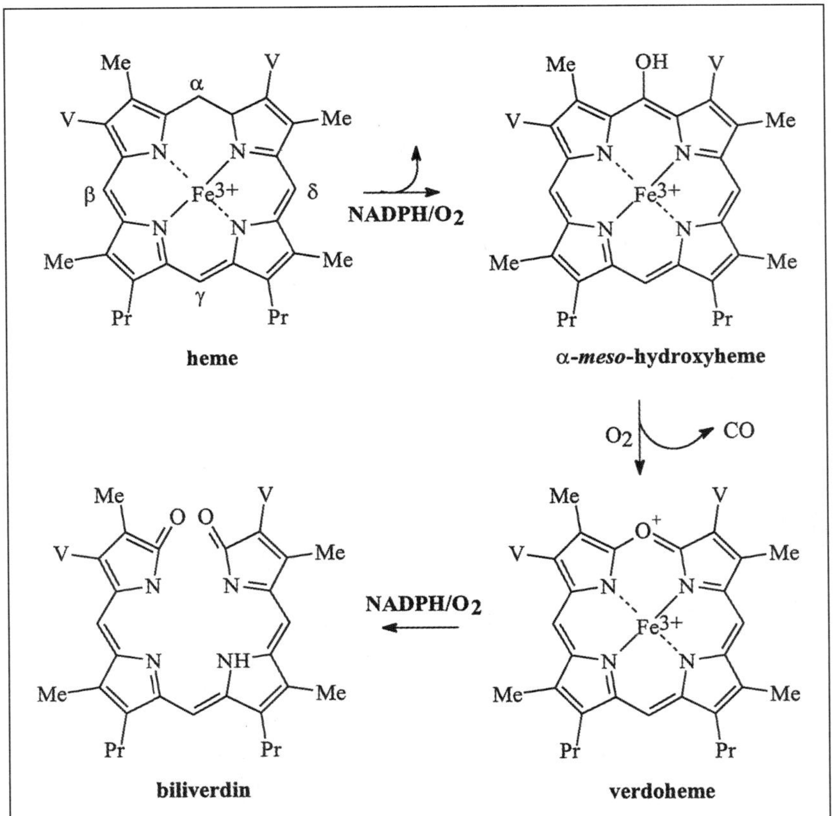

Figure 2. Conversion of heme to biliverdin by HO-1. The abbreviations are as follows; Me, methyl; V, vinyl; Pr, propionic acid.

of the band is also a strong indication of the ionization state of the proximal histidine. The band at 216 cm^{-1} in the heme:HO-1 complex (111) is close to that of myoglobin (221 cm^{-1}) (53,55), where the proximal histidine is only weakly hydrogen-bonded (Figure 5). The presence of strong hydrogen bonding to the proximal histidine in cytochrome c peroxidase shifts the iron–imidazole stretch to 233 cm^{-1}, while complete ionization gives a value closer to 246 cm^{-1} (105,115). In cytochrome c peroxidase, it has been shown that removal of the hydrogen bonding network by mutation of Asp-235 to an Ala shifts the Fe-His stretching mode from 246 to 206 cm^{-1}. The relatively low value for the heme:HO-1 complex can therefore be interpreted as a proximal histidine that is weakly or not hydrogen-bonded (106).

It is interesting that the structure of the heme:HO-1 complex resembles the oxygen-binding proteins more than the oxygen-activating peroxidases. The presence in the heme:HO-1 complex of a histidine proximal ligand that is neither ionized nor hydrogen-bonded has strong implications for the mechanism of action of HO-1. The ability of the proximal ligand to destabilize the dioxygen bond and assist in cleavage to form the activated ferryl species is largely dependent on the electron density present on the proximal ligand. As the electron density increases by partial or complete deprotonation of the proximal histidine, the ability to carry out dioxygen bond cleavage increases.

In previous studies, we have generated a ferryl species analogous to that of com-

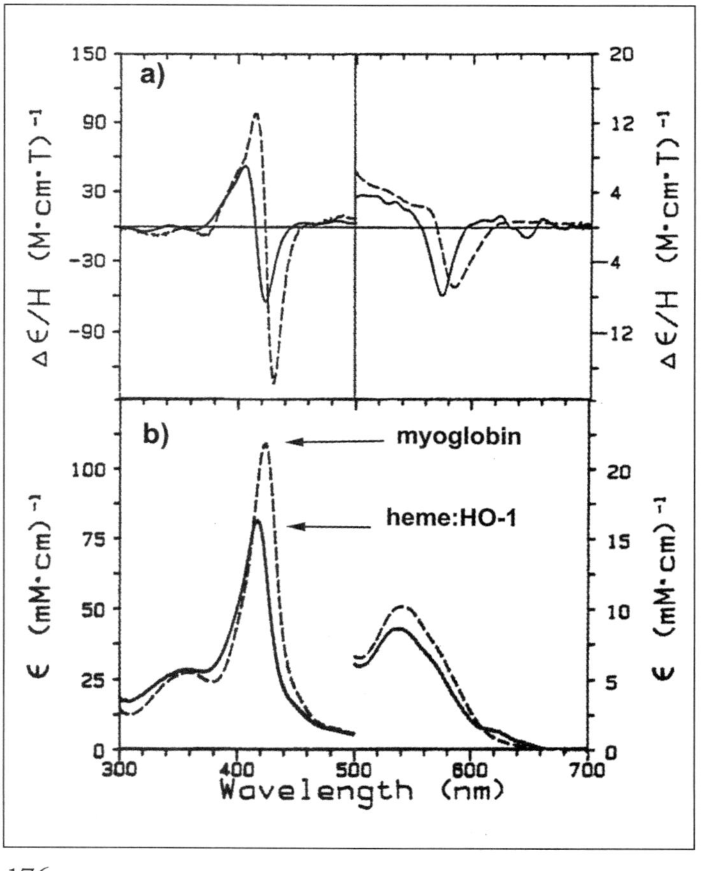

Figure 3. (a) MCD and (b) electronic absorption spectra of the cyanoferric heme:HO-1 complex and of the cyanoferric sperm whale myoglobin. The MCD spectra of the human HO-1 and sperm whale myoglobin have previously been reported in References 40 and 16, respectively.

pound II in peroxidase and shown that this species is not an intermediate in the normal physiological reaction of heme oxygenase (128). Subsequent studies with ethylhydroperoxide showed that the reaction proceeded by the normal pathway with the formation of α-meso-ethoxyheme, analogous to the formation of α-meso-hydroxyheme in the enzymatic conversion of heme to biliverdin (130). The ethylhydroperoxide reaction, therefore, favors an electrophilic mechanism with the addition of the terminal ethoxy or hydroxy (Fe-O-OH) moiety to the heme (Figure 6). The addition of the terminal oxygen of the ferric peroxo anion ($Fe\text{-}O\text{-}O^-$) to the α-meso-edge of the heme is ruled out as the terminal oxygen of the ethylhydroperoxide species is blocked. The proposed electrophilic mechanism is consistent with the presence of a proximal ligand that is neither ionized or hydrogen-bonded.

Isotopic labeling and 2-dimensional NMR studies on the rat HO-1 have revealed an unusual heme electronic structure (43). The proton contact shift patterns of the heme reveal large differences in the delocalized spin density in which the spin density is highest at positions 3 (methyl) and 2 (vinyl) and much smaller at positions 1 (methyl) and 4 (vinyl) (Figure 7). The asymmetric spin density within a pyrrole is unprecedented in any low-spin ferric hemoprotein, where it is more common to find asymmetry between different pyrroles. Based on studies with model hemes, the distribution of the spin density suggests a direct electronic perturbation of the heme

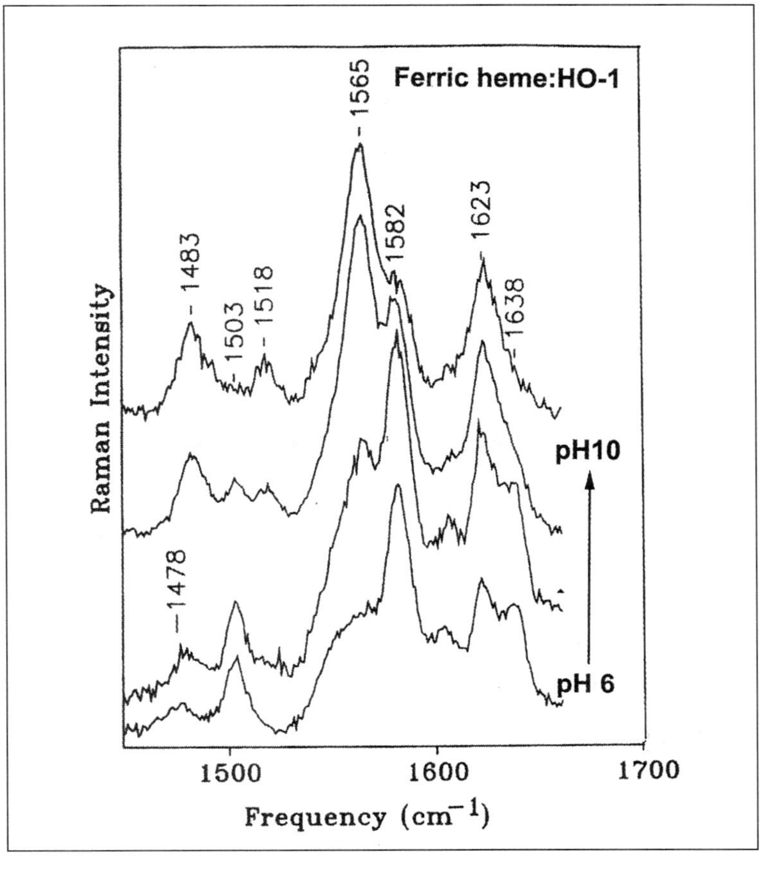

Figure 4. Resonance Raman spectra of the heme:HO-1 complex in the spin and coordination state frequency region as a function of pH (413.1 nm excitation). The resonance Raman data for the heme:HO-1 complex has previously been reported (111).

by the protein environment such that there is an increase in electron density at the α-meso-edge of the heme. The increased electron density at the α-meso-edge would facilitate the electrophilic addition of the protonated terminal oxygen to the α-meso-position (Figure 6).

The extensive characterization of the heme:HO-1 complex by spectroscopic techniques such as MCD, resonance Raman, and NMR have greatly aided the biochemical and enzymatic studies in elucidating the mechanism of action of HO-1. It is also gratifying that the recent 3-dimensional structure of the heme:HO-1 complex has in large part confirmed many aspects of the previous spectroscopic data (97).

6. SUMMARY

The area of hemoprotein research has advanced dramatically in recent years with the advent of molecular cloning techniques. The availability of high level expression systems for hemoproteins, such as the cytochrome P-450 enzymes, and the ability to generate site-directed mutants has advanced our knowledge of both mechanism and structure. The rapidly expanding role of hemoproteins in cell signaling processes typified by the eukaryotic nitric oxide synthases and soluble guanylate cyclase, and in gene regulation with transcription factors such as CooA, has expanded the repertoire of hemoprotein motifs. It

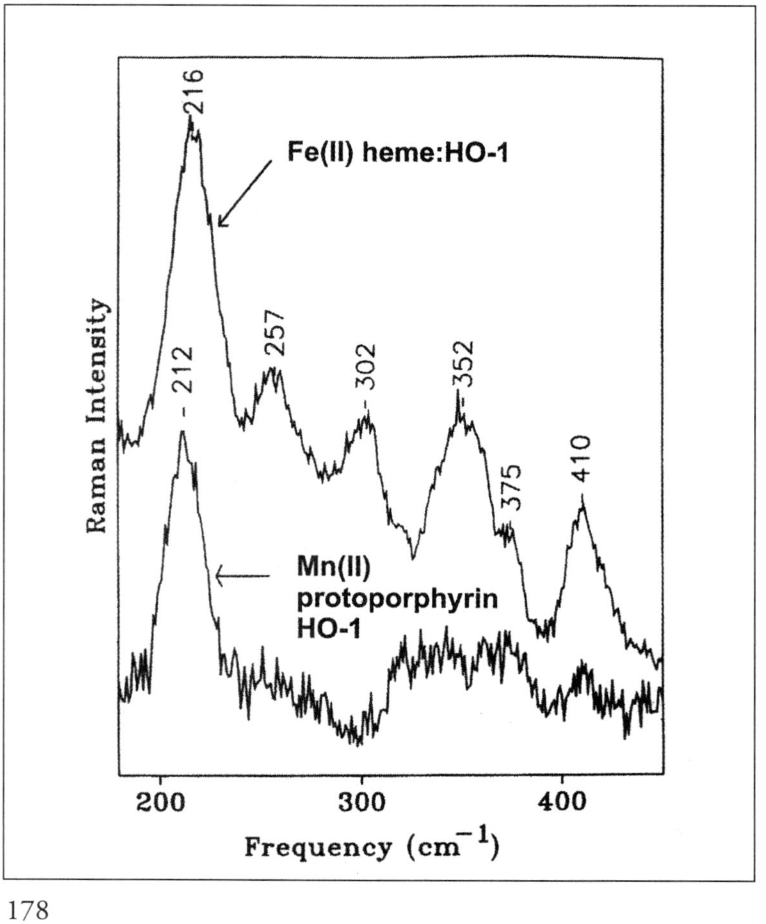

Figure 5. Resonance Raman spectra of the ferrous heme:HO-1 complex and the manganese (II) protoporphyrin:HO-1 complex in the low frequency region (441.6 nm excitation). The resonance Raman data for the heme:HO-1 complex has previously been reported (111).

is with great anticipation that we look forward to the discovery of new and unique hemoproteins and the critical role they may play in molecular and cellular processes.

ABBREVIATIONS

DTT, dithiothreitol; EDTA, ethylene diamine tetra acetic acid; HO-1, human heme oxygenase isozyme 1; HRP, horseradish peroxidase; HSAP, hemoprotein spectral analysis program; IEF, isoelectric focusing; H_2O_2, hydrogen peroxide; LDS, lithium dodecyl sulfate; MCD, magnetic circular dichroism; NMR, nuclear magnetic resonance; PAGE, polyacrylamide gel electrophoresis; PCA, principal component analysis; PMSF, phenylmethylsulfonyl fluoride; PVDF, polyvinylene difluoride membrane; SDS, sodium dodecyl sulfate; SVD; singular value decomposition analysis; TCEP, tris(2-carboxyethyl) phosphine; TMBZ, 3,3′,5,5′-tetramethylbenzidine.

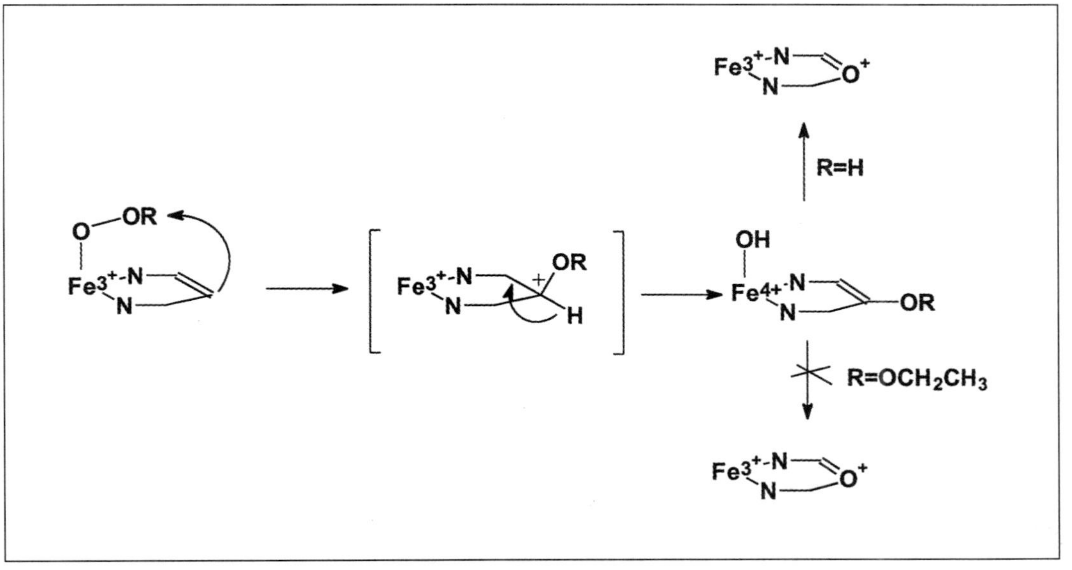

Figure 6. Electrophilic oxidation of the porphyrin macrocycle by the ferric peroxide complexes.

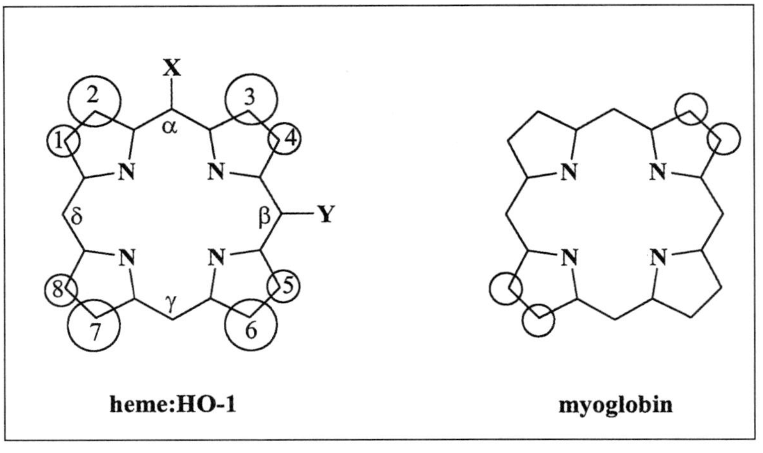

Figure 7. Schematic representation of the unpaired spin distribution about the heme periphery in HO-1 and myoglobin. The size of the circle represents the population of the spin distribution. Adapted from Reference 33.

ACKNOWLEDGMENTS

I would like to thank Paul Ortiz de Montellano for his continuous support and encouragement over the years. I also thank the National Institutes of Health for their financial support.

REFERENCES

1. **Agalidis, I., E. Jauneau, and F. Reiss-Husson.** 1974. Influence of iron deficiency on the cytochromes of *Rhodopseudomonas spheroides*. Eur. J. Biochem. *47*:573-580.
2. **Aono, S., H. Nakajima, K. Saito, and M. Okada.** 1996. A novel heme protein that acts as a carbon monoxide-dependent transcriptional activator in *Rhodospirillum rubrum*. Biochem. Biophys. Res. Commun. *228*:752-756.
3. **Barnes, H.J., M.P. Arlotto, and M.R. Waterman.** 1991. Expression and enzymatic activity of recombinant cytochrome P450 17 alpha-hydroxylase in *Escherichia coli*. Proc. Natl. Acad. Sci. USA *88*:5597-5601.
4. **Bartsch, R.G.** 1971. Cytochromes: bacterial. Methods Enzymol. *23*:344-363.
5. **Beck von Bodman, S., M.A. Schuler, D.R. Jollie, and S.G. Sligar.** 1986. Synthesis, bacterial expression, and mutagenesis of the gene coding for mammalian cytochrome b5. Proc. Natl. Acad. Sci. USA *83*:9443-9447.
6. **Black, S.M. and P.R. Ortiz de Montellano.** 1995. Characterization of rat neuronal nitric oxide synthase expressed in *Saccharomyces cerevisiae*. DNA Cell Biol. *14*:789-794.
7. **Boyle, R.C., A.L. Tappel, A.A. Tappel, H. Chen, and H.J. Andersen.** 1994. Quantitation of heme proteins from mixtures. J. Agric. Food Chem. *42*:100-104.
8. **Braun, V., K. Hantke, and W. Koster.** 1998. Bacterial iron transport: mechanisms, genetics, and regulation. Met. Ions Biol. Syst. *35*:67-145.
9. **Champiat, D., A. Roux, O. Lhomme, and G. Nosenzo.** 1994. Biochemiluminescence and biomedical applications. Cell Biol. Toxicol. *10*:345-351.
10. **Chance, B.** 1954. Spectrophotometry of intracellular respiratory pigments. Science *120*:767.
11. **Charles, I.G., A. Chubb, R. Gill, J. Clare, P.N. Lowe, L.S. Holmes, M. Page, J.G. Keeling, S. Moncada, and V. Riveros-Moreno.** 1993. Cloning and expression of a rat neuronal nitric oxide synthase coding sequence in a baculovirus/insect cell system. Biochem. Biophys. Res. Commun. *196*:1481-1489.
12. **Charles, I.G., C.A. Scorer, M.A. Moro, C. Fernandez, A. Chubb, J. Dawson, N. Foxwell, R.G. Knowles, and S.A. Baylis.** 1996. Expression of human nitric oxide synthase isozymes. Methods Enzymol. *268*:449-460.
13. **Chen, H., A.L. Tappel, and R.C. Boyle.** 1993. Oxidation of heme proteins as a measure of oxidative damage to liver tissue slices. Free Radic. Biol. Med. *14*:509-517.
14. **Cornejo, J., R.D. Willows, and S.I. Beale.** 1998. Phytobilin biosynthesis: cloning and expression of a gene encoding soluble ferredoxin-dependent heme oxygenase from *Synechocystis* sp. PCC 6803. Plant J. *15*:99-107.
15. **Dawson, J.H. and D.M. Dooley.** 1989. Iron Porphyrins. Part 3, p. 1-135. *In* A.P.B. Lever and H.B. Gray (Eds.). VCH publishers, New York.
16. **Dawson, J.H., S. Kadkhodayan, C. Zhuang, and M. Sono.** 1992. On the use of iron octa-alkylporphyrins as models for protoporphyrin IX-containing heme systems in studies employing magnetic circular dichroism spectroscopy. J. Inorg. Biochem. *45*:179-192.
17. **DePillis, G.D., S. Ozaki, J.M. Kuo, D.A. Maltby, and P.R. Ortiz de Montellano.** 1997. Autocatalytic processing of heme by lactoperoxidase produces the native protein-bound prosthetic group. J. Biol. Chem. *272*:8857-8860.
18. **Dierks, E.A., Z. Zhang, E.F. Johnson, and P.R. de Montellano.** 1998. The catalytic site of cytochrome P4504A11 (CYP4A11) and its L131F mutant. J. Biol. Chem. *273*:23055-23061.
19. **Dorward, D.W.** 1993. Detection and quantitation of heme-containing proteins by chemiluminescence. Anal. Biochem. *209*:219-223.
20. **Durante, W., M.H. Kroll, N. Christodoulides, K.J. Peyton, and A.I. Schafer.** 1997. Nitric oxide induces heme oxygenase-1 gene expression and carbon monoxide production in vascular smooth muscle cells. Circ. Res. *80*:557-564.
21. **Dutta, C. and H.L. Henry.** 1990. Detection of hemoprotein peroxidase activity on polyvinylidene difluoride membrane. Anal. Biochem. *184*:96-99.
22. **Elkins, C., C.J. Chen, and C.E. Thomas.** 1995. Characterization of the hgbA locus encoding a hemoglobin receptor from *Haemophilus ducreyi*. Infect. Immun. *63*:2194-2200.
23. **Estabrook, R.W., J. Peterson, J. Barn, and A. Hildebrant.** 1972. The spectrophotometric measurement of turbid suspensions of cytochromes associated with drug metabolism, p. 303-350. *In* C.F. Chignell (Ed.), Methods in Pharmacology. Vol. 2. Appleton-Century-Crofts, New York.
24. **Falk, J.E.** 1963. Part A. Pyrrole pigments: chemistry and biochemistry of porphyrins and metalloporphyrins, p. 3-33. *In* M. Florkin and E.H. Stotz (Eds.), Comprehensive Biochemistry, Vol. 9. Elsevier, Amsterdam.
25. **Falk, J.E.** 1964. Porphyrins and Metalloporphyrins. Elsevier, New York.
26. **Fishel, L.A., J.E. Villafranca, J.M. Mauro, and J. Kraut.** 1987. Yeast cytochrome c peroxidase: mutagenesis and expression in *Escherichia coli* show tryptophan-51 is not the radical site in compound I. Biochemistry *26*:351-360.
27. **Fronticelli, C., J.K. O'Donnell, and W.S. Brinigar.** 1991. Recombinant human hemoglobin: expression and refolding of beta-globin from *Escherichia coli*. J. Prot. Chem. *10*:495-501.
28. **Gegner, J.A. and F.W. Dahlquist.** 1991. Signal transduction in bacteria: CheW forms a reversible complex with the protein kinase CheA. Proc. Natl. Acad Sci. USA *88*:750-754.

29. **George, H.J., D.E. Van Dyk, R.A. Straney, J.M. Trzaskos, and R.A. Copeland.** 1996. Expression purification and characterization of recombinant human inducible prostaglandin G/H synthase from baculovirus-infected insect cells. Protein Expr. Purif. *7*:19-26.
30. **Gerber, N.C., C.R. Nishida, and P.R. Ortiz de Montellano.** 1997. Characterization of human liver inducible nitric oxide synthase expressed in *Escherichia coli*. Arch. Biochem. Biophys. *343*:249-253.
31. **Gerber, N.C. and P.R. Ortiz de Montellano.** 1995. Neuronal nitric oxide synthase. Expression in *Escherichia coli*, irreversible inhibition by phenyldiazene, and active site topology. J. Biol. Chem. *270*:17791-17796.
32. **Gilles-Gonzalez, M.A., G. Gonzalez, M.F. Perutz, L. Kiger, M.C. Marden, and C. Poyart.** 1994. Heme-based sensors, exemplified by the kinase FixL, are a new class of heme protein with distinctive ligand binding and autoxidation. Biochemistry *33*:8067-8073.
33. **Gonzalez, F.J. and K.R. Korzekwa.** 1995. Cytochromes P450 expression systems. Annu. Rev. Pharmacol. Toxicol. *35*:369-390.
34. **Guengerich, F.P.** 1983. Oxidation-reduction properties of rat liver cytochromes P-450 and NADPH-cytochrome p-450 reductase related to catalysis in reconstituted systems. Biochemistry *22*:2811-2820.
35. **Guengerich, F.P., N.A. Hosea, A. Parikh, L.C. Bell-Parikh, W.W. Johnson, E.M. Gillam, and T. Shimada.** 1998. Twenty years of biochemistry of human P450s: purification, expression, mechanism, and relevance to drugs. Drug Metab. Dispos. *26*:1175-1178.
36. **Guengerich, F.P., M.V. Martin, Z. Guo, and Y.J. Chun.** 1996. Purification of functional recombinant P450s from bacteria. Methods Enzymol. *272*:35-44.
37. **Halaka, F.G., G.T. Babcock, and J.L. Dye.** 1985. The use of principal component analysis to resolve the spectra and kinetics of cytochrome c oxidase reduction by 5, 10-dihydro-5-methyl phenazine. Biophys. J. *48*:209-219.
38. **Halpert, J.R., T.L. Domanski, O. Adali, C.P. Biagini, J. Cosme, E.A. Dierks, E.F. Johnson, J.P. Jones et al.** 1998. Structure-function of cytochromes P450 and flavin-containing monooxygenases: implications for drug metabolism. Drug Metab. Dispos. *26*:1223-1231.
39. **Hartmann, C. and P.R. Ortiz de Montellano.** 1992. Baculovirus expression and characterization of catalytically active horseradish peroxidase. Arch. Biochem. Biophys. *297*:61-72.
40. **Hawkins, B.K., A. Wilks, L.S. Powers, P.R. Ortiz de Montellano, and J.H. Dawson.** 1996. Ligation of the iron in the heme-heme oxygenase complex: X-ray absorption, electronic absorption and magnetic circular dichroism studies. Biochim. Biophys. Acta *1295*:165-173.
41. **Hernan, R.A., H.L. Hui, M.E. Andracki, R.W. Noble, S.G. Sligar, J.A. Walder, and R.Y. Walder.** 1992. Human hemoglobin expression in *Escherichia coli*: importance of optimal codon usage. Biochemistry *31*:8619-8628.
42. **Hernan, R.A. and S.G. Sligar.** 1995. Tetrameric hemoglobin expressed in *Escherichia coli*. Evidence of heterogeneous subunit assembly. J. Biol. Chem. *270*:26257-26264.
43. **Hernandez, G., A. Wilks, R. Paolesse, K.M. Smith, P.R. Ortiz de Montellano, and G.N. La Mar.** 1994. Proton NMR investigation of substrate-bound heme oxygenase: evidence for electronic and steric contributions to stereoselective heme cleavage. Biochemistry *33*:6631-6641.
44. **Hoffman, S.J., D.L. Looker, J.M. Roehrich, P.E. Cozart, S.L. Durfee, J.L. Tedesco, and G.L. Stetler.** 1990. Expression of fully functional tetrameric human hemoglobin in *Escherichia coli*. Proc. Natl. Acad. Sci. USA *87*:8521-8525.
45. **Hofrichter, J., J.H. Sommer, E.R. Herry, and W.A. Eaton.** 1983. Nanosecond absorption spectroscopy of hemoglobin. Proc. Natl. Acad. Sci. USA *80*:2235-2239.
46. **Holmans, P.L., M.S. Shet, C.A. Martin-Wixtrom, C.W. Fisher, and R.W. Estabrook.** 1994. The high-level expression in *Escherichia coli* of the membrane-bound form of human and rat cytochrome b5 and studies on their mechanism of function. Arch. Biochem. Biophys. *312*:554-565.
47. **Hui, H.L., J.S. Kavanaugh, M.L. Doyle, A. Wierzba, P.H. Rogers, A. Arnone, J.M. Holt, G.K. Ackers, and R.W. Noble.** 1999. Structural and functional properties of human hemoglobins reassembled after synthesis in *Escherichia coli*. Biochemistry *38*:1040-1049.
48. **Ishikawa, K., M. Sato, and T. Yoshida.** 1991. Expression of rat heme oxygenase in *Escherichia coli* as a catalytically active, full-length form that binds to bacterial membranes. Eur. J. Biochem. *202*:161-165.
49. **Jennings, P.A., M.J. Stone, and P.E. Wright.** 1995. Overexpression of myoglobin and assignment of its amide, C alpha and C beta resonances. J. Biomol. NMR. *6*:271-276.
50. **Jones, S., M. Tasab, J.E. Ogden, D.J. Ballance, and M.J. Powell.** 1996. Expression of rat neuronal nitric oxide synthase in *Saccharomyces cerevisiae*. J. Biotechnol. *48*:37-41.
51. **Jung, C., O. Ristau, and H. Rein.** 1991. The high-spin/low-spin equilibrium in cytochrome P-450—a new method for determination of the high-spin content. Biochim. Biophys. Acta *1076*:130-136.
52. **Kakuno, T., R.G. Bartsch, K. Nishikawa, and T. Horio.** 1981. Redox componenets associated with chromatophores from *Rhodospirillum rubrum*. J. Biochem. *70*:79-94.
53. **Kincaid, J., P. Stein, and T.G. Spiro.** 1979. Absence of heme-localized strain in T state hemoglobin: insensitivity of heme-imidazole resonance Raman frequencies to quaternary structure. Proc. Natl. Acad. Sci. USA *76*:549-552.
54. **King, M.T. and G. Drews.** 1975. The respiratory electron transport system of heterotrophically-grown *Rhodopseudomonas palustris*. Arch. Microbiol. *102*:219-231.
55. **Kitagawa, T., K. Nagai, and M. Tsubaki.** 1979. Assignment of the Fe-Nepsilon (His F8) stretching band in the resonance Raman spectra of deoxy myoglobin. FEBS Lett. *104*:376-378.
56. **Knaff, D.B. and B.B. Buchanan.** 1975. Cytochrome b and photosynthetic sulfur bacteria. Biochim. Biophys. Acta. *376*:549-560.

57. **Koenigs, L.L. and W.F. Trager.** 1998. Mechanism-based inactivation of cytochrome P450 2B1 by 8-methoxypsoralen and several other furanocoumarins. Biochemistry *37*:13184-13193.
58. **Koenigs, L.L. and W.F. Trager.** 1998. Mechanism-based inactivation of P450 2A6 by furanocoumarins. Biochemistry *37*:10047-10061.
59. **Laemmli, U.K.** 1970. Cleavage of structural proteins during the assembly of the head of bacteriophage T4. Nature *227*:680-685.
60. **Lee, B.C.** 1992. Isolation of an outer membrane hemin-binding protein of *Haemophilus influenzae* type b. Infect. Immun. *60*:810-816.
61. **Lindenmayer, A. and R.W. Estabrook.** 1958. Low temperature spectral studies on the biosynthesis of cytochromes in bakers yeast. Arch. Biochem. Biophy. *78*:66-82.
62. **List, B.M., P. Klatt, E.R. Werner, K. Schmidt, and B. Mayer.** 1996. Overexpression of neuronal nitric oxide synthase in insect cells reveals requirement of haem for tetrahydrobiopterin binding. Biochem. J. *315*:57-63.
63. **Maines, M.D.** 1997. The heme oxygenase system: a regulator of second messenger gases. Annu. Rev. Pharmacol. Toxicol. *35*:517-554.
64. **Mayer, B., P. Klatt, B.M. List, C. Harteneck, and K. Schmidt.** 1996. Large-scale purification of rat brain nitric oxide synthase from baculovirus overexpression system. Methods Enzymol. *268*:420-427.
65. **McMillan, K., D.S. Bredt, D.J. Hirsch, S.H. Snyder, J.E. Clark, and B.S. Masters.** 1992. Cloned, expressed rat cerebellar nitric oxide synthase contains stoichiometric amounts of heme, which binds carbon monoxide. Proc. Natl. Acad. Sci. USA *89*:11141-11145.
66. **McMillan, K., J.C. Salerno, and B.S. Masters.** 1996. Nitric oxide synthases: analogies to cytochrome P450 monooxygenases and characterization of recombinant rat neuronal nitric oxide synthase hemoprotein. Methods Enzymol. *268*:460-472.
67. **Miller, J.P. and R.E. White.** 1994. Photoaffinity labeling of cytochrome P450 2B4: capture of active site heme ligands by a photocarbene. Biochemistry *33*:807-817.
68. **Miyake, Y. and N. Takayama.** 1976. Spectral intermediates during the reduction of hepatic microsomal cytochrome P-450. J. Biochem. (Tokyo) *79*:1077-1087.
69. **Modi, S., M.J. Paine, M.J. Sutcliffe, L.Y. Lian, W.U. Primrose, C.R. Wolf, and G.C. Roberts.** 1996. A model for human cytochrome P450 2D6 based on homology modeling and NMR studies of substrate binding. Biochemistry *35*:4540-4550.
70. **Mok, T.C., P.A. Rickard, and F.J. Moss.** 1969. The carbon monoxide-reactive haemoproteins of yeast. Biochim. Biophys. Acta *172*:438-449.
71. **Muchmore, D.C., L.P. McIntosh, C.B. Russell, D.E. Anderson, and F.W. Dahlquist.** 1989. Expression and nitrogen-15 labeling of proteins for proton and nitrogen-15 nuclear magnetic resonance. Methods Enzymol. *177*:44-73.
72. **Nagai, K. and H.C. Thogersen.** 1984. Generation of beta-globin by sequence-specific proteolysis of a hybrid protein produced in *Escherichia coli*. Nature *309*:810-812.
73. **Nakane, M., J.S. Pollock, V. Klinghofer, F. Basha, P.A. Marsden, A. Hokari, T. Ogura, H. Esumi, and G.W. Carter.** 1995. Functional expression of three isoforms of human nitric oxide synthase in baculovirus-infected insect cells. Biochem. Biophys. Res. Commun. *206*:511-517.
74. **North, J.A., R. Dietrich, and A.L. Tappel.** 1996. Multicomponent analysis of heme protein spectra in biological materials. Anal. Biochem. *233*:115-123.
75. **Ohnishi, T., S. Miura, and Y. Ichikawa.** 1993. Photoaffinity labeling of cytochrome P-45011 beta with methyltrienolone as a probe for the substrate binding region. Biochim. Biophys. Acta *1161*:257-264.
76. **Ortiz de Montellano, P.R.** 1995. Arylhydrazines as probes of hemoprotein structure and function. Biochimie *77*:581-593.
77. **Ortiz de Montellano, P.R., C. Nishida, I. Rodriguez-Crespo, and N. Gerber.** 1998. Nitric oxide synthase structure and electron transfer. Drug Metab. Dispos. *26*:1185-1189.
78. **Ortiz de Montellano, P.R. and A. Wilks.** 2000. Structure and mechanism of heme oxygenase. Adv. Inorg. Chem. *51*:359-402.
79. **Osawa, Y., K. Nakatsuka, M.S. Williams, J.T. Kindt, and M. Nakatsuka.** 1996. Reactions of reactive metabolites with hemoproteins—toxicological implications: covalent alteration of hemoproteins. Adv. Exp. Med. Biol. *387*:37-45.
80. **Otto, B.R., S.J. van Dooren, J.H. Nuijens, J. Luirink, and B. Oudega.** 1998. Characterization of a hemoglobin protease secreted by the pathogenic *Escherichia coli* strain EB1. J. Exp. Med. *188*:1091-1103.
81. **Pollock, W.B., F.I. Rosell, M.B. Twitchett, M.E. Dumont, and A.G. Mauk.** 1998. Bacterial expression of a mitochondrial cytochrome c. Trimethylation of lys72 in yeast iso-1-cytochrome c and the alkaline conformational transition. Biochemistry *37*:6124-6131.
82. **Poole, R.K., R.I. Scott, and B. Chance.** 1980. Low-temperature spectral and kinetic properties of cytochromes in *Escherichia coli* K-12 grown at lowered oxygen tension. Biochim. Biophys. Acta *591*:471-482.
83. **Poulos, T.L.** 1991. Modeling of mammalian P450s on basis of P450cam X-ray structure. Methods Enzymol. *206*:11-30.
84. **Poulos, T.L.** 1993. Peroxidases. Curr. Opin. Biotechnol. *4*:484-489.
85. **Ramos, L.S., K.R. Beebe, W.P. Karey, M.E. Sanches, B.C. Erickson, B.E. Wilson, B.E. Wangen, and B.R. Kowalski.** 1986. Chemometrics. Anal. Chem. *58*: 294R-515R.
86. **Renaud, J.P., D.R. Davydov, K.P. Heirwegh, D. Mansuy, and G.H. Hui Bon Hoa.** 1996. Thermodynamic studies of substrate binding and spin transitions in human cytochrome P-450 3A4 expressed in yeast microsomes. Biochem. J. *319*:675-681.
87. **Rieske, J.S.** 1967. The quantitative determination of mitochondrial hemoproteins. Methods Enzymol. *10*:488-493.
88. **Ristau, O., H. Rein, S. Greschner, G.R. Janig, and K. Ruckpaul.** 1979. Quantitative analysis of the spin equilibrium of cytochrome P-450 LM2 fraction from rabbit liver microsomes. Acta Biol. Med. Ger. *38*:177-185.

89. **Rivera, M., C. Barillas-Mury, K.A. Christensen, J.W. Little, M.A. Wells, and F.A. Walker.** 1992. Gene synthesis, bacterial expression, and 1H NMR spectroscopic studies of the rat outer mitochondrial membrane cytochrome b5. Biochemistry *31*:12233-12240.
90. **Riveros-Moreno, V., B. Heffernan, B. Torres, A. Chubb, I. Charles, and S. Moncada.** 1995. Purification to homogeneity and characterisation of rat brain recombinant nitric oxide synthase. Eur. J. Biochem. *230*:52-57.
91. **Rodgers, K.R.** 1999. Heme-based sensors in biological systems [In Process Citation]. Curr. Opin. Chem. Biol. *3*:158-167.
92. **Rodriguez-Crespo, I., N.C. Gerber, and P.R. Ortiz de Montellano.** 1996. Endothelial nitric-oxide synthase. Expression in *Escherichia coli*, spectroscopic characterization, and role of tetrahydrobiopterin in dimer formation. J. Biol. Chem. *271*:11462-11467.
93. **Roman, L.J., E.A. Sheta, P. Martasek, S.S. Gross, Q. Liu, and B.S. Masters.** 1995. High-level expression of functional rat neuronal nitric oxide synthase in *Escherichia coli*. Proc. Natl. Acad. Sci. USA *92*:8428-8432.
94. **Sari, M.A., S. Booker, M. Jaouen, S. Vadon, J.I. Boucher, D. Pompon, and D. Mansuy.** 1996. Expression in yeast and purification of functional macrophage nitric oxide synthase. Evidence for cysteine-194 as iron proximal ligand. Biochemistry *35*:7204-7213.
95. **Sarma, S., R.J. DiGate, D.L. Banville, and R.D. Guiles.** 1996. 1H, 13C and 15N NMR assignments and secondary structure of the paramagnetic form of rat cytochrome b5. J. Biomol. NMR *8*:171-183.
96. **Schmidt, M.L. and J.Q. Trojanowski.** 1986. Enzymatic detection of native and derivatized horseradish peroxidase in sodium dodecyl sulfate polyacrylamide gels. Anal. Biochem. *155*:371-375.
97. **Schuller, D.J., A. Wilks, P.R. Ortiz de Montellano, and T.L. Poulos.** 1999. Crystal structure of human heme oxygenase-1. Nat. Struct. Biol. *6*:860-867.
98. **Seo, H.G., J. Fujii, H. Soejima, N. Niikawa, and N. Taniguchi.** 1995. Heme requirement for production of active endothelial nitric oxide synthase in baculovirus-infected insect cells. Biochem. Biophys. Res. Commun. *208*:10-18.
99. **Shelver, D., M.V. Thorsteinsson, R.L. Kerby, S.Y. Chung, G.P. Roberts, M.F. Reynolds, R.B. Parks, and J.N. Burstyn.** 1999. Identification of two important heme site residues (cysteine 75 and histidine 77) in CooA, the CO-sensing transcription factor of *Rhodospirillum rubrum*. Biochemistry *38*:2669-2678.
100. **Shen, T.J., N.T. Ho, V. Simplaceanu, M. Zou, B.N. Green, M.F. Tam, and C. Ho.** 1993. Production of unmodified human adult hemoglobin in *Escherichia coli*. Proc. Natl. Acad. Sci. USA *90*:8108-8112.
101. **Shioi, Y., K. Takamiya, and M. Nishimura.** 1972. Studies on energy and electron transfer systems in green photosynthetic bacterium *Chloropseudomonas ethylica* strain 2K. I. Isolation and characterization of cytochromes from *Chloropseudomonas ethylica* strain 2K. J. Biochem. (Tokyo) *71*:285-294.
102. **Shrager, R.I. and R.W. Hendler.** 1982. Titration of individual components in a mixture with resolution of difference spectra, pK's and redox transitions. Anal. Chem. *54*:1147-1152.
103. **Shu, F., V. Ramakrishnan, and B.P. Schoenborn.** 1996. High-level expression and deuteration of sperm whale myoglobin. A study of its solvent structure by X-ray and neutron diffraction methods. Basic Life Sci. *64*:309-323.
104. **Smith, A.T., N. Santama, S. Dacey, M. Edwards, R.C. Bray, R.N. Thorneley, and J.F. Burke.** 1990. Expression of a synthetic gene for horseradish peroxidase C in *Escherichia coli* and folding and activation of the recombinant enzyme with Ca^{2+} and heme. J. Biol. Chem. *265*:13335-13343.
105. **Smulevich, G., J.M. Mauro, L.A. Fishel, A.M. English, J. Kraut, and T.G. Spiro.** 1988. Heme pocket interactions in cytochrome c peroxidase studied by site-directed mutagenesis and resonance Raman spectroscopy. Biochemistry *27*:5477-5485.
106. **Smulevich, G., M.A. Miller, J. Kraut, and T.G. Spiro.** 1991. Conformational change and histidine control of heme chemistry in cytochrome c peroxidase: resonance Raman evidence from Leu-52 and Gly-181 mutants of cytochrome c peroxidase. Biochemistry *30*:9546-9558.
107. **Spiro, T.G.** 1978. Resonance Raman spectra of hemoproteins. Methods Enzymol. *54*:233-249.
108. **Spiro, T.G.** 1985. Resonance Raman spectroscopy as a probe of heme protein structure and dynamics. Adv. Prot. Chem. *37*:111-159.
109. **Spiro, T.G. and R.S. Czernuszewicz.** 1995. Resonance Raman spectroscopy of metalloproteins. Methods Enzymol. *246*:416-460.
110. **Springer, B.A. and S.G. Sligar.** 1987. High-level expression of sperm whale myoglobin in *Escherichia coli*. Proc. Natl. Acad. Sci. USA *84*:8961-8965.
111. **Sun, J., A. Wilks, P.R. Ortiz de Montellano, and T.M. Loehr.** 1993. Resonance Raman and EPR spectroscopic studies on heme-heme oxygenase complexes. Biochemistry *32*:14151-14157.
112. **Tamburini, P.P., G.G. Gibson, W.L. Backes, S.G. Sligar, and J.B. Schenkman.** 1984. Reduction kinetics of purified rat liver cytochrome P-450. Evidence for a sequential reaction mechanism dependent on the hemoprotein spin state. Biochemistry *23*:4526-4533.
113. **Taylor, K.L., D.J. Uhlinger, and J.M. Kinkade, Jr.** 1992. Expression of recombinant myeloperoxidase using a baculovirus expression system. Biochem. Biophys. Res. Commun. *187*:1572-1578.
114. **Teale, F.W.J.** 1959. Cleavage of the haem-protein link by acid methyl ethyl ketone. Biochim. Biophy. Acta. *35*:543.
115. **Teraoka, J. and T. Kitagawa.** 1981. Structural implication of the heme-linked ionization of horseradish peroxidase probed by the Fe-histidine stretching Raman line. J. Biol. Chem. *256*:3969-3977.
116. **Thomas, P.E., D. Ryan, and W. Levin.** 1976. An improved staining procedure for the detection of the peroxidase activity of cytochrome P-450 on sodium dodecyl sulfate polyacrylamide gels. Anal. Biochem. *75*:168-176.
117. **Tschirret-Guth, R.A., K.F. Medzihradszky, and P.R. Ortiz de Montellano.** 1998. Specific azidophenyldiazene hemoprotein active site probes. Cross-linking of the heme to His-64 in myoglobin. J. Am. Chem. Soc. *120*:7404-7410.

118.**Tschirret-Guth, R.A., K.F. Medzihradszky, and P.R. Ortiz de Montellano.** 1999. Trifluoromethyldiazirinylphenyldiazenes: new hemeprotein active-site probes. J. Amer. Chem. Soc. *121*:4731-4737.

119.**Tschirret-Guth, R.A. and P.R. Ortiz de Motellano.** 1998. Synthesis of photoaffinity probes for heme-containing proteins. J. Org. Chem. *63*:9711-9715.

120.**Tsutsui, K.** 1986. Affinity chromatography of heme-binding proteins: synthesis of hemin-agarose. Methods Enzymol. *123*:331-338.

121.**Tsutsui, K. and G.C. Mueller.** 1982. A protein with multiple heme-binding sites from rabbit serum. J. Biol. Chem. *257*:3925-3931.

122.**Varadarajan, R., A. Szabo, and S.G. Boxer.** 1985. Cloning, expression in *Escherichia coli*, and reconstitution of human myoglobin. Proc. Natl. Acad. Sci. USA *82*:5681-5684.

123.**Vickery, L.E.** 1978. Spin states of heme proteins by magnetic circular dichroism. Methods Enzymol. *54*: 284-302.

124.**von Wachenfeldt, C., T.H. Richardson, J. Cosme, and E.F. Johnson.** 1997. Microsomal P450 2C3 is expressed as a soluble dimer in *Escherichia coli* following modification of its N-terminus. Arch. Biochem. Biophys. *339*:107-114.

125.**Vuletich, J.L. and Y. Osawa.** 1998. Chemiluminescence assay for oxidatively modified myoglobin. Anal. Biochem. *265*:375-380.

126.**Wilks, A., S.M. Black, W.L. Miller, and P.R. Ortiz de Montellano.** 1995. Expression and characterization of truncated human heme oxygenase (hHO- 1) and a fusion protein of hHO-1 with human cytochrome P450 reductase. Biochemistry *34*:4421-4427.

127.**Wilks, A. and P.R. Ortiz de Montellano.** 1992. Intramolecular translocation of the protein radical formed in the reaction of recombinant sperm whale myoglobin with H_2O_2. J. Biol. Chem. *267*:8827-8833.

128.**Wilks, A. and P.R. Ortiz de Montellano.** 1993. Rat liver heme oxygenase. High level expression of a truncated soluble form and nature of the meso-hydroxylating species. J. Biol. Chem. *268*:22357-22362.

129.**Wilks, A. and M.P. Schmitt.** 1998. Expression and characterization of a heme oxygenase (Hmu O) from *Corynebacterium diphtheriae*. Iron acquisition requires oxidative cleavage of the heme macrocycle. J. Biol. Chem. *273*:837-841.

130.**Wilks, A., J. Torpey, and P.R. Ortiz de Montellano.** 1994. Heme oxygenase (HO-1). Evidence for electrophilic oxygen addition to the porphyrin ring in the formation of alpha-meso-hydroxyheme. J. Biol. Chem. *269*:29553-29556.

131.**Wu, C., J. Zhang, H. Abu-Soud, D.K. Ghosh, and D.J. Stuehr.** 1996. High-level expression of mouse inducible nitric oxide synthase in *Escherichia coli* requires coexpression with calmodulin. Biochem. Biophys. Res. Commun. *222*:439-444.

132.**Yoshida, T., K. Ishikawa, and M. Sato.** 1991. Degradation of heme by a soluble peptide of heme oxygenase obtained from rat liver microsomes by mild trypsinization. Eur. J. Biochem. *199*:729-733.

133.**Yoshida, T. and G. Kikuchi.** 1978. Purification and properties of heme oxygenase from pig spleen microsomes. J. Biol. Chem. *253*:4224-4229.

134.**Yun, C.H., G.J. Hammons, G. Jones, M.V. Martin, N.E. Hopkins, W.L. Alworth, and F.P. Guengerich.** 1992. Modification of cytochrome P450 1A2 enzymes by the mechanism-based inactivator 2-ethynylnaphthalene and the photoaffinity label 4-azidobiphenyl. Biochemistry *31*:10556-10563.

135.**Zabel, U., M. Weeger, M. La, and H.H. Schmidt.** 1998. Human soluble guanylate cyclase: functional expression and revised isoenzyme family. Biochem. J. *335*:51-57.

8 Hemoproteins Purification and Characterization by Using Aqueous Two-Phase Systems

Daniel Forciniti
Chemical Engineering Department, University of Missouri-Rolla, Rolla, MO, USA

1. INTRODUCTION

There are many techniques that can be used to purify and characterize proteins, but one especially caught my attention about 10 years ago. Is it possible to purify a protein using just water? A smart trick that does the job is aqueous 2-phase extraction. In 1958, Albertsson (2) found that when 2 aqueous solutions of hydrophilic polymers such as polyethyleneglycol (PEG) and dextran (Dx) are mixed above critical concentrations, a liquid–liquid phase separation occurs. Moreover, when he added these 2 polymers to a disrupted cell broth, he found that proteins or enzymes and cell debris tend to partition unequally between the 2 phases or between the phases and the interface, thus allowing for the extraction of a particular protein. The high water content of these systems, which gives the technique its name, provides a gentle environment for biologically active proteins, cells, and cell organelles. Many proteins have been partitioned in aqueous 2-phase and 3-phase systems, and the subject has been described in several reviews (2,22,50,52, 56). A comprehensive database containing all the publications in this topic since 1956 can be found in (http://bama.ua.edu/~rdrogers/aq2phase/). Detail recipes for most of the protocols used in different applications of aqueous 2-phase systems can be found in Reference 22.

Aqueous 2-phase extraction is a technique that is ideally suitable for the removal of cell debris and simultaneous concentration of the target protein; furthermore, it can be used as a very selective procedure by adding affinity ligands to one of the phase-forming polymers. Table 1 shows some examples of the use of aqueous 2-phase systems for the purification of various proteins. Worth mentioning is the isolation of *Vitreoscilla* hemoglobin from *Escherichia coli* lysate by 3 extraction steps with an overall yield of 47% and a purity higher than 95% (21). For either industrial or recombinant proteins, the method offers notable advantages (27,54). For industrial enzymes, the method's utility stems from the ease with which it can be adapted to continuous production and scaled up to meet industrial needs. For example, Genencor Intl. (Palo Alto, CA, USA) (27) currently uses a PEG/salt system coupled with ion exchange chro-

Heme, Chlorophyll, and Bilins: ***Methods and Protocols***
Edited by A.G. Smith and M. Witty

Table 1. Some Examples of the Use of Aqueous Two-Phase Systems for the Protein Purification

Substance	Source	Yield	Concentration Factor/Purity	Observations	Reference
Thaumatin	Cell homogenate of *E. coli*	92%	65%	Two extraction steps using PEG/phosphate systems.	5
Lactate dehydrogenase	Porcine muscle extract	54%	11.8-fold	PEG/Dx followed by affinity precipitation.	19
Ecdysone (hormone)	Spinach	88%	ND	UCON 50-hb-5111/Reppal PES-200[a].	33
3-phosphoglycerate kinase	Yeast homogenate	57%	5.2-fold	UCON 50-HB-5100/Dx T-500.	47
NAD-kinase	*Lactobacillus cellobiosus*	100%	3-fold	PEG/salt.	23
Leucine dehydrogenase	*Bacillus sphaericus*	98%	9.5-fold	PEG/crude Dx (unfractionated Dx of high, >1 000 000 molecular weight).	23
Beta interferon	Human fibroblasts		350-fold		31
Beta interferon	Recombinant *E. coli*	76%	ND	PEG/salt/PEG phosphate ester followed by back extraction.	18
Fractionation of proteins	Human serum	ND	ND	PEG/Dx and PEG/Aquaphase PPT[b] plus addition of various affinity ligands. Separation done by counter-current distribution.	7
Lactate dehydrogenases	Recombinant *B. stearothermophilus*	80%	8-fold	Cu(II)poly-VI-VCL and Dx-70.	30
Chymosin	Recombinant *A. awamori*	ND	ND	PEG-8000/NA_2SO_4 Industrial scale production (Genencor Intrl.).	27
Recombinant Protein	Mammalian cell culture	90%	100-fold	PEG/Pluronic F68/Ammonium sulfate. Industrial scale production (MedImmune, Gaithersburg, MD, USA)	4
Cutinase	Recombinant *Saccharomyces cerevisiae*	ND	26-fold	PEG-1500/sodium phosphate (plus Triton® X-100, or Tween® 20, or butyrate).	9
Vitreoscilla hemoglobin	*E. coli* lysate	47%	95%	Sequential extraction: *(i)* PEG/Dx, *(ii)* PEG/Na_2SO_4, *(iii)* PEG/$MgSO_4$	21

[a]Reppal PES-200 is a hydroxypropyl starch produced by CarbamyLab (Kristianstad, Sweden).
[b]Aquaphase PPT is a modified starch product produced by Perstop (Sweden).
ND: not determined by the authors.

matography for the purification of chymosin. For recombinant proteins, its utility is based on its attractiveness as a gentle first step or on its selectivity on a small scale by using affinity ligands. For example, Mattiasson and Galaev (30) purified lactate dehydrogenase (carrying a tag of 6 histidine residues) from recombinant *Bacillus stearothermophilus* with an 80% yield and concentrated 8-fold using an aqueous 2-phase system formed by Cu(II)poly-VI-VCL (copolymers of N-vinylcaprolactam and 1-vinyl imidazole) and Dx-70. One application of the method is aqueous 2-phase affinity partitioning chromatography, in which the stationary (support) phase is coated with the Dx-rich bottom phase of an aqueous 2-phase system, and the top phase is used as the mobile phase. Aqueous 2-phase systems are routinely used, among other places, in the Institute of Enzyme-Technology (Julich, Germany), at the Departments of Biotechnology, Biochemistry, and Center for Chemistry and Chemical Engineering Lund University, and at the Royal Free Hospital School of Medicine, University of London.

In addition to its application as a separation process, aqueous 2-phase partitioning often yields information on important physical properties, replacing other more cumbersome analytical techniques. For example, aqueous 2-phase partitioning can be used to determine the isoelectric point of proteins, to measure the hydrophobicity and charge of a protein, to calculate dissociation constants between enzymes and substrates, to fractionate cell populations, or to characterize cell surfaces. One of these applications is cross-partitioning (see section 3.5.1), which has been used to determine the isoelectric point of hemoglobins from different species (51).

The most common systems are created using either 2 incompatible polymers or one polymer and a salt. PEG/Dx and PEG/phosphate are by far the most popular ones. An extended list including some new phase systems is presented in Table 2. Recently, temperature-induced phase splitting systems, made with copolymers of ethylene oxide and propylene oxide, are gaining popularity (20).

Aqueous 2-phase systems have a very low interfacial tension (the reversible work to create unit area of surface at constant temperature, volume, and chemical potential) and densities very close to that of pure water. The interfacial tension, which ranges from 1 to 300 μN/m, of aqueous 2-phase systems depends on polymer molecular weight, polymer concentration, and temperature. We have found that the logarithm of the interfacial tension varies linearly with the logarithm of the difference in polymer concentration between the top and bottom phases. If systems of the same tie-line length (TLL; the length of the line joining the composition of both phases and the total system composition in a concentration–concentration plane) are compared, increasing the molecular weight of one of the two polymers increases the interfacial tension. A very complex temperature dependence of the interfacial tension was also found. An extensive collection of densities and interfacial tensions is available (15). The reader is referred to Table 3 for representative values. The very low density difference between top and bottom phases makes phase separation under gravity very slow. So, centrifugation is often used to achieve phase separation.

The preparation of aqueous 2-phase systems is straightforward. Stock solutions of all the phase components are mixed in appropriate amounts (by weight), and then the emulsion is allowed to separate into 2 phases under gravity or in a centrifuge. After complete phase separation is achieved, the phases are sampled. In the laboratory-scale experiments, aqueous 2-phase systems with a final mass of 5 to 10 g are usually prepared. In some systems, one of the phase polymers derivatized with charged groups

Table 2. Commonly Used Phase-Forming Species

Species 1	Species 2	References
PEG	Dx	2,16
PEG	Ficoll	2
Dx	Ficoll	2
Methyl Cellulose	Dx	2
PEG	Benzoil Dextran	28
P(EO-PO)	Water (NaCl)	47
P(EO-PO)	Reppal (Hydroxypropyl starch)	20
PEG	Sodium citrate/citric acid	48
PEG	NaH_2PO_4/K_2HPO_4	2,43
PEG	Ammonium Sulfate	2
Sodium Dextran Sulfate	PEG (NaCl)	2
PEG	NaCl	44
PEG	SCNNa/NaCl	10
PEG	GuHCl/NaCl	46
DX	PVP	56
Dx	PVA	56
PEG	Polyvinyl methyl ether	56
Methyl Cellulose	Hydroxypropyl Dextran	2

All systems contain a large amount of water (or buffer), up to 90% (wt/wt).

or affinity ligands is also included. The derivatized species often constitute a minor constituent (1%–5%) of the total system.

The equilibrium data for aqueous 2-phase systems are usually presented by plotting the amount of one phase forming species (A) versus the amount of the other (B) in both top and bottom phases. The locus of equilibrium compositions of top and bottom phase in the A-B plane is called the binodal (See Figure 1). Figure 1 shows the phase diagram for a polyethylene propylene random copolymer P(EO-PO)/salt system at 25°C. A few tie-lines are also indicated. Simple geometrical arguments show that the length of the tie-line is equal to $TLL = (\Delta A^2 + \Delta B^2)^{1/2}$, where $\Delta A = [A]^{top} - [A]^{bottom}$ and $\Delta B = (B)^{bottom} - (B)^{top}$. The ratio between segments BO and AO is equal to the volume ratio between the top and bottom phases (V_T/V_B). The working tie-line and the volume ratio between the top and the bottom phase can be chosen by using published equilibrium data (56). Since there are variations in the molecular weight of the polymers and, sometimes, the experimental conditions used to determine a binodal are not completely specified, it may be appropriate to create specific a phase diagram if a particular system is going to be used routinely.

The equilibrium curve or binodal is affected by:

1. Polymer molecular weight. The higher the molecular weight of the polymer used, the lower the polymer concentration (on a weight by weight basis) needed to obtain phase separation. The binodal curve becomes more asymmetrical as the molecular weight difference

Table 3. Densities and Interfacial Tensions of some Representative Systems

System	[PEG] % wt/wt	[Dx] % wt/wt	ρ^{top}	ρ^{bottom}	Interfacial Tension μN/m
Dx-40/PEG-4000	7.64	10.29	1.023	1.092	59.10
Dx-40/PEG-20 000	5.05	8.50	1.017	1.068	87.60
Dx-500/PEG-4000	7.64	10.46	1.014	1.098	154.5
Dx-500/PEG-20 000	4.97	8.46	1.016	1.074	145.2

[PEG] and [Dx] are the overall concentration of PEG and Dx (before phase separation). The densities (ρ in units of g/cm^3) of each phase were measured with U-shaped oscillator densitometer, and the interfacial tensions were measured with a spinning drop tensiometer (Krüss Site 04, Hamburg, Germany). All systems are at 25°C.

between the polymers increases.

2. Polymer hydrophobicity. The polymers become less compatible when the hydrophobicity of one of them is increased. This can be done by derivatizing the polymer. For example, two phases are formed by hydroxypropyl Dx and Dx.
3. Temperature. The effect of this variable is small. In PEG/Dx mixtures, less polymer is needed at low temperatures. On the other hand, the phase transition in PEG-salt systems is facilitated by high temperatures.

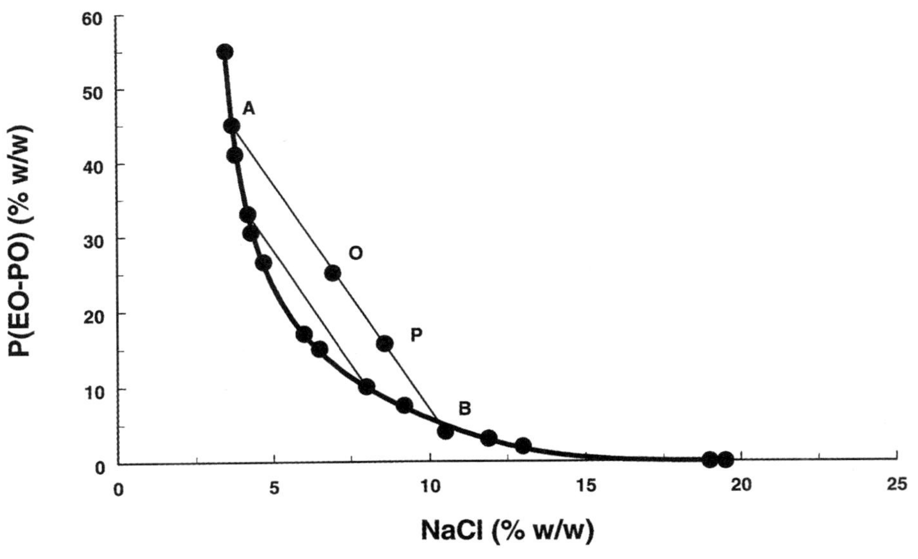

Figure 1. Equilibrium curve for a P(EO-PO)/NaCl system at 25°C. Points to the left the equilibrium curve correspond to one phase, and points to the right correspond to 2 liquid phases at equilibrium Point B representing the composition of the top phase, point A the composition of the bottom phase, and points O and P two possible overall concentrations of P(EO-PO) and NaCl. An overall concentration equal to O yields $V_T/V_B = 1$, and an overall concentration equal to P yields $V_T/V_B = 1/3$.

4. Polydispersity of the polymers. Commercial grade polymers have a wide distribution of molecular weights. Therefore, phase separation occurs in a small range of polymer concentrations changing the binodal curve to a binodal zone. If the polymer molecular weight distribution is too wide, more than two phases can be formed.
5. Ionic Strength. The length of the tieline increases when salts are added to PEG-salt systems. In PEG/Dx/water systems the ionic strength does not have an appreciable influence on the position of the binodal.

1.1. Distribution of Proteins in Aqueous Two-Phase Systems

The separation power of a given aqueous 2-phase system is given by the partition coefficient, K_p, of a protein. K_p is defined as the ratio of the protein concentration in the top and bottom phases. This partition coefficient depends on the difference in chemical potential of the protein between top and bottom phases, and therefore, it is a function of the chemical nature of the polymers, the protein, added electrolytes, and temperature. It is usually manipulated by changing the pH (13,14) or the temperature, by adding salts or affinity ligands, or by changing the molecular weight of the polymers or their concentration.

Manipulation of the pH is of primary importance in partitioning studies of proteins. It is convenient to split the partition coefficient into 2 contributions, one that is independent of pH, and the other that is proportional to the net charge of the protein. In so doing, electrostatic and nonelectrostatic contributions to the partition coefficient can be conveniently (but artificially) separated. This approach predicts that the logarithm of the partition coefficient is a linear function of the charge of the protein. In the analysis of partitioning data at different pHs, we must consider any change in the physical properties of the proteins due to changes in the pH and how these changes may affect the protein partition coefficient. For example, whereas lysozyme does not change its conformation over the pH range 1.2 to 11.3 in dilute salt solutions at moderate temperatures, it polymerizes reversibly at pH above 5 (45). This change in the molecular weight of the protein will change the partition coefficient.

The molecular weight and concentration of the phase-forming species also affect the partition coefficient strongly (13). For example, in a PEG/Dx system, low PEG molecular weight favors the partitioning of proteins into the PEG-rich phase. The effect of polymer concentration on the partition coefficient is also well known. If K_p is smaller than 1, an increase in either one of the phase-forming species decreases K_p. Similarly, if K_p is larger than 1, an increase in either of the phase-forming species increases K_p. The amount of the phase-forming species also affects the volume ratio between the phases. For example, in Figure 1, a system of total composition O has a top:bottom volume ratio of 1, whereas a system of total composition P has a top:bottom volume ratio of 1/3. Therefore, the targeted protein can be not only purified, but also concentrated in a single step.

Generally, the partition coefficient increases with increasing temperature in Dx/PEG systems between 4° and 40°C (12). The partition coefficient of small and hydrophilic proteins is only slightly affected by changes in temperature, whereas the partition coefficient of bigger and more hydrophobic proteins is strongly affected by temperature changes. High temperatures (around 40°C) may be used to minimize protein association, whereas low temperatures may be desirable to maintain protein stability.

Salts have unequal affinities for the top and bottom phases of aqueous 2-phase systems (25). This uneven partition of salts between the 2 phases affects the chemical potential of the protein in each phase and thus its partition coefficient. The following mental exercise helps to understand this phenomenon. Consider 2 phases at equilibrium in which a salt has been previously partitioned. The different affinity of the salt for the bottom and top phases creates 2 distinct ionic atmospheres in both phases. Picture a charged protein molecule at infinite distance from the phases. Bring the protein and try to insert it in each of the 2 phases. Since the protein is going to see different ionic atmospheres in both phases, the work needed to insert it in one or the other phase will be different. This difference in the work of insertion of a charged protein in each of the 2 phases at equilibrium is proportional to the electrochemical potential difference of the protein between them. Consequently, the partition coefficient, which is a function of that potential difference, is strongly affected by the type and concentration of salts and by the charge on the protein.

In general, the partition coefficient of proteins away from their isoelectric point depends on both the type and concentration of cation and anion (25). For example, for positively charged proteins the partition coefficient in PEG/Dx systems is higher in potassium chloride than in potassium phosphate; the reverse is true for negatively charged proteins. The effect of the cation on the partition coefficient of positively charged proteins is K approximately equal to Cs greater than Na greater than Li, whereas the reverse order holds for negatively charged proteins. It is possible to correlate the partition coefficient of a protein with its charge and with the type of salt. Johansson (25) found that,

$$\log K_p = \log K_0 + \gamma Z \qquad \text{[Eq. 1]}$$

where, K_0 depends on the particular protein and phase-system, but it is independent on protein's charge, Z is the charge of the protein, and γ depends on the types and concentration of polymers, temperature, and types of electrolytes. Values of γ are plotted in Figure 2 for various salts (values are for Dx-500/PEG-8000 at 4°C). The net effect of a salt can be calculated by $\gamma = \gamma_{cation} - \gamma_{anion}$. For example, the addition of tetrabutyl ammonium phosphate yields a $\gamma = -79$, whereas the addition of potassium perchlorate yields a $\gamma = 30$. This figure can be used to select the appropriate electrolyte to move the desired protein to either the top or bottom phases. A pH away from the isoelectric point of the protein must be selected, since the partition coefficient of proteins at their isoelectric point is quite insensitive to salt type and concentration (see Equation 1).

The partition coefficient of proteins can be manipulated by adding an affinity ligand or by using liquid–liquid chromatography, counter–current extraction, or by a combination of the first two. Affinity ligands, such as PEG-palmitate (42) or PEG-red (24), have been routinely used to improve the selectivity of aqueous 2-phase extraction. In affinity partitioning (26), it is customary to define $\Delta \ln K_p = \ln K^L - \ln K^0$ (K^L and K^0 are the partition coefficients of the protein with and without added affinity ligand, respectively) to quantify the enhancement in partitioning as a result of the addition of the affinity ligand. Liquid–liquid chromatography, with or without the addition of an affinity ligand, can be used to improve the selectivity of aqueous 2-phase systems. For example, we have used liquid–liquid chromatography with an immobilized Dx-rich phase to purify formate dehydrogenase (FDH) from *Candida boinidi* with and without the addition of PEG-red as an affinity ligand (50). Another attractive alternative technique is the use of metal affinity aqueous 2-phase extraction (1,40,55). In this technique, a small portion (less than 1%) of the total PEG is replaced by PEG-iminodiacetic acid (PEG-IDA), which chelates divalent cations like copper or zinc. Histidine groups on the protein surface recognize these metals, and the strength of the binding is proportional to the number of histidine groups on the protein surface.

One of the most common analytical uses of aqueous 2-phase systems is the determination of isoelectric points by determining cross-partitioning points (3, 41,51,53). The method uses the different sensitivity of the partition coefficient on the kind of salt used and determines the pH at which the partition coefficient is independent of salt type. If the protein of interest does not interact with contaminating materials in the solution, its isoelectric point can be determined without prior protein purification (other than desalting).

The determination of isoelectric points of proteins by cross-partitioning is straightforward. Two sets of 2-phase systems (usually Dx/PEG) covering a wide pH range are prepared. One set contains NaCl, whereas the other set contains Na_2SO_4. The pH dependence of the partition coefficient of proteins in the presence of NaCl is usually different from that in the presence of Na_2SO_4. So, the partition coefficient versus pH curves cross each other at a pH (cross-point) that usually agrees with the isoelectric point of the proteins. Discrepancies between the isoelectric points determined by cross-partitioning and by isoelectric focusing can be due to conformational changes of the protein in the phases system or by interactions between the polymers and the proteins.

The knowledge of the hydrophobicity of a protein is useful for the design of reverse phase chromatography units, for the understanding of protein–ligand interactions, and for the understanding of protein

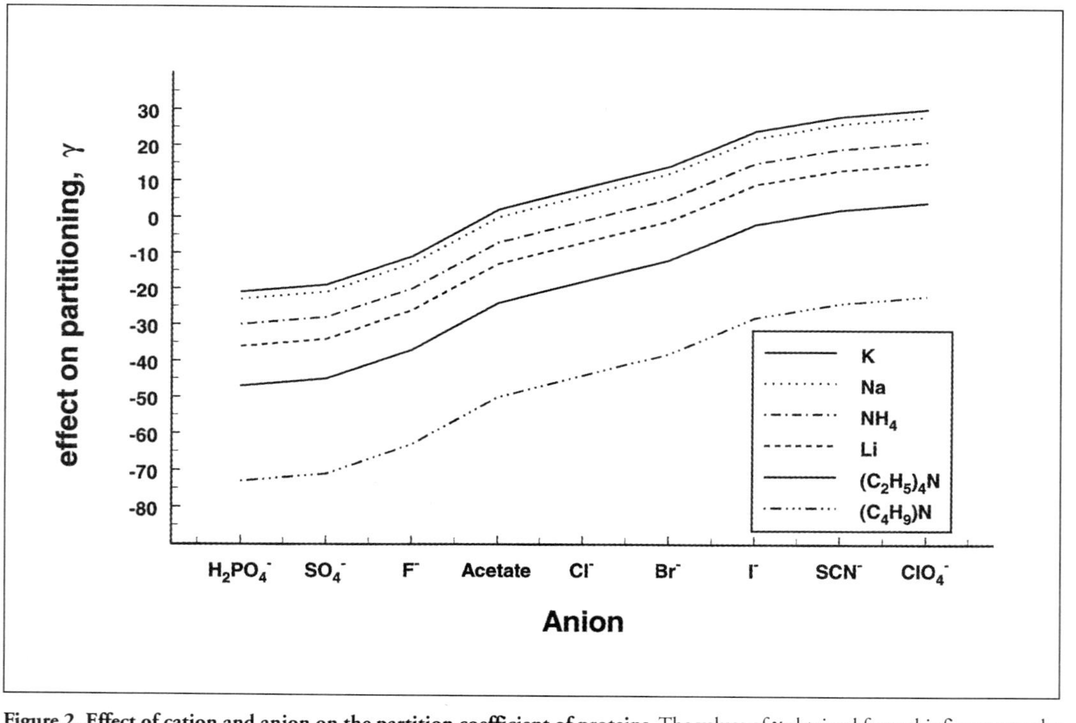

Figure 2. Effect of cation and anion on the partition coefficient of proteins. The values of γ obtained from this figure must be used with Equation 1.

folding and refolding. The usual way of measuring hydrophobicity of a solute is to measure the free energy of transfer of that solute from water into an organic solvent. Because of the need to use an organic solvent, these methods are not well suited to measure hydrophobicities of polar and flexible biological molecules. Aqueous 2-phase systems have been used to determine the relative hydrophobicity of various proteins (56). The overall idea is to "calibrate" a series of aqueous 2-phase systems (Dx/Ficoll® and Dx/PEG have been used) by partitioning a series of small solutes (amino acids) of increasing hydrophobicity. It has been found that the partition coefficient of amino acid i, correlates with the partition coefficient of glycine by (56):

$$\ln K_i = \ln K_{Gly} + (HF)(RH_i) \qquad \text{[Eq. 2]}$$

where HF is known as the hydrophobicity factor (a constant for a given 2-phase system) and RH_i is the hydrophobicity of solute i relative to glycine. For a protein, the surface hydrophobicity of the i protein relative to molecule j, HSF_i is given by:

$$HSF_i = \ln K_{ij} / HF_j \qquad \text{[Eq. 3]}$$

The technique has been used to determine the hydrophobicity of several hemoproteins including hemoglobin, apomyoglobin, and cytochrome c (56). The change of surface properties caused by pH-induced denaturation of some of these proteins has been investigated by this technique (56). For example, it has been found that the difference in hydrophobic index (expressed in equivalent CH_2 groups) for denatured and native cytochrome c is 146, which indicates the exposure, as the protein denatures, of hydrophobic residues otherwise buried in the interior of the polypeptide.

It also is possible to follow protein–protein and protein–ligand association by 2-phase partitioning (29). The basic principle here is the strong dependence of the partition coefficient with the molecular size of the solutes. For example, Petersen (38,39) found that the partition coefficient in a Dx/PEG system of cytochrome c oxidase was 20, whereas that of cytochrome c was 0.275 and used these differences to show that both oxidized and reduced cytochrome c formed a 1:1 complex with the oxidase. In another study, Middaugh and Lawson (32) determined the association constant of hemoglobin by using aqueous 2-phase systems. They found that the partition coefficients of oxyhemoglobin and methemoglobin produced a sigmoidal curve when they were plotted against protein concentration. From these plots, they determined the dimer–tetramer association constant for these proteins.

The potential uses of this technique are vividly pictured in the following examples. PEG-coated liposomes are a current alternative to increase the stability of liposomes. The behavior of the modified liposomes will depend on their surface properties. Because the partition behavior of a particle is a signature of its surface properties, partitioning of PEG-coated liposomes in aqueous 2-phase systems can be used to anticipate their behavior in the blood stream. For example, Moribe et al. (34) used aqueous 2-phase systems to detect surface differences of PEG-coated liposomes. They partitioned the coated and uncoated liposomes (after exposing them to plasma) in two kinds of systems, charge sensitive (5% PEG-8000, 5% Dx T-500, and 0.11 M NaPO4, pH 7.0) and charge insensitive (0.01 NaPO4, 5% PEG-8000, 5% Dx T-500, and 0.15 M NaCl). Here charge sensitive or insensitive refers to systems in which the charge of the substance to be partitioned either affects its partition behavior or not. They concluded that in spite of PEG being a steric barrier for the interaction between plasma proteins and the liposomes, a weak interaction remains between the PEG-coated liposomes and plasma proteins. Berggren et al. (6) used P(EO-PO)/Reppal 2-phase systems to study the hydrophobicity of a series of proteins. They partitioned in these systems several salts, Na_2PO_4, NaCl, and $NaClO_4$, and several proteins of different hydrophobicity, myoglobin, cytochrome c, lysozyme, bovine serum albumin, and β-lactoglobulin. They were able to correlate the partition coefficients of these proteins with their tryptophan content.

1.2. Batch and Continuous Partitioning

Most of the times, aqueous 2-phase partitioning is carried out in test tubes, so it is a batch operation. Attempts have been made, however, to evolve the technique into a continuous operation, mostly to improve the purification factor of a given protein. One that we personally recommend is liquid–liquid partitioning chromatography. In liquid–liquid partitioning chromatography, one phase (for example the Dx-rich phase in a PEG/Dx system) is immobilized on a convenient support, and the other phase (in this case the PEG-rich phase) is used as the mobile phase.

The column is made of agarose or silica diol beads whose surface is derivatized by growing a hydrophilic polymer (polyacrylamide) on it. The Dx-phase is retained, and the silica diol or agarose beads become impermeable to the proteins as determined in our laboratory. The PEG-rich phase is used as the mobile phase. The elution times suggest that the partitioning into Dx-phase is significant. For a column used by us, of 2.5 cm in diameter and 40 cm length (about 196.3-mL volume), the continuous phase (PEG) is about 10 mL, and the Dx-phase is about 5 mL. If the Dx-phase is assumed to coat the beads uniformly, then the ratio (radius of the bead with the coat)/(radius without the coat) is about 1.009. For a bead of radius 10 μm, the film thickness is about 900 nm.

2. MATERIALS

2.1. Polymers

A variety of polymers have been used to prepare 2-phase systems (see Table 2). PEG, a linear synthetic polymer of ethylene oxide units, and Dx, poly(α-1,6-glucose), are the most commonly used polymers in the preparation of aqueous 2-phase systems. Examples of other sugar polymers used are Ficoll (polysucrose), pullulan, and maltodextrins. Derivatized carbohydrate polymers have also been used; these include methylcellulose, hydroxyethylcellulose (HEC), Reppal PES (hydroxypropyl starch), benzoyl, dextran sulfate, and diethylaminoethyl (DEAE)/Dx. Examples of synthetic polymers, besides PEG, are polyvinyl alcohol (PVA), polyvinylpyrrolidone, pluronic, and random copolymers of ethylene oxide and propylene oxide: $EO_{50}PO_{50}$, $EO_{20}PO_{80}$ (UCON), etc. Several suppliers can be used worldwide: Sigma (St. Louis, MO, USA) (PEG, Dx, $Dx\text{-}SO_4$, Ficoll, Methylcellulose), Union Carbide (Bound Brook, NJ, USA) (PEG, UCON), Amersham Pharmacia Biotech (Piscataway, NJ, USA) (Dx and Ficoll), Polysciences (Warrington, PA, USA) (PEG), Shearwater Polymers (Huntsville, AL, USA) (PEG and PEG derivatives), etc. For most applications, the polymers are used as received. Multivalent ions in commercial Dx can be eliminated by dialysis, ultrafiltration, or by a desalting step. Impurities in PEG (antioxidants, ethylene glycol, and diethylene glycol) can be eliminated by ether or hexane precipitation of a PEG/acetone solution (1,11). Often, the molecular weight and the molecular weight distribution are given by the manufacturer. In the absence of accurate information, the molecular weight can be determined by a size exclusion chromatography-low angle light scattering tandem (no internal standards are needed) or by size exclusion chromatography using the appropriate standards. Some companies, like Wyatt Technology (Santa Barbara, CA, USA), provide determination of molecular weights for a fee. The molecular weights of PEG and Dx can be determined by using a Superose™ 12 column (Amersham Pharmacia Biotech) eluted with a 3% NaCl solution at room temperature (17). Figure 3 shows a typical molecular weight distribution curve for Dx. Molecular weight standards for PEG can be bought from Polysciences. Since molecular weight standards for Dx are difficult to obtain, narrow fractions of pullulan (Polysciences) can be used. Polydisperse polymers of good quality can be purchased from speciality chemical companies like Sigma. Polymer batches of narrow molecular weight distributions can also be purchased but at a much higher price. For example, less polydisperse Dx can be bought from Amersham Pharmacia Biotech, whereas narrow fractions of PEG can be purchased from PolySciences.

2.2. Buffers

A variety of buffers have been used to regulate the pH in aqueous 2-phase systems. The two most commonly used are phosphate and Tris buffers. The buffers must be kept in a refrigerator and used within 30 days.

2.3. Additives

A series of additives are normally used in aqueous 2-phase systems. Bacteriocides (either sodium azide or chloroacetamide) are conveniently added to the polymer stock solutions or in solid form to the 2-phase system. A series of salts (shown in Figure 2) are commonly used to drive the protein of interest into one or another phase.

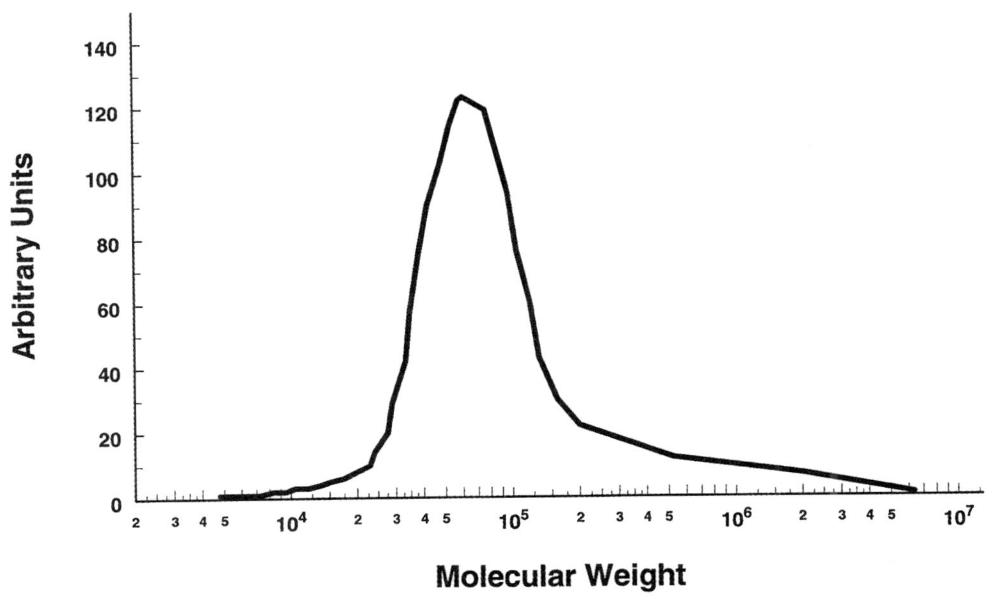

Figure 3. Molecular weight distribution of Dx T-500. This was obtained by running a Dx solution through a Superose 12 column eluted with NaCl (3%). The column was calibrated with pullulan standards.

2.4. Polymer Derivatives

Different PEG derivatives have been used in aqueous 2-phase partitioning experiments as affinity ligands. The three main kinds are PEG-dye, PEG-fatty acid, and PEG-imidazole compounds. Several dyes, including Cibacron Blue F3G-A (Ciba-Geigy, Basel, Switzerland), Procion Red HE-3B, Procion Green H-4G, and Procion Brown MX-5BR (I.C.I. Organic Division, Blackely, UK), and Remazol dyes (Hoechst, Frankfurt, Germany) can be conveniently attached to PEG-amine (from Shearwater Polymers). Some PEG-dye derivatives are available commercially (for example, Sigma commercializes PEG-red, MW 8000), but they can be easily prepared in the laboratory, as can other derivatives.

❖ Procedure 1. Preparation of PEG Derivatives

A. Preparation of PEG-Red

1. Dissolve aminated PEG and Procion Red HE 3B (1:1.8 ratio) in water.
2. Adjust the pH to 11.0 (5 M NaOH) and incubate the mixture at 60°C for 24 hours with constant stirring.
3. Remove the excess salts by dialysis.

B. Preparation of Iminodiacetate PEG (for Metal Affinity Two-Phase Partitioning)

1. The synthesis begins with PEG-chloride (either produced in the laboratory or obtained from Shearwater Polymers), to which iminodiacetic acid and potassium carbonate are added.
2. The solution is refluxed for 48 hours.
3. The reaction is stopped by adding sodium sulfate and allowed to separate into 2 phases.
4. The PEG-rich phase is diluted and dialyzed first against sodium bicarbonate and then against water.

PEG-fatty acids can be also easily produced in the laboratory by reacting PEG with the chloride or anhydride of the fatty acid in toluene

3. METHODS

3.1. Preparation of Stock Solutions

The concentration of each stock solution is only limited by the solubility of the polymers and by the increase in viscosity, which makes the solutions very cumbersome to handle. For a polymer–polymer system, stock solutions of polymers and the appropriate buffer are prepared; the polymer stocks can be prepared in water or buffer. For a polymer–salt system, the stock polymer solution, and a stock salt solution of the desired pH, are prepared. In the systems for affinity partitioning, stocks of derivatized polymers are also prepared. Sodium azide (1 mmol) or chloroacetamide (up to 5g/L) should be added to each stock solution as a bacteriocide. Stock solutions of the polymers must be stored at 4°C and used within 30 days of being prepared. Stock solutions of PEG must be stored in the dark to prevent UV-induced oxidation. Age and exposure to light induces the formation of acidic groups in spite of the addition of antioxidants. A decrease in pH and a yellowish color of a PEG solution are clear indications that oxidation has taken place. Preparation of some representative stock solutions is described below.

PEG stock of 30% to 50% (wt/wt) is prepared by accurately weighing the polymer and the water or buffer in a flask and stirring for an hour or more on a magnetic stirrer until a clear solution is obtained. Solid PEG, if properly stored, contains less than 0.5% water. Dx stock of 20% to 30%

(wt/wt) concentration is prepared by first making a paste of the powder with a small amount of water and then adding the rest of the water to reach the final mass. Because of the presence of water (5%–10%) in commercial Dx, an amount of Dx in excess to that needed may be weighted. Heating the Dx solution up to 95°C on a hot plate is highly recommended to facilitate the dissolution of the polymer and to reduce bacterial growth.

The final polymer concentration in the stock solution can be easily determined by refractive index measurements (either PEG or Dx), polarimetry (Dx), colorimetric essays (PEG), or freeze-drying (both polymers). To measure the concentration of PEG by refractive index measurements, it is necessary to measure the refractive index increment of the solution above the buffer (the refractive index increment above water of a 1% PEG solution is 0.00139). The density of the solution can be measured very precisely using a U-shaped oscillator densitometer (Anton PAAR USA, Asland, VA, USA) or estimated from ρ_{PEG} = [0.997 + 0.169 C_{PEG}/100] and ρ_{Dx} = [0.0997 + (0.391 C_{Dx}/100)], with C in g/100 mL and the densities of PEG and Dx in g/mL at 25°C. Alternatively, the concentration of PEG can be obtained by a colorimetric assay: *(i)* 5 mL of 0.5 M perchloric acid are added to 1 mL of PEG solution; *(ii)* after 15 minutes, the precipitate (if any) is discarded; *(iii)* 1 mL of 5% $BaCl_2$ and 0.4 mL of 0.1 M iodine are added to 4 mL of a PEG solution; *(iv)* after 15 minutes, the absorbance is measured at 525 nm against a blank of all the above chemicals except PEG; and *(v)* a calibration curve is prepared in the concentration range of 0.1 to 0.6 g/mL of PEG. The concentration of Dx in the stock solution can be measured using a polarimeter with a Na lamp at 589 nm and 25°C: *(i)* 5 g of the solution is diluted to 25 mL with water; and *(ii)* the optical rotation (*a*) is measured (the specific optical rotation of Dx is +199° mL/gdm). This method is applicable for the determination of concentration of other carbohydrate polymers as well.

Freeze-drying can be used to measure the concentration of polymers in the stock solutions: *(i)* a known amount (from 5 to 20 g) of polymer stock solution is added to a freeze-drying flask; *(ii)* the solution is freeze-dried for about 8 hours; and *(iii)* the dried polymer is dissolved in 2 mL of water, then freeze-dried for another 8 hours. The user should check for constant weight at least the first time that this technique is used. We have found that extensive freeze-drying followed by rehydration in a small amount of water and subsequent freeze-drying yields results that are identical to those obtained with other techniques.

Two to four times concentrated stock solutions of salts are prepared using reagent grade chemicals. The solutions may be adjusted to the required pH before making up the final mass of the solution. For example, in case of phosphate and citrate solutions, the acid and the basic salts are weighed in molar ratios determined by the desired pH. Salts can also be used directly in the solid form.

For cross-partitioning experiments, the following stocks are also required:

a. Stock solutions of a series of 0.04 M buffers (glycine or sodium phosphate) spanning the pH range from 3.5 to 11.5.
b. Stock solutions of alkali (i.e., lithium, sodium, potassium) chlorides (0.33 M).
c. Stock solutions of alkali sulfates (0.167 M).

Protein stock solution, 1 g/L. Proteins should be desalted before use. Dialysis against a buffer made in nanopure water, ultrafiltration operated in dialysis mode, or a desalting column can be used for this purpose. For example, to desalt a protein

Table 4. Some Convenient Aqueous Two-Phase Systems

Concentration of Species 1 in % (wt/wt)	Concentration of Species 2 in % (wt/wt)	Water (or Buffer) in % (wt/wt)	V_T:V_B
PEG-3400 6.5%	Dx-500 6.50%	87	1:1
PEG-6000 6.74%	Dx-70 10.82%	75.94	1:1
PEG-1500 13.66%	K_2PO_4 (pH 7.0) 13.12%	73.22	1:1
PEG-1000 14.5%	$MgSO_4$ 10%	75.5	1:1

The polymers molecular weights are nominal.

by ultrafiltration prepare 50 g of a 1 g/L protein solution and ultrafiltrate with 5 volumes of a 50 mmol Tris buffer, pH 7.0.

Stock solutions of affinity ligands, e.g., PEG-bound ligand, can be prepared at a concentration of hundredfold or more, as the final concentration of these species in the 2-phase system is quite low. The cost of the PEG derivatives limits the amount of stock solution to be prepared.

3.2. Selection and Preparation of Aqueous Two-Phase Systems

The selection of the appropriate phase system depends on the final application. PEG/Dx is by far the most common pair of polymers used. Other incompatible polymers were already mentioned and summarized in Table 2. High molecular weight Dx (Dx-500 000) is highly recommended, since it can be used with low molecular weight PEGs reducing the viscosity of the phases. Also popular are PEG/K_2HPO_4-KH_2PO_4 systems. After the phase forming species have been selected, the next step is to select a particular tie-line. Good sources of tie-lines are the monograph by P.Å. Albertsson (2), the book by Zaslavsky (56), which contains about 150 phase diagrams for Dx/PEG, Dx/Polyvinylpyrrolidine (PVP), Dx/Polyvinyl alcohol (PVA), Dx/Ficoll, PEG/Polyvinyl methyl ether, and PEG-salt, and several articles (16,43,44). Some convenient 2-phase systems are shown in Table 4. As a rule of thumb, those who are using aqueous 2-phase systems for the first time should choose equal volumes of top and bottom phases to facilitate sampling and protein partition coefficient determination. For analytical purposes, 5- or 10-g systems are very convenient. Although any buffer can be used, for acidic and neutral pHs, phosphate buffers are recommended, whereas for basic pHs, Tris buffers can be used. Buffer concentration should be kept between 20 to 50 mM. It is less cumbersome to work at room temperature since the mixing, equilibration, and sampling has to be done at the same temperature. Because protein partitioning is only marginally affected by temperature changes, low temperatures may be desirable to maintain protein stability. The final preparation of an aqueous 2-phase system is quite straightforward, and a detailed recipe can be found in Reference 11. An example is given in Procedure 2.

❖ Procedure 2. Preparation of a Dx-500 000/PEG-4000 System at Room Temperature

1. Shake the stock solutions well so that there are no density gradients.
2. Place a graduated centrifuge tube of 15 mL total volume on a weighing balance.
3. Weigh out the stock solutions into the tube in order of their increasing densities, and layer them carefully over each other. This facilitates the removal of portions of one stock solution in case of error during weighing. Because of

the problem of accurately pipetting the polymer stock solutions due to their high viscosity, they are best measured by weight and are easily transferred using a Pasteur pipet with a broken tip.

4. Mix the contents of the test tube thoroughly, first by hand, and then in a rotary shaker (20 min is enough) at the equilibration temperature.
5. Let the systems settle for a period of 30 minutes to 24 hours depending on the system composition, or centrifuge them for 2 to 15 minutes at 1500× *g*. Poor temperature control in centrifuges makes it more convenient to sediment the systems in a water bath or in a chromatographic chamber when working at a temperature other than ambient. In general, the time of phase separation depends on the distance of the working tie-line from the critical point. Close to the critical point, the phase separation time is long. At intermediate tie-lines the phase separation time is shorter. If the more viscous phase volume is larger than the volume of the less viscous phase, the phase separation time increases. If the system is to be used in a liquid–liquid partition chromatography system, one must chose a total concentration of polymers such that the PEG-rich phase constitutes most of the volume. This phase system must be allowed to settle for 24 to 48 hours before the PEG-rich phase is used as the mobile phase.

If instructions are followed carefully, the preparation of aqueous 2-phase systems should be routine laboratory work. Still, a common source of frustration for those using aqueous 2-phase systems for first time is their apparent lack of reproducibility. As indicated before, systems are normally prepared according to some published binodal. Often, the prepared system differs from the one published. Specifically, the ratio between top and bottom phase volumes of both published and prepared systems may not be the same. The appearance of only one phase after following a published recipe step-by-step is equally frustrating. These apparent inconsistencies cause people to believe that the lack of reproducibility is an inherent property of aqueous 2-phase systems. Fortunately, this is not true. The most common reasons for these inconsistencies are:

- The selected tie-line is too close to the critical point. So, small differences in the molecular weight or molecular weight distribution of the polymers, presence of additives, or differences in temperature moved the system into the one phase region. Addition of small amounts of one of the two polymers will move the system into the 2-phase region.
- The selected tie-line is too far from the critical point. When working at longer tie-lines poor mixing is normally the cause for the lack of production of 2 phases. Since the denser stock solution is added to the centrifuge tube first, it is quite difficult to mix the residue of stock solution that is trapped in the tip of the tube with the rest of the solution. So, the 2-phase system is actually prepared using a considerable smaller amount of one of the two polymers. To assure good mixing, mix the content of the tube in a vortex mixer and inspect the tip and walls of the tubes for stock solution residues. Continue mixing after no deposits are present, and place the tube in a rotary shaker.

3.3. Preparation of Liquid–Liquid Partitioning Chromatography Systems

Specific guidelines for the use of this method can be found in the various articles by Muller (35–37). The main steps are outlined in Procedure 3.

❖ Procedure 3. Preparation of Partitioning Chromatography Systems

1. Measure the partition coefficient of the raw materials in batch systems before attempting to run a liquid–liquid partition chromatography (LLPC) column. The partition coefficient of the target protein must be adjusted to be between 0.3 and 0.1 (the data shown in Figure 2 can be used for this purpose), and the salt concentration should be high enough to shield elec-

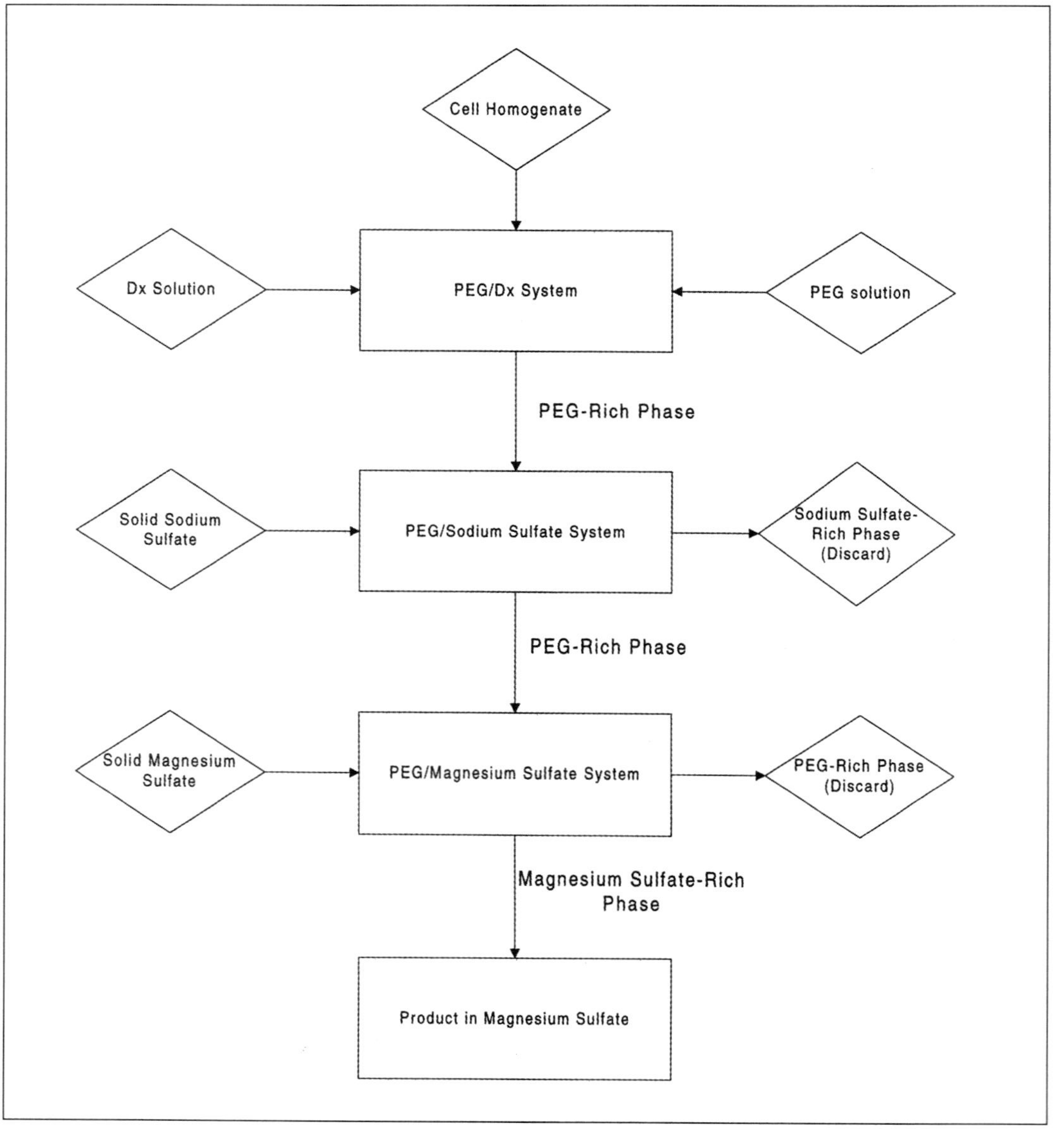

Figure 4. Purification of hemoglobin from *E. coli.* Cell homogenate is mixed with Dx and PEG solutions. The Dx-rich phase is discarded, and Na_2SO_4 is added to the PEG-rich phase. The Na_2SO_4-rich phase is discarded, and $MgSO_4$ is added to the PEG-rich phase. The protein is recovered from the salt-rich phase.

trostatic interactions between the proteins and the gel (ionic strength of 0.05 or higher).

2. After the optimum conditions to obtain an appropriate partition coefficient have been identified, prepare enough top phase at the right pH and at the right ionic strength to elute the column. The top (PEG-rich) phase is allowed to equilibrate for several days in the presence of small amount of bottom phase.
3. The column is packed according to Muller (36) and equilibrated until the UV noise of the effluent has dropped below 0.005 OD units at 280 nm. This ensures that all the Dx-rich phase that is not bound to the beads has been washed out.
4. The sample is dissolved in the mobile phase (it should not exceed 2%–3% of the bed volume for analytical runs and as twice as much for preparative runs) and injected into the column. The elution is started immediately.

As in any chromatographic separation, sample preparation is quite important. If the starting material is a cell homogenate, solids must be sedimented out by centrifugation for about 15 minutes at 2000× *g*. The clear supernatant is mixed with the appropriate amount of PEG that is going to be used as a mobile phase. If aggregates are observed, they must be eliminated by centrifugation. If no precipitation is observed, more PEG is added (to reach 30%), the liquid is cooled in an ice bath for 10 minutes, and the protein precipitate is removed and resuspended in buffer.

Some features of these type of systems are: *(i)* the partition coefficients must be sufficiently different from 1 to make an impact on the retention times. That is, they must be in the right range to make a multicomponent chromatographic separation possible; *(ii)* the elution volumes correlate quite well with the partition coefficients of the proteins obtained in batch experiments, so scale-up is straightforward; *(iii)* very low amounts of Dx are needed, which is of direct benefit as far as costs go. In addition, if the losses are proportional to the total Dx, then the losses are expected to be low as well; *(iv)* PEG precipitates large proteins (above 200 000 Da) at the stationary phase–mobile phase interface. This is avoided at all costs, as the precipitate clogs the column; and *(v)* the eluant contains significant amounts (around 10% wt/wt) PEG. Depending on the final application, this can be removed as described in the following section.

❖ Procedure 4. Determination of Partition Coefficients

1. Approximately 1 mL of the protein solution to be purified is added to the phase-forming species mixture replacing an equal amount of buffer. Mixing and phase separation are done as described above for systems that do not contain any protein.
2. Mixing has to be done carefully. It has to be vigorous enough to allow distribution of the proteins between the 2 phases but gentle enough to prevent protein denaturation (a rotary shaker is highly recommended, whereas the use of a vortex mixer is discouraged).
3. The phase systems are centrifuged at 1500× *g* for 20 minutes to speed phase settling.
4. Sampling is done by pipetting carefully 1 mL of top phase and 1 mL of bottom phase from each partitioning tube (the amount pipetted should be controlled by weighing for more precise sampling). Impurities may accumulate at the liquid–liquid or liquid–air interfaces. They do not constitute a problem unless they are pipetted during

sampling, so a positive pressure on the pipet as it enters the phases is always recommended. Blank phase systems are sampled in the same manner.

5. The samples are diluted with buffer. The actual dilution depends on the particular protein and on its partition coefficient. Since the viscosity of the phases is very high, improper mixing of the sample and the dilution buffer may result. Uncontrollable scattering from regions of different densities within the sample produces erroneous absorbance readings. As a general rule, mix the sample of the phase with the dilution buffer and stir in a vortex mixer.
6. Leave the solution resting and stir again. Inspect the solution to detect density differences along the axial direction of the test tube. Continue stirring until the solution is completely transparent. Because of the relatively high absorbance of the blanks at 280 nm, an absorbance reading of protein containing samples between 0.5 and 1 is recommended. Hemoproteins are conveniently measured at 540 nm. Standard protein tests like Bradford's test (8) can be also used.
7. The partition coefficient is calculated from $K = [\text{Absorbance}_{\text{sample}} - \text{Absorbance}_{\text{blank}}]^{\text{top}}/[\text{Absorbance}_{\text{sample}} - \text{Absorbance}_{\text{blank}}]^{\text{bottom}}$.
8. For preparative applications a precise mass balance is not necessary. For analytical purposes, a protein mass balance can easily be performed, since the volumes of the phases are very easy to measure, and the density of each phase is well correlated with polymer concentration. If the mass balance is not close (within 5%), check for the formation of a precipitate at the liquid–liquid interface. If a precipitate is present, one should use more diluted protein solutions (a decrease of 50% in protein concentration is usually enough). If no precipitate is present, poor sampling is probably the source of error.

3.4. Removal of PEG

Depending on the final application of the protein purified by aqueous 2-phase extraction, it may be desirable to eliminate all or most of the polymer that contaminates the protein of interest. Although the overall yield of the separation may be reduced, one of the easiest ways of eliminating the polymer (usually PEG) from the protein solution is to repartition the PEG-rich phase against a salt (phosphate or sulfate) rich phase. This is accomplished rather easily by first separating the top (PEG-rich) phase from the bottom (Dx-rich) phase and by adding either solid sodium phosphate or sulfate directly into the PEG-rich phase. By driving the protein into the salt-rich phase most of the PEG is eliminated. If the size of the protein is sufficiently different from the size of PEG, PEG-protein mixtures can be separated by ultrafiltration and by gel permeation chromatography. For example, we have used Ultra free-20 (Sigma) centrifuge tubes with a nominal molecular weight cut-off of 10 000 to separate lysozyme from PEG-4000. The samples were centrifuged at 12 000× *g* for 30 minutes. Up to 85% of the PEG present is eliminated in this way.

3.5. Methods for Characterization Experiments

3.5.1. Cross-Point Determination

Prepare 2 sets of Dx/PEG systems (Set A which contains alkali chloride and Set B which contains alkali sulfate) spanning a pH range from 3.5 to 11.5. Two runs are highly recommended for precision work. In the first run, 4 or 5 different pH values are enough. In the second run, 5 or 6 points should be obtained in the neighborhood of the cross-point.

❖ Procedure 5. Cross-Point Determination

1. Set A. Add to a 10-mL centrifuge tube (37): *(i)* 2.5 g of Dx stock solution; *(ii)* 1.0 g of PEG stock solution; *(iii)* 3 g of sodium chloride; *(iv)* 2.5 g of buffer; and *(v)* 1 g of the protein stock solution.
2. Set B. Add to a 10-mL centrifuge tube: *(i)* 2.5 g of Dx stock solution; *(ii)* 1.0 g of PEG stock solution; *(iii)* 3 g of or sodium sulfate solution; *(iv)* 2.5 g of buffer; and *(v)* 1 g of the protein stock solution.
3. The final phase system composition is 7.5% (wt/wt) Dx, 5.0 (wt/wt) PEG, 0.1 M alkali chloride or 0.05 M alkali sulfate, and 0.04 M glycine or sodium phosphate buffer.
4. Prepare blanks of the phases without added protein.
5. Mix, equilibrate, and sample the phases as explained in Section 3.4.6.
6. The pH in each phase is measured with a microelectrode directly on the undiluted phases. Because of the high viscosity of the phases, the pH measurements must be done over a relatively long period of time.
7. The partition coefficients of Sets A and B are plotted versus the pH. The pH and the partition coefficient values at which one K_p versus pH line (Set A) crosses the other one (Set B) and are read from the axes.

The lines of K_p versus pH may not cross each other because of errors in pH or in the values of the partition coefficients. One must be sure that the pH has been measured long enough to reach equilibrium and that the pH of both phases agrees within the experimental uncertainty (approximately 0.05 pH units). Erroneous values of K_p are generally due to poor sampling, and a protein mass balance should be done to assure that sampling has been done correctly.

The sensitivity of cross-partitioning depends upon the angle at which the 2 lines intersect. If the lines are perpendicular, the sensitivity is at a maximum, while parallel or nearly parallel lines yield no cross-point or a "cross-point range". The slope of the lines depends upon the type of salt, the change in net charge of the protein with pH, the specific interactions between the ions and the proteins, and the salt-induced changes in the interactions between polymer and protein. The sensitivity can be manipulated by varying the molecular weight of the polymers, the temperature, the concentration of the polymers, and the type of salt. So, cross-partitioning should be done using the lowest possible PEG molecular weight to minimize problems associated with the high viscosity of the phases. If the sensitivity is not good enough, the experimentalist needs to explore different conditions until a good sensitivity is found.

The pH and the partition coefficient at the cross-point are only marginally dependent on the combination of salts used and on their concentration. For example, NaCl can be replaced by potassium chloride and/or sodium sulfate by lithium sulfate without affecting the results. Still, some small differences in pH values at the cross-point with different salts have been observed. These differences are similar to those encountered in the electrophoretic determination of isoelectric points, which can also be slightly affected by the salt used.

This independence of cross-partitioning on the type and concentration of salt makes cross-partitioning a viable option for determining the isoelectric point of proteins that are stable only at high salt concentrations. In contrast, the type and concentration of salts have a strong influence on the shape of the ln K_p versus pH curves.

3.5.2. Surface Hydrophobicity

Dx/Ficoll-400 systems are prepared by weighting stock solutions of the polymers to a final concentration of 12.5% Ficoll and 10.8% Dx-70. The systems are prepared in sodium phosphate buffer at pH 7.4 at a concentration of (56):

$$C_{buffer} = 0.11 - 0.67 C_{NaCl} \qquad \text{[Eq. 4]}$$

where the concentration of the buffer is varied from 0.01 to 0.11 M and the concentration of NaCl is varied from 0 to 0.15 M. The protein(s) of interest is partitioned in this set of systems as indicated above. The logarithm of the partition coefficient is plotted versus the ionic strength. The zero intercept yields a parameter that represents the strength of all the interactions of the protein with an aqueous environment relative to that of a methylene group and the slope yields a parameter that reflects the strength of the hydration interactions of all the ionogenic groups of the protein relative to that of the α-carboxyl group of DNP-amino acid.

4. EXAMPLES

4.1. Isolation of Recombinant Hemoglobin from Cell Homogenates

The work by Hart and Bailey (21) is a good example of the use of aqueous 2-phase systems for the purification of a recombinant protein. They isolated *Vitreoscilla* hemoglobin from *E. coli* lysate. The purification was done in three extraction steps (see Figure 4) with an overall yield of 47% and a purity higher than 95%. In the first partition step, a PEG/Dx system was used. In this system, the contaminant proteins partitioned strongly into the Dx-rich phase, whereas the hemoglobin preferred the PEG-rich phase. Additional purification of the target protein is achieved by adding solid sodium sulfate to the PEG phase, thus forming another PEG-salt 2-phase system. Again, the hemoglobin preferred the PEG-rich phase. Solid magnesium sulfate was added to the PEG-rich phase to form a PEG/$MgSO_4$ 2-phase system. In contrast to the PEG/Na_2SO_4 system, the hemoglobin preferred the salt-rich phase. The use of Mg titration to separate the protein from PEG may be applicable to other proteins too.

4.2. Partitioning of Hemoproteins in PEG/Dx/Cu plus PEG Systems

Small amounts of metal chelate PEG (PEG-iminodiacetic acid loaded with Cu^{++}) added to conventional aqueous 2-phase systems have been used to extract proteins that contain histidine (40,55). A plot of $\Delta \ln K_p$ is linear with the number of exposed histidine groups on the protein molecules. They were able to separate cytochrome c, myoglobin, and hemoglobins. pH control is critical when using this technique, since at low pHs the free base of the imidazole form a noncoordinating imidazolium side chain that does not bind Cu^{++}. The enhancement in the partitioning of the proteins into the PEG-rich phase caused by the addition of PEG-IDA is remarkable. For example, at pH 8.0, the partition coefficient of human hemoglobin is 0.38 (7% PEG-8000, 4.4% Dx T-500, 0.1 M NaCl, and 0.01 M sodium phosphate), whereas upon addition of PEG-IDE the partition coefficient increases to 14.

4.3. Cross-Partitioning of Hemoglobins

Numerous hemoproteins have been studied by using cross-partitioning experiments. The pH at the cross-point agrees very well with the isoelectric point pH determined by using other techniques. For example, human A hemoglobin has a pH at the cross-point of 7.0 (isoelectric point: 7.0), and cytochrome c from horse yields a pH at the cross-point of 9.85 (isoelectric point: 9.8). The cross-partitioning of hemoproteins from cytochrome c (MW 12 000) to catalase (MW 240 000) in Dx/PEG systems (51) yields partition coefficients at the cross-point that do not show the clear dependence on protein molecular weight found with nonhemoproteins. Even though the molecular weights of human hemoglobin variants (A, F, S, and C) and hemoglobins from different species are essentially the same, the partition coefficient at the cross-point of hemoglobins A and F and those of hemoglobins from different mammalian species show measurable differences. Although the 4 human hemoglobin variants differ in charge, adult hemoglobins A, S, and C have the same the partition coefficient, while the fetal hemoglobin (F) has a lower partition coefficient.

4.4. Liquid–Liquid Partitioning Chromatography

There are numerous examples of the use of this technique for the purification of nucleic acids and proteins. In a few of them, the addition of affinity ligands to improve the separation has been explored.

Because the elution volume is extremely sensitive to the partition coefficient of the protein (if it is smaller than 3), even in the absence of affinity ligands considerable resolution can be obtained. For example, Muller (37) separated a synthetic mixture of lysozyme (retention volume, V = 17 mL), peroxidase (V = 19 mL), cytochrome c (V = 25 mL), myoglobin (V = 30 mL), β-lactoglobulin (V = 39 mL), ovotransferrin (V = 50 mL), ovalbumin (V = 60), and human serum albumin (V = 100) (0.6–1 mg each) in a 300 × 10 mm column packed with Lipargel coated with a solution of Dx-40 and eluted at a flow rate of 0.3 mL/minute with a PEG-6000 solution.

We have demonstrated that affinity ligands can be used in LLPC to increase the purification factor of a given protein (49). The tune-up of the separation is quite straight forward, and it includes optimization of the partition coefficient of the target protein as compared to the contaminant proteins, optimization of the amount of affinity ligand, and inhibition of other enzymes that compete with the target protein for the affinity ligand. For example, in the purification of FDH by using affinity liquid–liquid partition chromatography, we found that the optimum conditions were achieved when the stationary phase was Dx-500, the mobile phase PEG-20 000 (2.7% PEG and 4.5 Dx), 75 mM potassium bromide, 12 mM phosphate buffer at pH 7.5, and a concentration of PEG-red in the mobile phase of 5×10^{-5} M. The sample was 500 μL of crude extract of *C. boidinii* heated for 10 minutes at 55°C. We also showed that the separation of the ligand from the enzyme is quite straightforward. The eluate containing the FDH–ligand complex was treated with potassium phosphate forming a PEG-salt system. Using a concentration of potassium phosphate of 10% (wt/wt) the K value of FDH was 0.003, and the volume ratio was 1:4.22 (top to bottom). The yield of the enzyme in the lower phase was 99%. The column was stable for more than 1 year, and scale-up was straightforward.

5. CONCLUDING REMARKS

Aqueous 2-phase extraction is a very well-established technique that has been used in biochemical laboratories for the last 40 years. It is a versatile, easy to use, and low cost technique. Although the primary use of the method is the purification of proteins and other biological materials, it can be also used for protein characterization studies. Uncountable proteins have been purified using this technique either from cell homogenates or from previously fractionated mixtures. Although it has been used to separate cell debris from proteins providing a first cheap purification step, it has been also used to purify proteins in a single step from cell debris by making use of a variety of affinity ligands. The possibility of using the technique in continuous mode by immobilizing 1 phase opens even more possibilities. One of its main advantages is its low cost and easy use. The cost of chemicals is minimum (except if affinity ligands has to be used), and the hardware needed to implement it is available in any biochemistry laboratory. Members of the aqueous 2-phase systems community can be reached at our Web page for helping those who wish to use this technique. I hope that this chapter will encourage researchers in the heme and related compounds research field to consider aqueous 2-phase systems for their isolation procedures.

ABBREVIATIONS

Dx, dextran; HF, hydrophobicity factor; HSF_i, surface hydrophobicity of protein I; K_p, partition coefficient (= Ct/Cb); K_0, partition coefficient at zero charge; K^0, partition coefficient in the absence of affinity ligands; K_L, partition coefficient in the presence of affinity ligands; PEG, polyethyleneglycol; P(EO-PO), polyethylene propylene random copolymer (UCON); PVA, polyvinyl alcohol; PVP, polyvinylpyrrolidone; RH_i, hydrophobicity of solute i relative to glycerin; TLL, tie-line length.

REFERENCES

1. **Aguinada-Diaz, P.A. and R.Z. Guzman.** 1996. Affinity partitioning of metal ions in aqueous polyethylene glycol/salt two-phase systems with PEG-modified chelators. Sep. Sci. Technol. *31*:1483-1499.
2. **Albertsson, P.Å.** 1986. Partition of Cell Particles and Macromolecules, 3rd ed. Wiley (Interscience), New York.
3. **Albertsson, P.Å., S. Sasakawa, and H. Walter.** 1970. Cross partition and isoelectric points of proteins. Nature *228*:1329-1330.
4. **Alred, A.** 1999. Purification of a recombinant protein from mammalian cell culture: an industrial application. 11th International Conference on Partitioning in Aqueous Two-Phase Systems, Gulf Shore, Alabama.
5. **Asenjo, J.A., R.E. Turner, S.L. Mistry, and A. Kaul.** 1994. Separation and purification of recombinant proteins from *Escherichia coli* with aqueous two-phase systems. J. Chromatogr. A *668*:129-138.
6. **Bergreen, K., G. Johansson, and F. Tjerneld.** 1995. Effects of salts and the surface hydrophobicity of proteins on partitioning in aqueous two-phase systems containing thermoseparating ethylene oxide-propylene oxide copolymers. J. Chromatogr. A *718*:67-79.
7. **Birkenmeir, G., G. Kopperschlager, P.A. Albertsson, G. Johansson, F. Tjerneld, H.E. Akerlund, S. Berner, and H. Wickstroem.** 1987. Fractionation of proteins from human serum by counter-current distribution. J. Biotechnol. *5*:115-129.
8. **Bradford, M.M.** 1976. A rapid and sensitive method for the quantitation of microgram quantities of protein utilizing the principle of protein dye binding. Anal. Biochem. *72*:248-254.
9. **Fernandez, S., G. Johansson, and R. Hatti-Kaul.** 1999. Two-phase partitioning of recombinant Cutinase. 11th International Conference on Partitioning in Aqueous Two-Phase Systems, Gulf Shore, Alabama.
10. **Forciniti, D.** 1994. Protein refolding using aqueous two-phase systems. J. Chromatogr. A *668*:95-100.
11. **Forciniti, D.** 1999. Preparation of aqueous two-phase systems. *In* R. Hatti-Kaul (Ed.), Aqueous Two-Phase Systems–Methods and Protocols. Humana Press, New Jersey.
12. **Forciniti, D., C.K. Hall, and M.-R. Kula.** 1991. Effect of temperature on protein partitioning. Bioseparations *2*:115-128.
13. **Forciniti, D., C.K. Hall, M.-R. Kula.** 1991. Protein partitioning at the isoelectric point: effect of polymer concentration and polymer molecular weight. Biotechnol. Bioeng. *38*:986-994.
14. **Forciniti, D., C.K. Hall, and M.-R. Kula.** 1992. Protein partitioning. Effect of pH and polymer molecular weight. Chem. Eng. Sci. *47*:165-175.
15. **Forciniti, D., M.-R. Kula, and C.K. Hall.** 1990. Interfacial tension in aqueous two-phase systems. Influence of temperature and polymer molecular weight. J. Biotechnol. *16*:279-290.
16. **Forciniti, D., M.-R. Kula, and C.K. Hall.** 1991. Influence of polymer molecular weight and temperature on phase composition in aqueous two-phase systems. Fluid Phase Equilib. *61*:243-262.
17. **Forciniti, D., M.-R. Kula, and C.K. Hall.** 1991. Molecular weight distribution and aqueous two-phase systems. Biotechnology *20*:151-162.
18. **Guan, Y., T.H. Lilley, T.E. Treffry, C.-L. Zhou, and P.B. Wilkinson.** 1996. Use of aqueous two-phase systems in the purification of human interferon-alpha1 from recombinant *E. Coli*. Enzyme Microb. Technol. *19*:446-455.
19. **Guoqiang, D., R. Kaul, and B. Mattiasson.** 1994. Integration of aqueous two-phase extraction and affinity precipitation for the purification of lactate dehydrogenase. J. Chromatogr. A *668*:145-154.

20.**Harris, P.A., G. Karlström, and F. Tjerneld.** 1991. Enzyme purification using temperature- induced phase formation. Bioseparation *2*:237-246.
21.**Hart, R.A. and J.E. Bailey.** 1991. Purification and aqueous two-phase partitioning properties of recombinant *Vitreoscilla* hemoglobin. Enzyme Microb. Technol. *13*:788-795.
22.**Hatti-Kaul, R. (Ed.).** 1999. Aqueous Two-Phase Systems–Methods and Protocols, Chs. 2 and 3. Humana Press, New Jersey.
23.**Hustedt, H., K.H. Kroner, and M.-R. Kula.** 1984. Applications of phase partitioning in biotechnology, p. 529-589. *In* H. Walter, D.E. Brooks, and D. Fisher (Eds.), Partitioning in Aqueous Two-Phase Systems. Academic Press, New York.
24.**Johansson, G.** 1984. Partitioning of proteins, p. 161-226. *In* H. Walter, D.E. Brooks, and D. Fisher, (Eds.), Partitioning in Aqueous Two-Phase Systems. Academic Press, New York.
25.**Johansson, G.** 1994. Charge determination by partitioning. *In* H. Walter and G. Johansson (Eds.), Methods in Enzymology. Academic Press, New York.
26.**Kopperschlager, G.** 1994. Affinity extraction with dye ligands. *In* H. Walter and G. Johansson (Eds.), Methods in Enzymology. Academic Press, New York.
27.**Lorch, J.** 1999. Two-phase aqueous extraction as a process development tool. 11th International Conference on Partitioning in Aqueous Two-Phase Systems, Gulf Shore, Alabama.
28.**Lu, M., F. Tjernald, G. Johansson, and P.Å. Albertsson.** 1991. Preparation of benzoyl-Dx and its use in aqueous two-phase systems. Bioseparation *2*: 247-255.
29.**Lundeberg, S. and L. Backman.** 1994. Protein-protein and protein-ligand interactions, p. 241-254. *In* H. Walter and G. Johansson (Eds.), Methods in Enzymology. Academic Press, New York.
30.**Mattiasson, B. and I.Y. Galaev.** 1999. Affinity partitioning using aqueous two phase systems formed by thermosensitive polymers. 11th International Conference on Partitioning in Aqueous Two-Phase Systems, Gulf Shore, Alabama.
31.**Menge, U., M. Morr, U. Mayr, and M.-R. Kula.** 1983. Purification of human fibroblast interferon by extraction in aqueous two-phase systems. J. Appl. Biochem. *5*:75-90.
32.**Middaugh, C.R. and E.Q. Lawson.** 1980. Analysis of protein association by partitioning in aqueous two-phase polymer systems: applications to the tetramer-dimer association of hemoglobin. Anal. Biochem. *105*:364-368.
33.**Modlin, R.F., P.A. Alred, and F. Tjerneld.** 1994. Utilization of temperature-induced phase separation for the purification of ecdysone and 20-hydroxyecdysone from spinach. J. Chromatogr. A *668*:229-236.
34.**Moribe, K., K. Maruyama, and M. Iwatsuru.** 1997. Estimation of surface state of poly(ethylene glycol)-coated liposomes using an aqueous two-phase partitioning technique. Chem. Pharm. Bull. *45*:1683-1687.
35.**Muller, W.** 1989. Aqueous two-phase polymer systems for liquid/liquid partition-chromatography of biopolymers. Ber. Bunsen-Ges. Phys. Chem. *93*:956-961.
36.**Muller, W.** 1994. Columns using aqueous two-phase systems. *In* H. Walter and G. Johansson (Eds.), Methods in Enzymology. Academic Press, NY.
37.**Muller, W.** 1994. Separation of proteins and nucleic acids. *In* H. Walter and G. Johansson (Eds.), Methods in Enzymology. Academic Press, New York.
38.**Petersen, L.C.** 1978. Cytochrome c-cytochrome aa_3 complex formation a low ionic strength studied by aqueous two-phase partition. FEBS Lett. *94*:105-108.
39.**Petersen, L.C.** 1978. Measurements of cytochrome c-cytochrome aa_3 complex formation by aqueous two-phase partition. Biochem. Soc. Trans. *6*:1274-1275.
40.**Plunkett, S.D. and F.H. Arnold.** 1990. Metal affinity extraction of human hemoglobin in an aqueous polyethylene glycol-sodium sulfate two-phase system. Biotechnol. Tech. *4*:45-48.
41.**Sasakawa, S. and H. Walter.** 1972. Partition behavior of native proteins in aqueous Dx-poly(ethylene glycol) phase systems. Biochemistry *11*:2760-2765.
42.**Shanbhag, V.P. and P.E.H. Jensen.** 1999. Affinity partitioning using poly(ethylene glycol) with covalently coupled hydrophobic groups. *In* R. Hatti-Kaul (Ed.), Aqueous Two-Phase Systems—Methods and Protocols. Humana Press, New Jersey.
43.**Silva, L.H.M., J.S.R. Coimbra, and A.J.A. Meirelles.** 1997. Equilibrium phase behavior of poly(ethylene glycol) + potassium phosphate + water two-phase systems at various pH and temperatures. J. Chem. Eng. Data *42*:398-401.
44.**Snyder, S.M., K.D. Cole, and C.C. Szlag.** 1992. Phase composition viscosities, and densities for aqueous two-phase systems composed of polyethylene glycol and various salts at 25 C. J. Chem. Eng. Data *37*:268-274.
45.**Sophianopulos, A.J. and K.E. Van Holde.** 1964. Physical studies of muramidase (lysozyme). J. Biol. Chem. *239*:2516-2524.
46.**Spears, T. and D. Forciniti.** 1994. Protein refolding using chaotropic aqueous two-phase systems. *In* R.D. Rogers (Ed.), Aqueous Two-Phase Systems: From Metal Ions to Biomolecules. ACS Books, Washington.
47.**Tjerneld, F., P.A. Alred, R.F. Modlin, A. Kozlowski, and J.M. Harris.** 1995. Purification of biomolecules using temperature-induced phase separation. *In* R.D. Rogers and M.A. Eiteman (Eds.), Aqueous Biphasic Separations: Biomolecules to Metal Ions. Plenum Press, New York.
48.**Vernau, J. and M.-R. Kula.** 1990. Extraction of proteins from biological raw materials using aqueous PEG/citrate phase systems. Biotechnol. Appl. Biochem. *12*:397-404.
49.**Walsdorf, A., D. Forciniti, and M.-R. Kula.** 1990. Investigation of affinity partition chromatography using formate dehydrogenase as a model. J. Chromatogr. *523*:103-117.
50.**Walter, H., D.E. Brooks, and D. Fisher (Eds.).** 1984. Partitioning in Aqueous Two-Phase Systems, p. 498-528. Academic Press, New York.
51.**Walter, H. and D. Forciniti.** 1994. Cross-partitioning: determination of isoelectric point by partitioning, p. 223-233. *In* H. Walter and G. Johansson (Eds.), Methods in Enzymology. Academic Press, New York.
52.**Walter, H. and G. Johansson (Eds.).** 1994. Methods in Enzymology, Vol. 228. Academic Press, New York.

53. **Walter, H., S. Sasakawa, and P.Å Albertsson.** 1972. Cross partition of proteins. Effect of ionic composition and concentration. Biochemistry *11*:3880-3883.
54. **Winter, C., D. Ansaldi, J. Clifford, P. Lester, and B. Wolk.** 1999. Initial isolation of recombinant proteins from whole fermentation lysates using aqueous two phase systems. 11th International Conference on Partitioning in Aqueous Two-Phase Systems, Gulf Shore, Alabama.
55. **Wuenschell, G.E., E. Naranjo, and F.H. Arnold.** 1990. Aqueous two-phase metal affinity extraction of heme proteins. Bioprocess Eng. *5*:199-202.
56. **Zaslavsky, B.Y.** 1995. Aqueous Two-Phase Partitioning. Marcel Dekker, New York.

9 Structural Study of Heme Proteins by Electron Microscopy of 2-Dimensional Crystals

Terrence G. Frey
San Diego State University San Diego, CA, USA

1. INTRODUCTION

Characterization of membrane proteins has always lagged behind that of soluble proteins, because most techniques of protein purification and structural study depend upon having a water-soluble preparation. The development of myriad classes of detergents over the past decades has greatly facilitated the purification of integral membrane proteins and many of the techniques of structural characterization. Direct measurement of the 3-dimensional (3D) structures of integral membrane proteins has proceeded more slowly, but significant progress was made in the past 10 to 15 years, and the atomic structures of a number of membrane proteins are now known. Although X-ray crystallography and 2-dimensional (2D) nuclear magnetic resonance (NMR) spectroscopy have been used to elucidate the structures of thousands of soluble proteins, their utility in the structural study of membrane proteins has been limited by the difficulty of growing 3D crystals in the case of X-ray techniques or by the large size of protein–detergent micelles in the case of NMR spectroscopy. Thus, electron microscopy of 2D crystals with the application of image processing techniques, sometimes called electron crystallography because many of the computational tools and techniques are derived from X-ray crystallography, was the first technique to reveal the structure of a membrane protein in three dimensions (43). It continues to play a very significant role in the structural study of membrane proteins. In this paper, I will review the techniques of electron crystallography applied to integral membrane heme-containing proteins describing in general terms:

- Techniques for growth of 2D crystals.
- Specimen preparation for electron microscopy and collection of micrographs.
- Data processing to produce a 3D model.
- Comparison of results from electron crystallography and X-ray crystallography.

The reader should recognize that in the context of these methods, heme-containing membrane proteins are not intrinsically

Heme, Chlorophyll, and Bilins: ***Methods and Protocols***
Edited by A.G. Smith and M. Witty

different from other membrane proteins, and the techniques described and referenced here are applicable to other classes of membrane proteins (86). The methods used are in many cases very technical, and a full description is well beyond the scope of a single article or even a whole volume. Thus, I will outline the methods in relatively general terms and give the reader references for greater depth of description of the techniques involved. Amos et al. have written an excellent and comprehensive review of 3D structure determination of 2D crystals (4).

2. GROWTH OF 2D CRYSTALS

2D crystals can be defined as crystals in which a motif, a protein, or assembly of proteins in the case of a protein crystal, is repeated at the points of a 2D lattice. Thus, a 2D crystal is one that is characterized by a unit cell that periodically repeats at precise locations in two dimensions but not in the third. Note that a 2D crystal is not a 2D object but is 3D. The nomenclature used to describe space groups of 2D crystals was first described by Holser (45) and revived by Fuller et al. (37). The first symbol describes the type of lattice and the second symbol the class of symmetry perpendicular to the crystal. The following symmetry operations describe symmetry elements within the plane of the crystal. There are a number of excellent reviews of membrane protein crystallization, and some of these are devoted to growth of 2D crystals (50,55,60). Thus, I will not go into great detail describing these techniques here, but will offer my own perspective.

2.1. Naturally Occurring Crystals

One of the first membrane proteins studied by electron crystallography, bacteriorhodopsin, occurs naturally in patches of 2D crystalline "purple membrane" within the plasma membrane of *Halobacterium halobium*. The purple membrane can be readily purified and consists of large 2D arrays of bacteriorhodopsin molecules arranged on a 2D lattice and belonging to the 2-sided space group *p*3. This was one of the specimens exploited by Henderson and Unwin to demonstrate the efficacy of recording noisy electron micrographs at low electron doses and then recovering a high resolution image of a single unit cell by averaging thousands of unit cells within a crystal (78). They subsequently also calculated the first 3D image of a membrane protein and at a resolution sufficient to demonstrate transmembrane α-helices (43). A number of other membrane proteins occur naturally in 2D crystals; these include gap junctions and the photosynthetic membrane of the bacterium *Rhodopseudomonas viridis* (68,76). Others can be induced to form 2D crystals within their native membranes by the addition of specific ligands (62,71), by removing lipid enzymatically (59), or by other methods (61), and since membrane proteins are inserted the same way into their native bilayers, these crystals are all of the type shown in Figure 1a. It is also possible to form 2D crystals of a membrane protein by expressing a recombinant form in high concentration in another type of cell, as was the case with cardiac gap junction protein (76). The latter approach affords the opportunity to modify the protein using recombinant DNA technology, either by removing components that might inhibit crystallization or to identify the location of a specific sequence in the final structure.

2.2. Growth by Detergent Extraction

In some cases, it is possible to form 2D crystals of a protein by extracting native membranes in which it is present at high concentration using detergent to remove not only excess lipid but also other contaminating proteins. For example, cytochrome *c* oxidase constitutes approximately 10% of the protein of the mitochondrial inner membrane, and one can purify a membrane fraction of nearly pure cytochrome oxidase by treating beef heart mitochondria with appropriate concentrations of detergents followed by centrifugation to separate detergent solubilized components from the membrane residue (30,66,80). This requires 2 to 3 detergent treatments, and in each case, the pellet becomes enriched in cytochrome oxidase while other membrane proteins and excess phospholipid are removed by decanting the supernatants. Depending upon the type and concentration of detergent used, two different crystal forms have been obtained. Multiple extractions with nonionic Triton® X-114 and X-100 produces a vesicular preparation of nearly pure cytochrome oxidase and 25% by weight of residual phospholipid. The molecules of cytochrome oxidase are arranged as dimers related by a crystallographic 2-fold axis. The crystal is formed when a large vesicle collapses causing molecules from two sides of the vesicle to interdigitate in the center of the vesicle, as shown in Figure 1b, forming a 2D crystal in the space group $p22_12_1$ with unit cell dimensions of a = 100 Å, b = 125 Å, and a thickness of 210 Å (42). In this case, each unit cell of the crystal contains one molecule from each layer of the

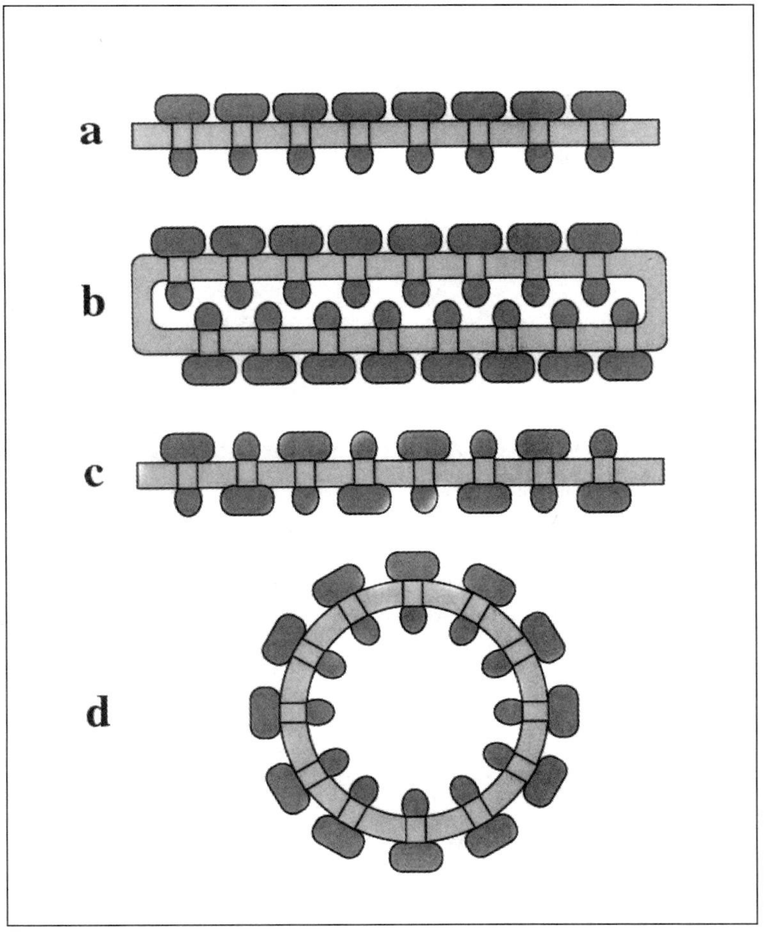

Figure 1. Molecular packing in four classes of 2D crystal. (a) All molecules oriented in the same direction in a single bilayer. (b) One crystal composed of two layers of a collapsed vesicle. (c) Molecules with alternating orientation in a single bilayer. (d) A crystal of class in panel a rolled into a cylinder producing a structure with helical symmetry.

collapsed vesicle, i.e., each unit cell contains two dimers. The space group describes a primitive orthorhombic lattice with a 2-fold axis of rotation perpendicular to the membrane. The symmetry operators 2_1 indicate 2-fold screw axes (rotation by 360°/2 followed by translation by half of a unit cell) parallel to the *a* and *b* crystal axes in the plane of the crystal.

Treatment with sodium deoxycholate removes a large proportion of phospholipid and dissociates cytochrome oxidase dimers into monomers. The resulting 2D crystal is not vesicular but consists of sheets of cytochrome oxidase monomers arranged on a primitive 2D lattice in the space group $p12_1$, denoting no symmetry perpendicular to the membrane and a 2_1 screw axis of symmetry along one of the crystal axes (37). These crystals are of the class shown in Figure 1c. The unit cell dimensions are a = 69 Å and b = 140 to 170 Å depending upon the preparation; the thickness is approximately 110 Å (35,37). In both cases, the structures of the preparations are more complex than described here. The vesicular $p22_12_1$ dimer crystals are predominantly multilayered vesicles with many small 3D crystals containing layers of the 2D crystal motif stacked in register (42,51,80). Single 2D crystals are more rare but can readily be found. The $p12_1$ crystal form contains single layer crystals as well as stacks of the simple 2D crystal motif in which successive layers are offset from one another, and electron micrographs show several differently appearing projections (35). Multilayered crystals of membrane proteins are commonly observed and often severely inhibit structural study because of the heterogeneity in the structures of different crystals (35,54,69).

2.3. Reconstitution of Purified Protein with Purified Lipids

The most general method for growing 2D crystals of membrane proteins begins with purified detergent solubilized protein and detergent solubilized lipids, either "native" lipids purified from the same type of membrane as the protein are derived or synthetic lipids using Procedure 1 (55,60,82,84). This is described in more detail in Chapter 11.

❖ Procedure 1. Growth of 2D Crystals

1. Dissolve lipids in an organic solvent (e.g., chloroform) and dry a measured amount on the surface of a suitable vessel under a stream of nitrogen.
2. Suspend the dried lipid in an appropriate volume of buffered detergent by sonication, producing mixed detergent–lipid micelles.
3. Mix the protein and lipid solutions approximately 1–10 mg/mL) to give a relatively high protein to lipid ratio (approximately 1:1 by weight).
4. Remove the detergent by dialysis or by adsorption.

As the detergent concentration decreases, the protein–detergent micelles and the protein–lipid micelles merge and eventually form bilayer membranes containing a high concentration of protein. Under the proper conditions, the protein molecules within the bilayers arrange themselves onto a 2D lattice forming 2D crystals (see Figure 2) (50,55,81). There are two common methods to remove detergent. One is to adsorb excess detergent by adding commercially available resin beads, e.g., Bio-Beads® (Bio-Rad Laboratories, Hercules, CA, USA) (81). A slower more controlled method is to dialyze the detergent–protein–lipid solution against detergent-free buffer. The speed of this process depends on the characteristics of the detergent employed, principally the critical micelle concentration (cmc)(50,55). In aqueous solution, detergent molecules aggregate to form micelles that are in equilibrium with individual detergent molecules; the cmc is essentially the concentration of individual molecules in the presence of micelles. Since detergent micelles are too large to pass through the pores of common dialysis tubing, dialysis of detergent solutions proceeds by the movement of individual detergent molecules through the dialysis tubing. Thus, the rate of dialysis depends on the cmc; the higher the cmc, the faster dialysis occurs. Of course dialysis also proceeds more rapidly at higher temperature. Triton detergents have low cmc's, so dialysis even at room temperature takes a relatively long period of time. For this reason, Weiss et al. used the Bio-Bead adsorption method to crystallize Complex III (cytochrome *c* oxidoreductase) purified in Triton X-100 from *Neurospora* mitochondria (81,84). Kim et al. grew crystals of purified cytochrome *c* oxidase essentially identical to the dimer crystals described above by reconstitution with purified phospholipids (53).

Many factors can affect the prospects of success in crystallizing a membrane protein. As with most crystallization experiments, a critical factor is the protein sample which should be pure and monodisperse. The latter property can be difficult to achieve in the case of membrane proteins which, owing to their hydrophobic surfaces, can adopt multiple aggregation states even in the presence of detergents. Thus, the choice of detergent can be critical and affects both the aggregation of the protein and the methods employed in removing it during crystallization trials. For example, Suarez et al. found that beef heart mitochondrial cytochrome *c* oxidase is polydisperse in many common detergents, but became monodisperse when transferred to dodecyl-maltoside (also known as lauryl maltoside), a detergent consisting of a maltose polar head and a 12 carbon saturated hydrocarbon tail that they synthesized for this purpose (now commercially available) (70). Choice of detergent also affects the crystallization process, since, as mentioned above, the cmc determines the dialysis rate, and other properties affect the efficacy of adsorption to beads. Frequently, two detergents within the same generic class can have significantly different cmcs. For example, dodecyl-maltoside has

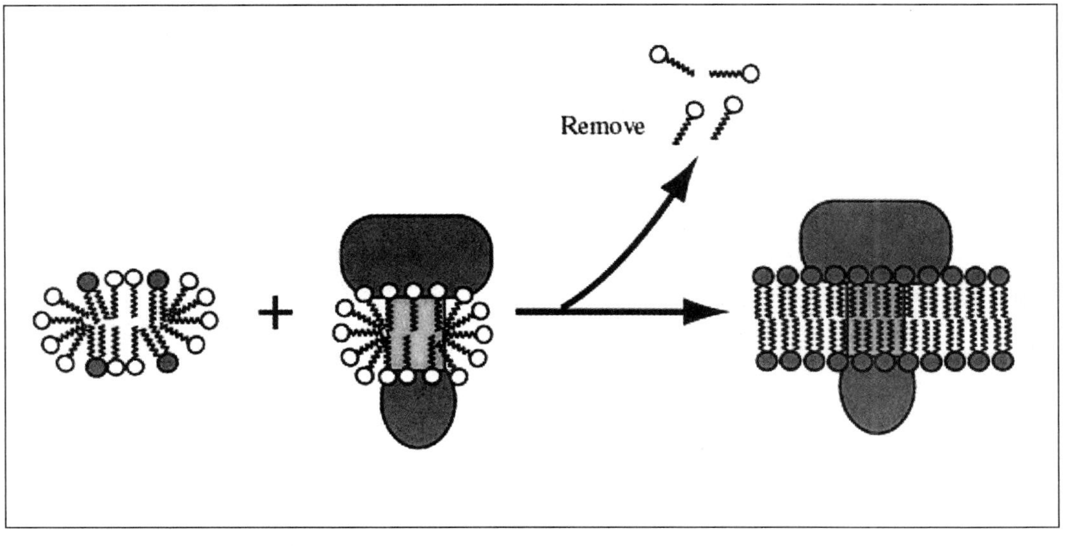

Figure 2. Reconstitution of purified phospholipids and purified membrane protein into a protein–lipid bilayer by mixing detergent solubilized lipid (left) and detergent solubilized protein (center) and then removing the detergent molecules.

a cmc of 0.15 mM, while decyl-maltoside has a cmc of 1.6 mM. In some cases it may be desirable to slow down dialysis of a detergent with a high cmc, and this can be accomplished by dialyzing the sample against a buffer containing the detergent at a concentration that is below its cmc and then reducing the detergent concentration when the dialysate is changed. Useful information on detergent properties can be found in Jap et al. (50) and in *Crystallization of Membrane Proteins* edited by Michel (60) as well as in a pamphlet published by Calbiochem-Novabiochem (6).

Other factors to be considered and varied are:

- pH.
- Ionic strength.
- Temperature.
- Choice of lipid.
- Protein concentration (generally several mg/mL).
- The presence of other solutes that might influence protein conformation (specific ligands, etc.) or solubility.

Given all of the conditions that might be varied, it is useful or essential to have a convenient system for microdialysis of many different samples. There are a number of commercial microdialysis cells suitable for this purpose. One can also construct microdialysis cells from glass tubing similar to those originally described by Zeppenzauer for crystallization of soluble proteins (87). Convenient and inexpensive microdialysis cells can also be constructed from a standard microcentrifuge tube by cutting off the conical tube leaving the tube cap and collar. In this case the compartment is formed by the tube cap over which one lays a small piece of dialysis tubing that is then sealed with the collar of the tube (Dr. Alok Mitra, personal communication).

3. SPECIMEN PREPARATION AND ELECTRON MICROSCOPY

Preparation of specimens for electron microscopy is a critical step that must take into consideration the goals of the project and the limitations imposed by the electron microscope. There are many books describing general techniques and methods for biological electron microscopy, and the reader should consult these for detailed protocols (7,22,40). Generally, the goal is to obtain the 2D or 3D structure of the protein at the highest possible resolution. The limitations imposed by the electron microscope are:

- The specimen should be thin, generally less than 1000 Å. A related goal is to obtain a preparation of 2D crystals that are a single unit cell thick, rather than multiple layers forming small 3D crystals.
- The specimen must be stable at very high vacuum, approximately 1 x 10^{-7} Torr or better. This usually requires removing most or all of the water, unless the specimen is maintained at very low temperature, -160°C or below. Removal of water will generally destabilize the specimen conformation, so one is then faced with finding some mechanism to stabilize and support the specimen structure.
- Biological specimens yield very low contrast, as they are composed of light elements. Contrast is commonly increased by adding some form of heavy atom contrast agent (e.g., stain), but adequate contrast can be achieved on unstained specimens by adjusting the focus of the objective lens (see below).

3.1. Negative Staining

One of the most common techniques of

specimen preparation is negative staining first developed in the 1960s. Negative staining works well with 2D membrane protein crystals, but the useful resolution that can be achieved is limited to about 10 Å, and 20 Å is more common. The specimen in an aqueous suspension of approximately 1 mg protein/mL is adsorbed onto a thin carbon film. Carbon films are generated by evaporation of carbon onto a suitable surface, for example freshly cleaved mica, from which it can be floated off onto a water surface and lowered onto the surface of electron microscope grids (commonly 400 mesh copper grids) previously placed on a support beneath the film. The carbon film can be lowered gently onto the grid surfaces by draining water out of the vessel. Alternatively, one can coat grids with a plastic film, generally Collodion, but Formvar is also used, after which a carbon film is evaporated onto the plastic film. These carbon–Collodion films can be used directly, but best results are usually obtained if the plastic is dissolved away with a solvent, since pure carbon films are thinner and more conductive giving less movement when irradiated with an electron beam. Ideally, one should use the thinnest carbon film possible in order to minimize its contribution to the image, but very thin carbon films are fragile, and 400 mesh grids are too coarse to support them adequately. In this case, one can first coat the grid with a holey film, a plastic film containing numerous circular or ellipsoidal holes of varying sizes. There are several methods by which holey plastic films can be formed (36,57,63), and they are generally stabilized to electron irradiation by evaporation of carbon onto them. Holey films make an excellent support for very thin carbon films and can even be used to support thin layers of stain over the holes without a carbon support film.

Freshly evaporated carbon films are generally hydrophilic and readily adsorb proteins and lipid vesicles. As they are stored, however, they quickly become more hydrophobic and then do not adsorb the specimen as well. This change is readily observed as one draws a liquid drop off of the grid with a piece of filter paper. If the liquid draws off evenly leaving a thin film, the grid is hydrophilic, but if it is all rapidly drawn off leaving none behind, the grid is hydrophobic and will probably not adsorb the specimen. One can overcome this change to some extent by increasing the concentration of the specimen, but it is often necessary to render the surface hydrophilic again by a process called glow discharge, in which the films are exposed to the ions formed by an electric discharge in the residual gases in a vacuum apparatus pumped down to several tens of microns of pressure. A glow discharge accessory is common in commercial vacuum evaporators found in electron microscopy laboratories, or an inexpensive glow discharge apparatus can be constructed from a plastic vacuum dessicator, a Tesla coil vacuum tester, and an inexpensive rotary pump (3). Procedure 2 describes a typical method to prepare negatively stained specimens.

❖ Procedure 2. Preparation of Negatively Stained Specimens

1. Allow a drop of the specimen to adsorb to a carbon-coated grid for 1 to 5 minutes.
2. Wash the grid with several drops of water or buffer followed by several drops of a suitable negative stain, commonly 1% to 2% uranyl acetate, but uranyl formate and phosphotungstic acid are also used. Uranyl stains generally provide higher resolution, but must be used at a pH of about 4.5,

which may disrupt the structures of some specimens.

3. Draw off excess stain with a piece of filter paper leaving a very thin layer that dries down forming a glass-like electron dense replica surrounding the specimen molecules supporting and contrasting them.

In the electron microscope, the actual protein structure appears light against a dark background formed by the stain, hence the term "negative" stain. It is important to realize that negative staining contrasts the 3D surface of the molecule with atoms that are approximately 7 Å in diameter. Thus, the resolution in negative-stained specimens is limited to approximately 10 Å at best and more commonly 20 Å. Furthermore, the internal structure of protein domains are not contrasted.

3.2. Other Methods

Images of negatively stained specimens are a projection of the electron density of the 3D stain replica onto the 2D image plane. Ultimately, one would wish to calculate a 3D structure from multiple images of tilted specimens (see below), but information on the 3D configuration of the molecule can often be quickly obtained by other common specimen preparation techniques. Indeed, this information is often essential to confirm the molecular packing of a new crystal form, information that is required in calculating the 3D structure. One of the most useful techniques is heavy metal shadowing of freeze-dried specimens (Procedure 3).

❖ Procedure 3. Heavy Metal Shadowing

1. Adsorb the specimen to a flat hydrophilic surface, such as a carbon film or freshly cleaved mica.
2. Rinse with water or volatile buffer to remove sample that is not adsorbed to the surface.
3. Freeze by plunging into a cryogen; liquid nitrogen is acceptable, but ethane or propane cooled in liquid nitrogen freezes much more rapidly.
4. Place the sample in a freeze fracture–etch instrument and remove ice by sublimation at approximately 70°C.
5. Contrast the surface by evaporating a heavy atom, generally platinum evaporated with carbon, but tungsten–tantalum may produce smaller metal grains yielding higher resolution (12). This creates a shadow effect that highlights the surface topography and can also be used to measure the thickness of the specimen.

This technique has been applied to contrast selectively the surface of vesicle crystals of cytochrome oxidase dimers indicating that the enzyme protrudes 20 to 30 Å beyond the bilayer surface on the exterior surface (corresponding to the matrix side of the inner mitochondrial membrane) (33) and to measure the thickness of a number of 2D crystals (37). Berriman, Leonard, and coworkers used shadowing to demonstrate that crystals of the mitochondrial cytochrome bc_1 complex thinned markedly during electron irradiation (5), and Smith and Ivanov have published a procedure to compute the surface relief structure from images of shadowed specimens (67). A related technique, freeze fracture–replication, can also be used to study the structure of membrane protein crystals in order to help determine the molecular packing (13). Heuser has adapted the technique of rapid slam freezing and freeze fracture–etch to look at molecules adsorbed to a slurry of small mica chips (44). Conventional plastic embedding and thin sectioning can also be used to evaluate the gross structure of a new crystal preparation of a membrane protein and to help confirm the molecular packing model (31,51,80).

3.3. Low Dose Electron Microscopy

The challenges of specimen preparation for electron microscopy are compounded by the sensitivity of biological specimens to electron irradiation. The exposure required to record a single image at moderate resolution, approximately 1 nm, can be as high as 100 to 300 electrons/Å^2. However, measurements of electron damage at much lower exposures paint a gloomy picture for prospects of achieving even this modest resolution. An exposure of even 20 to 30 electrons/Å^2 results in loss of 20% to 30% of the mass of a typical biological specimen (20), exposure of less than 10 electrons/Å^2 disrupts the higher order features of a protein crystal (38), and an electron dose of approximately 0.5 electrons/Å^2 is sufficient to inactivate enzymes (39). The primary function of stain is to increase the contrast of biological specimens so lower electron doses can be used to achieve useful images. Furthermore, heavy atom stains are more resistant to damage by electron irradiation. Nevertheless, studies in the 1970s demonstrated the efficacy of recording electron micrographs using minimal electron exposure even for negatively stained specimens (77,83). Now, most electron microscopes allow one to focus and correct astigmatism on an area of the specimen grid adjacent to the specimen and then record an image of the specimen, exposing it to only the electrons required to expose the photographic film. This procedure is absolutely essential when recording images of unstained specimens that are much more sensitive to electron irradiation than are stained specimens.

3.4. Unstained Specimens

Ideally, one would like to record electron micrographs of unstained specimens, since the resulting images would be of the actual biological molecules rather than the distribution of heavy atom stains around them. There are several problems in achieving this goal, however, beginning with the low contrast afforded by biological specimens and by their sensitivity to exposure to high energy electrons. Through the use of low dose techniques, one can record electron micrographs of unstained specimens at electron doses low enough to minimize radiation damage. The disadvantage is that although these images may technically be high resolution, the signal-to-noise ratio (S/N) is very low, often less than 1.0, and cannot be interpreted. The S/N can be increased by averaging many images of identical structures such as the unit cells of a crystal; statistically the S/N is increased by a factor equal to the square root of the number of structures averaged. This can be a very powerful tool in the case of 2D crystals, where a very small area might contain 100 unit cells giving an increase in the S/N of tenfold. A more typical situation would be a crystal containing several thousand unit cells giving and an increase in S/N of thirty- to fiftyfold. This was first demonstrated by Unwin and Henderson with their images of unstained purple membrane containing thousands of bacteriorhodopsin molecules; these electron micrographs appear featureless, but the averaged image was clearly interpreted at 7 Å resolution as resulting from the presence of transmembrane α-helices (78).

The remaining problem is how to prepare unstained specimens for the high vacuum conditions in an electron microscope. Unwin and Henderson dried their purple membrane specimens in a thin layer of 1% glucose in order to surround them with a hydrophilic substance that could also support them structurally. This approach has worked very well with purple membrane and with a number of other examples, but has the disadvantage that glucose has a density very similar to protein and thus actually reduces the already low contrast rather than increasing it. This was acceptable in the case of bacteriorhodopsin molecules as nearly all of the protein lies within the lipid bilayer surrounded by the lower density hydrocarbon tails of the lipid molecules. Cytochrome oxidase crystals, however, project much of their structure beyond the lipid bilayer surface, and these portions of the structure are virtually invisible above 10 Å resolution if the crystals are embedded in glucose (17,18,42). Better results have been obtained with aurothioglucose, a glucose derivative containing gold atoms, or with glucose mixed with uranyl acetate (79). These mixtures provide low resolution contrast of the hydrophilic domains of membrane proteins while embedding the structure in a hydrophilic substance. Kuhlbrandt and others have obtained excellent results using tannic acid rather than glucose (54).

3.5. Frozen Hydrated Specimens

The ideal method is to maintain an aqueous environment around the crystal as is the case with 3D protein crystals studied by X-ray diffraction. Although environmental chambers have been constructed that maintain significant partial pressure of water around the specimen by differential pumping, these proved too unstable for high resolution imaging. Taylor and Glaeser demonstrated another approach, freezing the specimen in a thin layer of ice and keeping it frozen at low temperature, -130°C or less, with a specially designed cryoelectron microscope stage (72). The early attempts, particularly by Dubochet's group and Unwin's group, provided very useful results but suffered from stage vibrations that limited resolution. This spurred efforts to construct more stable cryospecimen holders now marketed by Gatan (Pleasanton, CA, USA) and by Oxford Instruments (Concord, MA, USA) for the popular side entry electron microscopes. The techniques of cryoelectron microscopy have been described in an excellent review by Dubochet et al. (21), and I will only summarize the important points here. The key to this technique is to freeze the specimen very rapidly in a thin layer of water. With freezing velocities above approximately 10 000 degrees/second, the water is transformed to vitreous ice, a noncrystalline ice form that has a density and structure similar to liquid water. It is very difficult to freeze a thick specimen this rapidly, but a thin layer of water clinging to an electron microscope grid can be readily frozen to vitreous ice by plunging it into a suitable cryogen such as liquid propane or liquid ethane cooled by liquid nitrogen. The specimen grid is held in a pair of fine tweezers clamped into a simple device for plunging, and Procedure 4 is followed (see Figure 43 in Reference 21).

❖ Procedure 4. Preparation of Frozen Hydrated Specimens

1. Apply 1 to 5 µL of the specimen suspended in a suitable buffer at relatively high concentration, approximately 5 to 20 mg/mL, to a grid with a hydrophilic substrate, either continuous carbon film or holey film that has been recently glow discharged.
2. Blot the grid by pressing it firmly by hand between two layers of filter paper.
3. Plunge immediately into the cryogen; this step is facilitated if the plunge device has a foot pedal release.
4. Transfer very quickly to liquid nitrogen, quickly flicking off excess cryogen, and store under liquid nitrogen until use.

The result is a specimen embedded in a material very similar to its native environment at a very low temperature where movement is inhibited. The specimen may be adsorbed to a thin carbon film as for negative staining, or it may be suspended over the holes of a holey film. The latter method has the advantage that the specimen is not in contact with a solid support prior to freezing, but the holey film has a significantly smaller area suitable to record images. The specimen grid should be frozen immediately after blotting, but the thin film of water supporting the specimen may still dry significantly if the relative humidity of the environment is low. To minimize drying, one can maintain higher humidity around the specimen by:

- Blowing humidified air across it (21).
- Freeze in a cold room where humidity is high.
- Use a specially constructed freezing device that incorporates a humidity chamber (64).

Adrian et al. have adapted this procedure to incorporate heavy atom salts in the vitrified water layer in order to increase the contrast of the frozen specimens (1). A very important benefit of cryoelectron microscopy is the reduction of electron beam damage at low temperature. In measurements of loss of higher resolution information as a function of electron irradiation, specimens at -170°C can be exposed to 5 to 10 times the number of electrons as those at room temperature (38).

The low contrast provided by the relatively small density differences between vitreous ice and protein can be enhanced by appropriate choice of focus of the objective lens. In the brightfield mode of a transmission electron microscope, contrast is generated by two mechanisms: *(i)* amplitude contrast is generated when electrons are scattered by the specimen at a wide enough angle to cause them to be intercepted by the objective aperture subtracting them from the image. This is analogous to absorption contrast in the light microscope and contrasts relatively low resolution details; and *(ii)* phase contrast is generated when the phases of electrons are retarded as they pass through the specimen. The phase of these electrons are further modified by the objective lens, and the extent of this phase shift depends upon the:

- Angle of diffraction.
- Spherical aberration of the objective lens.
- Focus of the objective lens.

Thus, it is possible to control the amount of phase shift of the diffracted electrons by changing the focus of the objective lens, and with an appropriate level of underfocus, some of the diffracted electrons can be further phase-shifted by approximately 90°, generating appropriate phase contrast when combined with undiffracted electrons at the image plane. But for each choice of underfocus, only electrons diffracted at particular angles are phase-shifted by 90°, generating proper contrast. Electrons diffracted at other angles are phase-shifted by smaller amounts, generating less contrast, or are phase-shifted in the wrong direction, generating inverted contrast.

Diffraction angle correlates with resolution, and electrons diffracted at higher angles contain higher resolution information. The changes in the phase of electrons as a function of their diffraction angle is described by the contrast transfer function (CTF) (23,26). In order to visualize individual macromolecules, one must adjust the objective lens to a relatively large underfocus, one micron or more. This generates contrast of lower resolution, features making the molecules visible, but may introduce contrast reversals at high resolution. Figure 3c is an optical diffraction pattern (equivalent to a plot of Fourier transform intensities) of the cytochrome oxidase crystal in Figure 3a, and the effects of the CTF can be seen in the concentric rings of high background noise. The regions of the Fourier transform, where phases have been shifted in the wrong direction generating reversed contrast, are shown in the plot of the CTF displayed as an insert in Figure 3c on the same scale as the as the diffraction pattern. By definition, CTF values greater than zero represent incorrect phase shifts,

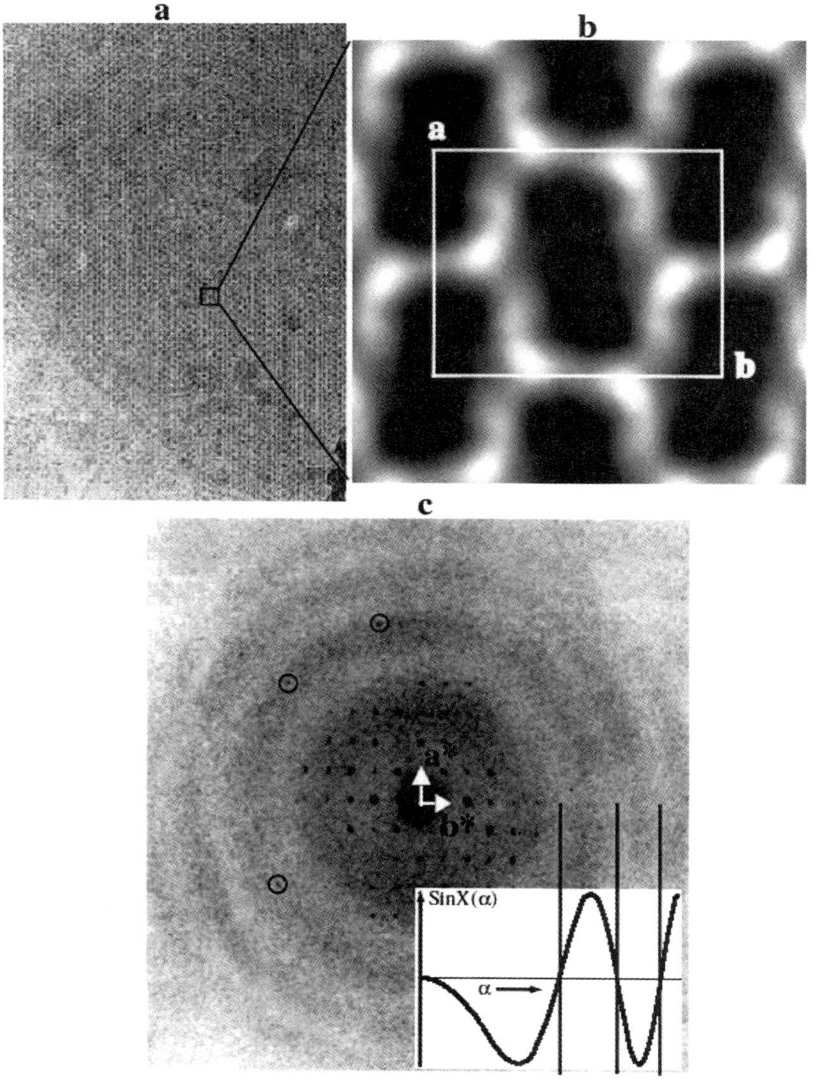

Figure 3. (a) An electron micrograph of a frozen hydrated crystal of cytochrome oxidase dimers; one unit cell is outlined. (b) A Fourier-filtered image with dramatically increased S/N calculated from 5 electron micrographs similar to panel a. One unit cell is outlined with unit cell axes of a = 100 Å and b = 125 Å. (c) Optical diffraction pattern of the crystal in panel a; the optical diffraction pattern is equivalent to a plot of the intensities of the Fourier transform. The reciprocal lattice vectors, a^* and b^*, are indicated. The inset is a plot of the phase CTF, $\chi(\alpha)$, for this defocus shown on the same scale, and the zeros in the CTF are indicated by horizontal lines showing regions of minimal contrast in the diffraction pattern.

and the circled diffraction spots lie within rings of the diffraction pattern that have been phase-shifted in the wrong direction. In order to calculate a high resolution image of a biological specimen, one must correct for the effects of the CTF. Since different values of underfocus optimally contrast features at different levels of resolution (different diffraction angles), it is sometimes advantageous to record more than one image of the same specimen, the first at lower defocus for high resolution information, and the second at greater defocus for lower resolution information (11,28,73).

3.6. Collecting Tilt Data

Transmission electron micrographs are 2D projections of the 3D electron density of the specimen. While these projections reveal important information about the structure of the specimen, much detail is lost when structural features are projected upon one another. If the specimen is tilted and another micrograph recorded, a different set of features will be superimposed, providing new information about structure. This is most readily seen if the two projections are viewed as a stereo pair. A full 3D reconstruction requires many images of the specimen tilted at different angles, greatly exceeding the number that can be recorded from a single specimen without very significant radiation damage to unstained specimens (14,43). Thus, low dose images of many different but identical specimens must be recorded in order to sample the 3D Fourier transform (see section 3.6 below and Figure 4). All images must be translated to bring them into alignment at a common origin, so it is usually necessary to record images at a variety of tilt angles and merge the data, beginning with images recorded with the smallest tilt and then adding images in order of increasing tilt angle. In the case of negatively stained specimens, one normally records several images at different tilt angles for each specimen, since these specimens are more resistant to radiation damage. Unstained specimens are much more sensitive to electron irradiation, and generally, only one high resolution image is recorded from each. Specimens are more stable to irradiation at low temperature (10,38), so it is possible to record more than one micrograph from a single specimen at low temperature, particularly if the highest resolution is not required.

3.7. Specific Labeling

Most studies of protein structure by electron crystallography do not yield resolution sufficient to construct an atomic model, and methods to identify the locations of functionally important sites and/or components are needed in order to exploit fully lower resolution structures. There are two general approaches to identify specific sites on low resolution structures: *(i)* specific labeling with molecules visible by electron microscopy; and *(ii)* comparison with structures lacking one or more components.

Both of these approaches have been applied with success to low resolution structures of heme proteins derived from electron microscopic data.

The most common method of specific labeling in electron microscopy of biological structures is the use of specific antibody molecules. Antibodies are most commonly used to determine the distribution proteins in cells and organelles, but can also be used to identify the position of antibody epitopes on molecular structures. Frey et al. used subunit-specific antibodies to determine that the surface exposed in vesicle crystals of cytochrome oxidase dimers corresponded to the matrix surface of the inner mitochondrial membrane, concluding that the interior surface of the vesicles corresponds to the surface exposed to the intermembrane space (30). They subsequently prepared monovalent antibody fragments, Fabs, to label subunit IV on the surface of these crystals, concluding that this subunit lies 20 to 30 Å from the 2-fold axis of the dimer and near the *a* crystal axis (32). Based upon the prediction of a transmembrane α-helix from residues 80 to 97, the volume of the N-terminal domain of subunit IV could account for the 20 to 30 Å structure projecting from the bilayer surface that was detected in freeze-dried and shadowed specimens (34). Fab fragments have also been used as bulky affinity labels to identify their corresponding epitope binding sites by electron microscopy in many other specimens (2,82).

In many cases, the protein being studied binds another protein with sufficient affinity and specificity that the protein ligand can be used to label its binding site. This is the case with cytochrome *c*, one of the substrates for cytochrome *c* oxidase. Frey and Murray (35) incubated crystals of cytochrome oxidase monomers with cytochrome *c*, which binds to cytochrome oxidase with an affinity comparable to that of specific antibodies. After extensive image processing, the site of cytochrome *c* binding to cytochrome oxidase monomers was deduced from difference images (Figure

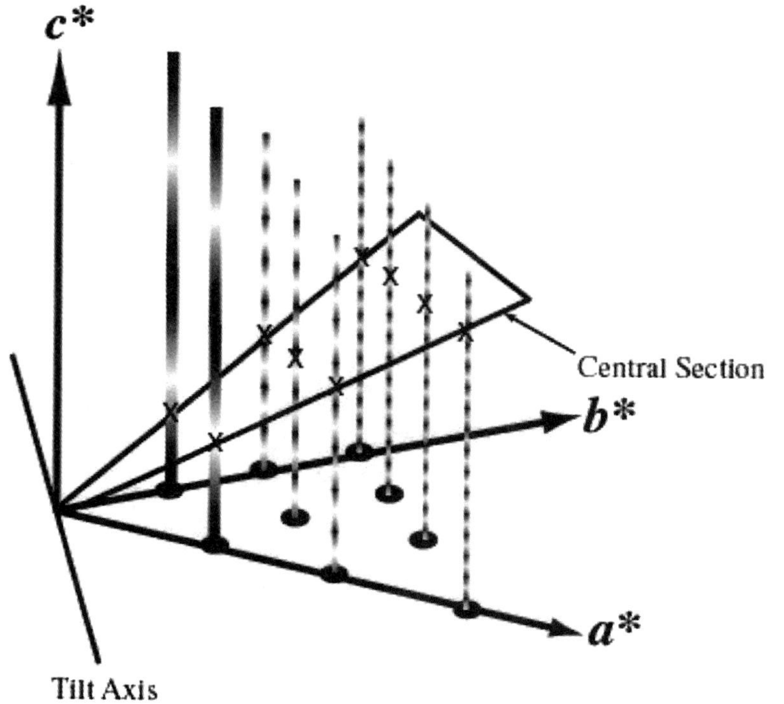

Figure 4. Lattice lines of the 3D Fourier transform of a 2D crystal. The positions of the reflections in the diffraction pattern of an untilted crystal are shown as black ellipses. In the 3D Fourier transform, these reflections extend perpendicular to the plane of the crystal as shown by the lines of periodically varying intensity. The Fourier transform of an electron micrograph of a crystal that has been tilted is a central section of the 3D Fourier transform that intercepts the lattice lines at the points shown by the X's.

5a), consistent with the cytochrome *c* binding site in the atomic structure, determined by X-ray diffraction and from biochemical studies (see section 4.1 and Figure 5b).

Other labeling studies have taken a different approach, using heavy atom cluster molecules that have been modified to react selectively with certain functional groups of proteins, generally reactive sulfhydryl groups of cysteine residues (29). A special issue of *The Journal of Structural Biology* is devoted to results from this approach using gold cluster compounds (1999, volume 127, issue 2). Crum et al. used a monomaleimide derivative of an undecagold cluster compound to label specifically Cys-115 of cytochrome oxidase subunit III in crystals of cytochrome oxidase dimers. They then identified the binding site by low dose cryoelectron microscopy of specimens embedded in glucose and uranyl acetate (16).

A different approach to identify the various components of a macromolecular complex is to compare structures of the intact complex with structures of subcomplexes. This approach was used in the study of the mitochondrial cytochrome bc_1 complex. Weiss, Leonard, and coworkers crystallized *Neurospora* mitochondrial cytochrome *c* reductase (cytochrome bc_1or Complex III) by reconstituting it with purified lipids and adsorbing excess detergent with Bio-Beads. This produced crystals of the type shown in Figure 1c, although these were generally formed in the two layers of a collapsed vesicle giving two overlapping crystalline layers (56,84). Their low resolution 3D reconstruction of the intact complex is shown in Figure 6b. Hovmoller et al. subsequently formed crystals of the purified subcomplex lacking two large "core" proteins, and comparison of the 3D structure with that of the intact complex allowed them to identify the functional components as shown in Figure 6 (46,47). The core subunits, which probably function in facilitating the assembly of the complex, were later purified, and their

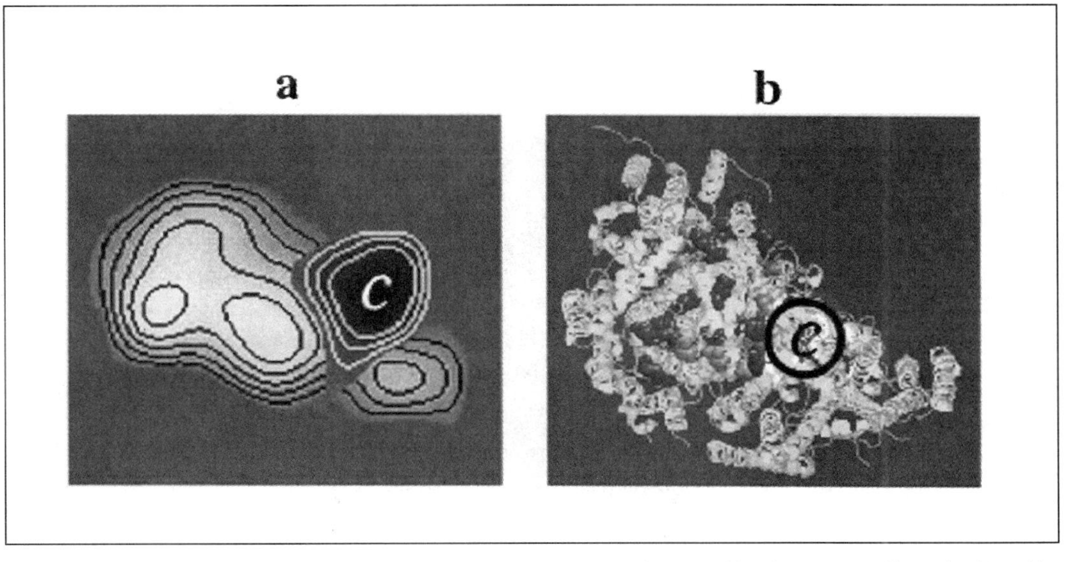

Figure 5. A comparison of the structure of cytochrome oxidase monomers determined by electron crystallography (a) and by X-ray crystallography (b). (a) A low resolution structure in projection calculated from electron micrographs of frozen hydrated crystals of cytochrome oxidase monomers. The dark peak outlined in white contour lines is the position of cytochrome *c* binding calculated from difference images. (b) A ribbon diagram produced from the atomic coordinants calculated from the high resolution X-ray structure and displayed by the program RasMol. The cytochrome *c* binding site is placed between Cys-115 of subunit III and the acidic residues of subunit II as determined biochemically.

low resolution structure was determined from helical aggregates confirming the assignment in Figure 6 (48). The cytochrome b_6f complex found in the thylakoid membrane of chloroplasts has a function in photosynthesis very similar to that of cytochrome bc_1 in mitochondria, but lacks the core subunits. The low resolution projection structure of the cytochrome b_6f complex purified from *Chlamydomonas reinhardtii* was determined by electron crystallography of 2D crystals grown by reconstitution and found to be very similar to the cytochrome bc_1 subcomplex lacking the core subunits (8).

4. DATA PROCESSING

4.1. Selecting Micrographs — Optical Diffraction

Although data collection by low dose electron microscopy is a critical element in determining the structure of a membrane protein, data processing is equally important. The first step is to identify which of the many micrographs recorded are suitable for further processing; generally, only a minority of micrographs contain high resolution information. The simplest method to screen micrographs for quality is optical diffraction, since the diffraction pattern is the Fourier transform of the object and is a diagnostic of several important image characteristics. Formation of an optical diffraction pattern is readily accomplished with an optical diffractometer consisting of a laser (generally a 1–5 mW He/Ne laser), beam expansion–spatial filter, and a single diffraction lens to collect the diverging beam and focus it to a point at some distance (24,65). When an electron micrograph is placed in the optical path just after the diffraction lens, the focused diffraction pattern of the illuminated portion of the micrograph can be observed at the focus of the lens. The undiffracted beam appears as a bright spot in the center of the diffraction pattern and is surrounded by the diffraction pattern of the object image. In the case of crystalline objects, the diffraction pattern consists of peaks of light at the points of a 2D lattice whose spacings are the reciprocal of the crystal lattice; thus, the lattice in the diffraction image is called the "reciprocal lattice". The structural information common to all unit cells within the illuminated area of the micrograph

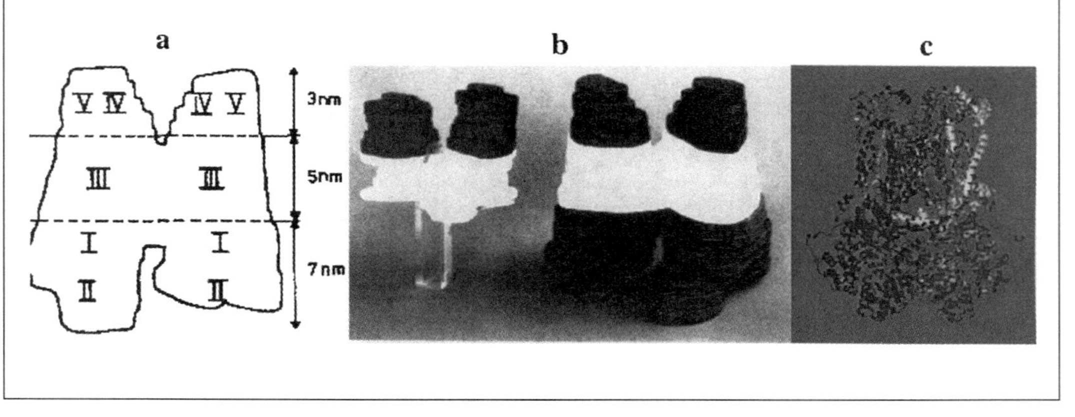

Figure 6. A comparison of the structure of cytochrome *c* reductase (cytochrome bc_1 complex) determined by electron crystallography and by X-ray crystallography. (a) A drawing interpreting the positions of 5 subunits in the dimeric complex with respect to the lipid bilayer in the center. Core subunits I and II face the matrix space. (b) Balsa wood models of the low resolution structures determined by electron microscopy–crystallography of: *(i)* a subcomplex lacking core subunits I and II on the left, and *(ii)* the intact complex. *(iii)* A ribbon diagram based upon the atomic coordinants of all subunits determined by X-ray crystallography.

is contained at the points of the diffraction pattern's reciprocal lattice, while nonperiodic noise is distributed throughout the diffraction pattern. Although one can obtain equivalent information by digitizing the electron micrograph and calculating its diffraction pattern, an optical diffractometer accomplishes this instantaneously, allowing one to move the micrograph around in the beam to select the best area quickly. One generally looks for two criteria revealed by the optical diffraction pattern:

1. Does the image contain high resolution information about the crystal structure? The diffraction pattern is the Fourier transform of the object illuminated, representing its structure in frequency space, with points furthest from the center representing higher frequency components contributing higher resolution information. One therefore looks for micrographs whose diffraction patterns display diffraction spots on the reciprocal lattice that extend relatively far from the origin of the diffraction pattern. Once the diffraction constant of a particular optical diffractometer is calibrated, usually with a diffraction object or grating of known spacing, the resolution of the information from each micrograph can be calculated based upon the distance from the origin of the furthest diffraction spot.
2. Is the micrograph properly focused with astigmatism corrected? Characteristics of the transfer of information from the object to the image are described by the CTF as described below. A plot of the CTF in diffraction space shows that it periodically passes through zero at points determined by the wavelength of the electron wave, the spherical aberration of the objective lens, and by the focus of the objective lens. The zeroes appear as concentric circles in the diffraction pattern (Figure 3c), and the radii of the circles can be used to calculate accurately the focus of the objective lens. If the objective lens has residual astigmatism, the focus is different in orthogonal directions, and the zeroes produce concentric ellipses rather than circles.

4.2. Digitizing

In order to process electron micrographs by computer, they must first be converted to digital form. Although it is now possible to purchase sensitive high resolution digital cameras for transmission electron microscopes, film is still the best media on which to record low dose high resolution images. The most accurate scanners have been mechanical, based upon wrapping the micrograph around a rotating drum or on precise movement in two dimensions, and these are generally quite expensive. More recently, high resolution digital cameras based upon charge-coupled device (CCD) technology have become available, and these represent a suitable lower cost alternative to mechanical scanners for many applications. Whatever the device used, there are several factors to consider in digitizing an image, and these have been covered in earlier publications (19,26). The first is the resolution one requires or expects in the digitized image. In digital sampling of a continuous function (analog signal), one must sample the function at an interval that is one half the interval or resolution one wants to obtain in the digitized image; this is called the Nyquist sampling rate. Thus, if one requires 10 Å information in a digital image, it must be sampled (digitized) at 5 Å or smaller intervals. In order to obtain very high resolution information from a low dose image, one must digitize the image of a large 2D crystal, at very small intervals, producing a very large data file. Electron micrographs of 2D crystals are typically recorded at about 40 000×, and 1 Å spacing in the specimen corresponds to 4 μm on the film. Thus, in order to record 10 Å information in the digitized electron micrograph, one would have to sample it at 20-μm (5 Å) intervals, and for 4 Å resolution, the sampling would have to be at 8-μm intervals. In practice, one actually samples more finely than strictly required by the Nyquist limit, since there is some fall off in the transfer of high resolution information that depends on the sampling aperture size.

4.3. Fourier Filtering

One of the principal goals of image processing applied to electron micrographs of 2D crystals is to extract an image with high S/N from a micrograph with a very low S/N; this is accomplished by signal averaging. The Fourier transform of a perfect 2D crystal of infinite extent would be nonzero only at the points of a 2D reciprocal lattice, but as seen in Figure 3, the Fourier transform of an actual image of a 2D protein crystal contains nonperiodic noise, and the peaks at the points of the reciprocal lattice are spread out somewhat as the crystal is finite and not perfectly ordered. If the image is reconstructed using only the information lying at the reciprocal lattice points of the Fourier transform, the result is the average of all of the unit cells within the digitized image, and the S/N is increased by the square root of the number of unit cells averaged. This is shown in Figure 3b compared with the original image in Figure 3a.

4.4. Correlation Alignment

The resolution of the averaged image depends upon the inherent resolution of the original electron micrograph (defined by the CTF) and upon the order of the crystal being averaged. As techniques of electron microscopy and computational averaging improved, leading to improvements in image resolution, it became apparent that disorder in membrane protein crystals was limiting the resolution that could be achieved after averaging the unit cells contained within an image. This problem was first recognized by Crowther and Sleytr who developed the first computer programs to attempt to correct for crystalline disorder (15). Henderson et al. later developed a method and software to correct the lattice disorder in images of very large 2D protein crystals in order to improve the S/Ns of their images at higher resolution, and their method is described in Procedure 5 (41). Although this may seem like a laborious process, the improvements in data can be dramatic, and most steps of the procedure are automated.

A similar approach to this problem developed out of efforts that began in Joachim Frank's group to create software tools to align electron microscope images of individual particles using correlation methods. The single particle correlation methods can also be used to align individual unit cells if the S/N is high enough to permit accurate alignment, or it can be applied to patches of unit cells in the case of low S/N images of unstained specimens. Once the patches of unit cells are aligned, they can be averaged to increase dramatically the S/N of the resulting images (27). The single particle averaging software can also be used for structural study of molecules that are not crystalline and have yielded dramatic results when applied to images of individual ribosomes and ribosomal subunits (58).

❖ Procedure 5. Resolution of the Electron Micrograph

1. Apply a mask to the Fourier transform that passes only the information that lies within a specified radius of each

reciprocal lattice point.

2. Calculate the inverse Fourier transform of the "masked" transform to produce a "coarsely" filtered image in which each unit cell is averaged with its nearest neighbors.
3. Select a reference image from the coarsely filtered image and calculate the crosscorrelation function of this reference image with the entire filtered image.

Identify the positions of each unit cell by searching for the peaks in the crosscorrelation function.

Reinterpolate the sampling of the original image based upon the positions of all of the unit cells identified in the crosscorrelation function in order to remove crystal lattice disorder.

4.5. Correcting the CTF

The performance of the objective lens of an electron microscope is defined by the CTF, which is the Fourier transform of the point-spread-function that describes how a point on the object (specimen) appears in the image (23,26). Figure 3c shows the effect of the CTF in modulating the intensities of the Fourier transform (displayed in the optical diffraction pattern) of the image in Figure 3a, and the inset graph shows the phase contrast component of the CTF for this defocus, demonstrating that it causes periodic phase reversals (phase shifts of 180°) within concentric bands of spatial frequencies

Note: When the CTF is plotted in this manner, correct transfer of contrast is indicated when sin c(a) = -1.

Not shown in the inset is the effect of amplitude contrast generated when electrons are scattered into the objective lens aperture removing them from the image; amplitude contrast is important at low spatial frequencies contributing contrast to low resolution features (approximately 50 Å or greater). Correction of the CTF is not absolutely required if all of the information contained in a micrograph lies within the first zero of the CTF; this is the region around the origin of the Fourier transform and within the first ring of low noise where the CTF goes through zero on the inset graph. However, if one wishes to obtain an accurate representation of the object, even at low resolution, the amplitudes of the Fourier transform must be increased by varying amounts to compensate for the fact that the CTF is not -1.0 across this frequency spectrum. The most significant correction is for those bands of the frequency spectrum of the Fourier transform where the CTF has caused a phase shift of 180°. In Figure 3c, the circled lattice points contain information about the crystal structure whose phases have been shifted by 180°, and these will contribute incorrect information to the image unless they are corrected. Once the defocus and residual astigmatism of a particular micrograph has been defined, the CTF can be calculated, and the phase changes are easily corrected. Correction of the amplitudes is more complicated because: *(i)* the amount by which amplitudes must be increased can be difficult to determine, since one must include contributions from amplitude contrast; and *(ii)* regions of the Fourier transform near the zeroes of the CTF require very large corrections, which can greatly magnify the contribution of noise in the image. The proper methods for making this correction are beyond the scope of this article, but more detailed information can be found in the literature (73,89).

4.6. 3D Reconstruction

The ultimate goal is to calculate a 3D structure of the protein under study. Electron micrographs are 2D projections of the 3D electron density. Most people are intuitively aware that one can gain a better knowledge of a complex 3D object's structure by viewing it from several angles. This intuitive approach is quantitatively achieved by a number 3D reconstruction algorithms that make use of 2D projections along different directions. In the case of 2D crystals, the most common of these algorithms make use of a property of Fourier transforms described by the central section theorem; this states that the Fourier transform of a 2D projection of a 3D object is a central section (a section that passes through the origin) of the 3D Fourier transform of the object. Thus, as one collects 2D projections along different angles, one can fill in the 3D Fourier transform and estimate its value at a resolution limited by:

1. The number of 2D projections and the angle between them. This is so because higher resolution information is contained further from the origin of the Fourier transform where the 2D central sections are further apart; eventually they diverge enough that the values of the 3D Fourier transform can no longer be estimated from the values on the 2D sections.
2. The size of the object. The reason for this limitation is more subtle and arises from the fact that larger objects have Fourier transforms that vary more rapidly than smaller objects with the same level of detail. Thus, the 2D sections for larger objects must be more closely spaced to achieve comparable resolution compared to smaller objects. Another way of viewing this is to recognize that defining the structure of a larger object requires more data (e.g., more projections).

The semiquantitative relationship between resolution, object size, and number of tilts was expressed by Crowther et al. in the equation:

$$m \cong \pi D/d$$

where m equals the number of views, D is the particle diameter, and d is the desired resolution (14).

Fourier transforms of 2D crystals have a special property that renders the use of Fourier transforms computationally efficient; they are sampled on a 2D lattice parallel to the plane of the crystal and are continuous along "lattice lines" perpendicular to the crystal plane at the points of the 2D reciprocal lattice as shown in Figure 4. For example, in the Fourier transform of the cytochrome oxidase crystal in Figure 3, the information about the crystalline structure is contained at the points of the reciprocal lattice defined by the lattice vectors a^* and b^*, while nonperiodic noise is distributed over the entire transform. If one could view the 3D Fourier transform, one would see that the information intersected by this central section varies continuously along lines, lattice lines, parallel to one another, and perpendicular to the plane of the transform in Figure 3c as diagrammed in Figure 4. The Fourier transforms of images of tilted crystals sample these lattice lines at the positions where the central section intersects them as shown in Figure 4, and collecting the information required for a complete 3D reconstruction of the crystal requires collecting central sections at different tilt angles in order to sample these lattice lines finely enough to estimate their value continuously out to the desired resolution. The higher the resolution and the larger the unit cell, the more projections required and the finer the angular intervals between them. Once the lattice lines have been measured from central sections, they can be sampled at appropriate regular intervals, and the 3D structure calculated by an inverse 3D Fourier transformation (4).

5. COMPARISON WITH STRUCTURES FROM X-RAY DIFFRACTION

5.1. Cytochrome *c* Oxidase

When the first atomic structure of a eukaryotic cytochrome oxidase determined by X-ray crystallography was published in 1995 (74), its structure had previously been determined in 2D projection at approximately 8 to 10 Å resolution (79) and in 3 dimensions at approximately 15 Å resolution (35). This is too low a resolution to discern subunit boundaries, let alone trace the polypeptide chains, but a number of structural features had been deduced by specimen preparation to contrast different domains selectively, by various labeling experiments, and by comparing the structures of both 2D crystal forms, the monomer form, and the dimer form. The transmembrane α-helices predicted by hydropathy plots based on amino acid sequences proved to be fairly accurate, and according to the X-ray model, they separate the molecule into two hydrophilic domains that protrude 35 to 40 Å beyond the lipid bilayer into the intermembrane space and into the matrix space of mitochondria. This is in contrast to the marked asymmetric distribution of protein mass, 60 Å into the intermembrane space and less than 10 Å into the matrix space, proposed from 3D reconstructions by electron crystallography (18,42,79). It is difficult to reconcile these and accept the X-ray model as being more accurate. On should note, however, that Frey et al. correctly determined that the matrix side domain projected 20 to 30 Å based upon the lengths of shadows cast in specimens that had been freeze-dried and shadowed with platinum–carbon (32–34). This highlights the importance of using different specimen preparation techniques in studying complex biological structures by electron microscopy.

Although in projection, the dimers observed in 2D crystals by cryoelectron microscopy appear very similar to those generated from the X-ray coordinants (obtained from the Protein Data Bank and displayed with the program RasMol; see Figure 7) (75), comparison of their sizes indicates that they must be different structures. The maximum dimension parallel to the membrane of dimers in the 2D crystals is approximately 100 Å, the length of the a crystal axis along which the molecules are aligned. The X-ray model, on the other hand, has a maximum dimension of approximately 150 Å. Thus, it appears that the dimer in 2D crystals must be a more compact structure with the individual monomers more closely aligned than the crystallographic dimers in the 3D crystals used to determine the structure by X-ray diffraction. This is also indicated by the fact that the dimer in 2D crystals has its highest concentration of mass around the 2-fold axis, while the dimer in the 3D crystals is less densely packed in this region. The size and shape of a cytochrome oxidase monomer observed in 2D crystals compares more favorably with the X-ray structure (Figure 5). At 95 x 53 Å, the structure determined by electron microscopy (35) is somewhat smaller than the 110 x 63 Å structure measured from X-ray coordinants, but this difference can be explained by the ambiguity in determining the molecular boundary in a 2D projection.

The sites identified by electron microscopy following specific labeling with Fabs, cytochrome *c*, and a monomaleimide undecagold cluster are, for the most part, confirmed by the atomic structure determined by X-ray crystallography. As shown in Figure 5, the cytochrome *c* binding site in images of cytochrome oxidase monomers is in essentially the same position as the binding site in the X-ray structure deduced from the positions of Cys-115 of subunit III and the acidic residues of subunit II that have been shown to bind to opposite surfaces of cytochrome *c* in the active site (9,25). Although the position of Cys-115 of subunit III in the X-ray structure appears to be quite distant from the peak identified for the undecagold cluster compound bound to Cys-115 in the low resolution structure determined by electron crystallography (16), one must remember that the dimer observed in 2D crystals by electron microscopy is much more compact than the dimer found in 3D crystals by X-ray diffraction. In order to compare these 2 structures, the monomers in the X-ray structure must each be moved approximately 25 Å towards one another placing Cys-115 of subunit III within the 15 Å length of the link between the undecagold cluster observed by electron microscopy and the maleimide group bound to the sulfhydryl of Cys-115.

5.2. Cytochrome *c* Reductase

The low resolution structure of the *Neurospora* cytochrome *c* reductase (cytochrome bc_1 complex) determined by electron microscopy is very similar the atomic structure of the beef heart mitochondrial enzyme determined by X-ray diffraction (49,85,88). As shown in Figure 6, the structure calculated from electron micrographs of 2D crystals displays an asymmetric distribution of mass protruding 30 Å on one side of the bilayer and 70 Å on the other with a 50 Å domain within the lipid bilayer (56,81). The smaller domain protruding 30 Å was identified as the

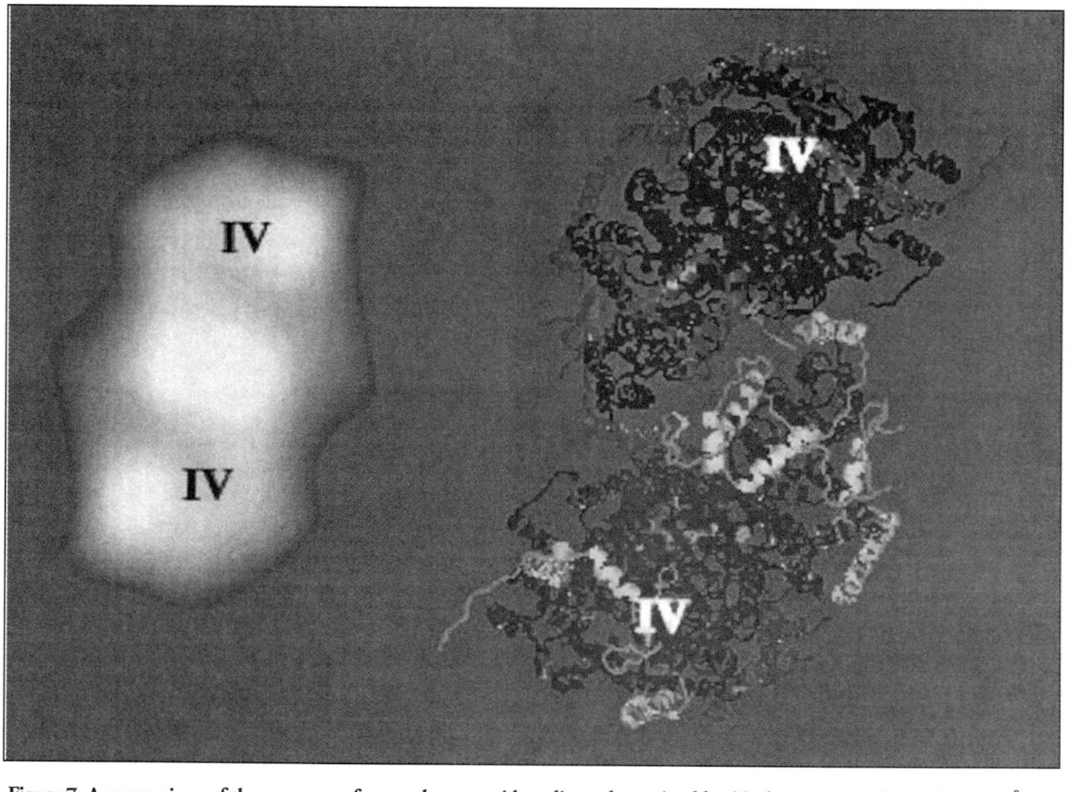

Figure 7. A comparison of the structure of a cytochrome oxidase dimer determined by (a) electron crystallography at 15 Å resolution and (b) a ribbon diagram based upon the atomic structure determined by X-ray crystallography. The position of subunit IV in the electron microscopy structure was determined by labeling with anti-IV Fabs. The structures are on the same scale.

hydrophilic subunits of cytochrome c_1 and the Rieske iron sulfur protein (subunits IV and V in Figure 7), and the larger domain as the core subunits I and II based upon the structure of a subcomplex lacking subunits I and II (47,52). These assignments are confirmed in the X-ray structure in which the cytochrome c1 and Rieske iron sulfur domains extend 30 Å beyond the bilayer and the core subunits 70 Å beyond. The dimensions of the cytochrome *c* reductase dimer are also similar: 120 x 75 Å in the electron microscopy structure versus 143 x 102 Å in the X-ray structure. Here, the somewhat smaller structure determined by electron microscopy can be attributed to: *(i)* shrinkage when the 2D crystals are prepared for electron microscopy by negative staining; and/or *(ii)* the fact that the structure determined by electron microscopy is of the *Neurospora* enzyme and that determined by X-ray diffraction is of the beef heart enzyme. The structure of cytochrome b_6f in projection (8) is very similar to the structure of the cytochrome bc_1 subcomplex calculated from atomic coordinants, but a 3D structure of the cytochrome b_6f complex is not yet available.

ABBREVIATIONS

cmc, critical micelle concentration; CCD, charge-coupled device; CTF, contrast transfer function; S/N, signal to noise ratio; 2D, two-dimensional; 3D, three-dimensional.

6. SUMMARY

Determination of the structures of membrane proteins by electron microscopy of 2D crystals has proven to be a very powerful method that, in the best cases, can yield atomic models of these proteins. The examples of heme proteins described here have yielded only low resolution structures, however, and the complete structures were obtained by X-ray crystallography of 3D crystals. The failure of electron crystallography to produce high resolution structures of these enzymes may be explained in part by their large sizes. Cytochrome *c* oxidase has a molecular weight of 200 000 (400 000 for the dimer form), and cytochrome *c* reductase has a molecular weight of 250 000 (500 000 for the dimer), and both contain many different polypeptide subunits. Nevertheless, low resolution models calculated from electron micrographs provided early insight into the structures of these critically important enzymes. Identification of functional sites and domains by specific labeling and by crystallization of subcomplexes also proved valuable in elucidating the structures of these enzymes and in explaining some aspects of their function. Furthermore, the technique of electron crystallography has been proven to be capable of determining the atomic structures of a number of proteins and will surely prove useful in elucidating the structures of other heme-containing membrane proteins.

REFERENCES

1. **Adrian, M., J. Dubochet, S.D. Fuller, and J.R. Harris.** 1998. Cryo-negative staining. Micron *29*:145-160.
2. **Aebi, U., B. Heggeler, L. Onorato, J. Kistler, and M.K. Showe.** 1977. New method for localizing proteins in periodic structures: Fab fragment labeling combined with image processing of electron micrographs. Proc. Natl. Acad. Sci. USA *74*:5514-5518.
3. **Aebi, U. and T.D. Pollard.** 1987. A glow discharge unit to render electron microscope grids and other surfaces hydrophilic. J. Electron Microsc. Tech. *7*:29-33.
4. **Amos, L.a., R. Henderson, and P.N.T. Unwin.** 1982. Three-dimensional structure determination by electron microscopy of two-dimensional crystals. Prog. Biophys. Mol. Biol. *39*:183-231.
5. **Berriman, J., R.K. Bryan, R. Freeman, and K.R. Leonard.** 1984. Methods for specimen thickness determination in electron microscopy. Ultramicroscopy *13*:351-364.

6. **Bhairi, S.M.** 1997. Detergents: A Guide to the Properties and Uses of Detergents in Biological Systems. Calbiochem-Novabiochem Corporation, San Diego.
7. **Bozzola, J.J. and L.D. Russell.** 1999. Electron Microscopy. Principles and Techniques for Biologists. Jones and Bartlett, Boston.
8. **Bron, P., J.J. Lacapere, C. Breyton, and G. Mosser.** 1999. The 9 Å projection structure of cytochrome b6f complex determined by electron crystallography. J. Mol. Biol. *287*:117-126.
9. **Capaldi, R.A., V. Darley-Usmar, S. Fuller, and F. Millett.** 1982. Structural and functional features of the interaction of cytochrome c with complex III and cytochrome c oxidase. FEBS Lett. *138*:1-7.
10. **Chiu, W. and T.W. Jeng.** 1982. Electron radiation sensitivity of protein crystals. Ultramicroscopy *10*:63-69.
11. **Conway, J.F. and A.C. Steven.** 1999. Methods for reconstructing density maps of "single" particles from cryoelectron micrographs to subnanometer resolution. J. Struct. Biol. *128*:106-118.
12. **Costello, M.J., J. Escaig, K. Matsushita, P.V. Viitanen, D.R. Menick, and H.R. Kaback.** 1987. Purified *lac* permease and cytochrome o oxidase are functional as monomers. J. Biol. Chem. *262*:17072-17082.
13. **Costello, M.J. and T.G. Frey.** 1982. Membranous cytochrome c oxidase. A freeze-fracture electron microscopic analysis. J. Mol. Biol. *162*:131-156.
14. **Crowther, R.A., D.J. De Rosier, and A. Klug.** 1970. The reconstruction of a three-dimensional structure from projections and its application to electron microscopy. Proc. R. Soc. A *317*:319.
15. **Crowther, R.A. and U.B. Sleytr.** 1977. An analysis of the fine structure of the surface layers from two strains of *Clostridia*, including correction for distorted images. J. Ultrastruct. Res. *58*:41-49.
16. **Crum, J., K.J. Gruys, and T.G. Frey.** 1994. Electron microscopy of cytochrome c oxidase crystals: labeling of subunit III with a monomaleimide undecagold cluster compound. Biochemistry *33*:13719-13726.
17. **Deatherage, J.F., R. Henderson, and R.A. Capaldi.** 1982. Relationship between membrane and cytoplasmic domains in cytochrome c oxidase by electron microscopy in media of different density. J. Mol. Biol. *158*:501-514.
18. **Deatherage, J.F., R. Henderson and R.A. Capaldi.** 1982. Three-dimensional structures of cytochrome c oxidase vesicle crystals in negative stain. J. Mol. Biol. *158*:487-499.
19. **De Rosier, D.J. and P.B. Moore.** 1970. Reconstruction of three-dimensional images from electron micrographs of structures with helical symmetry. J. Mol. Biol. *52*:355-369.
20. **Dubochet, J.** 1975. Carbon loss during irradiation of T4 bacteriophages and *E. coli* bacteria in electron microscopes. J. Ultrastruct. Res. *52*:276-288.
21. **Dubochet, J., M. Adrian, J.J. Chang, J.C. Homo, J. Lepault, A.W. McDowall, and P. Schultz.** 1988. Cryo-electron microscopy of vitrified specimens. Q. Rev. Biophys. *21*:129-228.
22. **Dykstra, M.J.** 1992. Biological Electron Microscopy. Theory, Techniques, and Troubleshooting. Plenum Press, New York.
23. **Erickson, H.P. and A. Klug.** 1971. Measurement and compensation of defocusing and aberrations by Fourier processing of electron micrographs. Phil. Trans. R. Soc. Lond. B *261*:105-118.
24. **Erickson, H.P., W.A. Voter, and K. Leonard.** 1978. Image reconstruction in electron microscopy: enhancement of periodic structure by optical filtering. Methods Enzymol. *49*:39-63.
25. **Ferguson-Miller, S., D.L. Brautigan, and E. Margoliash.** 1978. Definition of cytochrome c binding domains by chemical modification. III. Kinetics of reaction of carboxydinitrophenyl cytochromes c with cytochrome c oxidase. J. Biol. Chem. *253*:149-159.
26. **Frank, J.** 1973. Computer processing of electron micrographs, p. 215-274. *In* J.K. Koehler (Ed.), *Advanced Techniques in Biological Electron Microscopy.* Springer-Verlag, Berlin.
27. **Frank, J., W. Chiu, and L. Degn.** 1988. The characterization of structural variations within a crystal field. Ultramicroscopy *26*:345-360.
28. **Frank, J. and P. Penczek.** 1995. On the correction of the contrast function in biological electron microscopy. Optik *98*:125-129.
29. **Frey, P.A. and T.G. Frey.** 1999. Synthesis of undecagold labeling compounds and their applications in electron microscopic analysis of multiprotein complexes. J. Struct. Biol. *127*:94-100.
30. **Frey, T.G., S.H. Chan, and G. Schatz.** 1978. Structure and orientation of cytochrome c oxidase in crystalline membranes. Studies by electron microscopy and by labeling with subunit-specific antibodies. J. Biol. Chem. *253*:4389-4395.
31. **Frey, T.G. and T. Chang.** 1986. The structure of membrane bound cytochrome c oxidase. Ann. NY Acad. Sci. *483*:120-130.
32. **Frey, T.G., M.J. Costello, and S.H. Chan.** 1984. Selective contrast in electron microscopy of crystalline cytochrome oxidase. Ultramicroscopy *13*:85-91.
33. **Frey, T.G., M.J. Costello, B. Karlsson, J.C. Haselgrove, and J.S. Leigh.** 1982. Structure of the cytochrome c oxidase dimer. Electron microscopy of two-dimensional crystals. J. Mol. Biol. *162*:113-130.
34. **Frey, T.G., L.A. Kuhn, J.S. Leigh, Jr., M.J. Costello, and S.H. Chan.** 1985. Cytochrome oxidase: structural insights from electron microscopy and from secondary structure prediction. J. Inorg. Biochem. *23*:155-162.
35. **Frey, T.G. and J.M. Murray.** 1994. Electron microscopy of cytochrome c oxidase crystals. Monomer-dimer relationship and cytochrome c binding site. J. Mol. Biol. *237*:275-297.
36. **Fukami, A. and K. Adachi.** 1965. A new method of preparation of a self-perforated micro plastic grid and its application. J. Electron Microsc. *14*:112-118.
37. **Fuller, S.D., R.A. Capaldi, and R. Henderson.** 1979. Structure of cytochrome c oxidase in deoxycholate-derived two-dimensional crystals. J. Mol. Biol. *134*:305-327.
38. **Glaeser, R.M. and K.A. Taylor.** 1978. Radiation damage relative to transmission electron microscopy of biological specimens at low temperature: a review. J. Microsc. *112*:127-138.

39. **Hahn, M., J. Seredynski, and W. Baumeister.** 1976. Inactivation of catalase monolayers by irradiation with 100 keV electrons. Proc. Natl. Acad. Sci. USA *73*:823-827.
40. **Hayat, M.A.** 1986. Transmission Electron Microscopy. Academic Press, Orlando.
41. **Henderson, R., J.M. Baldwin, K.H. Downing, J. Lepault, and F. Zemlin.** 1986. Structure of purple membrane from *halobacterium halobium*: recording, measurement and evaluation of electron micrographs at 3.5 A resolution. Ultramicroscopy *19*:147-178.
42. **Henderson, R., R.A. Capaldi, and J.S. Leigh.** 1977. Arrangement of cytochrome oxidase molecules in two-dimensional vesicle crystals. J. Mol. Biol. *112*:631-648.
43. **Henderson, R. and P.N. Unwin.** 1975. Three-dimensional model of purple membrane obtained by electron microscopy. Nature *257*:28-32.
44. **Heuser, J.** 1989. Protocol for 3-D visualization of molecules on mica via the quick- freeze, deep-etch technique. J. Electron Microsc. Tech. *13*:244-263.
45. **Holser, W.T.Z.** 1958. Point groups and plane groups in a two-sided plane and their subgroups. Z. Kristallogr. *110*:266-281.
46. **Hovmoller, S., K. Leonard, and H. Weiss.** 1981. Membrane crystals of a subunit complex of mitochondrial cytochrome reductase containing the cytochromes *b* and c_1. FEBS Lett. *123*:118-122.
47. **Hovmoller, S., M. Slaughter, J. Berriman, B. Karlsson, H. Weiss, and K. Leonard.** 1983. Structural studies of cytochrome reductase. Improved membrane crystals of the enzyme complex and crystallization of a subcomplex. J. Mol. Biol. *165*:401-406.
48. **Hutchinson, E.G., W. Tichelaar, G. Hofhaus, H. Weiss, and K.R. Leonard.** 1989. Identification and electron microscopic analysis of a chaperonin oligomer from *Neurospora crassa* mitochondria. EMBO J. *8*:1485-1490.
49. **Iwata, S., J.W. Lee, K. Okada, J.K. Lee, M. Iwata, B. Rasmussen, T.A. Link, S. Ramaswamy, and B.K. Jap.** 1998. Complete structure of the 11-subunit bovine mitochondrial cytochrome bc1 complex (see comments]. Science *281*:64-71.
50. **Jap, B.K., M. Zulauf, T. Scheybani, A. Hefti, W. Baumeister, and U. Aebi.** 1992. 2D crystallization: from art to science. Ultramicroscopy *46*:45-84.
51. **Jayaraman, U., T. Chang, T.G. Frey, and J.K. Blasie.** 1987. Electron density profile of two-dimensionally crystalline membranous cytochrome c oxidase at low resolution. Biophys. J. *51*:475-486.
52. **Karlsson, B., S. Hovmoller, H. Weiss, and K. Leonard.** 1983. Structural studies of cytochrome reductase. Subunit topography determined by electron microscopy of membrane crystals of a subcomplex. J. Mol. Biol. *165*:287-302.
53. **Kim, C.H., M. Radermacher, M. Kessel, J. Frank, and T.E. King.** 1985. Three-dimensional reconstruction of cytochrome oxidase vesicle crystals prepared by cholate solubilization. J. Inorg. Biochem. *23*:163-169.
54. **Kühlbrandt, W.** 1988. Structure of light-harvesting chlorophyll a/b protein complex from plant photosynthetic membranes at 7 A resolution in projection. J. Mol. Biol. *202*:849-864.
55. **Kühlbrandt, W.** 1992. Two-dimensional crystallization of membrane proteins. Q. Rev. Biophys. *25*:1-49.
56. **Leonard, K., P. Wingfield, T. Arad, and H. Weiss.** 1981. Three-dimensional structure of ubiquinol:cytochrome c reductase from *Neurospora* mitochondria determined by electron microscopy of membrane crystals. J. Mol. Biol. *149*:259-274.
57. **Lunsdorf, H. and E. Spiess.** 1986. A rapid method of preparing perforated supporting foils for the thin carbon films used in high resolution transmission electron microscopy. J. Microsc. *144*:211-213.
58. **Malhotra, A., P. Penczek, R.K. Agrawal, I.S. Gabashvili, R.A. Grassucci, R. Junemann, N. Burkhardt, K.H. Nierhaus, and J. Frank.** 1998. *Escherichia coli* 70 S ribosome at 15 A resolution by cryo-electron microscopy: localization of fMet-tRNAfMet and fitting of L1 protein. J. Mol. Biol. *280*:103-116.
59. **Mannella, C.A.** 1984. Phospholipase-induced crystallization of channels in mitochondrial outer membranes. Science *224*:165-166.
60. **Michel, H.** 1991. Crystallization of Membrane Proteins. CRC Press, Boca Raton.
61. **Miyazawa, A., Y. Fujiyoshi, M. Stowell, and N. Unwin.** 1999. Nicotinic acetylcholine receptor at 4.6 A resolution: transverse tunnels in the channel wall. J. Mol. Biol. *288*:765-786.
62. **Mohraz, M. and P.R. Smith.** 1984. Structure of (Na^+,K^+)-ATPase as revealed by electron microscopy and image processing. J. Cell Biol. *98*:1836-1841.
63. **Murray, J.M. and R. Ward.** 1987. Preparation of holey carbon films suitable for cryo-electron microscopy. J. Elec. Microsc. Tech. *5*:285-290.
64. **Murray, J.M. and R. Ward.** 1987. Principles for the construction and operation of a device for rapidly freezing suspensions for cryo-electron microscopy. J. Electron Microsc. Tech. *5*:279-284.
65. **Salmon, E.D. and D. DeRosier.** 1981. A surveying optical diffractometer. J. Microsc. *123*:239-247.
66. **Seki, S., H. Hayashi, and T. Oda.** 1970. Studies on cytochrome oxidase. I. Fine structure of cytochrome oxidase-rich submitochondrial membrane. Arch. Biochem. Biophys. *138*:110-121.
67. **Smith, P.R. and I.E. Ivanov.** 1980. Surface reliefs computed from micrographs of isolated heavy metal shadowed particles. J. Ultrastruct. Res. *71*:25-36.
68. **Sosinsky, G.E. and G.A. Perkins.** 2000. Electron crystallographic methods for investigating gap junction structure [In Process Citation]. Methods *20*:140-155.
69. **Stokes, D.L. and N.M. Green.** 1990. Three-dimensional crystals of CaATPase from sarcoplasmic reticulum. Symmetry and molecular packing. Biophys. J. *57*:1-14.
70. **Suarez, M.D., A. Revzin, R. Narlock, E.S. Kempner, D.A. Thompson, and S. Ferguson-Miller.** 1984. The functional and physical form of mammalian cytochrome c oxidase determined by gel filtration, radiation inactivation, and sedimentation equilibrium analysis. J. Biol. Chem. 259:13791-13799.
71. **Taylor, K.A., L. Dux, S. Varga, H.P. Ting-Beall, and A. Martonosi.** 1988. Analysis of two-dimensional crys-

tals of Ca^{2+}-ATPase in sarcoplasmic reticulum. Methods Enzymol. *157*:271-289.

72. **Taylor, K.A. and R.M. Glaeser.** 1976. Electron microscopy of frozen hydrated biological specimens. J. Ultrastruct. Res. *55*:448-456.

73. **Toyoshima, C. and N. Unwin.** 1988. Contrast transfer for frozen-hydrated specimens: determination from pairs of defocused images. Ultramicroscopy *25*:279-291.

74. **Tsukihara, T., H. Aoyama, E. Yamashita, T. Tomizaki, H. Yamaguchi, K. Shinzawa-Itoh, R. Nakashima, R. Yaono, and S. Yoshikawa.** 1995. Structures of metal sites of oxidized bovine heart cytochrome c oxidase at 2.8 A [see comments]. Science *269*:1069-1074.

75. **Tsukihara, T., H. Aoyama, E. Yamashita, T. Tomizaki, H. Yamaguchi, K. Shinzawa-Itoh, R. Nakashima, R. Yaono, and S. Yoshikawa.** 1996. The whole structure of the 13-subunit oxidized cytochrome c oxidase at 2.8 A [see comments]. Science *272*:1136-1144.

76. **Unger, V.M., N.M. Kumar, N.B. Gilula, and M. Yeager.** 1999. Three-dimensional structure of a recombinant gap junction membrane channel. Science *283*: 1176-1180.

77. **Unwin, P.N.T.** 1974. Electron microscopy of the stacked disk aggregate of Tobacco Mosaic Virus protein. II. The influence of electron irradiation on the stain distribution. J. Mol. Biol. *87*:657-670.

78. **Unwin, P.N.T. and R. Henderson.** 1975. Molecular structure determination by electron microscopy of unstained crystalline specimens. J. Mol. Biol. *94*:425-440.

79. **Valpuesta, J.M., R. Henderson, and T.G. Frey.** 1990. Electron cryo-microscopic analysis of crystalline cytochrome oxidase. J. Mol. Biol. *214*:237-251.

80. **Vanderkooi, G., A.E. Senior, R.A. Capaldi, and H. Hayashi.** 1972. Biological membrane structure. 3. The lattice structure of membranous cytochrome oxidase. Biochim. Biophys. Acta *274*:38-48.

81. **Weiss, H., S. Hovmoller, and K. Leonard.** 1986. Preparation of membrane crystals of ubiquinol-cytochrome-c reductase from *Neurospora* mitochondria and structure analysis by electron microscopy. Methods Enzymol. *126*:191-201.

82. **Wikoff, W.R., G. Wang, C.R. Parrish, R.H. Cheng, M.L. Strassheim, T.S. Baker, and M.G. Rossmann.** 1994. The structure of a neutralized virus: canine parvovirus complexed with neutralizing antibody fragment. Structure *2*:595-607.

83. **Williams, R.C. and H.W. Fisher.** 1970. Electron microscopy of Tobacco Mosaic virus under conditions of minimal beam exposure. J. Mol. Biol. *52*: 121-123.

84. **Wingfield, P., T. Arad, K. Leonard, and H. Weiss.** 1979. Membrane crystals of ubiquinone:cytochrome c reductase from *Neurospora* mitochondria. Nature *280*:696-697.

85. **Xia, D., C.A. Yu, H. Kim, J.Z. Xia, A.M. Kachurin, L. Zhang, L. Yu, and J. Deisenhofer.** 1997. Crystal structure of the cytochrome bc1 complex from bovine heart mitochondria [published erratum appears in Science 1997 Dec 19;278(5346):2037]. Science *277*:60-66.

86. **Yeager, M., V.M. Unger, and A.K. Mitra.** 1999. Three-dimensional structure of membrane proteins determined by two-dimensional crystallization, electron cryomicroscopy, and image analysis. Methods Enzymol. *294*:135-180.

87. **Zeppenzauer, M.** 1971. Formation of large crystals, p. 253. *In* W.B. Jacoby (Ed.), Methods of Enzymology. Academic Press, New York.

88. **Zhang, Z., L. Huang, V.M. Shulmeister, Y.I. Chi, K.K. Kim, L.W. Hung, A.R. Crofts, E.A. Berry, and S.H. Kim.** 1998. Electron transfer by domain movement in cytochrome bc1. Nature *392*:677-684.

89. **Zhu, J., P.A. Penczek, R. Schroder, and J. Frank.** 1997. Three-dimensional reconstruction with contrast transfer function correction from energy-filtered cryoelectron micrographs: procedure and application to the 70S *Escherichia coli* ribosome. J. Struct. Biol. *118*:197-219.

10 Analysis and Reconstitution of Chlorophyll–Proteins

Harald Paulsen and Volkmar H.R. Schmid
Institut für Allgemeine Botanik der Johannes-Gutenberg Universität Mainz, Mainz, Germany

1. INTRODUCTION

It is hard to believe that only some 30 years ago, it was a matter of debate whether chlorophyll (Chl) and other photosynthesis pigments are protein-bound or just dissolved in plant membranes. Philip Thornber, who vividly described this debate in his recollection of photosynthesis research in the 1960s (90), was one of the exponents who finally convinced their colleagues that most, if not all, Chl in plants is in fact organized in protein complexes. It was his laboratory that devised quite a number of chromatographic and electrophoretic techniques for isolating (bacterio)chlorophyll-containing complexes from bacteria and plants. These isolation techniques later paved the way for structural analyses of, e.g., photosynthetic reaction centers of purple bacteria (24) as well as bacterial (53) and plant (49) light-harvesting complexes (LHC). Many of these protocols for isolating pigment–protein complexes are still, in a more or less modified version, being used in many laboratories today.

Isolating and analyzing Chl-protein complexes is not always an easy task, as several of these complexes are quite unstable and dissociate very easily. This may be illustrated by analyses of the major LHC (LHCII) of photosystem II (PSII). Although this is the most abundant Chl-containing complex and certainly one of the more stable ones, its Chl-protein stoichiometry reported by various laboratories has fluctuated between 6 (44) and 15 (14). One of the major problems is the fact that, with few exceptions, Chl-binding polypeptides are membrane proteins that need to be solubilized by detergents for isolation and analysis. The quantification of proteins in detergent solution is often difficult, which in turn tends to render Chl/protein ratios unreliable. Moreover, looking at the crystal structure of LHCII monomeric complexes (49) where most of the Chls are exposed at the surface of the protein complex, it may be not too surprising that some of these pigments are easily lost upon treatment of the complex by detergent. The major secret of solubilizing Chl-containing protein complexes without producing large amounts of unbound pigments has been the choice of the right detergent.

Heme, Chlorophyll, and Bilins: ***Methods and Protocols***
Edited by A.G. Smith and M. Witty

Unfortunately, however, the one optimal detergent for all Chl-protein complexes does not seem to exist. Optimized isolation protocols with individual detergents or detergent mixtures have been developed for just about every Chl-protein. Only a selection of these procedures can be referred to in this chapter.

A new approach to studying Chl-protein complexes was opened up when Plumley and Schmidt (72) showed that LHCII can be reconstituted in vitro upon denaturing it in detergent solution. Under renaturing conditions in detergent solution, the protein refolds and comcomitantly assembles its pigments. This astonishing capability of self-organization in vitro seems to extend to all Chl *a/b*-containing complexes in plants and also to bacterial LHCs (see section 3.1). On the other hand, no successful reconstitution of the Chl *a*-containing complexes of reaction centers and inner antennas has been reported. Possibly this indicates some structural feature that is common only among Chl-*a/b* proteins which enables them to refold in the presence of detergents and pigments.

The approach of reconstitution is so powerful, because it allows the generation of recombinant pigment–protein complexes in vitro. Not only can the pigment and lipid composition in these complexes be altered, in order to assess their contribution to structure and function, but also bacterially expressed apoproteins carrying directed mutations can be used. This way, the impact of individual protein domains, or even single amino acids, on pigment binding or functioning of the complex can be addressed. Such a detailed analysis of structure–function relationships is much more difficult to achieve in an in vivo system.

In this chapter, we will focus on the analysis and reconstitution of Chl-*a/b* proteins. We hope the reader will also find the selected references to the analysis and reconstitution of other proteins helpful.

2. ISOLATION AND ANALYSIS OF CHLOROPHYLL–PROTEIN COMPLEXES

As pointed out in the introduction to this chapter, there is a tremendous number of different isolation procedures described in the literature that have been devised for various Chl-containing complexes. It is impossible to give a complete overview of these techniques. Therefore, we will point to procedures that we believe are good examples of different separation principles used for Chl-proteins and focus on those techniques that are also useful for isolating reconstituted pigment–protein complexes (see below, section 3.4). Likewise, we will describe in some more detail only those analytical procedures that have successfully been used to characterize recombinant pigment–protein complexes.

2.1. Isolation of Chlorophyll–Protein Complexes

2.1.1. Solubilization

Most Chl-protein complexes are localized in photosynthetic membranes; therefore, the first step in the isolation of these complexes notoriously includes the solubilization of the membranes by use of detergents. The detergent employed for solubilization needs to be wisely chosen, as it will influence the aggregational state and the integrity of the pigment–protein complexes, and it will often limit the choice of techniques accessible for further purification. Ionic detergents for instance that are difficult to remove quantitatively from the protein complexes, such as sodium dodecylsulfate (SDS), usually preclude subsequent ion exchange chromatography.

Nonionic glycosidic detergents such as octylglucoside (OG) or dodecylmaltoside (LM) have proven useful to solubilize many pigment–protein complexes under mild nondenaturing conditions (Table 1). However, other nonionic detergents like Triton® X-100 or ionic detergents such as lauryldimethylamine oxide (LDAO) and SDS (or lithiumdodecylsulfate, LDS, where SDS would precipitate at low temperatures) are also in use. SDS or LDS, in combination with nonionic detergents, allows the control of the denaturing stringency in partially denaturing polyacrylamide gels and, thus, is useful in separating more stable complexes from less stable ones (70).

2.1.2. Purification

The overview of separation procedures given in Table 2 is far from being complete, but rather gives a few examples of where the procedures have been used.

The majority of the procedures used for the isolation of native Chl-protein complexes includes ultracentrifugation steps in a density gradient or various chromatographic techniques or both; these techniques are also useful to isolate recombinant Chl-protein complexes. Some of the gel electrophoretic techniques tend to have a more strongly denaturing effect on pigment–protein complexes and, therefore, may lead to the loss of pigments. However, gel electrophoresis provides a rapid means to separate in vitro reconstituted Chl-protein complexes from unbound pigments and, thus, is useful for analyzing reconstitution products. Therefore, gel electrophoresis along with sucrose gradient centrifugation and ion exchange techniques will be described in some more detail in section 3.4.

2.2. Analysis of Chlorophyll–Protein Complexes

2.2.1. Biochemical Characterization

The first biochemical characterization of native and recombinant Chl-protein complexes usually is the analysis of the pigment and protein components. Very often, the first step consists of an extraction of the complexes by adding 80% acetone, which denatures and precipitates the protein moiety, whereas the noncovalently bound pigments end up in the supernatant and can subsequently be separated and analyzed. If Chl-protein complexes have been separated by polyacrylamide gel electrophoresis, acetone does not extract pigments from gel slices. This can be done quantitatively by using 2-butanol (52).

The pigments in the 80% acetone extract can be analyzed and quantified either spectrophotometrically or by high-performance liquid chromatography (HPLC). For the quantification of mixtures of Chl *a* and Chl *b* in 80% acetone (buffered to pH 7.8), the algorithm by Porra et al. (73) is frequently used:

$$\text{concentration}_{\text{Chl } a}\ (\mu\text{g/mL}) = 12.25\ A_{663.6} - 2.55\ A_{646.6}$$

$$\text{concentration}_{\text{Chl } b}\ (\mu\text{g/mL}) = 20.31\ A_{646.6} - 4.91\ A_{663.6}$$

where $A_{646.6}$ and $A_{663.6}$ are the absorbances at the given wavelengths at 1 cm pathlength, minus the absorbance at 750 nm. Some alternative algorithms are discussed in the same reference. For the determination of bacteriochlorophyll *a* in acetone:methanol (7:2, vol/vol), a molar absorption coefficient of ε_{770} = 76 000/Mcm (19) has been measured. For the rough quantification of α- and β-carotenoids and their derivatives, the specific absorption coefficient of Davies (22) is very useful: ε_{440} = 240 L/gcm.

Numerous HPLC protocols for separating Chls and carotenoids have been developed (29,34,65,93). One of the simplest ones is a gradient from 80% to 100% acetone. Usually, the conversion of peak areas to pigment quantities is calibrated on the basis of absorption coefficients such as those given above.

Quantification of the precipitated and redissolved protein can be performed by

Table 1. Detergents Used for Solubilizing Chl-Proteins

Detergent	Used to Solubilize	References
OG	PSII core complexes	37
	LH1 from *Rhodospirillum rubrum*	64
	Chl-protein complexes of thylakoids	15,69
LM	PSI for crystallization	30
	PSII core complexes with antenna complexes attached	7,37
	PSII core and antenna subunits	20
	CP43 and CP47	1
	LHCII subunits	21
	LHCI-730	86
	PSI and PSII from red algae	89
	Chl *a,b,c* complex of *M. squamata*	97
Heptylthioglucoside	PSII for crystallization	74,75
	CP22 (PsbS)	31
Triton® X-100	LHCII	13,48
	PSI	56
	Chl *a,b,c* complex of *M. squamata*	76
Lauryldimethyl-amine oxide (LDAO)	LH1 and LH2 from *Rhodospirillum molischianum*	92
SDS or LDS	Chl-protein complexes of thylakoids	4,25
Digitonin	Chl *a,c* complexes from various algae	10,26,27,39
Zwittergent 14	PSI and PSII from *Prochlorothrix hollandica*	95
Zwittergent 16	LHCI	38,51,91

Table 2. Separation Procedures Used to Isolate Chl-Proteins

	Used for	References
Sucrose-gradient ultracentrifugation	PSI	57
	PSII core and antenna subunits	20
	PSII core complex and PSII-LHCII super complex	37
	LH1 and LH2 from *R. molischianum*	92
	LHCI	63
	LHCII	13,48
	Chl *a,c* complexes from various algae	11,26,27,39
	Chl *a,b,c* complex of *M. squamata*	76
Sucrose gradient with subsequent anion-exchange chromatography	PSI reaction center complex	80
	PSI	80
	CP43 and CP47	1
	D1-D2-cytochrome b_{559}, CP43 and CP 47 and LHCII subunits	5
	LHCII complexes	61
	Recombinant CP24	62
	Recombinant CP29	35
	PSI and PSII from red algae	89
	Chl *a,b,c* complex from *M. squamata*	97
	PSI and PSII from *P. hollandica*	95
Anion exchange chromatography	Water-soluble Chl-protein from cauliflower and Brussels sprouts	43,60
Gel filtration chromatography	PSII-LHCII super complex	7
	LH1 from *R. rubrum*	64
Perfusion chromatography	PSII	81
	LHCI	91
Nickel chelating chromatography	Recombinant His-tagged LHCII	28,79

Table 2, continued

	Used For	References
	His-tagged bacterial photosynthetic reaction center	36
Copper chelating chromatography	PSI and PSII	77,87
Precipitation by mono- and divalent cations	LHCII	46
Partially denaturing LDS (Laemmli) gel electrophoresis	Recombinant LHCII	68,72
	Native and reconstituted Chl *a,b,c* complex from *M. squamata*	54
Modified Laemmli gel	Chl-protein complexes of thylakoids	2,45
Deriphat gel	Chl-proteins from thylakoids	69,84
Blue native polyacrylamide gel electrophoresis	Chloroplast protein complexes	47
Preparative flat-bed isoelectric focussing	CP22 (PsbS)	31
	LHCII subunits	21
	Subunits of PSII	5
Fluid-phase isoelectric focussing	PSII LHCs	3

applying a number of generally used protein assays such as the bicinchoninic acid reaction (in Reference 88, see Chapter 10), the Lowry test (71), or the Bradford test (9). Detergents are a problem in many protein assays. If the protein is precipitated from a solution containing SDS or LDS, co-precipitation of the dodecylsulfate salt can be reduced by acidifying the solution to pH 4.0–5.0 with acetic acid. Note that acidification is incompatible with subsequent Chl analysis in the supernatant, as it will turn the Chls into their pheophytins. If the protein is precipitated from a very dilute solution, it may be necessary to collect it by extended ultracentrifugation in order to pellet it quantitatively.

It is often desirable to exchange or remove detergents from Chl-protein complexes before their analysis or further handling. Detergents having a critical micellar concentration in the millimolar range, such as OG, can most easily be removed by dialysis. A method to exchange detergents in protein solutions by hydrophobic interaction chromatography on phenyl sepharose has been described (78); however, detergents that stably interact with proteins such as dodecylsulfate are only inefficiently exchanged by this technique. Removal of these detergents, SDS or LDS, is often problematic, particularly with recombinant Chl *a/b*-protein complexes that are reconstituted at high concentrations of LDS. The bulk of LDS can be removed by precipitating the dodecylsulfate with 200 mM potassium salt at 0°C in the presence of a nonionic detergent. The remaining dodecylsulfate can be removed by binding the complexes to an ion exchange column and washing them with buffer containing a nonionic detergent (41).

2.2.2. Spectroscopic Characterization

The spectroscopy used to analyze isolated Chl-protein complexes very much depends on which complex is to be analyzed and on its properties to be examined. Techniques generally used for a first characterization of pigment–protein complexes are absorption, fluorescence, and circular dichroism (CD) spectroscopy (96).

An immediate test for the intactness of isolated Chl *a*/*b* LHCs is the measurement of energy transfer from Chl *b* to Chl *a*. This can easily be measured by using a steady-state fluorescence spectrophotometer. The excitation is set to an absorption wavelength of Chl *b* (usually at around 470 nm, at the long-wavelength side of the Soret absorption band in order to minimize simultaneous excitation of Chl *a*). Fluorescence emission is scanned over the Chl *a* and Chl *b* emission wavelength region, between 650 and 750 nm. Emission from Chl *a* exclusively (maximum at around 680 nm, no shoulder at 660 nm) indicates quantitative energy transfer from Chl *b* to Chl *a*. Alternatively, the emission wavelength is set to the Chl *a* emission wavelength around 680 nm, and excitation is scanned in the Chl and carotenoid absorption domain. Excitation signals at 455 nm and around 475 nm indicate energy transfer from Chl *b* and carotenoids, respectively.

Care must be taken to avoid deceptive energy transfer in dilute detergent solutions. For instance in 0.1% (wt/vol) LM solution, as is often used in sucrose density gradient centrifugations, efficient energy transfer from Chl *b* to Chl *a* is detected, even with unbound pigments, in the absence of any pigment–protein complexes. As the detergent concentration is lowered, Chl concentrations in the decreasing number of detergent micelles rise until energy transfer becomes possible. Therefore, if energy transfer from Chl *b* to Chl *a* is detected even upon heat-denaturing the Chl-protein complexes, it cannot be indicative of intact complexes and most likely the detergent concentration is too low.

CD spectroscopy is useful to characterize pigment–protein complexes, as both the protein and the pigment moieties give rise to CD signals in the UV and visible regions, respectively (96). CD signals in the visible region in which Chls and carotenoids absorb have been taken as a criterion to test the structural authenticity of recombinant Chl *a*/*b* complexes (68,72). Beyond this fingerprint comparison, CD spectra provide information about the pigment organization in pigment–protein complexes (94) and the state of oligomerization of such complexes (33,41). Signals of CD in the UV domain have been used to compare protein folding in a recombinant with that in wild-type LHCII (66).

3. RECONSTITUTION OF PIGMENT–PROTEIN COMPLEXES

Reconstitution represents the controlled folding of an LHC-apoprotein in the presence of detergents and pigments (and lipids), giving rise to complexes with very similar properties when compared to the authentic complexes isolated from leaf material.

3.1. Survey of Reconstituted Pigment–Protein Complexes

The first pigment–protein complexes which were reconstituted are the major LHCII of higher plants and the core LHC

(LH1) of photosynthetic bacteria. In these initial experiments, authentic proteins isolated from LH1 (64) and total thylakoid membrane proteins (72) were employed. New possibilities were opened up by the availability of cDNAs for several LHC-apoproteins of different origins, which allowed the use of bacterially overexpressed LHC proteins in reconstitution experiments (17,68). Since bacterially expressed proteins can easily be mutated, the introduction of various structural alterations into recombinant pigment–protein complexes was facilitated. Using C and N terminally truncated apoproteins, the significance of these protein domains for the formation of LHCII could be identified (18,67). Later, reconstitution of different LHCs of higher plants, CP29 (35), CP26 (82), and CP24 (62), as well as LHCI-730 (86) and LHCI-680a (Schmid and Paulsen, unpublished) were accomplished. Additionally, an LHCI (LhcaR1) of the red alga *Porphyridium cruentum* (32), a LHC of the green alga *Chlorella fusca,* and a Chl *a,b,c*-containing complex of the prasinophyte *Mantoniella squamata* were successfully reconstituted (54). Recently, the peripheral LHC (LH2) of a purple bacterium has also been reconstituted (92).

Moreover, it was shown that it is not only possible to reconstitute monomeric LHC but also the oligomeric form of the major LHCII complex and of LHCI-730 (41,79,86).

3.2. Comparison of Reconstitution Procedures

All the LHCII reconstitution experiments until 1992 were performed by the original freeze-thaw method (72). Later, a new method, based on detergent exchange was developed (66), which proved to be very powerful as it has allowed reconstitution of the more labile complexes in recent years. In Table 3, these two methods and their applications are summarized. Both procedures will be given in more detail in section 3.3.2.

The result of reconstitutions of LHCI and LHCII by these two methods is depicted in Figure 1. It is obvious that application of the detergent exchange method yields reconstituted LHCI and LHCII, whereas with the freeze-thaw method, only reconstituted LHCII is obtained.

Additionally, a reconstitution technique was developed, which is based on detergent mixing, which also prompts protein folding and pigment binding. This method allows protein refolding to initiate very quickly and, therefore, facilitates time-resolved spectroscopic measurements (8).

3.3. Reconstitution Procedures

Reconstitutions of plant LHCs usually start from isolated plant pigments and bacterially expressed apoprotein.

3.3.1. Pigment Isolation

Reconstitution is performed either with a total thylakoid pigment extract or with (a mixture of) individual pigments. Some pigments are commercially available [e.g., Chl *a* and *b*, lutein, α- and β-carotene from Sigma (St. Louis, MO, USA); lutein and zeaxanthin from Roth (Germany)] but others are not. Therefore, their isolation is described in the following protocol. All isolation steps should be performed in dim light. As a source for pigment isolation, homogenized whole leaves or thylakoids isolated as in, e.g., Reference16, can be used. Thylakoids are suspended in a small volume of dilute buffer as 10 mM Tricine-NaOH (pH 7.8) and extracted by addition of acetone to a final concentration of 80%. Proteins are removed by a 10-minute centrifugation at 15 000× *g*. For isolation of total pigment extract or individual pigments, the supernatant is treated in different ways.

❖ Procedure 1. Total Pigment Extract

1. The pigment solution in acetone is mixed with 0.25 volumes diethyl ether in a separating funnel.
2. To improve phase separation, solid NaCl (e.g., 35 g to 600-mL solution) is added and dissolved by gently moving the funnel. If the lower acetone phase remains colored, the ether extraction should be repeated, adding more NaCl if phase separation is poor.
3. Combine and dry the ether phases, either by the addition of solid NaCl or by placing the ether solution in a -20°C freezer for at least 1 hour. Then ice or NaCl can be removed by filtration through a sintered glass funnel (precooled to -20°C if ice crystals are to be removed).
4. Evaporate the ether in a rotary evaporator or nitrogen stream.
5. Pigments are dissolved in acetone, quantified on the basis of their Chl content (see section 2.2.1), and aliquoted.
6. The aliquots are dried in a nitrogen stream and can be stored for months to years at -20°C under nitrogen or argon.

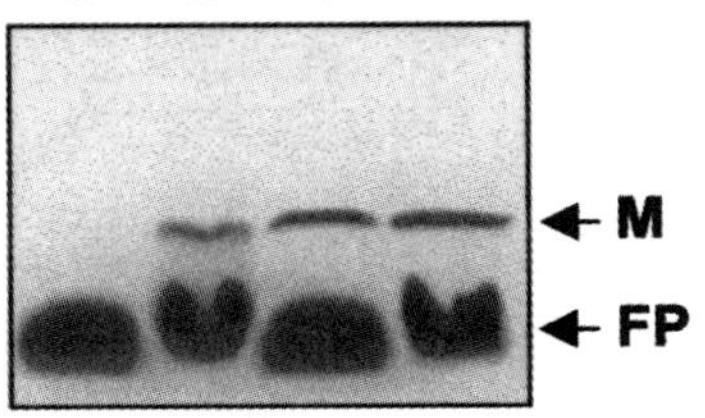

Figure 1. Partially denaturing gel electrophoresis of reconstitution mixtures with LHCI- (Lhca4; lanes a and b) or LHCII- (Lhcb1; lanes c and d) apoprotein. The mixtures were subjected to either the freeze-thaw (lanes a and c) or the detergent exchange method (lanes b and d). The resolution of bands with monomeric complexes (M) and free pigments (FP) is visible.

For the isolation of individual Chls and carotenoids, the following procedure is useful:

❖ Procedure 2. Isolation of Individual Pigments

1. The acetonic pigment solution is cooled to 0°C.
2. Dioxane is added to give a final concentration of about 15% (vol/vol) (42).
3. To the homogenous solution 0.16 volumes of water is added drop-wise under constant stirring, and the resultant solution is kept on ice for 1 hour without further stirring.
4. Aggregated Chls (as well as pheophytin and β-carotene) are collected by centrifugation (15 000× *g*, for 10 min), and the pelleted pigments are used for column chromatographic separation of the Chls and β-carotene. If the supernatant still contains a substantial amount of the Chl originally present, more water may be added drop-wise, and the additional precipitate collected. If the addition of water is too excessive or too fast, xanthophylls will also aggregate and contaminate the crude Chl preparation. The supernatant is kept for xanthophyll isolation.
5. For the separation of individual Chls (and xanthophylls), a reversed phase C18 material [e.g., 55–105 µ-Bondapak (Waters, Milford, MA, USA)] with acetone–water mixtures as the mobile phase is suitable. Chromatography can be done either at low pressure or by using an HPLC apparatus. A column volume of at least 4 mL (low pressure development) or 0.8 mL (high pressure processing) is recommended for each milligram of raw pigments applied.

Table 3. Comparison of the Experimental Steps of the Two Most Commonly Used Reconstitution Techniques

	Freeze-Thaw Method	Detergent Exchange Method
References that the method is used in:	17,18,54,55,68,72	32,35,40,41,62,66,82,86
Protein denaturation by LDS and heating	+	+
Addition of OG	-	+
Addition of pigments (and lipids)	+	+
Freeze at -20°C and thaw at 22°C 3 times	+	-/(+)
Removal of LDS by KCl addition	-	+
Application to nondenaturing polyacrylamide gel electrophoresis or sucrose density gradient centrifugation	+	+

6. Preequilibrate the column with 86% acetone.
7. Dissolve the Chl pellet in a small volume of 86% acetone.
8. To remove any particulate material the solution is centrifuged (15 000× *g* for 10 min).
9. The supernatant is loaded on the column.
10. Elution of Chl *b* (green) and Chl *a* (blue-green) is done with acetone of that concentration. For the elution of pheophytin (brown) and β-carotene (orange), the acetone concentration has to be raised to 90% and 95%, respectively. To obtain pure pigments, only the central part of the individual pigment fractions should be collected.
11. The eluted pigments are transferred to ether, dried, quantitated, and stored as described for total pigment extracts (steps 1–6).
12. For isolation of individual xanthophylls, the pigments in the supernatant of the dioxane precipitation (step 4) are ether-extracted and dried as described for total pigment extracts (steps 1–4).
13. The residue is dissolved in ethanol and made up to 8% KOH by addition of 0.1 volumes from an 80% (wt/vol) stock solution in water (22). The mixture is overlaid with nitrogen and kept overnight in a tightly capped bottle at 55°C in the dark. The saponification with KOH converts residual Chl and lipids into more hydrophilic products.
14. Xanthophylls are extracted by ether as described for total pigment extract (steps 1–2).
15. The ether fraction is washed twice with 3 volumes of water.
16. The xanthophylls are then obtained from the ether solution as described

above (total pigment extracts, steps 3–4). They can be either used as "total xanthophylls" (see section 2.2.1 for absorption coefficient for carotenoids) for reconstitution or, alternatively, subjected to a column chromatography procedure as outlined for the Chls (step 5) to isolate the individual xanthophylls.

17. Proceed as in steps 6 to 9, but use 74% acetone for preequilibration and as solvent.
18. Isocratic elution is started with 74% acetone. Following elution of neoxanthin and violaxanthin the acetone concentration is raised to 80% acetone for the elution of lutein.

The purity of the individual pigments is most conveniently tested by analytical HPLC or by thin-layer chromatography (TLC) with, e.g., RP18-plates (Merck, Darmstadt, Germany) and methanol as solvent. For the quantification of Chls see section 2.2.1; carotenoids can be quantified by means of the absorption coefficients of Davies (23). The specific absorption coefficients, for example for an ethanolic solution, are ε_{439} = 224.3 L/gcm (neoxanthin), ε_{443} = 255 L/gcm (violaxanthin), ε_{445} = 255 L/gcm (lutein), and ε_{453} = 262 L/gcm (β-carotene).

3.3.2. Protein Isolation

Originally, authentic proteins isolated from the membranes of the corresponding species were extracted and used for reconstitution experiments (72). Meanwhile, however, cDNAs derived from the genes are cloned into expression vectors, e.g., pDS-vectors (12), in order to obtain large quantities of the desired protein. All bacterially expressed LHC-apoproteins are accumulated in the form of insoluble inclusion bodies. Therefore, inclusion body isolation, which follows mainly the procedure described by Nagai and Thøgersen (58), is described here in detail.

❖ Procedure 3. Isolation of Recombinant LHC-Apoproteins

1. Start with an 5 mL overnight culture [Luria-Bertani medium supplemented with 100 µg ampicillin/mL (LB-Amp)] of *Eschericia coli* that harbors the respective expression plasmid.
2. Inoculate a 250-mL Erlenmeyer flask containing 100 mL LB-Amp with 1 mL of the overnight culture and grow the cells on a rotary shaker at 170 rounds/minute and 37°C to mid-log phase (OD_{550} of around 0.5), which takes about 2 hours.
3. Induce overexpression by addition of a 1 M isopropyl-β-D-thiogalactoside solution to 1 mM final concentration. The cells are cultivated for another 4 to 5 hours under the same conditions.
4. Harvest the cells by centrifugation (5 min at 5 000× *g*). The cell pellets are either stored at -20°C or processed further immediately.
5. Suspend the cell pellet in 500 µL lysis buffer [50 mM Tris-HCl (pH 8.0), 25% (wt/vol) sucrose, and 1 mM EDTA) and bring it to 800 µL with lysis buffer.
6. Add 200 µL lysozyme from a 1% (wt/vol, in lysis buffer) solution which is freshly prepared each time. The solution is mixed well and incubated at room temperature for 30 minutes.
7. Add 10 µL DNase I solution [0.1% (wt/vol) solution in 20 mM Tris-HCl (pH 8.0), 50 mM NaCl, 1 mM dithiothreitol (DTT), and 50% (vol/vol) glycerol; this solution can be stored at -20°C], together with 10 µL of 0.1 M $MnCl_2$ and 1 M $MgCl_2$. Incubation for another 30 minutes at room temperature follows.

8. Add 2 mL of detergent solution A [1% (wt/vol) deoxycholic acid (sodium salt), 1% (vol/vol) Nonidet® P-40, 0.2 M NaCl, 20 mM Tris-HCl (pH 7.5), 2 mM EDTA, 30 mM DTT). The solution is mixed well and kept at room temperature for 5 minutes.
9. Centrifuge the solution for 10 minutes at 10 000× *g*.
10. Suspend the pellet in 2 mL detergent solution B [0.5% (wt/wt) Triton® X-100, 1 mM EDTA-NaOH (pH 7.8), and 20 mM DTT] and keep the solution at room temperature for 5 minutes.
11. Collect inclusion bodies by centrifugation as in step 5. Sometimes the over-expressed protein does not form very stable inclusion bodies. In this case, the volume of the detergent solutions should be reduced (e.g., by 50%, but the appropriate amount has to be determined empirically).
12. Suspend the pellet in storage buffer [50 mM Tris-HCl (pH 8.0), 1 mM EDTA, 20 mM DTT]. If the protein pellet appears slimy at this point and is not easily resuspended, it is advisable to repeat steps 4–7 once again.
13. Assess the protein content by, e.g., the Bradford protein assay (9), which is compatible with DTT. Aliquots of the protein solution can be stored at -20°C.

3.3.3. Reconstitution Procedures

In the following, we describe reconstitution of Chl-protein complexes by *(i)* freeze-thaw cycles, *(ii)* detergent exchange, and *(iii)* detergent mixing. Quantities and volumes given are for subsequent separation in analytical polyacrylamide slab gels. For sucrose gradient ultracentrifugation, the amounts should be scaled up 4-fold [SW60 rotor, Beckman Coulter (Fullerton, CA, USA)] or 7-fold (SW40 or SW41 rotors, Beckman Coulter).

Procedure 4. Freeze-Thaw Method

1. Suspend 8 μg of LHCII or 25 μg of LHCI inclusion body protein in 16 μL storage buffer (section 3.3.1).
2. Mix the protein solution with an equivalent volume of 2× reconstitution buffer [200 mM Tris-HCl (pH 9), 4% (wt/vol) LDS, 100 mM DTT, 2 mM benzamidine, 10 mM ε-aminocaproic acid, and — for gel separation — 25% (wt/vol) sucrose].
3. The protein solution is heated for 1 minute in a boiling water bath and cooled on ice.
4. Dried Chls and total xanthophylls corresponding to 24 μg and 5 μg, respectively, (LHCI: total pigment extract equivalent to 30 μg Chl) are dissolved in 1.5 μL ethanol (the final ethanol content must not exceed 8% as otherwise protein precipitation may occur). Pigments must be dissolved completely, which is best done by vigorously vortex mixing followed by an incubation for 30 seconds in an ultrasonic bath. When pigments sediment during subsequent centrifugation of the pigment solution (1 min at 15 800× *g*), this step has to be repeated.
5. Add the protein solution to the pigment solution under vortex mixing.
6. The resultant solution is placed in a -20°C freezer for 2 hours and then thawed at room temperature for 15 minutes.
7. Repeat step 6 twice.
8. The solution is ready to be analyzed on a partially denaturing gel or in sucrose density gradients (see section 3.4).

❖ Procedure 5. Detergent Exchange Method

1. Prepare a protein solution as in step 1 of Procedure 4.

2. Reconstitution is continued by the addition of 3.7 µL of 10% (LHCII) or 20% (LHCI) OG to the protein solution.
3. Boil the solution for 1 minute and cool it down on ice.
4. Add 1.2 µL 1 M DTT to the sample.
5. Prepare a pigment solution as in step 4 of Procedure 4.
6. Transfer the protein solution to the pigments during mixing.
7. Add 4.27 µL 2 M (LHCII) or 6.78 µL 1 M (LHCI) KCl solution.
8. The resultant solution is mixed and kept at 4°C for 20 minutes.
9. Precipitated potassium dodecylsulfate is sedimented by centrifugation (15 800× *g* for 5 min at 4°C).
10. The supernatant is loaded on a gel or sucrose gradient (section 3.4).

❖ Procedure 6. Detergent Mixing Method

1. Twenty-three micrograms LHCII inclusion body protein is dissolved in 116 µL of 0.2% (wt/vol) SDS, 100 mM lithium borate (pH 9.0), 12.5% (wt/vol) sucrose, and 5 mM DTT.
2. Transfer the solution to a cuvette.
3. Twelve micrograms Chl and 3 µg total xanthophylls are dissolved in 3 µL ethanol (step 4 of Procedure 4).
4. Dissolved pigments are mixed with 116 µL of a solution with 2% (wt/vol) OG, 0.075% (wt/vol) phosphatidylglycerol, 100 mM lithium borate (pH 9.0), 12.5% (wt/vol) sucrose, and 5 mM DTT.
5. The pigment solution is added to the protein solution in the cuvette and mixed rapidly by stirring. Alternatively, these 2 solutions are rapidly mixed by means of a stopped-flow device which allows time resolved measurements down to the millisecond range. The dilution brought about by the mixing of the 2 samples results in folding of the protein (8).
6. The success of reconstitution can be examined by gel electrophoretic analysis or by following spectroscopic signals, e.g., the energy transfer from Chl *b* to Chl *a* (section 2.2.2).

Besides monomeric complexes, oligomeric complexes can also be reconstituted. Very convenient is the generation of the heterodimeric LHCI-730, which requires equal amounts of the apoproteins (12.5 µg of both) in the starting protein solution (86). Using a sophisticated, multistep procedure, trimerization of LHCII was achieved which allowed crystallization of the reconstituted complex. This method is described in detail in Hobe et al. (41). Meanwhile, another reconstitution method for trimeric LHCII has been developed, where refolding of the protein occurs while it is immobilized, via a His-tag, on a metal-chelate column material. This method facilitates faster production and isolation of trimeric complexes (79).

3.4. Isolation of Reconstituted Complexes

All the techniques described below should be performed at 0° to 4°C in dim light.

3.4.1. Partially Denaturing Polyacrylamide Gel Electrophoresis

Most of the commonly used partially denaturing gel systems go back to the recipes given by either Laemmli (50) or Neville (59). They proved to be also suitable for isolation of reconstituted complexes. For reconstituted LHCII, we prefer the

Laemmli system (68), and for LHCI, we prefer a modified Neville system (85). For most applications, analytical slab minigels are a good choice. The 30% (wt/vol) acrylamide stock solution we use has an acrylamide: N,N′-methylenebisacrylamide ratio of 30.

Laemmli Gel

Prepare the required volume of the resolving gel solution with 12% (wt/vol) acrylamide, 400 mM Tris-HCl (pH 8.8) and 10% (vol/vol) glycerol. While the solution is stirred, ammonium persulfate (APS) [10% (wt/vol) stock solution] and N,N,N′,N′-tetramethylethylenediamine (Temed) are added to final concentrations of 0.07% (wt/vol) and 0.05% (vol/vol), respectively. The gel solution is poured between the assembled glass plates up to about 7 mm below that point where the bottom of the comb will be located. The gel surface is overlaid with a thin layer of water to obtain homogenous polymerization and a plane gel surface. After 1 hour, the gel should be polymerized, the water is poured off, and, if necessary, the gel surface is dabbed with filter paper. The stacking gel solution with 4.5% (wt/vol) acrylamide, 130 mM Tris-HCl (pH 6.8), 10% (vol/vol) glycerol, 0.05% (wt/vol) APS, and 0.05% (vol/vol) Temed is poured on top of the resolving gel, and the comb is inserted by gently pushing it down starting from one side.

Neville Gel

The resolving gel is composed of 12% (wt/vol) acrylamide, 424 mM Tris-HCl (pH 9.1), and 10% (wt/vol) sucrose. Polymerization is initiated by the addition of 10% APS solution and Temed to final concentrations of 0.03% (wt/vol) and 0.075% (vol/vol), respectively. The stacking gel is composed of 4% (wt/vol) acrylamide, 54 mM Tris-H_2SO_4 (pH 6.1), 10% (wt/vol) sucrose, and polymerization is initiated by final concentrations of 0.06% (wt/vol) APS and 0.075% (vol/vol) Temed.

For both gel types, the same running buffer with 25 mM Tris, 196 mM glycine, and 0.1% (wt/vol) LDS (only required in the cathode buffer) is used. The buffer is best prepared as a tenfold stock solution and diluted before use.

Prior to electrophoresis, the gel and the buffer should be cooled to 4°C. After removal of the comb, the gel pockets are rinsed with running buffer. Samples equivalent to 10 µg Chl are applied to 4-mm-wide wells of a 1-mm-thick gel. Electrophoresis is either conducted with a constant voltage of 60 V (Laemmli) or with a constant current of 0.1 mA/ mm^2 of gel cross-section (Neville gel) for about 2.5 hours. Afterwards, the gel sandwich is disassembled, the gel documented, and individual bands can be excised and the protein eluted for further characterization (see section 2.2).

3.4.2. Sucrose Density Gradients

Compared to nondenaturing gel electrophoresis, ultracentrifugation through sucrose density gradients is a more gentle method for isolating labile LHC. Moreover, this method allows the isolation of sufficient material for further analyses and has the advantage that the green band collected from the centrifuge tube can be used immediately without the need to extract it from a gel.

Sucrose gradients can be formed either by means of a gradient mixer in combination with a peristaltic pump or, more conveniently, by the freeze-thaw method described by Bassi and Simpson (6). For the latter method, a solution with 0.5 M sucrose, 5 mM Tricine-NaOH (pH 7.8), and 0.1% (wt/vol) LM is filled in the centrifuge tubes. The tubes are placed in a -20°C freezer. Three hours before sample application, the tubes are transferred to a refrigerator and kept there until completely thawed. Subsequently, the upper tenth of the gradient solution is carefully removed, which results in gradients with a sucrose concentration of about 0.1 to 1 M sucrose. Depending on the rotor used, centrifugation at 4°C is performed at 450 000× *g* for 16 hours (SW60, Beckman Coulter) or 260 000× *g* for 23 hours (SW40 or 41, Beckman Coulter). Subjecting a reconstitution mixture containing the 2 apoproteins of LHCI-730 to density-gradient ultracentrifugation results in a separation as is shown in Figure 2. The resolution of zones with free pigments, monomeric complexes, and the heterodimeric LHCI-730 is clearly visible. The bands of interest are collected with a flat-tipped syringe needle from the top.

3.4.3. Anion Exchange Chromatography

As a consequence of the low isoelectric points of the higher plant pigment proteins, ranging from 4 to 5 (21), anion exchange chromatography is suitable for the isolation of these proteins. This method is a good choice if one intends large-scale isolation. Furthermore, the advantages described for sucrose density gradients also apply to this method.

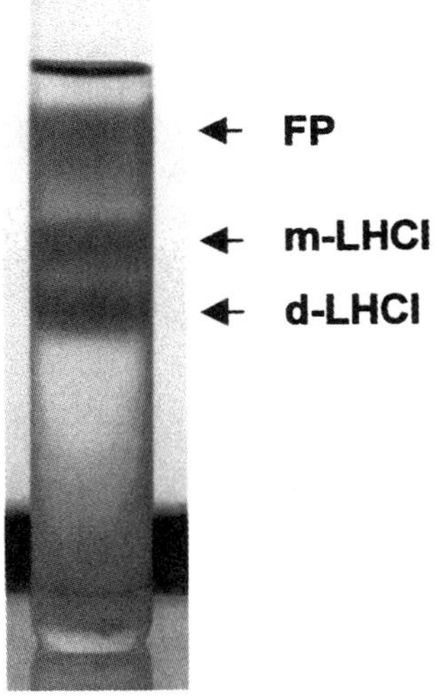

Figure 2. Sucrose gradient fractionation of reconstitution mixtures containing the two apoproteins of LHCI-730. FP, free pigments; m-LHCI, monomeric LHCI; d-LHCI, dimeric LHCI (LHCI-730).

Diethylaminoethyl (DEAE)-cellulose is mostly used as stationary phase for column development by gravity, fast flow Q-Sepharose® or Poros Q (both from Amersham Pharmacia Biotech, Piscataway, NJ, USA) for pump-mediated column processing (41,91,97). The mobile phase is usually composed of a slightly alkaline buffer (e.g., phosphate buffer, Tris), and a detergent such as LM, both in low concentrations [10 mM and 0.05% (wt/vol), respectively]. Prior to sample application, the column is washed first with 4 column volumes of, e.g., 10 mM sodium phosphate buffer (pH 7.4) and then with 2 volumes of the buffer supplemented with the detergent, which is used for the solubilization of the pigment–protein complexes, e.g., 0.05% (wt/vol) LM. Then the sample, adjusted to the same phosphate buffer (pH 7.4) and detergent concentration, is applied. After the sample has completely entered the column bed, the column is washed with 2 column volumes of buffer including detergent. Elution is achieved by the buffer plus detergent supplemented with NaCl. Mostly, a gradient of 0 to 400 mM NaCl works well. The steepness of the NaCl gradient has to be determined individually. Eluted bands can be characterized with regard to apoprotein composition (section 3.4.1), pigment composition (section 2.2.1), and spectral properties (section 2.2.2).

4. CONCLUDING REMARKS

Several aspects have been mentioned in the previous paragraphs of how reconstitution of Chl *a*/*b*-protein complexes can be used as a fine and sophisticated surgical tool in functional analyses of these complexes. Single Chl-binding amino acids, for instance, may be exchanged simply by mutating the bacterially expressed apoprotein (83). If the corresponding Chl then is lacking in the reconstituted complex, its individual spectroscopic properties may be deduced from spectral differences between this and the native complex.

It should, however, be kept in mind that reconstitution itself is quite a striking example of self-organization of a biological structure. A cue as simple as the mixing of 2 different detergents is sufficient to initiate the correct folding of a medium-size membrane protein and the binding of up to 15 or so pigment molecules of several different kinds to their correct binding sites. The effort seems worthwhile to try and understand this self-organization process itself: what is the sequence of events between the prereconstitution mixture of components and the fully formed stable complex, what are the structural features involved, and which is the driving force? Once we know the answer to these questions, we may understand why Chl *a/b*-protein complexes appear to reconstitute more easily than other pigment–protein complexes. We may then learn to design biomimetic structures that autonomously form in vitro.

ABBREVIATIONS

APS, ammonium persulfate; CD, circular dichroism; Chl, chlorophyll; CP, chlorophyll protein; DTT, dithiothreitol; HPLC, high-performance liquid chromatography; LB-Amp, Luria-Bertani medium supplemented with ampicillin; LDAO, lauryldimethylamine oxide; LDS, lithiumdodecylsulfate; LHC, light-harvesting complex; LH1 and LH2, light-harvesting complexes of purple bacteria; LM, dodecylmaltoside; OG, octylglucoside; PS, photosystem; SDS, sodium dodecylsulfate; Temed, N,N,N′,N′-tetramethylethylenediamine.

ACKNOWLEDGMENTS

We thank Martin Lohr (Universität Mainz, Germany) for carefully reading the manuscript. Work in the authors' laboratory was supported in part by grants from the Deutsche Forschungsgemeinschaft (Pa 324/6-1, Schm 1203/2-1).

REFERENCES

1. **Alfonso, M., G. Montoya, R. Cases, R. Rodriguez, and R. Picorel.** 1994. Core antenna complexes, CP43 and CP47, of higher plant photosystem II. Spectral properties, pigment stoichiometry, and amino acid composition. Biochemistry *33*:10494-10500.
2. **Allen, K.D. and L.A. Staehelin.** 1991. Resolution of 16 to 20 chlorophyll protein complexes using a low ionic strength native green gel system. Anal. Biochem. *194*:214-222.
3. **Allen, K.D. and L.A. Staehelin.** 1992. Biochemical characterization of photosystem-II antenna polypeptides in grana and stroma membranes of spinach. Plant Physiol. *100*:1517-1526.
4. **Anderson, J.M., J.C. Waldron, and S.W. Thorne.** 1978. Chlorophyll-protein complexes of spinach and barley thylakoids. FEBS Lett. *92*:227-233.
5. **Bassi, R., B. Pineau, P. Dainese, and J. Marquardt.** 1993. Carotenoid-binding proteins of photosystem II. Eur. J. Biochem. *212*:297-303.
6. **Bassi, R. and D. Simpson.** 1987. Chlorophyll-protein complexes of barley photosystem I. Eur. J. Biochem. *163*:221-230.
7. **Boekema, E.J., H. van Roon, F. Calkoen, R. Bassi, and J.P. Dekker.** 1999. Multiple types of association of photosystem II and its light-harvesting antenna in partially solubilized photosystem II membranes. Biochemistry *38*:2233-2239.
8. **Booth, P.J. and H. Paulsen.** 1996. Assembly of light-harvesting chlorophyll *a/b* complex *in vitro*. Time-resolved fluorescence measurements. Biochemistry *35*:5103-5108.
9. **Bradford, M.M.** 1976. A rapid and sensitive method for the quantitation of proteins utilizing the principle of protein-dye binding. Anal. Biochem. *72*:248-254.
10. **Büchel, C. and G. Garab.** 1997. Organization of the pigment molecules in the chlorophyll *a/c* light-harvesting complex of *Pleurochloris meiringensis* (*Xanthophyceae*): characterization with circular dichroism and absorbance spectroscopy. J. Photochem. Photobiol. B Biol. *37*:118-124.

11. **Büchel, C. and C. Wilhelm.** 1993. Isolation and characterization of a photosystem I-associated antenna (LHC I) and a photosystem I-core complex from the chlorophyll *c*-containing alga *Pleurochloris meiringensis* (*Xanthophyceae*). J. Photochem. Photobiol. B Biol. *20*:87-93.
12. **Bujard, H., R. Gentz, M. Lanzer, D. Stueber, M. Mueller, I. Ibrahimi, M.T. Haeuptle, and B. Dobberstein.** 1987. A T5-promoter-based transcription-translation system for the analysis of protein expression *in vivo* and *in vitro*. Methods Enzymol. *155*:416-433.
13. **Burke, J.J., C.L. Ditto, and C.J. Arntzen.** 1978. Involvement of the light-harvesting complex in cation regulation of excitation energy distribution in chloroplasts. Arch. Biochem. Biophys. *187*:252-263.
14. **Butler, P.J.G. and W. Kühlbrandt.** 1988. Determination of the aggregate size in detergent solution of the light-harvesting chlorophyll *a/b*-protein complex from chloroplast membranes. Proc. Natl. Acad. Sci. USA *85*:3797-3801.
15. **Camm, E.L. and B.R. Green.** 1980. Fractionation of thylakoid membranes with the nonionic detergent octyl-β-D-glucopyranoside. Plant Physiol. *66*:428-432.
16. **Camm, E.L. and B.R. Green.** 1982. The effects of cations and trypsin on extraction of chlorophyll-protein complexes by octyl glucoside. Arch. Biochem. Biophys. *214*:563-572.
17. **Cammarata, K.V., F. Plumley, and G.W. Schmidt.** 1990. Reconstitution of light-harvesting complexes: a single apoprotein binds Chl*a*, Chl*b* and xanthophylls, p. 341-344. *In* M. Baltscheffsky (Ed.), Current Research in Photosynthesis, Vol. 2. Kluwer Academic Publishers, Dordrecht.
18. **Cammarata, K.V. and G.W. Schmidt.** 1992. *In-vitro* reconstitution of a light-harvesting gene product–deletion mutagenesis and analyses of pigment binding. Biochemistry *31*:2779-2789.
19. **Clayton, R.K. and B.J. Clayton.** 1981. B 850 pigment–protein complex of *Rhodopseudomonas sphaeroides*. Extinction coefficients, circular dichroism, and the reversible binding of bacteriochlorophyll. Proc. Natl. Acad. Sci. USA *78*:5583-5587.
20. **Dainese, P. and R. Bassi.** 1991. Subunit stoichiometry of the chloroplast photosystem II antenna system and aggregation state of the component chlorophyll *a/b* binding proteins. J. Biol. Chem. *266*: 8136-8142.
21. **Dainese, P., G. Hoyer-Hansen, and R. Bassi.** 1990. The resolution of chlorophyll *a/b*-binding proteins by a preparative method based on flat bed isoelectric focusing. Photochem. Photobiol. *51*:693-703.
22. **Davies, B.H.** 1965. Analysis of carotenoid pigments, p. 489-532. *In* T.W. Goodwin (Ed.), Chemistry and Biochemistry of Plant Pigments. Academic Press, New York.
23. **Davies, B.H.** 1976. Carotenoids, p. 38-165. *In* T.W. Goodwin (Ed.), Chemistry and Biochemistry of Plant Pigments, Vol. 2. Academic Press, London.
24. **Deisenhofer, J., O. Epp, K. Miki, R. Huber, and H. Michel.** 1985. Structure of the protein subunits in the photosynthetic reaction center of *Rhodopseudomonas viridis* at 3 Å resolution. Nature *318*:618-624.
25. **Delepelaire, P. and N.H. Chua.** 1981. Electrophoretic purification of chlorophyll *a/b* protein complexes from *Chlamydomonas reinhardtii* and spinach and analysis of their polypeptides. J. Biol. Chem. *256*:9300-9307.
26. **Douady, D., B. Rousseau, and L. Caron.** 1994. Fucoxanthin chlorophyll *a/c* light-harvesting complexes of *Laminaria saccharina* — partial amino acid sequences and arrangement in thylakoid membranes. Biochemistry *33*:3165-3170.
27. **Durnford, D.G. and B.R. Green.** 1994. Characterization of the light harvesting proteins of the chromophytic alga, *Olisthodiscus luteus* (*Heterosigma carterae*). Biochim. Biophys. Acta *1184*:118-126.
28. **Flachmann, R. and W. Kühlbrandt.** 1996. Crystallization and identification of an assembly defect of recombinant antenna complexes produced in transgenic tobacco plants. Proc. Natl. Acad. Sci. USA *93*:14966-14971.
29. **Francis, G.W., B. Huseby, and O.M. Andersen.** 1993. An improved HPLC system for the analysis of photosynthetic pigments. Chromatographia *35*:189-192.
30. **Fromme, P., H.T. Witt, W.D. Schubert, O. Klukas, W. Saenger, and N. Krauss.** 1996. Structure of photosystem I at 4.5 angstrom resolution: a short review including evolutionary aspects. Biochim. Biophys. Acta *1275*:76-83.
31. **Funk, C., W.P. Schröder, A. Napiwotzki, S.E. Tjus, G. Renger, and B. Andersson.** 1995. The PSII-S protein of higher plants: a new type of pigment-binding protein. Biochemistry *34*:11133-11141.
32. **Gantt, E., F.X. Cunningham, B. Grabowski, and S. Tan.** 1998. Relatedness of caroteno-chlorophyll antenna complexes in algae and plants, p. 239-247. *In* G. Garab (Ed.), Photosynthesis: Mechanism and Effects, Vol. I. Kluwer Academic Publishers, Dordrecht.
33. **Garab, G., J. Kieleczawa, J.C. Sutherland, C. Bustamante, and G. Hind.** 1991. Organization of pigment–protein complexes into macrodomains in the thylakoid membranes of wild-type and chlorophyll b-less mutant of barley as revealed by circular dichroism. Photochem. Photobiol. *54*:273-281.
34. **Gilmore, A.M. and H.Y. Yamamoto.** 1991. Resolution of lutein and zeaxanthin using a non-endcapped, lightly carbon-loaded C18 high-performance liquid chromatographic column. J. Chromatogr. *543*:137-145.
35. **Giuffra, E., D. Cugini, R. Croce, and R. Bassi.** 1996. Reconstitution and pigment-binding properties of recombinant CP29. Eur. J. Biochem. *238*:112-120.
36. **Goldsmith, J.O. and S.G. Boxer.** 1996. Rapid isolation of bacterial photosynthetic reaction centers with an engineered poly-histidine tag. Biochim. Biophys. Acta *1276*:171-175.
37. **Hankamer, B., J. Nield, D. Zheleva, E. Boekema, S. Jansson, and J. Barber.** 1997. Isolation and biochemical characterisation of monomeric and dimeric photosystem II complexes from spinach and their relevance to the organisation of photosystem II *in vivo*. Eur. J. Biochem. *243*:422-429.
38. **Haworth, P., J.L. Watson, and C.J. Arntzen.** 1983. The detection, isolation, and characterisation of a light-harvesting complex which is specifically associated with photosystem I. Biochim. Biophys. Acta *724*:151-158.

39. **Hiller, R.G., C.D. Scaramuzzi, and J. Breton.** 1992. The organisation of photosynthetic pigments in a Cryptophyte alga — a linear dichroism study. Biochim. Biophys. Acta *1102*:360-364.
40. **Hobe, S., R. Förster, J. Klingler, and H. Paulsen.** 1995. N-proximal sequence motif in light-harvesting chlorophyll *a/b*-binding protein is essential for the trimerization of light-harvesting chlorophyll *a/b* complex. Biochemistry *34*:10224-10228.
41. **Hobe, S., S. Prytulla, W. Kühlbrandt, and H. Paulsen.** 1994. Trimerization and crystallization of reconstituted light-harvesting chlorophyll *a/b* complex. EMBO J. *13*:3423-3429.
42. **Iriyama, K., N. Ogura, and A. Takamiya.** 1974. A simple method for extraction and partial purification of chlorophyll from plant material, using dioxane. J. Biochem. *76*:901-904.
43. **Kamimura, Y., T. Mori, T. Yamasaki, and S. Katoh.** 1997. Isolation, properties and a possible function of a water soluble chlorophyll *a/b* protein from Brussels sprouts. Plant Cell Physiol. *38*:133-138.
44. **Kan, K.S. and J.P. Thornber.** 1976. The light-harvesting chlorophyll *a/b*-protein of *Chlamydomonas reinhardtii*. Plant Physiol. *57*:47-53.
45. **Knötzel, J., I. Svendsen, and D.J. Simpson.** 1992. Identification of the photosystem-I antenna polypeptides in barley — isolation of 3 pigment-binding antenna complexes. Eur. J. Biochem. *206*:209-215.
46. **Krupa, Z., N.P.A. Huner, J.P. Williams, E. Maissan, and D.R. James.** 1987. Development at cold-hardening temperatures. The structure and composition of purified rye light harvesting complex II. Plant Physiol. *84*:19-24.
47. **Kügler, M., L. Jansch, V. Kruft, U.K. Schmitz, and H.P. Braun.** 1997. Analysis of the chloroplast protein complexes by blue-native polyacrylamide gel electrophoresis (BN-PAGE). Photosynth. Res. *53*:35-44.
48. **Kühlbrandt, W.** 1984. Three-dimensional structure of the light-harvesting chlorophyll *a/b*-protein complex. Nature *307*:478-480.
49. **Kühlbrandt, W., D.N. Wang, and Y. Fujiyoshi.** 1994. Atomic model of plant light-harvesting complex by electron crystallography. Nature *367*:614-621.
50. **Laemmli, U.K.** 1970. Cleavage of structural proteins during the assembly of the head of bacteriophage T4. Nature *227*:680-685.
51. **Lam, E., W. Ortiz, and R. Malkin.** 1984. Chlorophyll *a/b* proteins of photosystem I. FEBS Lett. *168*:10-14.
52. **Martinson, T.A. and F.G. Plumley.** 1995. One-step extraction and concentration of pigments and acyl lipids by *sec*-butanol from *in vitro* and *in vivo* samples. Anal. Biochem. *228*:123-130.
53. **McDermott, G., S.M. Prince, A.A. Freer, A.M. Hawthornthwaite-Lawless, M.Z. Papiz, R.J. Cogdell, and N.W. Isaacs.** 1995. Crystal structure of an integral membrane light-harvesting complex from photosynthetic bacteria. Nature *374*:517-521.
54. **Meyer, M. and C. Wilhelm.** 1993. Reconstitution of light-harvesting complexes from *Chlorella fusca* (*Chlorophyceae*) and *Mantoniella squamata* (*Prasinophyceae*). Z. Naturforsch. C *48*:461-473.
55. **Meyer, M., C. Wilhelm and G. Garab.** 1996. Pigment-pigment interactions and secondary structure of reconstituted algal chlorophyll *a/b*-binding light-harvesting complexes of *Chlorella fusca* with different pigment compositions and pigment–protein stoichiometries. Photosynth. Res. *49*:71-81.
56. **Mullet, J.E. and C.J. Arntzen.** 1980. Simulation of grana stacking in a model membrane system. Mediation by a purified light-harvesting pigment–protein complex from chloroplasts. Biochim. Biophys. Acta *589*:100-117.
57. **Mullet, J.E., J.J. Burke, and C.J. Arntzen.** 1980. Chlorophyll proteins of photosystem I. Plant Physiol. *65*:814-822.
58. **Nagai, K. and H.C. Thøgersen.** 1987. Synthesis and sequence-specific proteolysis of hybrid proteins produced in *E. coli*. Methods Enzymol. *153*:461-481.
59. **Neville, D.M.** 1971. Molecular weight determination of protein-dodecyl sulfate complexes by gel electrophoresis in a discontinous buffer system. J. Biol. Chem. *246*:6328-6334.
60. **Nishio, N. and H. Satoh.** 1997. A water-soluble chlorophyll protein in cauliflower may be identical to Bnd22, a drought-induced, 22-kilodalton protein in rapeseed. Plant. Physiol. *115*:841-846.
61. **Nußberger, S., K. Dörr, D.N. Wang, and W. Kühlbrandt.** 1993. Lipid-protein interactions in crystals of plant light-harvesting complex. J. Mol. Biol. *234*:347-356.
62. **Pagano, A., G. Cinque, and R. Bassi.** 1998. *In vitro* reconstitution of the recombinant photosystem II light-harvesting complex CP24 and its spectroscopic characterization. J. Biol. Chem. *273*:17154-17165.
63. **Palsson, L.O., S.E. Tjus, B. Andersson, and T. Gillbro.** 1995. Ultrafast energy transfer dynamics resolved in isolated spinach light-harvesting complex I and the LHC I-730 subpopulation. Biochim. Biophys. Acta *1230*:1-9.
64. **Parkes-Loach, P.S., J.R. Sprinkle, and P.A. Loach.** 1988. Reconstitution of the B873 light-harvesting complex of *Rhodospirillum rubrum* from the separately isolated alpha and beta-polypeptides and bacteriochlorophyll *a*. Biochemistry *27*:2718-2727.
65. **Patzlaff, J.S. and B.A. Barry.** 1996. Pigment quantitation and analysis by HPLC reverse phase chromatography: a characterization of antenna size in oxygen-evolving photosystem II preparations from cyanobacteria and plants. Biochemistry *35*:7802-7811.
66. **Paulsen, H., B. Finkenzeller, and N. Kühlein.** 1993. Pigments induce folding of light-harvesting chlorophyll *a/b*-binding protein. Eur. J. Biochem. *215*:809-816.
67. **Paulsen, H. and S. Hobe.** 1992. Pigment-binding properties of mutant light-harvesting chlorophyll *a/b*-binding protein. Eur. J. Biochem. *205*:71-76.
68. **Paulsen, H., U. Rümler, and W. Rüdiger.** 1990. Reconstitution of pigment-containing complexes from light-harvesting chlorophyll *a/b*-binding protein overexpressed in *E. coli*. Planta *181*:204-211.
69. **Peter, G.F. and J.P. Thornber.** 1991. Biochemical composition and organization of higher plant photosystem II light-harvesting pigment–proteins. J. Biol. Chem. *266*:16745-16754.
70. **Peter, G.F. and J.P. Thornber.** 1991. Electrophoretic procedures for fractionation of photosystems I and II pigment proteins of higher plants and for determina-

tion of their subunit composition, p. 195-210. *In* L.G. Rogers (Ed.), Methods in Plant Biochemistry, Vol. 5. Academic Press, New York.

71. **Peterson, G.L.** 1977. A simplification of the protein assay method of Lowry et al. which is more generally applicable. Anal. Biochem. *83*:346-356.

72. **Plumley, F.G. and G.W. Schmidt.** 1987. Reconstitution of chlorophyll *a/b* light-harvesting complexes: xanthophyll-dependent assembly and energy transfer. Proc. Natl. Acad. Sci. USA *84*:146-150.

73. **Porra, R.J., W.A. Thompson, and P.E. Kriedemann.** 1989. Determination of accurate extinction coefficients and simultaneous equations for assaying chlorophylls *a* and *b* extracted with four different solvents: verification of the concentration of chlorophyll standards by atomic absorption spectroscopy. Biochim. Biophys. Acta *975*:384-394.

74. **Rhee, K.H., E.P. Morris, J. Barber, and W. Kühlbrandt.** 1998. Three-dimensional structure of the plant photosystem II reaction centre at 8 Å resolution. Nature *396*:283-286.

75. **Rhee, K.H., E.P. Morris, D. Zheleva, B. Hankamer, W. Kühlbrandt (Reprint Author), and J. Barber.** 1997. Two dimensional structure of plant photosystem II at 8 Å resolution. Nature *389*:522-526.

76. **Rhiel, E., W. Lange, and E. Mörschel.** 1993. The unusual light-harvesting complex of *Mantoniella squamata* — supramolecular composition and assembly. Biochim. Biophys. Acta *1143*:163-172.

77. **Ritter, S., J. Komenda, E. Setlikova, I. Setlik, and W. Welte.** 1992. Immobilized metal affinity chromatography for the separation of photosystem I and photosystem II from the thermophilic cyanobacterium *Synechococcus elongatus*. J. Chromatogr. *625*:21-31.

78. **Robinson, N.C., D. Wiginton, and L. Talbert.** 1984. Phenyl sepharose-mediated detergent-exchange chromatography: its application to exchange of detergents bound to membrane proteins. Biochemistry *23*:6121-6126.

79. **Rogl, H., K. Kosemund, W. Kühlbrandt, and I. Collinson.** 1998. Refolding of *Escherichia coli* produced membrane protein inclusion bodies immobilised by nickel chelating chromatography. FEBS Lett. *432*:21-26.

80. **Rögner, M., U. Mühlenhoff, E.J. Boekema, and H.T. Witt.** 1990. Mono-, di- and trimeric PS I reaction center complexes isolated from the thermophilic cyanobacterium *Synechococcus* sp.* — size, shape and activity. Biochim. Biophys. Acta *1015*:415-424.

81. **Roobol-Boza, M. and B. Andersson.** 1996. Isolation of hydrophobic membrane proteins by perfusion chromatography — purification of photosystem II reaction centers from spinach chloroplasts. Anal. Biochem. *235*:127-133.

82. **Ros, F., R. Bassi, and H. Paulsen.** 1998. Pigment-binding properties of the recombinant photosystem II subunit CP26 reconstituted *in vitro*. Eur. J. Biochem. *253*:653-658.

83. **Sandoná, D., R. Croce, A. Pagano, M. Crimi, and R. Bassi.** 1998. Higher plants light harvesting proteins. Structure and function as revealed by mutation analysis of either protein or chromophore moieties. Biochim. Biophys. Acta *1365*:207-214.

84. **Sarvari, E. and P. Nyitrai.** 1994. Separation of chlorophyll-protein complexes by Deriphat polyacrylamide gradient gel electrophoresis. Electrophoresis *15*:1068-1071.

85. **Schmid, V. and C. Schäfer.** 1994. Alterations of the chlorophyll-protein pattern in chronically photoinhibited *Chenopodium rubrum* cells. Planta *192*:473-479.

86. **Schmid, V.H.R., K.V. Cammarata, B.U. Bruns, and G.W. Schmidt.** 1997. *In vitro* reconstitution of the photosystem I light-harvesting complex LHCI-730: heterodimerization is required for antenna pigment organization. Proc. Nat. Acad. Sci. USA *94*:7667-7672.

87. **Setlikova, E., S. Ritter, R. Hienerwadel, J. Kopecky, J. Komenda, W. Welte, and I. Setlik.** 1995. Purification of a photosystem II reaction center from a thermophilic cyanobacterium using immobilized metal affinity chromatography. Photosynth. Res. *43*:201-211.

88. **Smith, P.K., R.I. Krohn, G.T. Hermanson, A.K. Mallia, F.H. Gartner, M.D. Provenzano, E.K. Fujimoto, N.M. Goeke, B.J. Olson, and D.C. Klenk.** 1985. Measurement of protein using bicinchoninic acid. Anal. Biochem. *79*:76-85.

89. **Tan, S., G.R. Wolfe, F.X. Cunningham, and E. Gantt.** 1995. Decrease of polypeptides in the PS I antenna complex with increasing growth irradiance in the red alga *Porphyridium cruentum*. Photosynth. Res. *45*:1-10.

90. **Thornber, J.P.** 1995. Thirty years of fun with antenna pigment–proteins and photochemical reaction centers: a tribute to the people who have influenced my career. Photosynth. Res. *44*:3-22.

91. **Tjus, S.E., M. Roobol-Boza, L.O. Palsson, and B. Andersson.** 1995. Rapid isolation of Photosystem I chlorophyll-binding proteins by anion exchange perfusion chromatography. Photosynth. Res. *45*:41-49.

92. **Todd, J.B., P.S. Parkes-Loach, J.F. Leykam, and P.A. Loach.** 1998. *In vitro* reconstitution of the core and peripheral light-harvesting complexes of *Rhodospirillum molischianum* from separately isolated components. Biochemistry *37*:17458-17468.

93. **Val, J., E. Monge, and N.R. Baker.** 1994. An improved HPLC method for rapid analysis of the xanthophyll cycle pigments. J. Chromatogr. Sci. *32*:286-289.

94. **van Amerongen, H. and W.S. Struve.** 1995. Polarized optical spectroscopy of chromoproteins. Biochem. Spectroscopy *246*:259-283.

95. **van der Staay, G.W.M., A. Brouwer, R.L. Baard, F. Vanmourik, and H.C.P. Matthijs.** 1992. Separation of photosystem I and photosystem II from the oxychlorobacterium (prochlorophyte) *Prochlorothrix hollandica* and association of chlorophyll-*b* binding antennae with photosystem II. Biochim. Biophys. Acta *1102*:220-228.

96. **van Holde, K.E., C.E. Johnson, and P.S. Ho.** 1998. Physical Biochemistry. Prentice Hall, London.

97. **Welte, C., R. Nickel, and A. Wild.** 1995. Three-dimensional crystallization of the light-harvesting complex from *Mantoniella squamata* (*Prasinophyceae*) requires an adequate purification procedure. Biochim. Biophys. Acta *1231*:265-274.

11 Two-Dimensional Crystallization of Chlorophyll Proteins

Georgios Tsiotis
Department of Chemistry, University of Crete, Heraklion, Greece

1. INTRODUCTION

1.1. Photosynthesis Begins with the Chlorophylls

The conversion of light energy into the free energy of organic compounds is the quintessence of photosynthesis. The primary process is the absorption of light by the antenna complexes and its transfer to the reaction center (RC) where photochemical energy storage take places. These are the most important proteins that catalyze the conversion of light energy into chemical energy and are known as type-I and type-II RCs. Type-I RCs use iron–sulfur clusters and are exemplified by photosystem I (PSI), found in plants, eukaryotic algae, and cyanobacteria. This complex catalyses the light-driven electron transfer from reduced plastocyanin to oxidized ferredoxin. Type-II RCs use quinones as electron acceptors, the best well-known example being photosystem II (PSII), which is found in all oxygen-evolving photosynthetic organisms, where it carries out the photochemical oxidation of water and produces reduced plastoquinone. The two types of RCs probably have a common evolutionary origin, as they apparently use a special pair of chlorophyll (Chl) or bacteriochlorophyll (BChl) molecules as the primary electron donor and a Chl or pheophytin as the primary electron acceptor (2). The photosystems are large multisubunit protein complexes integral to the thylakoid membrane.

The initial step in the energy conversion process of photosynthesis is absorption of light by Chls (or BChls). Most of the Chl (or BChl) molecules in the photosynthetic organisms serve as an antenna for light harvesting. Only one Chl molecule out of approximately 500, and one BChl molecule out of 50 to 100, is directly connected to the electron transfer chain through the reaction centers that trap excitation energy by rapid electron transfer to an acceptor. The pigment molecules involved in antenna light gathering function are bound to proteins of relatively small size (10–30 kDa), known as light-harvesting (LH) proteins, which are again integral membrane proteins (21).

In higher plants, algae, and cyanobacteria, the two photosystems are linked by the

Heme, Chlorophyll, and Bilins: *Methods and Protocols*
Edited by A.G. Smith and M. Witty

cytochrome b_6f complex. This is a plastoquinol:plastocyanin oxidoreductase, which is an integral membrane protein complex that participates in electron transfer and the generation of an electrochemical proton gradient in oxygenic photosynthesis.

In this chapter, various techniques for obtaining two-dimensional (2D) crystals of chlorophyll a-containing membrane proteins and a cytochrome b_6f complex are discussed.

1.2. Structural Analysis of Membrane Proteins

Membrane proteins are amphiphilic macromolecules incorporated vectorially in the fluid lipid membrane. Upon disruption of the membrane and removal of the lipids by detergents, such proteins tend to aggregate nonspecifically and to precipitate. This can be avoided, and the proteins can be solubilized in aqueous solutions, when detergents replace the membrane lipids. While structure determination of soluble proteins by X-ray crystallography and nuclear magnetic resonance (NMR) has progressed at a remarkable rate, this has been far less successful for membrane proteins, as manifested by the small number solved to atomic resolution. Three-dimensional (3D) crystals that allowed atomic resolution by X-ray crystallography have been obtained for the bacterial porins (25, 44,50), the photosynthetic RCs (1,34), two cytochrome c oxidases (20,48), the purple bacterium light-harvesting complex (LHC) II (32), and bacteriorhodopsin (40). The requirement that membrane proteins must be solubilized in detergents leads to large complexes not suitable for study by NMR.

In most cases, solubilized membrane proteins have been crystallized by conventional methods that foster interaction of hydrophilic protein surfaces. The initial hopes that detergent-solubilized membrane proteins could be handled analogously to soluble proteins with respect to crystallization have not been realized. For this reason, different attempts were undertaken to resolve this problem, such as the successful co-crystallization of the cytochrome c oxidase with antibody fragments (39) and the use of the cubic lipid phase to obtain high quality 3D crystals of bacteriorhodopsin (29).

Electron crystallography of 2D crystals offers a viable alternative to X-ray crystallography. The inherent adaptation of membrane proteins for the 2D environment of the lipid bilayer suggests that the most appropriate geometry for a regularly packed protein would be a 2D crystal. Suitable 2D crystals are more easily produced than 3D crystals, because membrane proteins have a tendency to pack within a lipid bilayer. The ultimate goal of electron microscopy-based techniques applied to proteins is the production of images from which a clear depiction of the internal atomic structure can be derived. Methods and instruments have been developed which potentially allow analysis at near atomic resolution (16,17).

There are several inherent advantages of electron microscopy when compared to X-ray crystallography for the study of membrane proteins: (*i*) 2D crystals of membrane proteins form in a continuous lipid bilayer. Thus, the protein environment in the crystal is thought to be similar to that experienced in vivo; (*ii*) during 2D crystallization, the time taken to effect crystallization is short, hence opening up the possibility of structural analysis on membrane proteins not sufficiently stable for 3-D crystallization trials; (*iii*) generally, 2D crystallization is more readily achievable than 3D crystallization; and (*iv*) phase information can be measured directly, eliminating the need to produce "heavy atom derivatives", and is generally of higher accuracy than those obtained in 3D crystallography.

Successful imaging of biological molecules requires the adoption of experimental protocols that take into account the fragility of such molecules when exposed to the hostile environment encountered in an electron microscope. Initially, due to the high vacuum, it is essential to provide protection against dehydration and stabilization against molecular collapse. Methods such as freezing in vitreous ice (8), sugar embedding (16), and negative staining (14) have been developed for this purpose. Frozen-hydrated and sugar-embedded techniques yield information on internal protein density, whereas negative staining provides information only on the molecular environment which is stain accessible.

2. TWO-DIMENSIONAL CRYSTALLIZATION

2D crystals of membrane proteins have been obtained in three different ways: (*i*) reconstitution of the purified membrane protein into a lipid bilayer at high protein density; (*ii*) improvement of the packing of a highly abundant protein into regular arrays in its native membrane; and (*iii*) addition of precipitants to promote protein–protein interactions, analogous to 3D crystallization, but yielding 2D crystals.

2.1. Crystallization of Purified Membrane Protein by Reconstitution into Lipid Bilayers

The crystallization of membrane proteins in 2D arrays requires membrane proteins which are solubilized and purified to homogeneity. A unique oligomeric species possessing a high intrinsic molecular symmetry is likely to favor the growth of crystalline arrays. Indeed, heterogeneous protein solutions and the presence of denatured proteins greatly hinder the formation of crystalline patches. The choice of detergent is critical. Ideally, detergent and isolation protocols should be selected not only to yield a homogeneous structural state but also to preserve the protein in a unique functional state (11,22). The pure protein is obtained in a detergent solution, often with residual lipids. In fact, the latter often contribute to the stability of the membrane protein and may be essential for successful 2D crystallization.

2.1.1. Detergent Choice and Concentration

Many membrane proteins are destabilized on extraction from their native membranes, especially when short-chain [high critical micellar concentration (CMC)] detergents are used. Although the proteins can be effectively solubilized with detergents that replace the lipid and keep the hydrophobic surfaces of the protein shielded from water, delipidation may destabilize the protein. The choice of detergent is critical; there is a fine balance between disruption of the membrane to solubilize a membrane protein and preserving its structural integrity. As discussed above, reconstitution is closely linked to the properties of the detergents used both during purification and the reconstitution itself. Therefore, the detergents used for purification can be exchanged for a different detergent used for reconstitution. This makes possible the use of mild detergents for isolation (most with long alkyl chain and low CMC) and then to change to detergents with high CMC for the crystallization. Among the detergents with an alkyl chain, those with a charged head group (e.g., dodecyl sulfate and hexatrimethylammonium) are generally not suited for isolation of "native" membrane proteins. Zwitterionic alkyl detergents (e.g., sulfobetains) and those with an N-oxide as polar head group can be used with many proteins, but they are still too harsh for most of the membrane proteins. Among the detergents with an alkyl chain, those with polyoxyethylene (e.g., $C_{12}E_8$, Lubrol-series, Triton® X-series, and Brij-series) or a disaccharide head group are the mildest (33). Different protocols have been developed for the exchange of detergent using ultrafiltration (e.g., with Centricon-100™ concentrator devices [Amicon Division, W.R. Grace, MA, USA]), gel filtration, anion or cation exchange chromatography, sucrose density gradient, isoelectric focusing, and selective absorption onto hydrophobic resin beads.

2.1.2. Determination of Chlorophyll and Protein Content

To set up optimal 2D crystallization trials, accurate knowledge of both lipid and protein is required, and the following procedures have been used with success in my laboratory. An advantage of Chl proteins is that the chromophores that are present (see Procedure 1 and Chapter 10 in this book by Paulsen and Schmid) correlate with the protein content (see Procedures 2 or 3 in this chapter).

❖ Procedure 1. Determination of Chlorophyll Concentration

Absorption spectroscopy is a quick and easy method for the estimation of chromophore content of Chl binding proteins. It is potentially nondestructive, although Chls are usually extracted from the protein using organic solvents.

1. Dilute 0.1 mL of the sample to 20 mL (200-fold) with 80% acetone.
2. Mix well and centrifuge at 3000× g for 5 minutes.
3. Measure the absorbance (A) of the acetone extract at 664 and 647 nm in a 1-cm path length cuvette.
4. Use the following equation to calculate the Chl *a*, Chl *b,* and total Chl derived from the known extinction coefficients of Chl *a* and *b* at 664 and 647 nm, respectively.

$$\text{Chl } a\ (\mu\text{M}) = 13.19 \times A_{664} - 2.57 \times A_{647}$$
$$\text{Chl } b\ (\mu\text{M}) = 21.10 \times A_{647} - 5.26 \times A_{664}$$
$$\text{Total Chl}\ (\mu\text{M}) = 7.93 \times A_{664} + 19.53 \times A_{647}$$

For cyanobacteria, which contain only Chl *a*, the concentration of an 80% acetone extract can be calculated from A_{663} using the formula:

$$\text{Chl } a\ (\mu\text{g/mL}) = 12.2 \times A_{663}\ (\varepsilon_{663} = 82\ \text{mg}^{-1}.\text{Chl } a.\text{L}^{-1})$$

❖ Procedure 2. Bio-Rad Protein Assay

This assay is based on the observation that the absorbance maximum for an acidic solution of Coomassie® Brilliant Blue G-250 changes from 465 to 595 nm when bound to protein (5).

1. Prepare a set of protein standards by diluting a stock of 2 mg/mL bovine serum albumin (BSA) standard in the buffer used for the experimental samples. Include a blank with no BSA.
2. Place 0.1 mL of each standard concentration of BSA and appropriately diluted experimental samples in test tubes.
3. Add 5.0 mL of the Bio-Rad protein assay dye reagent concentrate (Bio-Rad Laboratories, Hercules, CA, USA) (diluted 1:4 with water) to each test tube and vortex mix.
4. After a standard period from 5 minutes to 1 hour, measure the absorbance at 595 nm.
5. Plot the absorbance at 595 nm versus the protein concentration of standards.
6. Estimate the protein concentration of experimental samples from the standard curve.

❖ Procedure 3. Bicinchoninic Acid Assay to Determine Protein Concentration

The water-soluble salt of bicinchoninic acid (BCA) is a specific reagent for Cu^+.

Peptide bonds and 4 amino acids (cysteine, cystine, tryptophan, and tyrosine) reduce cupric (Cu^{2+}) ions in alkaline medium to Cu^{+} (Biuret reaction). The reaction product of 2 molecules BCA and 1 Cu^{+} exhibits a strong absorbance at 562 nm. Using BCA protein assay reagent available from Pierce Chemical (Rockford, IL, USA), the method below should be followed.

1. Prepare a set of protein standards of known concentration by diluting a stock 2 mg/mL BSA standard in the buffer used for the experimental samples. Include a blank with no BSA.
2. Prepare a working reagent by mixing 50 parts Reagent A with 1 part Reagent B.
3. Dilute 0.1 mL of each standard concentration of BSA and each experimental sample to 2 mL in a working reagent and incubate for 30 minutes at 37°C or room temperature for 2 hours.
4. Allow the samples to cool and measure the absorbance at 562 nm.
5. Prepare a standard curve by plotting the absorbance at 562 nm versus the protein concentration. Using the standard curve, determine the protein concentration for each unknown protein sample.

Since color development will continue slowly, it is necessary that all absorbance readings be done immediately. Care must be taken to avoid the presence of reducing agents such as thiols [dithiothreitol (DTT) or mercaptoethanol] or large amounts of sugars, which will interfere with the assay.

2.1.3. Lipid Mixture for Reconstitution

The lipid mixture used for reconstitution has an influence on crystallization results. Crystallization is more likely to occur when the lipid bilayer is in the fluid phase and thus allows some lateral mobility of the inserted membrane proteins. Native lipids are often ideal in terms of stability and transition temperatures, and they also provide mixtures of head group charges and molecular geometries similar to membranes from which the protein originated. However, the complexity of such preparations may preclude crystal formation. Nevertheless, since synthetic lipids, *Escherichia coli* lipids, soybean lecithin, and egg lecithin have all been successfully used for 2D crystallization, no general recommendations can be made on which lipid or lipid mixture is most suitable for any one particular membrane protein. Commercially available lipids (from Sigma [St. Louis, MO, USA] or Lipid Products [South Nutheld Redhill, Surrey RH1 5 PG, UK]) are often stored as chloroform solutions or a lyophilized powder at -20°C where they are stable for long periods. Prior to use for 2D crystallization trials, lipids have to be transferred to detergent-containing buffer solutions. The lipid content of the reconstitution mixture is, in general, a well-controlled parameter. Lyophilized lipids can easily be weighed and redissolved in buffer at 1 to 10 mg/mL containing a high concentration of detergent. Lipids from chloroform can be handled in the same way after removal of the organic solvent. The amount of detergent to solubilize completely a lipid stock can be calculated by use the following equation:

Concentration of detergent (mol/L) = CMC (mol/L) + 3 × concentration of the lipid (mol/L).

All organic solvent should be removed before solubilization in detergent buffer, as it interferes with the crystallization.

2.1.4. Lipid to Protein Ratio

The reconstitution of membrane proteins into bilayers is achieved by mixing lipids and protein, both solubilized in detergents, and decreasing the detergent concentration. Figure 1 shows an example where the concentration of octyl-polyoxyethylene (8-POE) was decreased by dilution, and the formation of structures of different sizes was monitored using dynamic light scattering (7). The dilution experiment led to the formation of vesicles with egg phosphatidylcholine (EggPC) or vesicles and 2D crystals with EggPC and the porin OmpF. The latter assembled only if the dilution rate was slow. The relationship between detergent concentration and structure sizes can be described as the "3-stage" model of Lichtenberg et al. (30). Stage I (crystals/vesicles area) is characterized by a detergent concentration too low to disrupt the lipid bilayer. Stage II (intermediate structures area) is the region of detergent concentration where lipid bilayer and mixed micellar structures coexist. The micelle-bilayer transition region (Stage II) was found to be the key to reconstitution and, by implication, to 2D crystallization (7,11). Cryo-electron microscopy (9) has shown for several lipid–detergent systems that this transition involves the formation of worm-like extended lipid micelles, probably capped by detergents that must convert to vesicles on detergent removal. Such rod-like structures are therefore thought to be important intermediates in the formation of 2D crystals.

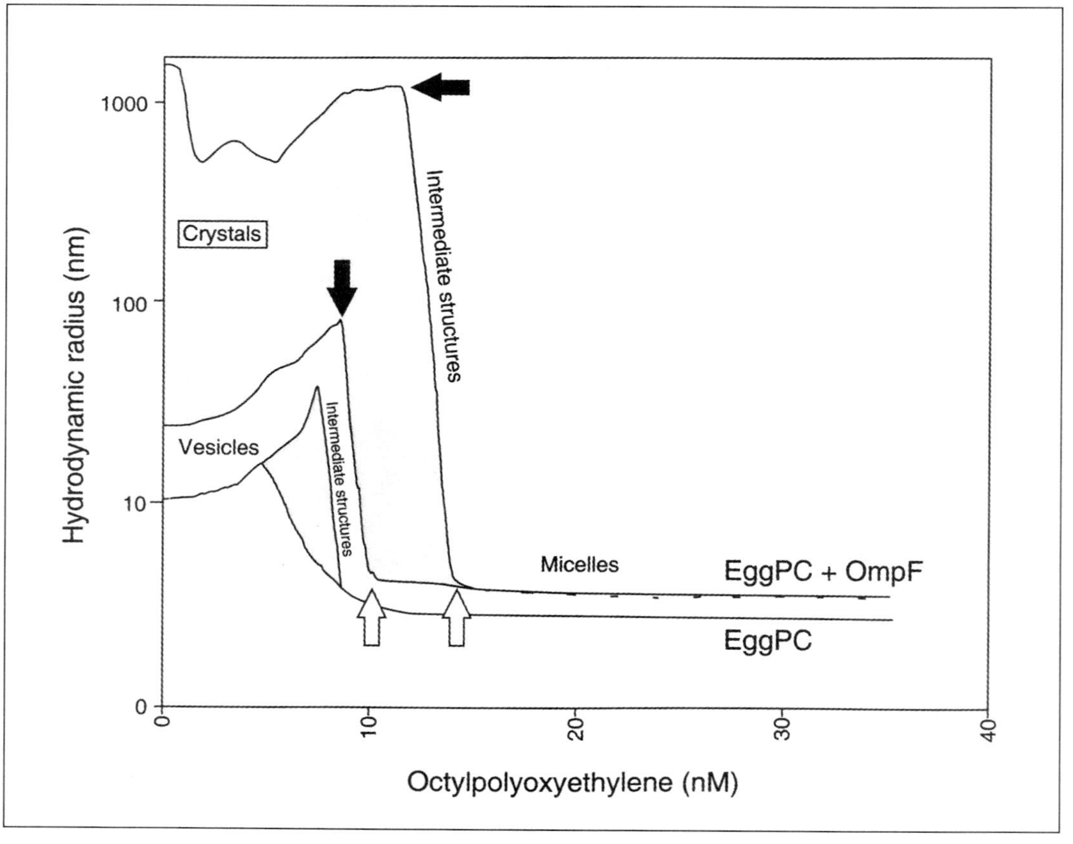

Figure 1. Hydrodynamic radii on dilution of a mixed micellar suspension containing a detergent (8-POE) and either lipid (EggPC) or lipid and a membrane protein (porin OmpF). Above 15 mM 8-POE, uniform micelles are seen. Below 15 mM 8-POE, each treatment produces a range of structure radii with large crystals seen in the presence of OmpF. The dilution is represented as a function of detergent concentration to illustrate the 3-stage model. Large bilayer structures (crystals/vesicles) at low detergent concentration (Stage I). Small micelles at high detergent concentration (Stage III). Mixture of structures at intermediate detergent concentration (Stage II). Black arrows indicate the saturation points, and white arrows indicate the solubilization points. Data from Reference 7. Figure was modified from Reference 15.

At the start of a typical reconstitution experiment, an excess of detergent ensures a homogenous distribution of protein and lipid in micelles. As detergent concentration is decreased, lipids and proteins interact due to the exposure of their hydrophobic surfaces. With an excess of lipid over protein, the protein is mainly incorporated into lipid bilayers, similar to its native state. In an excess of protein over lipid, the protein mostly ends up in amorphous aggregates, perhaps denatured. An important parameter is, therefore, the lipid:protein ratio (LPR), which should be low enough to promote crystal contacts between protein molecules, but not so low that the protein is lost to aggregation. When the membrane protein is reconstituted from a mixture of solubilized components, crystal ordering of proteins may occur during reconstitution. For crystal packing during reconstitution, the LPR of the reconstitution experiment must be as low as possible to ensure close packing without leading to excessive aggregation. While the lipid content of the reconstitution mixture is, in general, a well-controlled parameter, the content of monodisperse protein is sometimes unknown, because protein assays do not indicate the amount of aggregates.

2.2. Crystallization Methods

The manner in which the detergent concentration is decreased for reconstitution and subsequent 2D crystallization is an important consideration. The commonly used techniques for detergent removal are dilution (7,47), dialysis (11,22), and selective adsorption of the detergent on solid supports such as the hydrophobic resin beads (43).

2.2.1. Dilution Method

Diluting a solution of protein, lipid, and detergent decreases the concentrations of all components by equal factors, until the free detergent concentration drops below saturation. Crystallization by the dilution method requires a significant dilution of the protein, and therefore, rather high initial concentrations are required. On the other hand, the dilution method allows the process to be arrested when the saturation point is reached, extending the time in which an ordered assembly of the components can take place. As an illustration of this, the following 2 procedures describe the preparation of 2D crystals of PSI complexes from the thermophilic cyanobacterium *Synechococcus* sp. OD 24 (Procedure 4) and tubular crystals of spinach PSII (Procedure 5).

❖ Procedure 4. 2D Crystals of the PSI Complex from *Synechococcus* sp. OD 24 (7)

PSI complexes are isolated from *Synechococcus* sp. OD 24 according to Reference 24.

1. Make 600 μL of a starting mixture containing 1 mg/mL PSI, 1 mg/mL lecithin, and 7.5 mg/mL octyl-β-thioglucoside (OTG).
2. Dilute this mixture by the addition of sequential 25-μL aliquots of 10 mM 2-[N-morpholino]ethanesulfonic acid (MES), pH 6.0, 100 mM NaCl, 10 mM $MgCl_2$ to achieve a slow dilution of PSI.
3. After the addition of 4 aliquots (i.e., 100 μL, thus reducing the concentration of OTG to about 6 mM), large interconnected protein–lipid aggregates can be observed.
4. Upon further dilution of detergent (addition of 12 aliquots, i.e., 300 μL), distinct vesicles can be seen. Crystalline packing of PSI complexes should be obtained after the addition of 20 aliquots (i.e., 500 μL).

❖ Procedure 5. Tubular Crystals of PSII Complex from Spinach (47)

1. Isolate PSII complexes from spinach leaves as described in Reference 36.
2. Resuspend the complex in 50 mM MES, pH 6.0, 0.4 M sucrose, 10 mM NaCl, 0.4% OTG.
3. Adjust the Chl concentration to 1 mg/mL.
4. Mix with dimyristoyl phosphatidyl choline (DMPC), which has been solubilized in 1.3% OTG to obtain a ratio of 1:1 with Chl.
5. Dilute with 50 mM N-[2-hydroxyethyl] piperazine-N′-[2-ethanesulfonic acid] (HEPES) over a 4× range.
6. Incubate at 22°C for 1 day in the dark. A green pellet indicates the formation of aggregates.
7. Centrifuge for 5 minutes (4000× *g*) and wash the pellet with the same buffer to separate the crystals from the remaining material.
8. Repeat the wash step twice more.

The tubular crystals obtained have a length of 1 to 2 µm and a diameter of 72.9 nm. The rhombic unit cell ($a = 16.2$ nm, $b = 13.7$ nm, $\gamma = 142.4°$) contains one PSII complex.

2.2.2. Dialysis Method

Dialysis is the most widely used technique in 2D crystallization trials, usually in the form of small sample compartments dialyzed against large buffer volumes. To improve the reproducibility of crystallization conditions, a temperature-controlled continuous flow dialysis apparatus was developed (22) (see Figure 2). The advantage of this system is a precise control of the temperature profile that was found to be quite critical in the 2D crystallization of membrane proteins. Additionally, a maximal gradient of detergent concentration is maintained across the dialysis membrane, which improves reproducibility. A drawback of the method is the long dialysis times needed to remove low CMC detergents, making it only practical for medium to high CMC detergents (typically CMC > 1 mM) and for proteins with high structural stability. Before use, it is necessary to carry out pretreatment of the dialysis membranes (Procedure 6).

❖ Procedure 6. Pretreatment of the Dialysis Membranes

1. Heat the dialysis tubing (molecular weight cutoff 6000–8000, Spectra/Por 1) in a 2-L beaker of boiling 50% ethanol for 1 hour. Use a 1-L beaker containing water to weigh down the dialysis tubing.
2. Rinse the dialysis tubing well with several changes of distilled water.
3. Heat the dialysis tubing in boiling 10 mM Na_2CO_3, 1 mM EDTA for 1 hour.
4. Rinse with distilled water as before.
5. Heat the dialysis tubing in boiling distilled water for 1 hour.
6. Store in distilled water containing 0.05% NaN_3 at 4°C.

❖ Procedure 7. 2D Crystals of PSI *Synechococcus* sp. OD 24 (24)

Trimer PSI complexes were isolated from the thermophilic cyanobacterium *Synechococcus* sp. OD 24 as described in Reference 12.

1. Change the detergent Triton X-100 by polyethyleneglycol (PEG) 6000/$MgCl_2$ precipitation.
2. Repeat this step 3 times.
3. Resuspend the precipitated PSI in 10 mM HEPES, pH 7.0, containing 0.5% OTG to a protein concentration of 2 mg/mL.

4. Add DMPC solution to achieve a LPR of 1.
5. Dialyze the mixture against a detergent-free buffer containing 25 mM ammonium ferric citrate.

Dialysis cell temperature: 26°C for 24 hours; increase to 37°C over 12 hours; 37°C for 24 hours; and decrease to 26°C over 10 hours. Total dialysis time is 70 hours.

Digital image processing of negatively stained and frozen-hydrated specimens prepared using Procedure 7 revealed orthorhombic crystals with unit cell dimensions a = 13.8 nm, b = 14.5 nm, and $p12_1$ symmetry. The same procedure has also been used with trimeric PSI isolated from mesophilic cyanobacterium *Synechococcus* PCC 7002 by isoelectric focusing (46) to provide 2D crystals (Tsiotis, unpublished results).

Purple sulfur and nonsulfur bacteria possess membrane-bound LH complexes, which serve to transfer energy to the RC, where charge separation occurs. LH complex (170 μg), isolated from a carotenoid-less mutant of the purple nonsulfur bacterium *Rhodospirillum rubrum* G9 as described previously (13), was dissolved in 100 μL of 50 mM NH_4HCO_3, 1% octyl-β-glucoside (OG), pH 7.8. After dialysis against buffer containing 0.8% OG, 5 mM $MgCl_2$, and 50 mM NaCl in the dark at 4°C for 5 days, 2D crystals were obtained, which had a hexagonal pattern with a lattice constant of 12.3 nm. Modification of this procedure, by the use of sonicated vesicles of dioleoyl-9-10 phosphatidylcholine lipid solubilized in OG in a ratio 1:1 to proteins, allowed the formation of crystals which diffracted at 8.5 Å (23). Digital image processing of frozen-hydrated specimens revealed crystals with a $p22_12_1$ symmetry and unit cell with dimensions of a = 12.8 nm, b = 19.4 nm.

As an alternative to the dialysis apparatus shown in Figure 2, an inexpensive microdialysis arrangement with Eppendorf® tubes or dialysis buttons may be used (Figure 3). This method enables the dialysis of small volumes (<50 μL) of protein–lipid–detergent. Another interesting microdialysis device using a bent glass capillary tube, shown in Figure 4 has been described by Kuhlbrandt (26). A glass tube of 2.5 mm inner diameter and about 6.5 mm outer diameter is bent by 90° near to one end. The end is melted into a smooth surface, which forms a tight seal with the dialysis

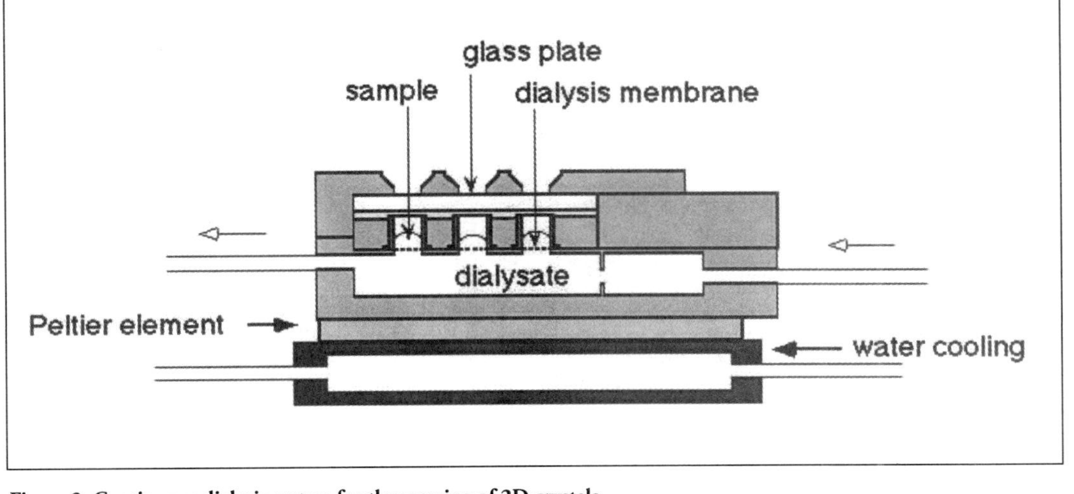

Figure 2. Continuous dialysis system for the growing of 2D crystals.

membrane. The dialysis membrane is fixed with a ring of silicon tubing, and 20 to 50 mL dialysate is fed into the capillary from the open end using a syringe (Figure 4). The dialysate is shaken down against the dialysis membrane to remove air bubbles, and the device is placed in a glass beaker with dialysis buffer. One interesting advantage of this device is that a sample for electron microscopy can be taken at anytime with a finer glass capillary without disturbing the progress of the experiment.

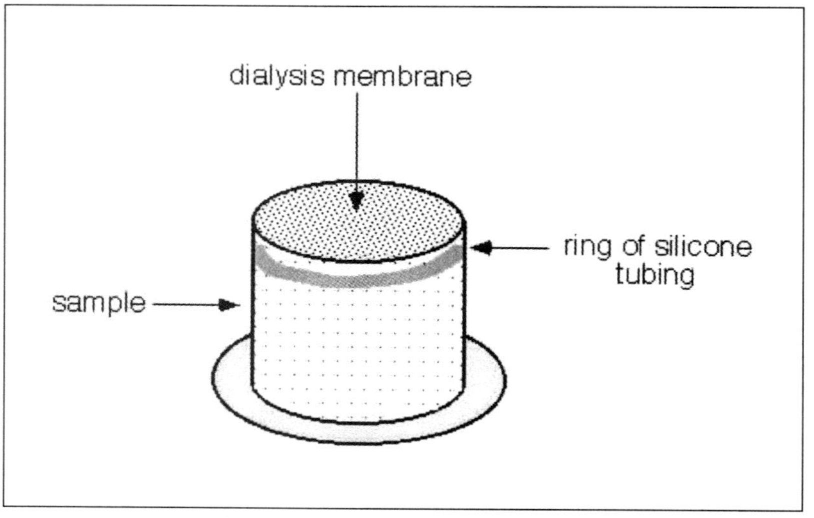

Figure 3. An Eppendorf cap for the growing of 2D crystals by microdialysis.

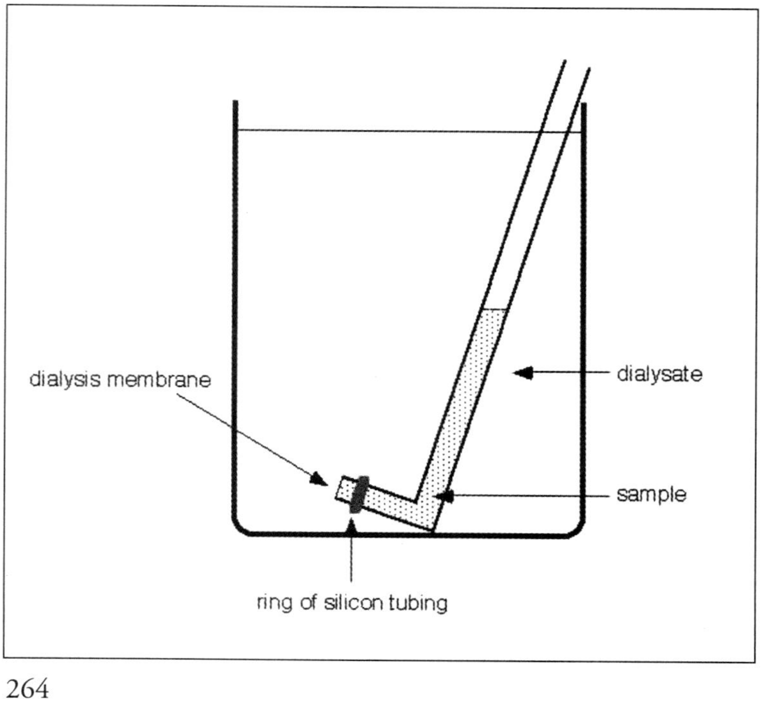

Figure 4. Glass capillary tube for the growing of 2D crystals by microdialysis.

❖ Procedure 8. 2D Crystals of a Spinach PSII Subcomplex (38)

PSII membranes were isolated according the method of Reference 10.

1. Concentrate the PSII complex to 0.5 mg Chl/mL by ultrafiltration (Centriprep-10,™ MW cut off 10 kDa [Amicon Division, W.R. Grace, MA, USA]).
2. Microdialyze (12–14 kDa MW cut off) 30 µL against a buffer containing: 40 mM MES, 20 mM NaCl, 1 mM $CaCl_2$, 1 mM zinc acetate, 1 mM NaN_3, 1 mM sodium ascorbate, 1 mg/mL butylated hydroxytoluene, pH 6.0, for 4 days at 20°C.
3. After dialysis, layer the crystalline suspension onto 7 mL 0.1 M sucrose, 40 mM MES, 20 mM NaCl, 1 mM NaN_3, pH 6.5.
4. Centrifuge at 2400× *g* for 5 minutes and collect the supernatant.
5. Centrifuge for 20 minutes at 35000× *g* at 2°C to collect the 2D crystals from supernatant.
6. Resuspend the pellet in a buffer containing 40 mM MES, 20 mM NaCl, 1 mM NaN_3, pH 6.5.

The 2D crystals contain CP47, D1, D2, cytochrome b-559, and the psbI subunit. The unit cell is rectangular with dimensions of 16.7 and 15.3 nm containing 4 monomers and $p22_12_1$ symmetry.

Procedure 8 and its modification, using higher Chl concentration (1 mg/mL) and including of 30% glycerol in the dialysis buffer and 1 to 3 weeks dialysis time, allowed the formation of well-ordered large tubular crystals of PSII complexes (42). The unit cell is rectangular with dimensions of 16.6 and 16.3 nm containing 4 monomers and belong to the same symmetry as the above mentioned crystals.

2.2.3. Bio-Beads Method

Adsorption of detergents to polystyrene beads (Bio-Beads® SM2; Bio-Rad Laboratories) is a simple alternative to conventional dialysis for obtaining 2D crystals of integral membrane proteins (43). Hydrophobic adsorption of detergents onto polystyrene beads can be used for detergent removal independent of the respective CMC and has been used to generate highly ordered 2D crystals of membrane proteins (43). Bio-Beads can be added to very small sample volumes with almost no dilution of the protein or lipids. This property is of great interest for membrane proteins for which purifying large quantities is particularly difficult. However, it should be noted that handling of small quantities of Bio-Beads is difficult, because the beads must not dry in the procedure. According to Rigaud's studies (43), the specific adsorption of lipids is about 100 to 200 times lower than the specific adsorption of detergents (2–4 mg phospholipids/g Bio-Beads SM2). Since crystallization reactions are generally performed at low LPRs (0.2–0.5), weak lipid adsorption may have some effects. However, lipid adsorption can be reduced by preincubation of the beads with an excess of sonicated liposomes prior to their use for detergent removal. 2D crystallization trials can be performed using either a 1-step addition of Bio-Beads, resulting in quick and straightforward removal of the detergent, or by addition of the same mass of Bio-Beads in several steps, which will slow down the process. The rate of detergent removal is directly linked to the weight of Bio-Beads used and also to the working temperature. The rate of adsorption of detergents doubles every 15°C. Quantitation of the Bio-Beads can present some difficulties, because the beads should stay wet to keep a reproducible adsorption property, and precise weighting has to be performed on freshly blotted beads. Prior to first use, wash the Bio-Beads with methanol and then with buffer (either in a batch procedure or packed on a column). Washed beads can be stored at 4°C as a

50% slurry in water supplemented with 0.2% NaN_3. Procedure 9 describes the use of this technique to crystallize the *Chlamydomonas reinhardii* cytochrome b_6f complex, purified as in Reference 41.

❖ Procedure 9. 2D Crystals of Cytochrome b_6f Complex from *C. reinhardii* (37)

1. Resuspend the purified cytochrome b_6f complex in buffer containing: 6.8 mM Tricine, 245 mM ammonium phosphate, 0.3 mM NaN_3, 2 mM $CaCl_2$, 1.1 mM benzamidine, 5.6 mM Â-aminocaproic acid, 0.2 mM phenylmethylsulfonyl fluoride (PMSF), 0.3% glycerol, 20 mM Hecameg, pH 8.0.
2. Add a mixture of EggPC and di-$C_{18:1}$-phosphatidylglycerol (1:1 ratio). The final protein concentration in reconstitution mixture should be 0.5 mg/mL and the LPR 0.2.
3. Preincubate the sample overnight in the cold room under gentle stirring.
4. Treat with 200 mg/mL SM2 Bio-Beads.
5. Incubate for 12 hours.
6. Pipet off the reconstituted material and keep for 24 hours at 4°C.
7. Freeze in liquid nitrogen, then thaw the sample at 37°C 3 times.

The crystals have a unit cell (a = 17.5 nm, b = 6.8 nm, and γ= 90°) and a $p22_12_1$ symmetry.

2.3. 2D Crystallization in Native Membranes

Some membrane proteins have a natural propensity to form regular arrays within the native membrane. Since the membrane protein does not dissociate from the lipid bilayer, its native orientation is maintained. Additionally, the proteins are not exposed to harsh detergents and, therefore, are kept under stabilizing conditions. One disadvantage is that these spontaneously formed 2D crystals are rarely highly ordered because other components can be trapped and hinder the crystal growth. Another disadvantage is that these methods are limited to cases where the membrane protein occurs at high density. However, the quality of naturally occurring 2D crystals can be improved by either detergent extraction, fusion of crystalline patches, or incubation with additives. Examples of naturally occurring crystalline membranes are the photosynthetic membranes from purple bacteria (35). PSII from higher plants is found in the thylakoid membrane at high density and sometimes in a regular lattice form (45). Detergent extraction of the thylakoid membrane under nonsolubilizing conditions promotes or improves the crystallinity (3,19,31).

For example, crystals of PSII, which provided evidence for the presence of a PSII dimer (3) were obtained as follows. PSII membranes, isolated as in Reference 4, were resuspended in 150 mM NaCl, 5 mM $MgCl_2$, and 20 mM Tricine, pH 7.5, at a final Chl concentration of 2 mg/mL. Triton X-100 was added to a concentration of 4% (wt/vol), and the mixture was incubated at 20°C for 20 minutes in the dark. It was then centrifuged for 30 minutes (45 000× *g*), and the pellet was washed once with buffer containing 10 mM NaCl, 5 mM $MgCl_2$, and 10 mM HEPES, pH 7.6. The crystals obtained had a rectangular unit cell (a = 17.8 nm and b = 26.7 nm). In contrast, similar crystals obtained using dodecyl maltoside (0.06%) instead of Triton X-100, and a subsequent sucrose gradient (0–2 M) (18) had unit cell dimensions of a = 16.8 nm, b = 18.9 nm, with γ = 91°, and a p1 symmetry group, and led to the proposal of the presence of a PSII monomer (19). Lastly, using a similar approach but with altered conditions, tubular PSII crystals have been obtained from spinach, which indicate the presence of 2 monomeric PSII complexes (31) (see Procedure 10).

❖ Procedure 10. Tubular Crystals of PSII from Spinach (31)

Thylakoid membranes were isolated as in Reference 35.

1. Resuspend thylakoid membranes in 15 mM NaCl, 50 mM sucrose, 5 mM $MgCl_2$, and 20 mM HEPES, pH 7.5.
2. Adjust Chl concentration to 0.8 mg/mL.
3. Add Triton X-100 to give a detergent to Chl ratio of 4.5:1.
4. Incubate for 20 minutes in ice with stirring.
5. Centrifuge for 20 minutes (20 000× *g*) and resuspend the pellet in 15 mM NaCl, 5 mM $MgCl_2$, 20 mM HEPES, pH 7.5
6. Determine the Chl concentration.
7. Add Triton X-100 to give a ratio of 3:1 of detergent to Chl.
8. Incubate for 10 minutes on ice.
9. Centrifuge for 15 minutes (15 000× *g*) and wash the pellet with the same buffer.

Additional washing steps can be included to separate the crystals from the remaining material.

The tubular crystals have a length of 1 to 2 μm and a diameter of 0.2 μm, and the unit cell has dimensions of a = 11.2 nm and b = 16.1 nm.

2.4. Crystallization of Purified Membrane Protein by Precipitation

In a few cases, a solubilized membrane protein has been crystallized into 2D sheets without detergent removal under conditions similar to those used in 3D crystallization experiments. Pea thylakoid LHC-II, the most abundant membrane protein in the chloroplast, was resolved to atomic resolution after formation of 2D crystals in a "batch method" (49) (see Procedure 11). The temperature profile proved to be critical for the crystallization of LHC-II. In addition, 2D crystals of some proteins with large extramembrane domains can be obtained at a pH close to the isoelectric point (35) or by increasing the ionic strength (6,12). These methods are best interpreted as variations of 3D crystallization and not as proper reconstitution of membrane proteins into a native-like environment.

❖ Procedure 11. 2D Crystals of LHC-II Complex from Pea Chloroplasts

LHC-II complexes from pea chloroplasts are isolated as described in Reference 27.

1. An aliquot of LHC-II complex (with a Chl concentration of 4 mg/mL in 0.4% Triton X-100) is diluted 50 times with distilled water and KCl added to a final concentration of 300 mM.
2. Centrifuge and collect the pellet. Proceed with either Method A or B.

Method A

(*i*) Dissolve the pellet to a Chl concentration of 0.7 mg/mL in Triton X-100 and glycerol at final concentrations of 0.23% and 40%, respectively.

(*ii*) Incubate at 35° to 40°C for 2 hours.

Method B

(*i*) Dissolve the pellet with 0.11% Triton X-100 and 0.24% n-nonyl-β-glucopyranoside.

(*ii*) Add glycine buffer (100 mM, pH 7.0) and glycerol to a final concentration of 10 mM and 40%, respectively. The Chl concentration should be 0.75 mg/mL.

(*iii*) Incubate for 2 days at 25°C and then for 2 hours at 40°C.

The crystals have a unit cell of a = 12.7 nm, b = 12.7 nm, with γ = 60°. The plane group has a p321 symmetry (28).

3. ANALYSIS OF 2D CRYSTALLIZATION REACTIONS

Even if 2D crystals of large membrane proteins can sometimes be observed by light microscopy, the shape of the crystals and their degree of order can best be assessed by electron microscopy. Thus, screening of 2D crystallization experiments requires access to an electron microscope. The quickest specimen preparation method for screening for 2D crystals is negative staining. This method is rapid, needs a small amount of sample (less than 5 µL), is remarkably simple, and can provide structural information to a resolution of about 2 nm. A negatively stained 2D crystal is embedded in a dry microcrystalline heavy atom replica. As heavy atoms used for negative staining scatter electrons much more than biological atoms (C, H, O, N, P, and S), the contrast is drastically increased but also inverted (hence the term "negative" staining). In addition, the heavy atom salts partially substitute the water in the native environment of the molecules, thus embedding the specimen and protecting it from collapse upon drying in air. The most commonly used heavy atom salts are uranyl acetate or formate and sodium or potassium phosphotungstate. Uranyl salts are more suitable for more protein samples, whereas tungstate is useful for lipid structures. The pH of tungstate solutions can be modified, whereas uranyl salts precipitate at pH greater than 4.5.

After negative staining, examination of the samples is carried out by transmission electron microscopy. The presence of large vesicles or sheets can be checked at low magnification (2000×–5000×) because of the high contrast obtained when operating at large underfocus even for very thin objects (Figure 5A). Further observations are done at higher magnification (Figure

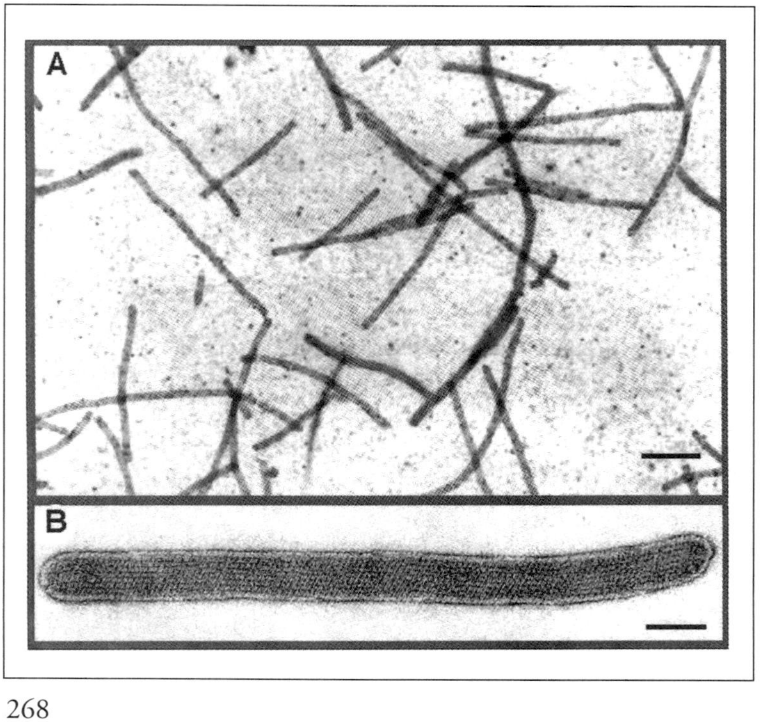

Figure 5. Negatively stained tubular crystals of PSII complex obtained as described in Reference 47. (A) Low magnification (5000×). Scale bar represents 500 nm. (B) High magnification (50 000×). Scale bar represents 100 nm.

5B). If no crystals are found, the presence of remaining detergent, single protein particles incorporated in lipid vesicles, solubilized proteins, or protein aggregates can all be identified. This information may be valuable for designing further crystallization trials.

4. CONCLUSIONS

Progress with the structural analysis of membrane proteins is limited by difficulties in handling protein–detergent and protein–lipid complexes. An alternative approach to X-ray crystallography of 3D crystals is the analysis of 2D crystals. 2D crystals have been obtained for many membrane proteins, indicating the natural tendency of these proteins to integrate into the 2D environment of the lipid bilayer. While technologies for data acquisition and data analysis are rapidly improving and are able to provide atomic resolution, the bottle-neck still remains the crystallization process.

ACKNOWLEDGMENTS

The author is indebted to A. Engel for helpful discussions. Financial support from Maurice E. Müller Foundation of Switzerland and the Swiss National Foundation for Scientific Research (Grant No. 31-46972.96) are gratefully acknowledged.

ABBREVIATIONS

2D and 3D, two- and three-dimensional; BChl, bacteriochlorophyll; BCA, bicinchoninic acid; BSA, bovine serum albumin; Chl, chlorophyll; CMC, critical micellar concentration; DMPC, dimyristoyl phosphatidyl choline; EggPC, egg phosphatidylcholine; HEPES, N-[2-hydroxyethyl]piperazine-N′-[2-ethanesulfonic acid]; LHC, light-harvesting complex; LPR, lipid:protein ratio; MES, 2-[N-morpholino]ethanesulfonic acid; OG, octyl-β-glucoside; OTG, octyl-β-thioglucoside; 8-POE, octyl-polyoxyethylene; PSI and PSII, photosystem I and II; RC, reaction center.

REFERENCES

1. **Allen, J.P., G. Feher, T.O. Yeates, H. Komiya, and D.C. Rees.** 1987. Structure of the reaction center from *Rhodobacter sphaeroides* R-26: the protein subunits. Proc. Natl. Acad. Sci. USA *84*:6162.
2. **Barber, J. and B. Andersson.** 1994. Revealing the blueprint of photosynthesis. Nature *370*:31-34.
3. **Bassi, R., A.G. Magaldi, G. Tognon, G.M. Giacometti, and K.R. Miller.** 1989. Two-dimensional crystals of the photosystem II reaction center complex from higher plants. Eur. J. Cell Biol. *50*:84-93.
4. **Berthold, D.A., G.T. Babcock, and C.F. Yocum.** 1981. A highly resolved, oxygen-evolving photosystem II preparation from spinach thylakoid membranes: EPR and electron transport properties. FEBS Lett. *134*:231-234.
5. **Bradford, M.** 1976. A rapid and sensitive method for the quantitation of microgram quantities of protein utilizing the principle of protein-dye binding. Anal. Biochem. *72*:248-254.
6. **Dekker, J.P., S.D. Betts, C.F. Yocum, and E.J. Boekema.** 1990. Characterization by electron microscopy of isolated particles and two-dimensional crystals of the CP47-D1-D2-cytochrome b-559 complex of photosystem II. Biochemistry *29*:3220-3225.
7. **Dolder, M., A. Engel, and M. Zulauf.** 1996. The micelle to vesicle transition of lipids and detergents in the presence of a membrane protein: towards a rationale for 2D crystallization. FEBS Lett. *382*:203-208.
8. **Doubochet, J., J. Lepault, R. Freeman, J.A. Berriman, and J.C. Homo.** 1982. Electron microscopy of frozen water and aqueous solutions. J. Microsc. *128*:219-237.
9. **Egelhaaf, S.U. and P. Schurtenberger.** 1994. Shape transformations in the lecithin-bile salt system: from cylinders to vesicles. J. Phys. Chem. *98*:8560-8573.
10. **Enami, I., K. Kamino, J.-R. Shen, K. Satoh, and S. Katoh.** 1989. Isolation and characterisation of photosystem II complex which lack light-harvesting chlorophyll a/b proteins but retain three extrinsic proteins related to oxygen evolution from spinach. Biochim. Biophys. Acta *977*:33-39.
11. **Engel, A., A. Hoenger, A. Hefti, C. Henn, R.C. Ford, J. Kistler, and M. Zulauf.** 1992. Assembly of 2-D membrane protein crystals: dynamics, crystal order, and fidelity of structure analysis by electron microscopy. J. Struct. Biol. *109*:219-234.
12. **Ford, R.C., A. Hefti, and A. Engel.** 1990. Ordered arrays of the photosystem I reaction centre after reconstitution: projections and surface reliefs of the complex at 2 nm resolution. EMBO J. *9*:3067-3075.

13.**Ghosh, R., A. Hoenger, A. Hardmeyer, D. Mihailescu, R. Bachofen, A. Engel, and J.P. Rosenbusch.** 1993. Two-dimensional crystallization of the light-harvesting complex from *Rhodospirillum rubrum*. J. Mol. Biol. *231*:501-504.

14.**Haschenmeyer, R.H. and R.J. Meyers.** 1972. Negative staining, p. 101-107. *In* M.A. Hayat (Ed.), Principles and Techniques of Electron Microscopy. Vol. 2, Van Nostrand, New York.

15.**Hasler, L., J.B. Heymann, T. Walz, J. Kistler, and A. Engel.** 1998. 2D crystallization of membrane proteins: rationales and examples. J. Struct. Biol. *121*:162-172.

16.**Henderson, R. and P.N.T. Unwin.** 1975. Three-dimensional model of purple membrane obtained by electron microscopy. Nature *257*:28-32.

17.**Henderson, R., J.M. Baldwin, K.H. Downing, J. Lepault, and F. Zemlin.** 1986. Structure of purple membrane from *Halobacterium halobium*: recording, measurement and evaluation of electron micrographs at 3.5 Å resolution. Ultramicroscopy *19*:147-178.

18.**Holzenburg, A., F.H. Wilson, M.E. Finbow, and R.C. Ford.** 1992. Structural investigations of membrane proteins: the versality of electron microscopy. Biochem. Soc. Trans. *20*:591-597.

19.**Holzenburg, A., M.C. Bewley, F.H. Wilson, W.V. Nicholson, and R.C. Ford.** 1993. 3-dimensional structure of photosystem-II. Nature *363*:470-472.

20.**Iwata, S., C. Ostermeier, B. Ludwig, and H. Michel.** 1995. Structure at 2.8 Å resolution of cytochrome c oxidase from *Paracoccus denitrificans*. Nature *376*:660-669.

21.**Jansson, S.** 1994. The light-harvesting chlorophyll a/b-binding proteins. Biochim. Biophys. Acta *1184*:1-19.

22.**Jap, B.K., M. Zulauf, T. Scheybani, A. Hefti, W. Baumeister, U. Aebi, and A. Engel.** 1992. 2D crystallization: from art to science. Ultramicroscopy *46*:45-84.

23.**Karrasch, S., P.A. Bullough, and R. Ghosh.** 1995. The 8.5 Å projection map of the light-harvesting complex I from *Rhodospirillum rubrum* reveals a ring composed of 16 subunits. EMBO J. *14*:631-638.

24.**Karrasch, S., D. Typke, T. Walz, M. Miller, G. Tsiotis, and A. Engel.** 1996. Highly ordered two-dimensional crystals of photosystem I reaction center from Synechococcus sp.: functional and structural analyses. J. Mol. Biol. *262*:336-348.

25.**Kreusch, A., A. Neubuser, E. Schiltz, J. Weckesser, and G. Schulz.** 1994. Structure of the membrane channel porin from *Rhodopseudomonas blastica* at 2.0 Å resolution. Prot. Sci. *3*:58-63.

26.**Kühlbrandt, W.** 1992. Two-dimensional crystallization of membrane proteins. Q. Rev. Biophys. *25*:1-49.

27.**Kühlbrandt, W., T. Thaler, and E. Wehrli.** 1983. The structure of membrane crystals of the light-harvesting chlorophyll a/b protein complex. J. Cell Biol. *96*:1414-1424.

28.**Kühlbrandt, W. and D.N. Wang.** 1991. Three-dimensional structure of plant light-harvesting complex determined by electron crystallography. Nature *350*:130-134.

29.**Landau, E.M. and J.P. Rosenbusch.** 1996. Lipidic cubic phases: a novel concept for the crystallization of membrane proteins. Proc. Natl. Acad. Sci. USA *93*: 14532-14535.

30.**Lichtenberg, D., R.J. Robson, and E.A. Dennis.** 1983. Solubilization of phospholipids by detergents. Structural and kinetic aspects. Biochim. Biophys. Acta *737*:285-304.

31.**Lyon, M.K., K.M. Marr, and P.S. Furcinitti.** 1993. Formation and characterization of 2-dimensional crystals of photosystem-II. J. Struct. Biol. *110*:133-140.

32.**McDermott, G., S.M. Prince, A.A. Freer, A.M. Hawthornthwaite-Lawless, M.Z. Papiz, R.J. Cogdell, and N.W. Isaacs.** 1995. Crystal structure of an integral membrane light-harvesting complex from photosynthetic bacteria. Nature *374*:517-521.

33.**Michel, H.** 1991. General and practical aspects of membrane protein crystallization, pp. 74-88. *In* H. Michel (Ed.), Crystallization of Membrane Proteins, CRC Press, Boca Raton.

34.**Michel, H., K.A. Weyer, H. Gruenberg, and F. Lottspeich.** 1985. The 'heavy' subunit of the photosynthetic reaction centre from *Rhodopseudomonas vidris*: isolation of the gene, nucleotide and amino acid sequence. The EMBO Journal *4*:1667-1672.

35.**Miller, K.R. and J.S. Jacob.** 1983. Two-dimensional crystals formed from the photosynthetic reaction centers. J. Cell Biol. *97*:1266-1270.

36.**Mishra, R.K. and D.F. Ghanotakis.** 1994. Selective extraction of CP 26 and CP 29 proteins without affecting the binding of the extrinsic proteins (33, 23 and 17 kDa) and the DCMU sensitivity of a photosystem II core complex. Photosynth. Res. *42*:37-42.

37.**Mosser, G., C. Breyton, A. Olofsson, J.L. Popot, and J.L. Rigaud.** 1997. Projection map of cytochrome b_6f complex at 8 Å resolution. J. Biol. Chem. *272*:20263-8.

38.**Nakazato, K., C. Toyoshima, I. Enami, and Y. Inoue.** 1996. Two-dimensional crystallization and cryo-electron microscopy of photosystem II. J. Mol. Biol. *257*: 225-232.

39.**Ostermeier, C., S. Iwata, B. Ludwig, and H. Michel.** 1995. Fv fragment-mediated crystallization of the membrane protein bacterial cytochrome c oxidase. Nat. Struct. Biol. *2*:842-6.

40.**Pebay-Peyroula, E., G. Rummel, J.P. Rosenbusch, and E.M. Landau.** 1997. X-ray structure of bacteriorhodopsin at 2.5 angstroms from microcrystals grown in lipidic cubic phases [see comments]. Science *277*:1676-81.

41.**Pierre, Y., C. Breyton, D. Kramer, and J.-L. Popot.** 1995. Purification and characterization of the cytochrome b_6f complex from *Chlamydomonas reinhardtii*. J. Biol. Chem. *270*:29342-29349.

42.**Rhee, K.H., E.P. Morris, D. Zheleva, B. Hankamer, W. Kuhlbrandt, and J. Barber.** 1997. Two-dimensional structure of plant photosystem II at 8-angstrom resolution. Nature *389*:522-526.

43.**Rigaud, J.L., G. Mosser, J.J. Lacapere, A. Olofsson, D. Levy, and J.L. Ranck.** 1997. Bio-Beads: an efficient strategy for two-dimensional crystallization of membrane proteins. J. Struct. Biol. *118*:226-235.

44.**Schirmer, T., T. Keller, Y. Wang, and J. Rosenbusch.** 1995. Structural basis for sugar translocation through maltoporin channels at 3.1 Å resolution [see comments]. Science *267*:512-514.

45.**Staehelin, L.A.** 1975. Chloroplast membrane structure. Intramembrenous particles of different sizes make contact in stacked membrane regions. Biochim. Biophys. Acta *408*:1-11.

46.**Tsiotis, G., W. Nitschke, W. Haase, and H. Michel.** 1993. Purification and crystallization of photosystem I complex from a phycobilisome-less mutant of the cyanobacterium *Synechococcus* PCC 7002. Photosynth. Res. *35*:285-297.

47.**Tsiotis, G., T. Walz, A. Spyridaki, A. Lustig, A. Engel, and D. Ghanotakis.** 1996. Tubular crystals of a photosystem II core complex. J. Mol. Biol. *259*: 241-248.

48.**Tsukihara, T., H. Aoyama, E. Yamashita, T. Tomizaki, H. Yamaguchi, K. Shinzawa-Itoh, R. Nakashima, R. Yaono, and S. Yoshikawa.** 1995. Structure of metal sites of oxidized bovine heart cytochrome c oxidase at 2.8 Å. Science *269*:1069-1074.

49.**Wang, D.N. and W. Kühlbrandt.** 1991. High-resolution electron crystallography of light-harvesting chlorophyll a/b-protein complex in three different media. J. Mol. Biol. *217*:691-699.

50.**Weiss, M.S., A. Kreusch, E. Schiltz, U. Nestel, W. Welte, J. Weckesser, and G.E. Schulz.** 1991. The structure of porin from *Rhodobacter capsulatus* at 1.8 Å resolution. FEBS Lett. *280*:379-382.

12 Biosynthesis and Analysis of Bilins

Matthew J. Terry
University of Southampton, Southampton, England, UK

1. THE DIVERSITY AND BIOSYNTHESIS OF NATURALLY OCCURRING BILINS

The term bilin is a collective one to describe a broad group of open chain tetrapyrroles and derives from the name "bile pigments" as the first of these compounds to be characterized were isolated from animal bile. These bilins, biliverdin (BV) and bilirubin (BR), are the sequential products of heme degradation (their green and yellow pigmentation can be detected during the discoloration of a bruise), with BR being conjugated to glucuronic acid to expedite excretion. The structures of BV and BR are shown in Figure 1, and their biochemistry is still the best understood of the bilins today. However, we now know that there is a great diversity of naturally occurring bilins that have a wide range of different functions. In cyanobacteria and two groups of algae, the rhodophytes (red algae) and the cryptomonads, a tremendous variety of bilins are utilized for light harvesting through covalent attachment to the phycobiliproteins, which comprise the photosynthetic apparatus of these organisms (25,26). The structures of two of the most common of these bilins, phycocyanobilin (PCB) and phycoerythrobilin (PEB), are also shown in Figure 1. Other light-harvesting pigments include phycoviolobilin, phycourobilin, 15,16-dihydrobiliverdin (DHBV), and mesobiliverdin (MBV) (25,26). In higher plants, the related bilin, phytochromobilin (PΦB) (Figure 1), serves as the chromophore of the phytochromes (17,58). This family of photoreceptors is important throughout plant development and regulates such diverse processes as germination, growth, flowering, and the synthesis of the photosynthetic apparatus. Also, BV and BR may not just be waste products, since both have been demonstrated to have potent antioxidant activities (51), and BR has even been implicated in having a role in circadian regulation in humans (41). BV is used for pigmentation in reptiles (4), fishes (23), and insects, where the insecticyanin protein is bound to BV IX rather than the IXα isomer more commonly found in nature (27). BV is also found in the eggshells of a wide variety of birds (31). More remarkably still, the marine snail *Aplysia californica* uses

Heme, Chlorophyll, and Bilins: ***Methods and Protocols***
Edited by A.G. Smith and M. Witty

PEB, which it obtains from a diet of red seaweed, as a defensive ink pigment (43).

The biosynthetic relationship between the bilins discussed in this chapter is shown in Figure 2. The first committed step in the pathway for the synthesis of all bilins is the oxidation of protoheme to BV IXα by the enzyme heme oxygenase (5,34,42,53). Following this universal reaction, BV IXα can be reduced at one of three different positions in a somewhat phylogenetic-dependent manner. In animals, BV IXα is reduced at the C-10 position (see Figure 1) by the enzyme BV reductase to give BR IXα (32,50). In higher plants, the plastid-localized enzyme PΦB synthase, which is a bilin 2,3-reductase, catalyses the formation of the A-ring ethylidene group required for covalent assembly to apophytochrome (see Reference 55 and Chapter 13 in this text by McDowell and Lagarias). The product of this reaction, 3(*Z*)-PΦB (57), is subsequently converted to 3(*E*)-PΦB, the immediate precursor of the bound phytochrome chromophore, by an, as yet unidentified, bilin $3^1,3^2$ *cis-trans* isomerase. In red algae, the pathways for the synthesis of bilins are quite complex, but have been recently been reviewed in some depth (5). Concentrating just on the central pathway, BV IXα is first reduced by a bilin 15,16-reductase to give 15,16-DHBV IXα (10) and subsequently

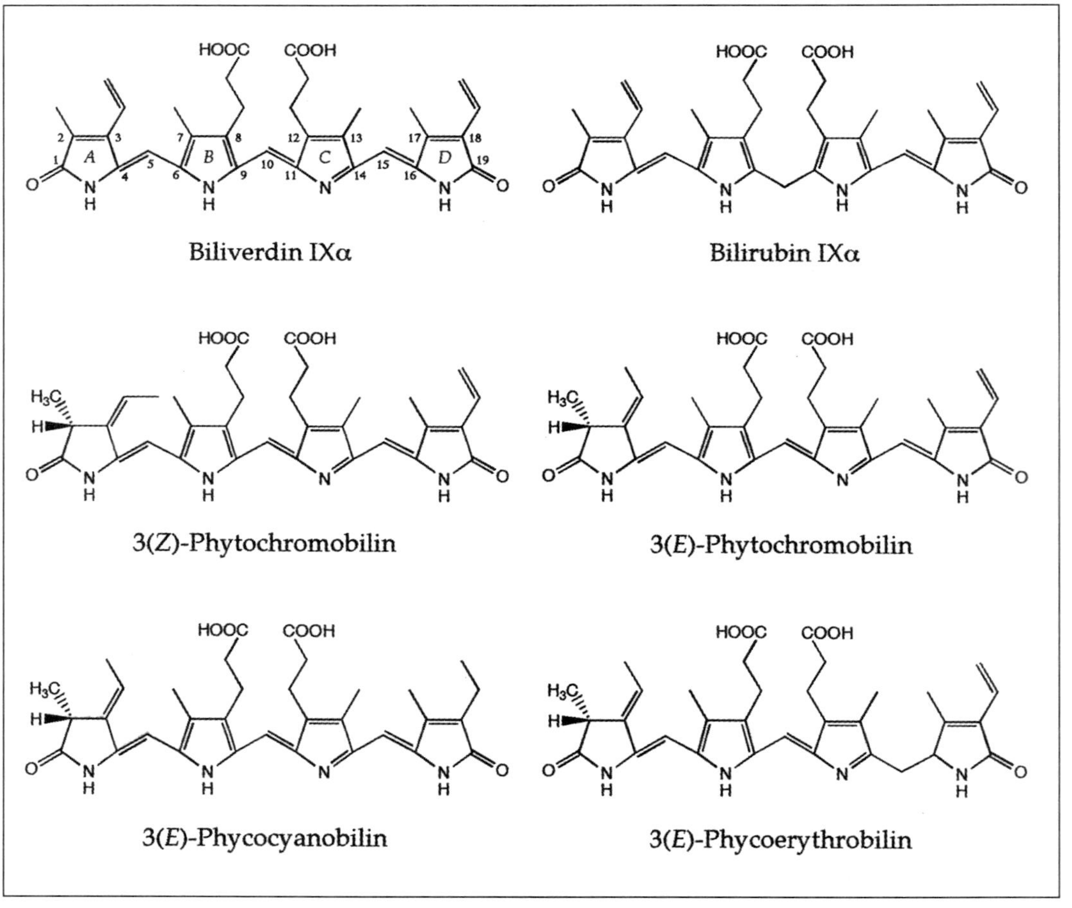

Figure 1. Chemical structures of the major bilins.

by a bilin 2,3-reductase to 3(*Z*)-PEB (9). This later enzyme is presumably related to PΦB synthase, as both enzymes accomplish the same reaction. 3(*Z*)-PEB then undergoes a series of isomerizations to produce the 3(*E*)-isomers of both PEB and PCB (5,9).

The complexity of bilin synthesis is further increased by recent evidence that the reduction of BV IXα is not simply related to broad phylogenetic group. The green alga *Mesotaenium caldariorum* is able to synthesize 3(*Z*)-PCB directly from 3(*Z*)-PΦB (67). Such a pathway suggests a reasonable biosynthetic alternative to the red algal pathway for PCB synthesis, and

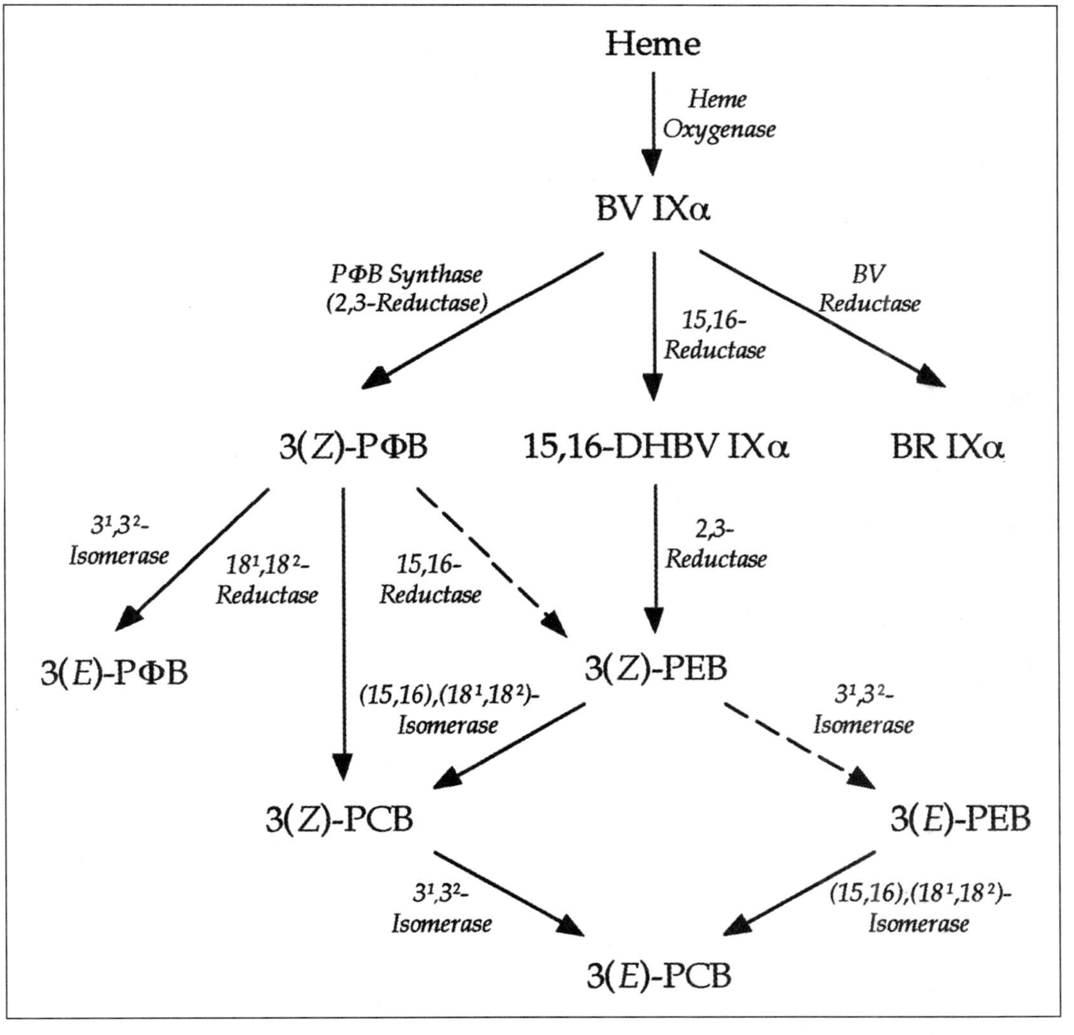

Figure 2. The biosynthetic relationship of the bilins discussed in the text. The full pathway shown has been constructed from biosynthetic pathways identified in a number of different organisms with no single organism containing all pathways. Heme oxygenase is present in animals, plants, algae, and cyanobacteria. Of the four reductases, BV reductase has been detected in animals and cyanobacteria, PΦB synthase (bilin 2,3-reductase) in plants and green algae, bilin 15,16-reductase in red algae and possibly cyanobacteria, and bilin 18^1,18^2-reductase in green algae alone. There are two isomerase activities. 3(*Z*),3(*E*) *cis-trans*-isomerase (3^1,3^2-isomerase) appears to be present in plants and algae, while (15,16),(18^1,18^2)-isomerase activity has only been detected in red algae. Broken arrows indicate possible biosynthetic routes that have yet to be shown experimentally. Abbreviations: BR, bilirubin; BV, biliverdin; DHBV, dihydrobiliverdin; PCB, phycocyanobilin; PEB, phycoerythrobilin; PΦB, phytochromobilin.

indeed the authors report the unpublished observation that *Cyanidium caldarium* has PΦB synthase activity. Since the yeast *Pichia pastoris* also has this activity (66), bilin 2,3-reductases may be considerably more widespread than first thought. To complicate matters further, it now appears that cyanobacteria also possess a BV (bilin 10,11-)reductase (49), while two more novel enzymes (a bilin 4,5-reductase and a bilin $12^1,12^2$-dehydrogenase) have been proposed to account for the synthesis of the five additional phycobilins identified to date (5).

While the diversity of bilins and their biosynthesis is a fascinating topic, this chapter is primarily concerned with the methods required to study bilin biosynthesis in general. In particular, I will focus on a few specific steps in bilin synthesis in photosynthetic organisms, only addressing the biochemistry of these pathways in animals for comparative purposes. It is intended that the methods described here could be applied equally well to the biosynthetic steps that have yet to be characterized.

2. ISOLATION AND PREPARATION OF BILINS

2.1. Biliverdins

BV can be purchased as a dihydrochloride of about 80% purity from ICN Pharmaceuticals (Costa Mesa, CA, USA) and, until very recently, from Sigma (St. Louis, MO, USA). One problem with these samples is that they contain a mixture of isomers making them unsuitable for many purposes. An alternative is to buy a more pure sample of the IXα isomer (the isomer that predominates in biological systems) from Porphyrin Products (Logan, UT, USA). The IXα isomer can then be further purified by high-performance liquid chromatography (HPLC) (see section 3.1). A more cost-effective method of obtaining BV is to purify it from a mixture of bile acids, which can be purchased quite cheaply from both Sigma and ICN Pharmaceuticals. An alternative method of obtaining BV IXα is by oxidation of the BR IXα with 2,3-dichloro-5,6-dicyano-1,4-benzoquinone (22,36). This method is also cost-effective if large quantities of BV IXα are required and is the most suitable for obtaining (^{14}C)-labeled BV IXα. In addition, it can be used for the synthesis of the BV IIIα and XIIIα isomers from the corresponding BR isomers or for MBV IXα from mesobiliverdin (MBR) IXα. Extensive methods for the synthesis of a wide range of BV dimethyl esters, including DHBV dimethyl ester, have also been published (52).

2.2. Phycobilins

The phycobilins can most easily be obtained by cleavage from the appropriate phycobiliprotein using a $HgCl_2$-assisted methanolysis protocol. A method for the preparation of PCB from lyophilized, mixed *Spirulina* species is shown in Procedure 1. This method has been described in Terry et al. (56) and is based on the earlier methods of Beale and Cornejo (9) for the isolation of phycobilins from *Porphyridium cruentum* and Arciero et al. (1) who used *Synechococcus.*

❖ Procedure 1. Isolation and Purification of PCB from Cyanobacteria

Reagents

- Lyophilized mixed *Spirulina* species (Sigma)
- C18 cartridge e.g., Sep-Pak (Waters-Millipore, Milford, MA, USA) or Bond Elut (Varian Instruments, Santa Clarita, CA, USA).

- Acetonitrile, 2-mercaptoethanol, $HgCl_2$, methanol, and trichloroacetic acid.
- 0.1% (vol/vol) trifluoroacetic acid (TFA).
- Acetonitrile:0.1% TFA (20:80, vol/vol).
- Acetonitrile:0.1% TFA (60:40, vol/vol).

Method

1. Rehydrate the *Spirulina* powder in 30 mL/g dry weight of deionized water and incubate for 10 minutes at room temperature.
2. Centrifuge at 30 000× *g* for 20 minutes and collect the deep-blue supernatant containing the phycocyanin protein.
3. Add trichloroacetic acid to give a final concentration of 1% (wt/vol) and incubate for 1 hour at 4°C in the dark to precipitate the protein.
4. Centrifuge at 30 000× *g* for 20 minutes and discard the supernatant.
5. Wash the pellet repeatedly in methanol (20 mL/g *Spirulina* powder), discarding the supernatant each time, until all of the chlorophyll has been removed.
6. Resuspend the pellet in methanol (2 mL/g *Spirulina* powder) containing 1 mg/mL $HgCl_2$, and incubate the sample for 20 hours at 42°C in darkness to cleave the covalently-bound PCB chromophore from the phycocyanin protein.

   Note: Samples containing free bilins should be kept as far as possible in low light (or even safe light) conditions to prevent photodegradation. This is a general rule that is applicable to all work with pigments.
7. Remove the protein by centrifugation at 10 000× *g* for 10 minutes.
8. Add 1 µL/mL 2-mercaptoethanol to the supernatant to precipitate the $HgCl_2$ and centrifuge at 30 000× *g* for 10 minutes. The PCB, which remains in the supernatant, can be further purified and concentrated using a C18 cartridge.

   Note: Alternatively, PCB can be extracted into methylene chloride:1-butanol (2:1 vol/vol) and purified by diethylaminoethyl (DEAE)-Sepharose® chromatography (Amersham Pharmacia Biotech, Piscataway, NJ, USA) (8).
9. Precondition the C18 cartridge with the sequential addition of acetonitrile (to aid hydration and remove any contaminating compounds that might otherwise be present in the final elution step), water, and 0.1% (vol/vol) TFA (to equilibrate the column to the sample pH).
10. Dilute the crude PCB solution 10-fold in 0.1% TFA and apply to the preconditioned cartridge.
11. Wash cartridge twice with 0.1% TFA followed by acetonitrile:0.1% TFA (20:80 vol/vol).
12. Elute PCB with a minimum, but known, volume of acetonitrile:0.1% TFA (60:40 vol/vol).
13. Aliquot and remove solvent under vacuum (without heat).
14. Store sample at -20°C, or below, in the dark until required.

HPLC analysis of the PCB sample isolated from *Spirulina* using Procedure 1 (see section 3.1.) indicates that it is predominantly 3(*E*)-PCB (>90%) with the 3(*Z*)-isomer comprising the major additional product. These isomers can be collected and reconcentrated by diluting 10-fold in 0.1% (vol/vol) TFA and applying to a C18 cartridge exactly as described above (Procedure 1, steps 9–14).

PEB has been isolated both by whole cell methanolysis (17) and by methanolysis of phycobiliprotein fractions (3,40). For whole cell methanolysis, either the red alga, *P. cruentum*, or other phycoerythrin-containing organisms, such as the cyanobacterium *Calothrix* sp. PCC 7601 can been used (17). *P. cruentum* is grown at 27°C in a liquid suspension culture continuously provided with 1% (vol/vol) CO_2. It is generally slow growing, but we have found that the ASP2 medium described in Provasoli et al. (44) greatly improves the growth rate compared with other reported media (17). *Calothrix* sp. PCC 7601 can be grown in BG-11 medium (47) containing 15 mM N-tris(hydroxymethyl)methyl-2-aminoethanesulfonic acid (TES) (pH 8.2) and 50 mM dextrose at 23°C (17). In both cases, an equal mixture of cool white and red fluorescent lamps was used (17). For *Porphyridium*, cells were extracted with dimethyl sulfoxide (DMSO), and the solution was diluted 8-fold with acetone. Following centrifugation, the cells were extracted 5 times with DMSO:acetone (1:8 vol/vol) and then a further 3 times with methanol, until the supernatant was colorless (17). However, extracting with acetone alone followed by methanol was sufficient. *Calothrix* cells can be extracted directly with methanol (17). The $HgCl_2$-assisted methanolysis procedure is essentially identical to that described for PCB in Procedure 1, but using a ratio of 8 mL methanol/g cells. The method outlined in Procedure 1 (steps 9–14) is also suitable for the purification of PEB, but requires a further HPLC step to separate PEB from other bilins in the sample. Details of HPLC purification methods for bilins are given below in section 3.1. As with the isolation of PCB, the 3(*E*)-isomer of PEB is the principal product of this procedure, but the 3(*Z*)-isomer can also be obtained. HPLC-purified samples can be concentrated by diluting 10-fold in 0.1% (vol/vol) TFA and applying to a C18 cartridge exactly as described for PCB (Procedure 1, steps 9–14). An alternative procedure for isolating PEB has recently been reported, in which the freeze-dried seaweed nori (*Porphyra yezoensis ueda*) was used as the starting material (40). The powdered material was extracted twice with deionized water, and the suspension was filtered through cheesecloth before the phycobiliproteins were precipitated with 65% ammonium sulfate. Standard methanolysis and purification protocols were then used (40).

Although methanolysis has proved to be the most widely adopted method of bilin isolation, other procedures have been reported. One interesting method that has been described in the literature comes from the observation that *C. caldarium* cells excrete large quantities of PCB when fed the tetrapyrrole precursor 5-aminolevulinic acid (ALA) (59). Another alternative is chemical synthesis, which has now been accomplished for the free acid (30). However, this is not a trivial procedure, and a better approach for most laboratories will be to let the enzymes themselves accomplish the required chemistry. Many of these enzymes have now been partially purified (5) (described later in text) and could already be used to synthesize phycobilins from BV IXα or heme. However, perhaps the real breakthrough will come from the cloning and expression of these enzymes, which will eventually permit the production of specific phycobilins to order. A similar strategy for the intermediates of heme and corrin synthesis is described in Chapter 4 in this text by Warren and Shoolingin-Jordan.

2.3. Phytochromobilins

As it is difficult to obtain enough phytochrome protein for methanolytic cleavage of the bound chromophore, there are really only two practical approaches for purifying PΦB. The first takes advantage of the observation that the acetone treatment required for extraction of *P. cruentum* cells prior to phycoerythrin methanolysis leads to oxidation of the bound PEB to give PΦB, which is then cleaved by the normal methanolysis procedure (17). Phycoerythrin-containing cells extracted with methanol do not produce PΦB. The methanolysis and purification procedures are exactly as described above, and the yield of, predominantly, 3(*E*)-PΦB was estimated to be approximately one third the yield of PEB (17).

A reasonable alternative is to synthesize PΦB enzymatically by incubating the BV IXa precursor with isolated plastids (57, 62). In principle, plastid preparations from a wide variety of tissues or species could be used, but in my experience, there is considerable variation in the yields of PΦB that can be obtained. Of the plants that have been investigated, the best starting material appears to be pea and oat seedlings, while tomato, cucumber, and *Arabidopsis* are much less suitable. It is also generally better to use etioplasts than chloroplasts, as bilin recovery is greater from these preparations. The procedure is relatively straightforward and only requires crude etioplast preparations. Indeed, attempts to purify plastids further with Percoll gradients led to decreased bilin yields, possibly because of interactions between the bilins and the Percoll itself. Procedure 2 shows a simple procedure for obtaining PΦB from BV using isolated pea (*Pisum sativum* L.) etioplasts (55,62).

❖ Procedure 2. Synthesis of PΦB from BV Using Isolated Pea Etioplasts

Reagents

- Homogenization stock solution [1 M sorbitol, 40 mM TES, 20 mM 4-(2-hydroxyethyl)-1-piperazineethanesulfonic acid (HEPES)-NaOH, pH 7.7, 1% (wt/vol) polyvinylpyrrolidone (soluble PVP), 2 mM $MgCl_2$, 2 mM EDTA (free acid), 2 mM EDTA (di-Na salt)].
- Bovine serum albumin (BSA) and cysteine.
- Assay buffer [500 mM sorbitol, 20 mM TES, 10 mM HEPES-NaOH, pH 7.7, with 1 mM phenylmethylsulphonyl fluoride (PMSF), 2 M leupeptin, 0.5 mM dithiothreitol (DTT) added fresh] 300 000 U/mL catalase in 5 mM citrate buffer, pH 7.5.
- NADPH regenerating system (12 mM $NADP^+$, 100 mM glucose-6-phosphate, and 15 U/mL glucose-6-phosphate dehydrogenase).
- 1 mM BV IXα stock solution in DMSO.

Method

1. Grow pea seedlings in moist vermiculite in the dark for 8 to 10 days at 25°C.
2. Prepare homogenization buffer from stock solution by diluting 2-fold with water. Add BSA and cysteine to final concentrations of 0.2% (wt/vol) and 5 mM, respectively, and adjust back to pH 7.7.
3. Under green safelight, harvest approximately 30 g apical tissue (top 3–4 cm of about 200 seedlings) and homogenize in a prechilled mortar and pestle in 2 mL/g fresh weight ice-cold homogenization buffer. The sample should be kept chilled (4°C) throughout the isolation procedure.
4. Filter homogenate through 4 layers of muslin.
5. Centrifuge for 1 minute at 8000× *g*.
6. Wipe away any starch from the side of the tube with a tissue and gently sus-

pend pellet in homogenization medium (approximately 1 mL/g fresh weight) using a paintbrush to tease the pellet away from the centrifuge tube.

7. Centrifuge for 1 minute at 100× *g* to remove unbroken cells and other debris.
8. Centrifuge supernatant for 2 minutes at 1500× *g*.
9. Wash the plastids by resuspending the pellet in a small volume of the assay buffer stock solution and centrifuge again for 2 minutes at 1500× *g*.
10. Resuspend the crude etioplast pellet in 1 mL assay buffer.
11. Prepare a 2 mL reaction mixture containing 1 mL etioplasts (the excess should be saved for protein determination; a final plastid protein concentration of about 1 mg/mL is recommended), 20 µL catalase, 200 µL NADPH regenerating system, and 560 µL assay buffer in a 25-mL Erlenmeyer flask.
12. Start the reaction by adding 20 µL of 1 mM BV IXα (final concentration 10 µM), deplete oxygen from the flask by replacing air with argon, and seal the flask.
13. Incubate in the dark for 3 hours at 28°C with gentle shaking.

    Note: In order to maximize the yield of PΦB, it is important to drive the reaction to completion. However, excessively long incubation periods can result in nonspecific bilin degradation and care needs to be taken when choosing incubation times.
14. Remove the etioplasts by centrifugation at top speed in a benchtop microfuge for 10 minutes.
15. Partially purify bilins on a C18 cartridge as described in Procedure 1 (steps 9–14) except that the cartridge is conditioned with 50 mM *N*-methylmorpholine–acetate buffer, pH 7.7, prior to the application of the sample.
16. Purify 3(*Z*)- and 3(*E*)-PΦB by HPLC as described in section 3.1.

The method outlined in Procedure 2 also produces a mixture of isomers, although in this case, the yield of 3(*Z*)-PΦB greatly exceeds that of 3(*E*)-PΦB. Again, purification of both isomers can be readily accomplished by HPLC.

With all bilin samples, it will generally be necessary to prepare stock solutions which can then be used for various biochemical experiments such as assembly of phytochrome (see Chapter 13 in this text by McDowell and Lagarias) or phycobiliproteins (see Chapter 14 in this text by Bryant and Schluchter) or as standards for biosynthesis studies. These solutions are routinely prepared by dissolving the dried bilin samples in DMSO. The concentration should be adjusted to 1 mM following spectrophotometric determination of the sample concentration (see section 3.2. and Table 1), and samples should then be stored at -80°C in the dark.

3. ANALYSIS OF BILINS

A wide range of techniques has been employed for the analysis of bilins, and the methods employed will depend on the information required. Bilins prepared by the large-scale isolation procedures described above will need to be analyzed for purity. HPLC can be used to determine the purity of samples and also to confirm the identity of the sample by co-injection of known standards. The absorption spectrum of the sample should also be taken, as the absorption properties of bilins are characteristically different and can also be used to confirm the identity of samples. In addition, absorption spectroscopy can be used to quantify purified bilin samples. If the sample to be analyzed contains an unknown bilin(s), then a wider range of techniques may be appropriate. Although HPLC and absorption spectroscopy will again prove useful, it may be necessary to determine the chemical structure in order to confirm or elucidate the identity of the unknown bilin. In this case, ^{1}H nuclear magnetic resonance (NMR) spectroscopy is the most appropriate technique.

3.1. HPLC Analysis and Purification

Bilins are readily separated by C18 reverse phase HPLC using isocratic solvent systems, and their distinctive absorption spectra make them easy to detect in the visible wavelength range. The most commonly used solvent system for analyzing most bilins is that of Beale and Cornejo (8), who separated phycobilins on an Altex Ultrasphere octadecylsilane (ODS) column (250 mm long × 46 mm diameter, 5 µm particle size). Separation was achieved using a mobile phase of acetone:ethanol:water: acetic acid (38:50:11:1 vol/vol/vol/vol) with a flow rate of 4 mL/minute at 30°C (8,17). An almost identical system (acetone:ethanol:water:acetic acid, 34:48:17:1 vol/vol/vol/vol) has also proved suitable for separating phytochromobilins, in this case using an Ultrasphere ODS column (250 × 10 mm, 5 µm particle size; Beckman Coulter, Fullerton, CA, USA) at room temperature (57). However, one potentially serious problem with this solvent system is that there is almost no resolution of BV IXα and 3(*Z*)-PΦB. To solve this problem, some improvements on this system have been developed in the Lagarias laboratory. The first of these, changing the mobile phase to acetone:ethanol:100 mM formic acid (25:65:10 vol/vol/vol), has been used successfully for the analysis of PΦB synthesis using a Supercosil LC-18 ODS column (250 × 4.6 mm, 5 µm particle size; Supelco, Bellefonte, PA, USA) with a flow rate of 1.5 mL/minute (54,62). However, even with this system BV IXα and 3(*Z*)-PΦB still elute close to each other, and further improvements in separation have been achieved using a mobile phase of acetone:20 mM formic acid (50:50 vol/vol). Interestingly, this fairly small change in solvent system leads to a complete reversal in the order that the bilins elute (67).

Another problem associated with most of these solvent systems is that retention times vary considerably from day to day and even during a series of runs. This probably reflects a pronounced temperature sensitivity of these systems, but may also be related in part to the condition of the column itself. This variation highlights the need to identify peaks by co-injection of known standards rather than by retention time. Perhaps the single biggest difficulty of using HPLC for bilin analysis is that there is tremendous variation between columns, with new columns, apparently identical to columns already in use, being completely unsuitable for bilin separation. This normally manifests itself as an inability to retain the bilins on the column, and this difference in performance is even apparent between columns from the same company. However, ODS columns from a number of companies (some of which have been named above) have been used successfully, and researchers new to the field should persevere in finding a suitable one.

The solvent systems described above have not only been used for separating phycobilins, but are also suitable for analyzing both BV IXα and MBV IXα synthesized from heme (54,62) and mesoheme (unpublished results), respectively. However, a simpler system comprising 95% (vol/vol) aqueous ethanol:acetic acid:water (92:1:7 vol/vol/vol) has also commonly been used (16,45). Both systems are also capable of separating the 4 BV IX isomers. For the analysis of BR IXα formation from BV IXα by cyanobacterial BV reductase, Schluchter and Glazer (49) used a linear

Table 1. Spectroscopic Properties of Bilins Discussed in the Text

Bilin[a]	λ_{max} (nm)[b]	ε ($M^{-1}cm^{-1}$)[c]	^{1}H-NMR spectra[d]
biliverdin IXα	377, 696 (37)	ε_{377} = 66, 200 (37)	9, 57
15,16-dihydrobiliverdin IXα	335, 560 (10)	ND	10
mesobiliverdin IXα	359, 685 (7)	ε_{359} = 78, 600 (7)	2, 13[e]
3(*Z*)-phycocyanobilin	369, 686 (9)	ε_{368} = 46, 774 (63)[e]	63[e,f], 9[e,f]
3(*E*)-phycocyanobilin	375, 692 (9)	ε_{374} = 47, 900 (13)[e]	19, 2, 30, 28[e]
3(*Z*)-phycoerythrobilin	327, 591 (9)	ND	9
3(*E*)-phycoerythrobilin	329, 592 (9)	ε_{594} = 25, 200 (12)[e]	20, 9
3(*Z*)-phytochromobilin	381, 698 (67)	ε_{382} = 38, 019 (63)[e]	57
3(*E*)-phytochromobilin	386, 700 (17)	ε_{386} = 64, 565 (63)[e]	17

[a]Values are given for the free acids unless otherwise stated.
[b]Absorption maxima are in HCl/methanol unless otherwise stated, with references in parentheses.
[c]Molar absorption (extinction) coefficients are in HCl/methanol unless otherwise stated, with references in parentheses.
[d]References are for ^{1}H-NMR spectra recorded in pyridine-D_5 unless otherwise stated.
[e]Dimethyl ester.
[f]Spectrum recorded in $CDCl_3$.
ND, not determined.

gradient from 50% solvent A (water), 50% solvent B (acetone:ethanol:water:acetic acid, 50:38:11:1 vol/vol/vol/vol) to 100% solvent B over 30 minutes.

One further consideration for the HPLC analysis of bilins is the most suitable wavelength for detection. As mentioned above, all these compounds absorb strongly in visible regions of the radiation spectrum making detection relatively straightforward. The wavelength of choice will depend on the range of bilins to be detected. Information about the absorption spectra of the common bilins is given in Table 1, but in general, detection at 370 or 380 nm is suitable for most bilins. The exception is PEB, which does not absorb strongly in this region. For detection of a mixture of phycobilins that includes PEB, 600 nm is more suitable (8,17).

3.2. Absorption Spectroscopy

The standard solvent for recording visible absorption spectra is methanol:36% (wt/vol) aqueous HCl (49:1 vol/vol), although other solvents such as acetonitrile:0.1% TFA (60:40 vol/vol) can and have been used. Absorption maxima for a range of common bilins in methanol/HCl are shown in Table 1. BV, PΦB, and PCB all have a Soret peak between about 370 to 390 nm and a broader peak at approximately 700 nm. The shorter wavelength peak is more reliable for quantitation, as its position and height are less dependent on factors such as solvent composition and temperature. Table 1 also gives molar absorption (extinction) coefficients for this peak, and these can be used for quantitation of bilin samples using Beer's law:

$$A = \varepsilon c l$$

where A is absorbance or optical density at a specific wavelength, ε is the molar absorption coefficient at the same wavelength, *c* is the concentration of the sample in M, and *l* is the light path in centimeters. It should be noted that for quantitation, absorption spectra should be recorded in the same solvent used to calculate the molar absorption coefficient, as absorption properties are highly solvent dependent.

PEB has a different spectrum with a major peak at about 590 nm and a smaller one near 330 nm. In this case, the longer wavelength peak is more suitable for quantitation as absorbance is much greater at this wavelength. BR is not soluble in methanol, but has a single major peak of 454 nm in chloroform (ε_{454} = 62 600 M^{-1} cm^{-1}) (36).

3.3. ^{1}H NMR Spectroscopy

^{1}H NMR spectroscopy is the most appropriate technique for determining the chemical structure of bilins. Obtaining and interpreting ^{1}H NMR spectra are specialized procedures that require considerable experience and are outside the scope of this review. However, help with these aspects of the technique can generally be provided by an NMR facility. For the researcher, the most important factor in obtaining good spectra is the quality of the sample, and great care should be taken during sample preparation. Small amounts of contamination by other compounds, or even dirt or grease, can severely affect the quality of the spectra. If glassware is involved, this should be washed in nitric acid and thoroughly rinsed in deionized water before use. Purified samples should be dissolved in pyridine-D_5 (100.0 atom % D; Sigma), although other solvents such as $CDCl_3$ have also been used. The interpretation of spectra is aided by comparison with known spectra. For this reason, spectra of related compounds for which the structure is known should be analyzed at the same time, and these can be used, together with published spectra, for the determination of the unknown structure. Table 1 provides a list of references containing information on ^{1}H NMR spectra for a range of bilins.

3.4. Other Spectroscopic Techniques

A number of other spectroscopic techniques have also proved useful. Structural information can also be obtained by mass spectrometry, and this technique has been used in the analysis of BV IXα (37), PCB (24), and the dimethyl esters of PCB (14), PEB (12), and various BVs (14,52). Circular dichroism spectroscopy can provide information on the configuration of chiral carbons. Circular dichroism spectra are also generally taken in HCl/methanol, as fully protonated bilins are free from helical conformations that can hinder the interpretation of the data. 3(*E*)-PΦB has a single chiral carbon at position 2 (see Figure 1), and this was shown to be in the *R* configuration (17). 3(*E*)-PCB (11) and 3(*E*)-PEB (29) isolated from phycobiliproteins also have a 2(*R*) configuration.

Fluorescence spectroscopy has not been widely used, as free bilins fluoresce very poorly. However, chelation with zinc results in a highly fluorescent compound, and this property has been used to study holophytochrome assembly (see Chapter 13 in this text by McDowell and Lagarias). More imaginative still has been the exploitation of the fluorescent properties of PEB bound to phytochrome apoprotein to generate a new series of fluorescent protein probes, the phytofluors (see Reference 40 and Chapter 13 in this text by McDowell and Lagarias).

4. BILIN BIOSYNTHESIS

Much of the original work that established the pathway for bilin biosynthesis in photosynthetic organisms has utilized the classical biochemical approach of feeding putative bilin precursors. Using these techniques, and in particular taking advantage of radiolabeled precursors, it was established that the phycobilins were derived from heme synthesized from ALA by the main tetrapyrrole pathway. These experiments are discussed by Beale (5). Similarly, ALA and BV IXα were also shown to be precursors of phytochromobilin (reviewed in Reference 58), though only recently was the intermediacy of heme confirmed (62). While this type of approach will continue to be important in bilin research, this section will concentrate on assaying the enzymes committed to bilin synthesis.

4.1. Heme Oxygenase

Heme oxygenase catalyses the synthesis of BV IXα from protoheme in a reaction that requires O_2 and reducing power, and liberates CO and Fe^{2+} (Figure 3) (34,42). In photosynthetic organisms, the most extensively characterized heme oxygenase system is that of *C. caldarium* (5). Heme oxygenase activity was first measured in *Cyanidium* extracts by Beale and Cornejo (7). The enzyme is soluble and has been partially purified using differential ammonium sulfate precipitation and DEAE-cellulose and blue-Sepharose affinity chromatography steps (16). This procedure was subsequently extended to include a ferredoxin-Sepharose affinity chromatography step to give a 200-fold purification of heme oxygenase activity (46).

The employment of a ferredoxin affinity purification step highlights one of the distinguishing features of heme oxygenases from photosynthetic organisms: the requirement for reduced ferredoxin, produced by ferredoxin-$NADP^+$ oxidoreductase and NADPH, for enzyme activity. In contrast, mammalian heme oxygenase (first described in Reference 53), found associated with microsomal membranes, utilizes an NADPH-cytochrome P450 reductase (35, 69). A second distinguishing feature is that the algal enzyme requires a second reductant in addition to ferredoxin (16): ascorbate appears to be the most effective with Trolox® (6-hydroxy-2,5,7,8-tetramethylchroman-2-carboxylic acid) also proving suitable (46). The reason that the reaction requires a second reductant is currently unknown.

The heme oxygenase assay in extracts or partially purified fractions of *Cyanidium* is an HPLC-based assay that takes advantage

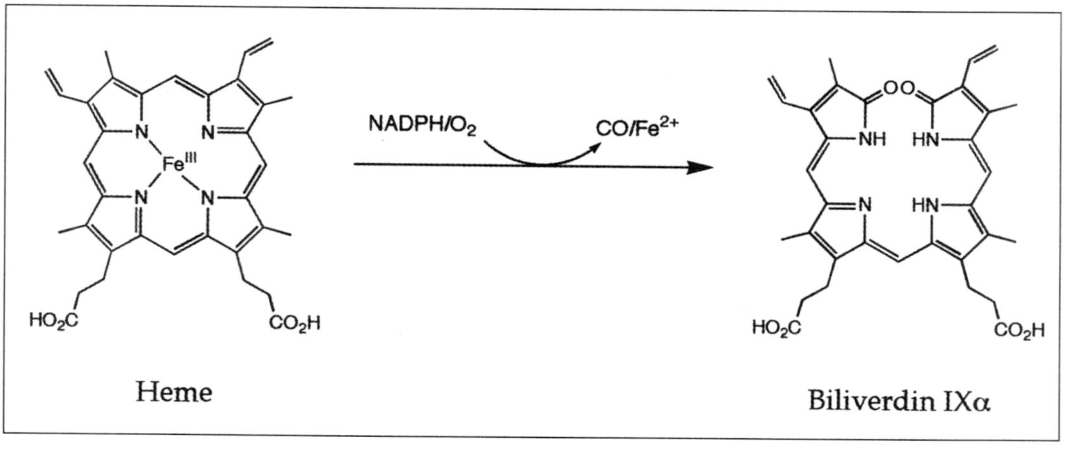

Figure 3. The reaction catalyzed by the enzyme heme oxygenase. Substrates and products are not shown stoichiometrically.

of the observation that MBV IXα, in contrast to BV IXα, is not further metabolized in these preparations. A typical reaction buffer contains 10 μM mesohemin IX in 20 mM Tris-HCl, pH 7.6, with 5 mM L-ascorbate, 0.1 mg/mL catalase, 0.5 mg/mL BSA, 10% (wt/vol) glycerol, and reducing power provided by an NADPH-regenerating system (10 mM glucose 6-phosphate, 0.5 mM $NADP^+$, and 0.3 U/mL glucose-6-phosphate dehydrogenase), together with 5 μg/mL *Porphyra umbilicalis* ferredoxin, and 0.01 U/mL spinach ferredoxin-$NADP^+$ reductase (46). O_2 is also required for the reaction to proceed. Mesohemin can be obtained from Porphyrin Products, while all the other components are readily available from Sigma. In the Beale laboratory, assays are performed at 42°C for 1 hour in the dark, and the bilin products are extracted into dichloromethane:1-butanol (2:1 vol/vol) and concentrated on DEAE-Sepharose prior to HPLC analysis (46).

The electrons required for the reaction are transferred directly from the ferredoxin, as evidenced by the observation that light-driven ferredoxin reduction via a spinach photosystem I fraction can support heme oxygenase activity (45). As there is likely to be a direct interaction between heme oxygenase and the ferredoxin, it is unsurprising that the source of ferredoxin is important for maximum activity. *Porphyra* ferredoxin is almost as good as a ferredoxin-containing fraction from *Cyanidium* (86%), but spinach ferredoxin only resulted in 20% of activity (16). There is also considerable variation in the effectiveness of different second reductants. L- and D-ascorbate were equally effective, while dehydroascorbate could not support any activity. Trolox (29% of the activity with ascorbate), dihydroquinone (16%), and DTT (15%) were all partially active (46). In the cyanobacterium *Synechocystis* sp. PCC 6701, Trolox was actually more effective as a second reductant than ascorbate (18). However, in most other respects, the cyanobacterial heme oxygenase appears to be similar to its algal counterpart (15,18).

One important consideration when assaying heme oxygenase is product identification. Heme readily undergoes coupled oxidation to a mixture of BV IXα, β, γ, and δ (particularly in the presence of ascorbate), and it is necessary to establish that the reaction product is exclusively the IXα isomer in order to demonstrate that BV synthesis is enzymatic. This can be done by HPLC as described in section 3.1. To prevent the formation of additional BV IX isomers from H_2O_2 generated by the reaction of ascorbate and O_2, catalase is routinely included in the reaction mixture (7). Heme oxygenase can be stimulated by the addition of strong iron chelators such as desferrioxamine or Tiron (4,5-dihydroxy-1,3-benzene disulphonic acid). These are thought to aid the dissociation of an Fe(III)-BV IXα complex that may be the primary product in vitro (46). A number of inhibitors are also available. Sn-protoporphyrin IX is a potent inhibitor of the algal enzyme (16), while the animal enzyme is inhibited by a wide range of metal porphyrins (60) that have yet to be tested on heme oxygenases from photosynthetic organisms. Heme oxygenase is also inhibited by diethylpyrocarbonate, probably through binding to active-site histidines, but is insensitive to the sulfhydryl reagent *p*-hydroxymercuribenzoate (16).

In preparations from higher plants, heme oxygenase has only been assayed in isolated plastids (62). The method is essentially the same as that described in Procedure 2, except that 10 μM heme replaces the BV, and the O_2 depletion step is not required. Under these assay conditions, BV IXα is completely metabolized to 3(*Z*)-PΦB. However, as for the algal system, mesoheme is also a substrate for the plant heme oxygenase, while the product MBV IXα is not further metabolized (unpublished results). In tomato etioplasts, BV IXα is the main product after incubation with heme (54). In this case, activities are quite low, and the reaction can only proceed if 5 mM sodium ascorbate is included in the assay. This leads to problems from coupled oxidation, and care needs to be taken to establish that BV IXα is the primary isomer produced.

Substantial progress has recently been made in our understanding of heme oxygenases by the cloning and expression of recombinant heme oxygenases from both *Synechocystis* sp. PCC 6803 (18) and *Arabidopsis* (39). Two genes were identified in *Synechocystis*, only one of which appeared to be expressed under the conditions tested (18). This gene encoded a protein of 27 kDa that was designated HO1 (18). Two genes have also been identified in higher plants (21,39). One gene, designated *HY1* after the mutant that led to its identification, but also known as *At*HO1, encodes a protein of 32.6 kDa (including a chloroplast target sequence of 6 kDa) and is plastid localized (39). This protein shows substantial similarity to specific conserved regions of mammalian heme oxygenases and has been shown to have heme oxygenase activity (39). Heme oxygenase activity of *At*HO2, encoded by the second gene with sequence similarity to known heme oxygenases, could not be established (21).

Genes encoding heme oxygenases have now been expressed in *Escherichia coli* resulting in large quantities of enzyme that can be readily purified by conventional methods (64,65) or by attaching a purification tag such as polyhistidine (18) or glutathione S-transferase (39). One benefit from the overexpression of plant heme oxygenase is that it has permitted the use of a spectrophotometric assay for heme oxygenase which had not been possible previously because of interference by cellular pigments. This assay has the advantage that the reaction can be followed in real time, thereby allowing the determination of initial reaction rates and is commonly performed as a coupled assay with excess BV reductase allowing BR synthesis to be measured. However, it is also possible to detect BV IXα formation directly by using the longer wavelength peak.

Using a coupled-assay approach, recombinant plant heme oxygenase was assayed under the following conditions: 100 mM Tris-HCl, pH 7.8, containing 15 μM hemin, 0.15 mg/mL BSA, recombinant cyanobacterial BV reductase (with an activity of 45 nmol BR/hour), 250 μM NADPH, 50 μg/mL spinach ferredoxin, 0.025 U/mL spinach ferredoxin-$NADP^+$ reductase, and 2 mM Tiron (39). The assay was performed at 25°C with reaction rates determined by following the formation of BR at 450 nm. It can be seen from the composition of the assay buffer that the plastidic plant enzyme has similar properties to the enzymes from algae and cyanobacteria. This similarity extends to the requirement for a second reductant such as ascorbate for maximal activity (T. Muramoto, M.J. Terry, A. Yokota, and T. Kohchi, unpublished results). One interesting difference is that there is a fairly strict requirement for an iron chelator for the reaction to proceed (39). Recombinant heme oxygenase has also been studied from the pathogenic bacterium *Corynebacterium diptheriae* (65). This heme oxygenase appears to utilize a third type of redox partner, as activity was most efficient with an NADH-dependent putidaredoxin–putidaredoxin reductase system (65).

There are also a number of additional assays for heme oxygenase activity. An alternative HPLC-based assay was recently described in some detail (48). This assay is suitable for measuring heme oxygenase in crude samples from mammalian tissues and therefore provides a suitable alternative to the standard spectrophotometric assay (see above) as it overcomes problems such as spectral interference and the effect of protein composition on the molar absorption coefficient of BR. In addition, it is also possible to measure heme oxygenase activity by the release of CO instead of BV IXα synthesis. The production of CO in crude samples has been measured by gas chromatography (61), although this method has obvious problems of specificity. Alternatively, the release of ^{14}CO can be measured following the incubation of purified heme oxygenase with ^{14}C-heme (70). However, a simpler method of CO detection in purified samples is to follow the change in the absorbance maximum of myoglobin following CO binding (65).

4.2. Bilin Reductases

Bilin reductases catalyzing four different reactions have been reported to date (see Figure 2). By far the most extensively characterized of these enzymes is mammalian BV reductase, which reduces BV IXα to BR IXα. This enzyme was first assayed directly by Singleton and Laster (50) who followed the disappearance of BV IXα spectrophotometrically. However, it is now standard practice to assay BV reductase by following the appearance of BR IXα. The enzyme is unusual in that it has two different pH optima for NADPH and NADH, both of which it uses directly (32). A standard assay for mammalian BV reductase, adapted by Terry et al. (56) from Kutty and Maines (32), is shown in Procedure 3.

❖ Procedure 3. Spectrophotometric Assay for Mammalian BV Reductase

Reagents

- Assay buffer A [0.1 M Tris-HCl, pH 8.7, containing 1 mg/mL BSA, 10 μM BV IXα, and an NADPH-regenerating system (0.1 mM $NADP^+$, 1 mM glucose-6-phosphate, and 0.1 U/mL glucose-6-phosphate dehydrogenase, type XII from Sigma)].
- Assay buffer B [0.1 M potassium phosphate buffer, pH 7.0, containing 1 mg/mL BSA, 10 μM BV IXα, and an NADH-regenerating system (0.2 mM NAD^+, 1 mM glucose-6-phosphate, and 0.2 U/mL glucose-6-phosphate dehydrogenase, type XXIV from Sigma)].
- 20 to 80 μg/mL BV reductase.

Method

1. Place 495 μL of assay buffer A or B into a cuvette and zero the baseline absorbance.
2. Start the reaction by adding 5 μL BV reductase. Mix thoroughly and quickly using a pipet.
3. Follow the appearance of BR by recording the absorbance changes at 466 nm (pH 8.7) or 458 nm (pH 7.0).
4. Determine the rate of BV reductase activity using the derived absorption coefficients for BR in these buffers of ε_{466} = 64 100 M^{-1} cm^{-1} at pH 8.7 and ε_{458} = 55 800 M^{-1} cm^{-1} at pH 7.0 (see Reference 56 and section 3.2.).

Interestingly, in addition to BV IXα, PΦB, PCB, and PEB are all substrates for BV reductase (56), and this property has been used creatively to produce phytochrome chromophore-deficient plants through overexpression of the mammalian BV reductase gene (33). The enzyme from human liver is not, however, capable of reducing BV IXβ, and it now appears that a second enzyme is present which has substantial activity with BV IXβ, γ, and δ, but cannot reduce the IXα isomer (68). Surprisingly, BV IXα reductase is also present in the cyanobacteria, *Synechocystis* sp. PCC 6803 (49), although the significance of this is not entirely clear, and it is unknown whether algae or higher plants also have this enzyme.

Algae appear to have at least four different bilin-reducing enzymes for which the substrate specificity has yet to be clearly defined (Figure 2). The best characterized of these are the enzymes from the red alga *C. caldarium*. Beale and Cornejo (6) showed that cell-free extracts of *Cyanidium* reduced BV IXα to give the final products of 3(*Z*)- and 3(*E*)-PCB. Subsequently, these authors identified two soluble activities, one of which reduced BV IXα to 15,16 DHBV, while a second reduced 15,16 DHBV to 3(*Z*)-PEB (9,10). 3(*Z*)-PEB is then further isomerized to PCB (see section 4.3.). In contrast to BV reductase, the two bilin reductases required reduced ferredoxin for activity (8,45). The assay conditions used were as follows: 25 mM HEPES buffer, pH 7.3, 10 μM BV IXα, 10% (vol/vol) glycerol, 1 mM $MgCl_2$, approximately 1 mg/mL BSA, 3000 U/mL catalase, and reducing power in the form of an NADPH-regenerating system together with ferredoxin and ferredoxin $NADP^+$ reductase (see conditions for the heme oxygenase assay in section 4.1). Following incubation at 40°C for 30 minutes, the bilin products were extracted and analyzed by HPLC as described in section 3.1.

Incubation of BV IXα with soluble protein extracts from plastids of the green alga *M. caldariorum* under similar assay conditions resulted in completely different products. In this case, 3(*Z*)-PΦB and 3(*Z*)-PCB were synthesized sequentially (67). This result suggests the presence of a homolog of the plant enzyme, PΦB synthase, and a previously undescribed $18^1,18^2$-reductase (see Figures 1 and 2). PΦB synthase activity was first detected in cucumber using a coupled assay to measure PΦB synthesis by assembly with apophytochrome (55). This assay has now been superceded by an HPLC-based assay and PΦB synthase activity has been measured in isolated plastids from oat (57), pea (62), and tomato (54). The conditions for measuring this enzyme in isolated plastids have been described in Procedure 2. Recently, progress has been made in purifying PΦB synthase (38). This work has revealed that the enzyme is soluble, has an apparent molecular mass of 29 kDa, and functions in the same ferredoxin-dependent manner as the algal (45) and cyanobacterial (15) reductases.

4.3. Bilin Isomerases

Little is known about the bilin isomerases. Two types of isomerase activity have been reported. A glutathione-dependent 3(*Z*) to 3(*E*) *cis-trans* isomerase was identified in *Cyanidium* extracts (9), and a similar activity has been proposed to exist in plants as incubation of BV IXα with plastids results in the synthesis, first of 3(*Z*)-PΦB and then of 3(*E*)-PΦB (57). It is still not entirely clear whether this is an enzyme-catalyzed reaction, as it may simply reflect the fact that the (*Z*)-isomer is chemically less stable than the (*E*)-isomer. However, cyanobacterial extracts synthesized 3(*Z*)-PCB, but not 3(*E*)-PCB, from BV IXα supporting a role for a specific 3(*Z*) to 3(*E*) *cis-trans* isomerase in other organisms (15). The second isomerase activity converts PEB to PCB with both the (*Z*)- and (*E*)-isomers serving as substrates (9). In this case, the reaction was catalyzed by a high molecular weight (>60 kDa) protein fraction, but did not require NADPH or reduced ferredoxin for activity (9).

ABBREVIATIONS

ALA, 5-aminolevulinic acid; BV, biliverdin; BR, bilirubin; BSA, bovine serum albumin; DHBV, dihydrobiliverdin;

DMSO, dimethyl sulfoxide; MBV, mesobiliverdin; PCB, phycocyanobilin; PEB, phycoerythrobilin; PΦB, phytochromobilin; TFA, trifluoroacetic acid.

ACKNOWLEDGMENTS

I would like to thank the Royal Society for their support through a Royal Society University Research Fellowship, Professors Sam Beale and Peter Shoolingin-Jordan for reading this manuscript prior to publication, Mark Milford for checking the Procedures, and Professor Clark Lagarias for the opportunity to learn about the wonderful world of bilins.

REFERENCES

1. **Arciero, D.M., D.A. Bryant, and A.N. Glazer.** 1988. *In vitro* attachment of bilins to apophycocyanin. I. Specific covalent adduct formation at cysteinyl residues involved in phycocyanobilin binding in C-phycocyanin. J. Biol. Chem. *263*:18343-18349.
2. **Arciero, D.M., J.L. Dallas, and A.N. Glazer.** 1988. *In vitro* attachment of bilins to apophycocyanin. II. Determination of the structures of tryptic bilin peptides derived from the phycocyanobilin adduct. J. Biol. Chem. *263*:18350-18357.
3. **Arciero, D.M., J.L. Dallas, and A.N. Glazer.** 1988. *In vitro* attachment of bilins to apophycocyanin. III. Properties of the phycoerythrobilin adduct. J. Biol. Chem. *263*:18358-18363.
4. **Austin, C.C. and K.W. Jessing.** 1994. Green-blood pigmentation in lizards. Comp. Biochem. Physiol. *109A*:619-626.
5. **Beale, S.I.** 1993. Biosynthesis of phycobilins. Chem. Rev. *93*:785-802.
6. **Beale, S.I. and J. Cornejo.** 1984. Enzymic Transformation of biliverdin to phycocyanobilin by extracts of the unicellular red alga *Cyanidium caldarium.* Plant Physiol. *76*:7-15.
7. **Beale, S.I. and J. Cornejo.** 1984. Enzymatic heme oxygenase activity in soluble extracts of the unicellular red alga, *Cyanidium caldarium.* Arch. Biochem. Biophys. *235*:371-384.
8. **Beale, S.I. and J. Cornejo.** 1991. Biosynthesis of phycobilins. Ferredoxin-mediated reduction of biliverdin catalyzed by extracts of *Cyanidium caldarium.* J. Biol. Chem. *266*:22328-22332.
9. **Beale, S.I. and J. Cornejo.** 1991. Biosynthesis of phycobilins. 3(*Z*)-phycoerythrobilin and 3(*Z*)-phycocyanobilin are intermediates in the formation of 3(*E*)-phycocyanobilin from biliverdin IXα. J. Biol. Chem. *266*:22333-22340.
10. **Beale, S.I. and J. Cornejo.** 1991. Biosynthesis of phycobilins. 15,16-dihydrobiliverdin IXα is a partially reduced intermediate in the formation of phycobilins from biliverdin IXα. J. Biol. Chem. *266*:22341-22345.
11. **Brockmann, H., Jr. and G. Knobloch.** 1973. Die absolute Konfiguration des 2E-Äthyliden-3-methylsuccinimids. Ein Beitrag zur Bestimmung der absoluten Konfiguration von Phycobilinen und Phytochrom. Chem. Ber. *106*:803-811.
12. **Chapman, D.J., W.J. Cole, and H.W. Siegelman.** 1967. The structure of phycoerythrobilin. J. Am. Chem. Soc. *89*:5976-5977.
13. **Cole, W.J., D.J. Chapman, and H.W. Siegelman.** 1967. The structure of phycocyanobilin. J. Am. Chem. Soc. *89*:3643-3645.
14. **Cole, W.J., D.J. Chapman, and H.W. Siegelman.** 1968. The structure and properties of phycocyanobilin and related bilatrienes. Biochem. *7*:2929-2935.
15. **Cornejo, J. and S.I. Beale.** 1997. Phycobilin biosynthetic reactions in extracts of cyanobacteria. Photosyn. Res. *51*:223-230.
16. **Cornejo, J. and S.I. Beale.** 1988. Algal heme oxygenase from *Cyanidium caldarium.* Partial purification and fractionation into three required protein components. J. Biol. Chem. *263*:11915-11921.
17. **Cornejo, J., S.I. Beale, M.J. Terry, and J.C. Lagarias.** 1992. Phytochrome assembly. The structure and biological activity of 2(*R*),3(*E*)-phytochromobilin derived from phycobiliproteins. J. Biol. Chem. *267*:14790-14798.
18. **Cornejo, J., R.D. Willows, and S.I. Beale.** 1998. Phytobilin biosynthesis: cloning and expression of a gene encoding soluble ferredoxin-dependent heme oxygenase from *Synechocystis* sp. PCC 6803. Plant J. *15*:99-107.
19. **Crespi, H.L., L.J. Boucher, G.D. Norman, J.J. Katz, and R.C. Dougherty.** 1967. Structure of phycocyanobilin. J. Am. Chem. Soc. *89*:3642-3643.
20. **Crespi, H.L. and J.J. Katz.** 1969. Exchangeable hydrogen in phycoerythrobilin. Phytochem. *8*:759-761.
21. **Davis, S.J., J. Kurepa, and R.D. Vierstra.** 1999. The *Arabidopsis thaliana* HY1 locus, required for phytochrome-chromophore biosynthesis, encodes a protein related to heme oxygenases. Proc. Natl. Acad. Sci. USA *96*:6541-6546.
22. **Elich, T.D., A.F. McDonagh, L.A. Palma, and J.C. Lagarias.** 1989. Phytochrome chromophore biosynthesis. Treatment of tetrapyrrole-deficient *Avena* explants with natural and non-natural bilatrienes leads to formation of spectrally active holoproteins. J. Biol. Chem. *264*:183-189.
23. **Fang, L.-S. and J.L. Bada.** 1990. The blue-green blood plasma of marine fish. Comp. Biochem. Physiol. *97B*:37-45.
24. **Fu, E., L. Friedman, and H.W. Siegelman.** 1979. Mass-spectral identification and purification of phycoerythrobilin and phycocyanobilin. Biochem. J. *179*: 1-6.
25. **Glazer, A.N.** 1989. Light guides. Directional energy transfer in a photosynthetic antenna. J. Biol. Chem. *264*:1-4.

26. **Glazer, A.N. and G.J. Wedemayer.** 1995. Cryptomonad biliproteins—an evolutionary perspective. Photosyn. Res. *46*:93-105.

27. **Goodman, W.G., B. Adams, and J.T. Trost.** 1985. Purification and characterization of a biliverdin-associated protein from the hemolymph of *Manduca Sexta*. Biochem. *24*:1168-1175.

28. **Gossauer, A. and W. Hirsch.** 1974. Totalsynthese des racemischen Phycocyanobilins (Phycobiliverdins) sowie eines "Homophycobiliverdins". Liebigs Ann. Chem. *1974*:1496-1513.

29. **Gossauer, A. and J.-P. Weller.** 1978. Chemical total synthesis of (+)-(2*R*,16*R*)- and (+)-(2*S*, 16*R*)-phycoerythrobilin dimethyl ester. J. Am. Chem. Soc. *100*:5928-5933.

30. **Kakiuchi, T., H. Kato, K.P. Jayasundera, T. Higashi, K. Watabe, D. Sawamoto, H. Kinoshita, and K. Inomata.** 1998. Total syntheses of (±)-phycocyanobilin and its derivatives bearing a photoreactive group at D-ring. Chem. Lett. *1998*:1001-1002.

31. **Kennedy, G.Y. and H.G. Vevers.** 1976. A survey of avian eggshell pigments. Comp. Biochem. Physiol. *55B*:117-123.

32. **Kutty, R.K. and M.D. Maines.** 1981. Purification and characterization of biliverdin reductase from rat liver. J. Biol. Chem. *256*:3956-3962.

33. **Lagarias, D.M., M.W. Crepeau, M.D. Maines, and J.C. Lagarias.** 1997. Regulation of photomorphogenesis by expression of mammalian biliverdin reductase in transgenic *Arabidopsis* plants. Plant Cell *9*:675-688.

34. **Maines, M.D.** 1988. Heme oxygenase: function, multiplicity, regulatory mechanisms, and clinical applications. FASEB J. *2*:2557-2568.

35. **Maines, M.D., N.G. Ibrahim, and A. Kappas.** 1977. Solubilization and partial purification of heme oxygenase from rat liver. J. Biol. Chem. *252*:5900-5903.

36. **McDonagh, A.F. and F. Assisi.** 1971. Commercial bilirubin: a trinity of isomers. FEBS Lett. *18*:315-317.

37. **McDonagh, A.F. and L.A. Palma.** 1980. Preparation and properties of crystalline biliverdin IXα. Simple methods for preparing isomerically homogenous biliverdin and (^{14}C) biliverdin by using 2,3-dichloro-5,6-dicyanobenzoquinone. Biochem. J. *189*:193-208.

38. **McDowell, M.D. and J.C. Lagarias.** 1997. Partial purification, photoaffinity labeling, and characterization of phytochromobilin synthase. Plant Physiol. *114*:S739.

39. **Muramoto, T., T. Kohchi, A. Yokota, I. Hwang, and H.M. Goodman.** 1999. The *Arabidopsis* photomorphogenic mutant *hy1* is deficient in phytochrome chromophore biosynthesis as a result of a mutation in a plastid heme oxygenase. Plant Cell *11*:335-347.

40. **Murphy, J.T. and J.C. Lagarias.** 1997. The phytofluors: a new class of fluorescent protein probes. Curr. Biol. *7*:870-876.

41. **Oren, D.A.** 1997. Bilirubin, rem sleep, and phototransduction of environmental time cues. A Hypothesis. Chronobiol. Int. *14*:319-329.

42. **Ortiz de Montellano, P.R.** 1998. Heme oxygenase mechanism: evidence for an electrophilic, ferric peroxide species. Acc. Chem. Res. *31*:543-549.

43. **Prince, J., T.G. Nolen, and L. Coelho.** 1998. Defensive ink pigment processing and secretion in *Aplysia californica*: concentration and storage of phycoerythrobilin in the ink gland. J. Exp. Biol. *201*:1595-1613.

44. **Provasoli, L., J.J.A. McLaughlin, and M.R. Droop.** 1957. The development of artificial media for marine algae. Archiv. Mikrobiol. *25*:392-428.

45. **Rhie, G. and S.I. Beale.** 1992. Biosynthesis of Phycobilins. Ferredoxin-supported NADPH-independent heme oxygenase and phycobilin-forming activities from *Cyanidium caldarium*. J. Biol. Chem. *267*:16088-16093.

46. **Rhie, G. and S.I. Beale.** 1995. Phycobilin biosynthesis: reductant requirements and product identification for heme oxygenase from *Cyanidium caldarium*. Arch. Biochem. Biophys. *320*:182-194.

47. **Rippka, R., J. Deruelles, J.B. Waterbury, M. Herdman, and R.Y. Stanier.** 1979. Generic assignments, strain histories and properties of pure cultures of cyanobacteria. J. Gen. Microbiol. *111*:1-61.

48. **Ryter, S., E. Kvam, and R.M. Tyrell.** 1999. Heme oxygenase activity determination by high-performance liquid chromatography. Methods Enzymol. *300*:322-336.

49. **Schluchter, W.M. and A.N. Glazer.** 1997. Characterization of cyanobacterial biliverdin reductase. J. Biol. Chem. *272*:13562-13569.

50. **Singleton, J.W. and L. Laster.** 1965. Biliverdin reductase of guinea pig liver. J. Biol. Chem. *240*:4780-4789.

51. **Stocker, R., Y. Yamamoto, A.F. McDonagh, A.N. Glazer, and B.N. Ames.** 1987. Science *235*:1043-1046.

52. **Stoll, M.S. and C.H. Gray.** 1977. The preparation and characterization of bile pigments. Biochem. J. *163*:59-101.

53. **Tenhunen, R., H.S. Marver, and R. Schmid.** 1968. The enzymatic conversion of heme to bilirubin by microsomal heme oxygenase. Proc. Natl. Acad. Sci. USA *61*:748-755.

54. **Terry M.J. and R.E. Kendrick.** 1996. The *aurea* and *yellow-green-2* mutants of tomato are deficient in phytochrome chromophore synthesis. J. Biol. Chem. *271*:21681-21686.

55. **Terry, M.J. and J.C. Lagarias.** 1991. Holophytochrome assembly. Coupled assay for phytochromobilin synthesis *in organello*. J. Biol. Chem. *266*:22215-22221.

56. **Terry, M.J., M.D. Maines, and J.C. Lagarias.** 1993. Inactivation of phytochrome- and phycobiliprotein-chromophore precursors by rat liver biliverdin reductase. J. Biol. Chem. *268*:26099-26106.

57. **Terry, M.J., M.D. McDowell, and J.C. Lagarias.** 1995. (3*Z*)- and (3*E*)-phytochromobilin are intermediates in the biosynthesis of the phytochrome chromophore. J. Biol. Chem. *270*:11111-11119.

58. **Terry M.J., J.A. Wahleithner, and J.C. Lagarias.** 1993. Biosynthesis of the plant photoreceptor phytochrome. Arch. Biochem. Biophys. *306*:1-15.

59. **Turner, L., J.D. Houghton, and S.B. Brown.** 1992. Isolation and partial purification of phycocyanin apoprotein and its role in studies of bilin-apoprotein attachment. Plant Physiol. Biochem. *30*:309-314.

60. **Vreman, H.J., D.A. Cipkala, and D.K. Stevenson.** 1996. Characterization of porphyrin heme oxygenase inhibitors. Can. J. Physiol. Pharmacol. *74*:278-285.

61. **Vreman, H.J. and D.K. Stevenson.** 1988. Heme oxygenase activity as measured by carbon monoxide production. Anal. Biochem. *168*:31-38.
62. **Weller, J.L., M.J. Terry, C. Rameau, J.B. Reid, and R.E. Kendrick.** 1996. The phytochrome-deficient *pcd1* mutant of pea is unable to convert heme to biliverdin IXa. Plant Cell *8*:55-67.
63. **Weller, J.-P. and A. Gossauer.** 1980. Synthese und photoisomerisierung des racem. Phytochromobilin-dimethylesters. Chem. Ber. *113*:1603-1611.
64. **Wilks, A. and P.R. Ortiz de Montellano.** 1993. Rat liver heme oxygenase. High level expression of a truncated soluble form and nature of the meso-hydroxylating species. J. Biol. Chem. *268*:22357-22362.
65. **Wilks, A. and M.P. Schmitt.** 1998. Expression and characterization of a heme oxygenase (Hmu O) from *Corynebacterium diphtheriae.* Iron acquisition requires oxidative cleavage of the heme macrocycle. J. Biol. Chem. *273*:837-841.
66. **Wu, S.-H. and J.C. Lagarias.** 1996. The methylotrophic yeast *Pichia pastoris* synthesizes a functionally active chromophore precursor of the plant photoreceptor phytochrome. Proc. Natl. Acad. Sci. USA *93*:8989-8994.
67. **Wu, S.-H., M.T. McDowell, and J.C. Lagarias.** 1997. Phycocyanobilin is the natural precursor of the phytochrome chromophore in the green alga *Mesotaenium caldariorum*. J. Biol. Chem. *272*:25700-25705.
68. **Yamaguchi, T., Y. Komoda, and H. Nakajima.** 1994. Biliverdin-IXα and biliverdin IXβ reductase from human liver. J. Biol. Chem. *269*:24343-24348.
69. **Yoshida, T. and G. Kikuchi.** 1978. Features of the reaction of heme degradation catalyzed by the reconstituted microsomal heme oxygenase system. J. Biol. Chem. *253*:4230-4236.
70. **Yoshida, T., M. Noguchi, and G. Kikuchi.** 1982. The step of carbon monoxide liberation in the sequence of heme degradation catalyzed by the reconstituted microsomal heme oxygenase system. J. Biol. Chem. *257*:9345-9348.

13 Analysis and Reconstitution of Phytochromes

Michael T. McDowell and J. Clark Lagarias
University of California-Davis, Davis, CA, USA

1. THE PHYTOCHROME FAMILY

Photosynthetic organisms, from bacteria to higher plants, possess light sensing molecules that enable adaptation to fluctuations in intensity, direction, duration, polarization and spectral quality of light from their environment (26). The most well known of these photoreceptors are the phytochromes, which sense the ambient light conditions via their ability to photointerconvert between red (Pr) and far-red (Pfr) light absorbing forms (17,47,51,57). This unique property of phytochromes is conferred by a linear tetrapyrrole (bilin) prosthetic group that is covalently linked to a large polypeptide. First discovered in plants, phytochrome-like molecules also have been identified in lower eukaryotic plant species (i.e. green algae, mosses, and ferns) (27,64) and, more recently, in cyanobacteria (21,25,66,70). It is well established that higher plants possess multiple phytochromes that are encoded by a small nuclear gene family termed *PHYA-F* (5, 45). All phytochrome proteins share a highly conserved photosensory domain, in which the bilin prosthetic group is linked via a thioether to an invariant cysteine residue. Biochemical and molecular cloning studies indicate that the basic architecture of eukaryotic phytochromes has been preserved, while a growing family of phytochrome-related genes in cyanobacteria encode polypeptides considerably more divergent in structure. Experimental methods outlined in this chapter discuss 2 major tools, difference spectroscopy and holophytochrome assembly, necessary for establishing whether candidate genes encode bonafide phytochromes.

With only 2 known exceptions, eukaryotic phytochromes are soluble homodimeric proteins with a subunit roughly 1100 amino acids in length (Figure 1). The bilin prosthetic group is associated with a highly conserved photosensory domain at the protein's N terminus, which is readily cleaved from the C-terminal region by limited proteolysis to yield a photochemically active 60 to 70 kDa monomer (24). The more diverged 500 amino acid C terminus of eukaryotic phytochromes specifies the high affinity subunit–subunit interaction (23) and also possesses 2 regulatory subdomains that genetic studies have established to be critical

Heme, Chlorophyll, and Bilins: ***Methods and Protocols***
Edited by A.G. Smith and M. Witty

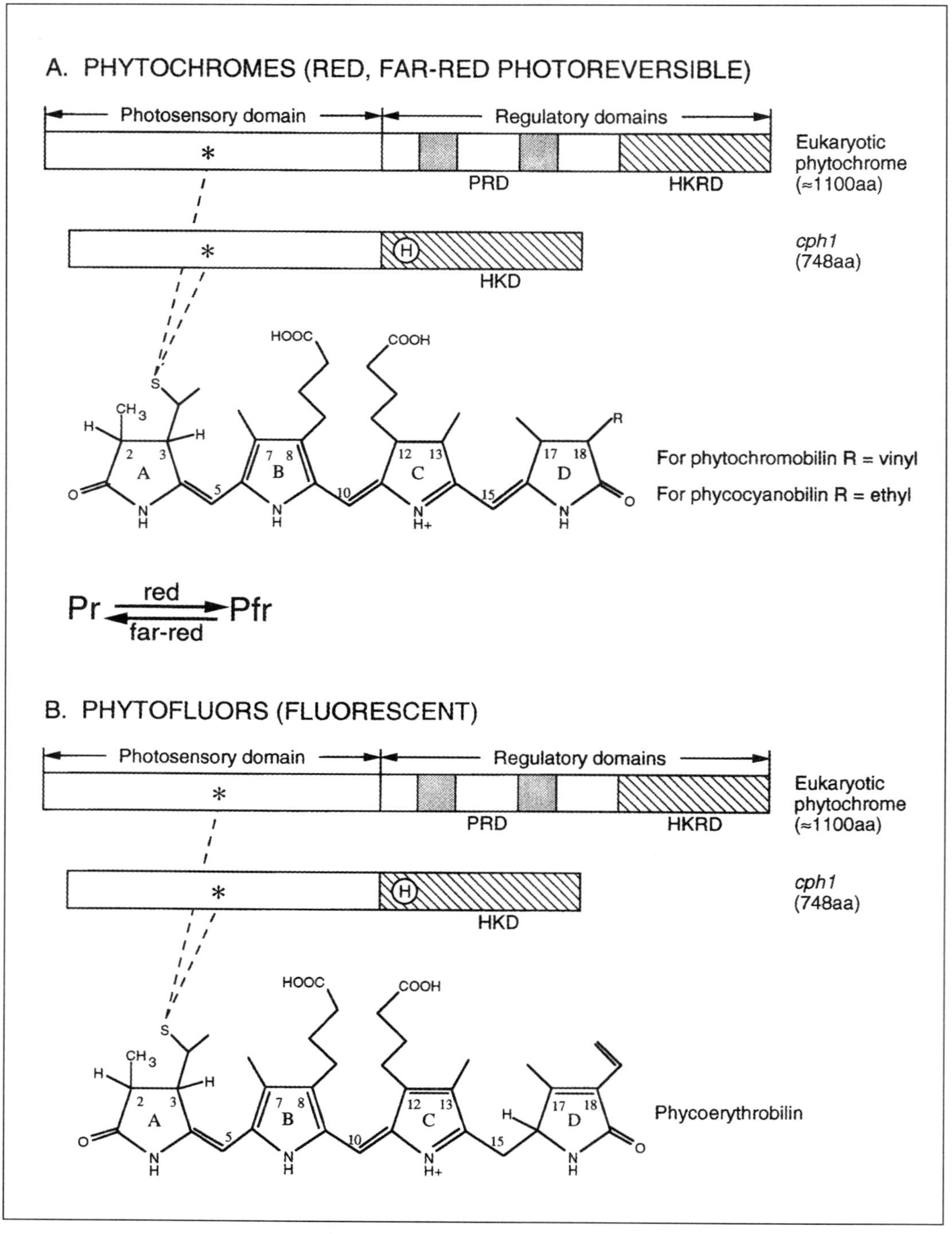

Figure 1. Domain structure of eukaryotic and cyanobacterial phytochromes and phytofluors. (A) Phytochromes exist in photointerconvertible red light absorbing (Pr) and far-red light absorbing (Pfr) forms. Phytochromes possess either phytochromobilin or phycocyanobilin prosthetic groups bound to a conserved cysteine residue indicated with an asterisk. (B) Phytofluors are orange fluorescent biliproteins consisting of phycoerythrobilin thioether-linked to the conserved cysteine residue on an apophytochrome. Regulatory domains include the PAS-related domain (PRD), which contain 2 direct repeats shown as dark boxes, and the histidine kinase domain (HKD) and the histidine kinase-related domain (HKRD), which are depicted with cross hatching.

for transducing the light signal (reviewed in References 47 and 63). These include the PRD, a domain related to the PAS domain found on eukaryotic regulatory proteins (30,59), and the HKRD, a domain related to histidine kinase transmitter domains of 2 component sensor proteins from bacteria (56). Like eukaryotic phytochromes, the cyanobacterial phytochrome Cph1 from *Synechocystis* sp PCC 6803 (21,70) possesses the conserved N-terminal photosensory domain and histidine kinase domain (HKD), but lacks the PRD regulatory subdomain (Figure 1). Recent investigations support the hypothesis that both Cph1 and eukaryotic phytochromes transduce the light signal perceived by the photosensory domain via changes in protein kinase activity of the regulatory domains (69,70).

It was not until the purification of native phytochrome from plants that a chemical examination of the chromophore of phytochrome took place. ^{1}H-Nuclear magnetic resonance (NMR) analysis of chromopeptides from oat phytochrome A revealed that its chromophore and linkage to the apoprotein were very similar to those found in phycobiliproteins (34,52). These studies also revealed that oat phytochrome possessed a phytochromobilin (PΦB) chromophore (Figure 1), confirming earlier investigations of bilins obtained from phytochromes by chemical cleavage (50). The precursors of the chromophores of the other phytochrome genes within a given plant species (i.e., *PHYB-E*) have not been directly determined, but they are assumed to be PΦB. One exception is phytochrome from the green alga *Mesotaenium caldariorum*, which possesses the phycobiliprotein chromophore precursor phycocyanobilin (PCB) (68). The natural chromophore precursor of Cph1 has not yet been determined. PCB has been proposed as a likely candidate owing to its intermediacy in the phycobiliprotein chromophore biosynthetic pathway in *Synechocystis* sp. PCC 6803 (6).

2. DIFFERENCE SPECTROSCOPY FOR PHYTOCHROME QUANTITATION

2.1. History and Development

The application of difference spectroscopy dates back to the early stages of phytochrome research with the discovery of the photoreversible nature of the biology of phytochrome. Action spectroscopy helped to define further the key wavelengths of light that phytochrome uses for its light switching mechanism (53). The key observation that led ultimately to the development of difference spectroscopy was that the physiological effects of red light irradiation on higher plants and algae could, in many cases, be reversed or nullified by a subsequent irradiation with far-red light. The result of these observations has become a key diagnostic test for phytochrome-mediated responses: red, far-red light photoreversibility. For the purification and characterization of phytochrome proteins, this led to the development of the difference spectrum. The two forms of phytochrome, Pr and Pfr, are spectrally different (Figure 2A). When the Pfr absorption spectrum is subtracted from the Pr absorption spectrum, the result is a difference spectrum with a very characteristic wave appearance (Figure 2B). The difference between the maximum and the minimum is then used to quantitate the amount of spectrally active phytochrome present in the sample.

2.2. Difference Spectral Assay for Holophytochrome

The following methodology refers only to the measurement of phytochrome in extracts and is based on the methods developed in this laboratory (31). For a discussion of in vivo phytochrome photoassay, we refer the reader to the following references (19,46). Two major caveats about phytochrome measurements need to be made at this point. First, to avoid continuous photocycling, all measurements should be performed in a laboratory and kept under dim green safelight (54). Second, the validity of the phytochrome difference assay depends upon the lack of other pig-

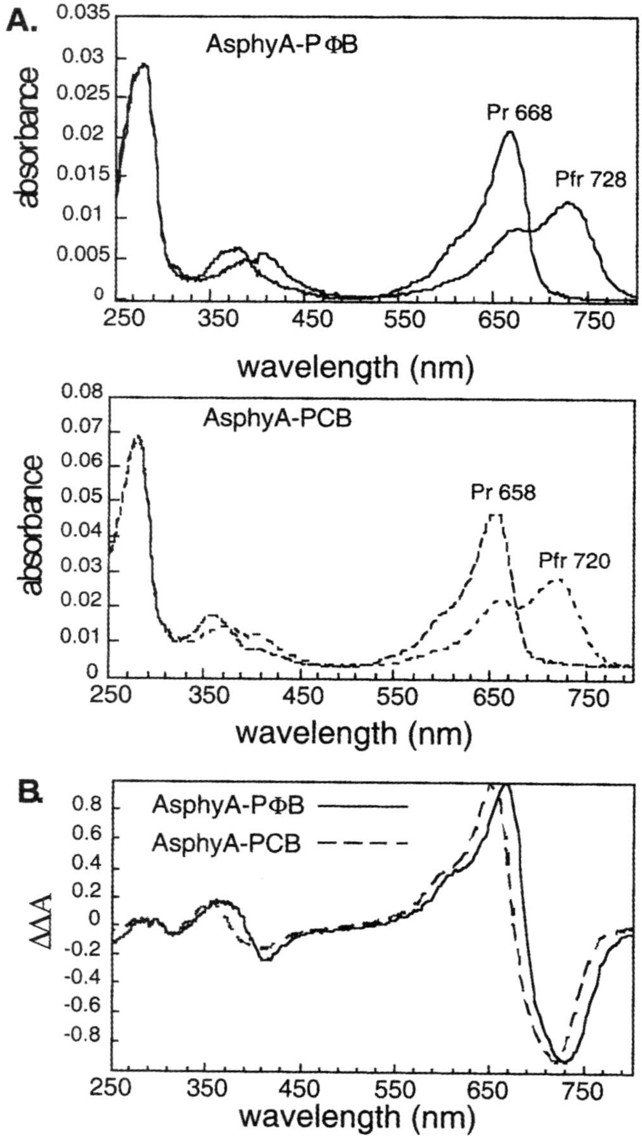

Figure 2. Absorption and difference spectra of purified recombinant oat phytochrome. (A) Absorption spectra of Pr and Pfr forms of PΦB and PCB adducts of recombinant oat apophytochrome A (AsphyA). (B) Difference spectra of PΦB and PCB adducts of recombinant AsphyA. Adapted from Murphy and Lagarias (44).

ments that absorb in the 500 to 800 nm wavelength range. For this reason, chlorophyll must be removed from extracts, from light-grown plant extracts which can be accomplished by precipitation with polyethyleneimine (41). The latter is not a concern with recombinant phytochromes prepared from bacterial, yeast, or insect cell extracts. Our standard method for phytochrome photoassay is outlined in Procedure 1. Because actinic illumination and sample measurement can be performed simultaneously, we recommend the use of a diode array spectrophotometer, such as the Model 8453 from Hewlett Packard (Wilmington, DE, USA). The length of irradiation will also depend on the light source. Typically, for a focused 250 W quartz halogen lamp, a 200-second irradiation is sufficient. The limit of sensitivity for this assay is approximately 10 nM phytochrome (i.e., 1 μg/mL protein).

❖ Procedure 1. Phytochrome Photoassay

1. The protein sample, either from plant tissue or reconstituted recombinant phytochrome (see later section), is diluted or reconstituted in the desired amount of TEGE buffer [25 mM Tris-HCl, pH 7.8, containing 25% (vol/vol) ethylene glycol, 1 mM EDTA, 2 mM phenylmethylsulfonyl fluoride (PMSF) and 1 mM dithiothreitol (DTT)], typically 500 μL. During irradiation and the subsequent analysis, the sample is kept at 10° to 16°C.
2. The sample is transferred to a 1-cm pathlength spectrophotometric cuvette, placed in the spectrophotometer, irradiated with saturating red light, and an absorption spectrum is taken. Saturating illumination is defined as the amount of illumination needed to produce a photostationary state and is determined empirically when the absorbance at 650 to 665 nm (i.e., Pr's absorption maximum) no longer changes with further illumination. For purified phytochrome preparations, absorbances between 250 and 800 nm are recorded; for crude samples, a 500 to 800 nm range is used. The wavelength of red light needed is dependent on the species of phytochrome under examination. For phytochromes containing the PΦB chromophore, the red actinic light should be 660 nm (± 5 nm), and for PCB-containing phytochrome, the actinic light should be less than 650 nm (e.g., we use 636 ± 5 nm). Actinic light is typically provided by a 250 W quartz halogen lamp filtered through an appropriate interference filter (Corion, Franklin, MA, USA).
3. The sample is then irradiated with saturating far-red light, and another absorption spectrum is taken. Far-red light is obtained using a 250 W quartz halogen lamp filtered through an FRS 700 black plexiglas filter (Rohm and Hass, Philadelphia, PA, USA).
4. The difference spectrum and quantitation of photoactive phytochrome is obtained by mathematically subtracting the spectrum of the red light-irradiated sample from the spectrum of the far-red light-irradiated sample (see Figure 2B). The quantitation is obtained by measuring the ΔΔA from the A_{max} and the A_{min} of the difference spectrum. To correct for the incomplete photoconversion of Pr to Pfr and back, the ΔΔA is multiplied by 0.86. To calculate the molar concentration of phytochrome, the corrected ΔΔA value is entered into the Lambert-Beers Law equation of $A_{corr} = \varepsilon cl$, where $\varepsilon_{665} = 132\,000\ M^{-1}cm^{-1}$ for purified oat phytochrome A (31).
5. To determine the relative purity for a particular phytochrome preparation, the specific absorbance ratio (SAR) is determined. To do this, the full spec-

trum of the Pr form of phytochrome (i.e., sample irradiated with saturating far-red light) is obtained, and the absorption of the red light maximum (i.e., 660 nm) is divided by the absorption of the protein maximum (i.e., 280 nm). Purified full-length phytochrome A preparations typically exhibit SARs of 1.0 to 1.1 (31,36).

2.3. Coupled Difference Spectral Assay for PΦB Synthase Activity

Another use for the phytochrome difference spectrum was first described by Elich and Lagarias (13,15) and more recently was exploited by Terry and Lagarias (60). This application was developed for the study of phytochrome chromophore biosynthesis. At the time of its development, there was no reliable way to assay directly the bilin intermediates in the biosynthesis of the phytochrome chromophore. This application suffers from two drawbacks, however. The first is that the reaction cannot be used to study the kinetics of the biosynthetic reactions. The other is, until recently, the lack of an abundant source of recombinant apophytochrome, as the systems previously developed for expression of recombinant higher plant apophytochrome could only produce small amounts of protein. Subsequently, high-performance liquid chromatography (HPLC) methods were developed to assay linear tetrapyrrole metabolism more directly (see Chapter 12 in this text by Terry). While HPLC analyses have facilitated the purification and biochemical characterization of phytochromobilin synthase (42), they are somewhat tedious and time-consuming, making analysis of column fractions a chore. The recent discovery of a Cph1 from *Synechocystis* sp. PCC 6803 had a significant side benefit; recombinant Cph1 apoprotein could be expressed in *Escherichia coli* and obtained in much larger amounts than higher plant apophytochrome (21,70). This discovery led us to develop the coupled assay for the purification of PΦB synthase which follows. This method is essentially that described by Terry and Lagarias (60), with the exceptions that Cph1 apophytochrome is substituted for oat apophytochrome A, and apophytochrome is added only after the assay is complete.

❖ Procedure 2. PΦB Synthase Assay Coupled to Phytochrome

1. Crude plastid preparations from dark grown oats were typically used as a source of phytochromobilin synthase activity (42,60). Dilute protein samples in a 1-mL final volume of 50 mM TES/KOH (pH 7.3) containing an NADPH regenerating system (6.5 mM glucose 6-phosphate, 0.82 mM $NADP^+$, 1.1 U/mL Torula yeast Type XII glucose-6-phosphate dehydrogenase EC 1.1.1.49), a ferredoxin, ferredoxin-reducing system (4.6 μM purified spinach ferredoxin, 0.025 U/mL ferredoxin:$NADP^+$ oxidoreductase EC 1.18.1.2), and 10 μM bovine serum albumin (BSA) (fraction V, heat shock).
2. Assays are initiated by adding biliverdin (BV) IXα in 10 μL of dimethyl sulfoxide (DMSO). Usually, the final concentration of BV in an assay was 5 μM.
3. Assay mixtures were incubated in a 28°C water bath under green safelight or subdued light for the desired amount of time and stopped by placing them on ice. For assays of intact plastid fractions or membrane fractions, the assays were clarified by centrifugation at 12000× *g* for 15 minutes at 4°C prior to analysis.
4. PΦB synthase assays can be analyzed either by HPLC (as described in Chapter 12 in this text by Terry) or by difference spectroscopy. To obtain a difference

spectrum, a small aliquot of concentrated recombinant Cph1 is added to the PΦB synthase assay mixture.

5. The assay was incubated an additional 20 to 30 minutes at room temperature to facilitate assembly, and then a difference spectrum is obtained as outlined above.

3. ASSAYS FOR HOLOPHYTOCHROME ASSEMBLY

3.1. History and Development

Based on the pioneering work of Gardner and collaborators (22), it is well established that the chromophore and apophytochrome biosynthetic pathways are essentially independent. Plants treated with inhibitors of 4-aminolevulinic acid (ALA) biosynthesis, such as gabaculine or amino-5-hexynoic acid, possess reduced amount of photoactive holophytochrome, but accumulate near normal amounts of apophytochrome (12,14,22). Using these inhibitors, it was shown that co-incubating the plants with ALA and BV IXα, as well as the unnatural isomer BV XIIIα, could restore the levels of spectrally active phytochrome (15). The phycobiliprotein chromophore precursor PCB was also found to restore the levels of spectrally active phytochrome, albeit with a blue-shifted difference spectrum (see Figure 2B). Follow-up investigations utilizing apophytochrome isolated from inhibitor-treated plants indicated that either the lyase enzyme copurifies with apophytochrome, or that apophytochrome is itself a bilin lyase (13). The development of recombinant apophytochrome expression systems by many laboratories (7,10,13,18,20,21,29,32,33,35,55, 62,65) have corroborated the latter hypothesis that apophytochrome is a bilin lyase that catalyzes thioether linkage formation with bilins. All bonafide apophytochromes examined to date can catalyze thioether formation with PΦB, as well as the phycobilins, PCB, and phycoerythrobilin (PEB), both precursors of the phycobiliproteins (37). This work is in striking contrast with phycobiliprotein assembly, whose bilin ligation requires separate lyases for proper assembly (see Chapter 14 in this text by Schluchter and Bryant). Analyses of bilin assembly to apophytochrome have relied on 3 major tools: (*i*) difference spectroscopy (described above); (*ii*) zinc blot analysis; and (*iii*) fluorescence spectroscopy. In the sections that follow, the latter 2 tools will be highlighted.

3.2. Zinc Blot Assay for Bilin Attachment

Early examination of bile pigments revealed that bilatrienes form intensely fluorescent complexes with zinc ions in the presence of iodine (1). This and related observations have led to the examination of fluorescent zinc complexes of bilin-linked polypeptides. The initial work was performed using standard sodium dodecyl sulfate polyacrylamide gel electrophoresis (SDS-PAGE) gels with the fluorescent products being observed following UV illumination (2). The purpose of SDS-PAGE is to remove the unbound bilin from the protein-bound bilin. This methodology, which can be used for phycobiliproteins as well (48), has been further improved by the extension to electroblotted protein samples or the "zinc blot" (37). The zinc blot is used as a general diagnostic technique to assess the ligation competency of a particular apophytochrome. The limit of sensitivity of this technique is approximately 50 ng/cm of phytochrome per lane for gels and 12 ng/cm of phytochrome per lane for blots. Procedure 2, for the zinc blot assay, is based on 2 references (2,37).

❖ Procedure 3. The Zinc Blot Assay for Bilins

1. Protein samples to be analyzed are electrophoresed in a SDS-polyacrylamide gel using any standard procedure. After electrophoresis, the gel is transblotted to a polyvinylidene difluoride (PVDF) membrane using any standard procedure for electroblotting.

   NOTE: Nylon and nitrocellulose are considerably less effective owing to their greater autofluorescence background and large UV absorption, respectively.
2. After electroblotting, the membrane is washed briefly with deionized water.
3. After the water wash, the blot is transferred to 1.3 M zinc acetate in deionized water and incubated at room temperature for roughly 30 minutes under reduced light.
4. Just prior to visualization, the blot is rinsed with deionized water to remove excess Zn^{2+} ion.
5. The blot is visualized by placing on a long wavelength UV transilluminator and photographing with Technical Pan film (Type 4415; Eastman Kodak, Rochester, NY, USA) using an RG-630 cutoff filter (Schott, Laurel, NJ, USA) (2-min exposure). Alternatively, the blot can be imaged using a Storm™ 860 PhosphorImager® (Molecular Dynamics, Sunnyvale, CA, USA) with the red laser in fluorescence mode. The image obtained using the Storm can be analyzed using the ImageQuant™ software or converted to tagged image format files (TIFF) and analyzed using a program such as National Institutes of Health (NIH) Image. For quantitative analyses, a dilution series of known quantities of holophytochrome (or phycobiliprotein) should be included on each blot.

3.3. Fluorescence Assay for Holophytochrome Assembly

The natural biological function of phytochrome is to act as a light switch due to its ability to photointerconvert between the Pr and Pfr light-absorbing forms. Based on NMR and resonance Raman spectroscopic analysis, this photoconversion has been proposed to involve the Z to E isomerization of the C15 methine bridge double bond (16,52). While the 2 known phytochrome chromophore precursors, PΦB and PCB, possess this double bond, PEB does not. Indeed, PEB adducts of apophytochromes are inactive photochemically, a result that led to the hypothesis that such adducts might be fluorescent. That PEB incubation with apophytochromes produce intensely fluorescent adducts was first documented in 1995 (39). This finding led to the development of a real-time kinetic assay for the study of phytochrome assembly that is described below. This assay has not only enabled the determination of both bilin binding and catalytic rate constants for the reconstitution of holophytochrome with its natural chromophore (i.e., $K_d \sim 1\ \mu M$, $k_{cat} \sim 0.25–0.3\ s^{-1}$), but it has also facilitated the analysis of potential inhibitors of this process, such as the PΦB precursor BV (i.e., $K_i \sim 1$ mM). The procedures for these assays, summarized below, are based on the work of Li, Murphy, and Lagarias (39). Two major fluorescent assay methodologies are described below, the standard and the competitive assays. The only difference between the kinetic assays is the data analysis. The data analysis for the standard assay is outlined in Scheme 1. The analyses of the 2 types of competitive fluorescence assays are outlined in Schemes 2 and 3, respectively.

❖ Procedure 4. Fluorescence Assay for Holophytochrome Assembly

1. The formation of the fluorescent PEB-

phytochrome adduct is initiated in a semimicrofluorescence cuvette by addition of apophytochrome (10 nM final concentration) from a concentrated stock solution to a greater-than or equal to 70-fold molar excess of PEB (typically 0.5 to 15 μM final concentration) in TEGE buffer [10 mM Tris-HCl, pH 8.0, 25% (vol/vol) ethylene glycol, 1 mM EDTA, 1 mM PMSF, and 1 mM DTT]. The PEB is dissolved in Me_2SO, with the final Me_2SO concentration in the assay no greater than 2% (vol/vol). The assay mixture is rapidly mixed and placed in a fluorescence spectrophotometer. Typically, measurements are performed with samples maintained at room temperature (22°–25°C).

$$\text{bilin} + \text{apoPC} \underset{k_{-1}}{\overset{k_1}{\rightleftharpoons}} \text{bilin:apoPC} \xrightarrow{k_2} \text{bilin-PC}$$

$$\frac{-d[\text{apoPC}]_T}{dt} = k_2[\text{PEB}:\text{apoPC}] \qquad \textbf{[Eq. 1]}$$

$$[\text{apoPC}]_T = [\text{apoPC}] + [\text{bilin:apoPC}]$$

$$K_{bilin} = \frac{[\text{bilin}][\text{apoPC}]}{[\text{bilin}:\text{apoPC}]}$$

Combining the above equations,

$$k_{app} = k_2\left[\frac{[\text{PEB}]K_{\text{BV}}}{K_{\text{PEB}}K_{\text{BV}} + K_{\text{BV}}[\text{PEB}] + K_{\text{PEB}}[\text{BV}]}\right] \qquad \textbf{[Eq. 2]}$$

When [bilin] >> [apoPC],

$$= k_{\text{app}}[\text{apoPC}]_T$$

Integrating,

$$\ln\left[\frac{[\text{apoPC}]_t}{[\text{apoPC}]_0}\right] = -k_{app}t \qquad \textbf{[Eq. 3]}$$

where

$$k_{app} = k_2\left[\frac{[\text{bilin}]}{\text{K}_{\text{bilin}} + [\text{bilin}]}\right] \qquad \textbf{[Eq. 4]}$$

Rearranging,

$$\frac{1}{k_{app}} = \frac{\text{K}_{\text{bilin}}}{k_2}\frac{1}{[\text{bilin}]} + \frac{1}{k_2} \qquad \textbf{[Eq.5]}$$

Scheme 1. Kinetic analysis of holophytochrome assembly (adapted from Reference 39). The enzyme reaction of a bilin, typically PEB, with apoPC to produce a fluorescent product is shown in the diagram. Equation 1 is the integrated rate equation describing this reaction. Equation 2 defines the kinetics of bilin–apophytochrome adduct formation. If the bilin precursor concentration is kept essentially constant, by providing a large excess, semilog plots of the fraction of apophytochrome remaining are expected to be linear as described in Equation 3. Equation 4 and its reciprocal, Equation 5, provide a means for determining the affinity of apophytochrome for a particular bilin, K_{bilin}, and the rate constant, or turnover, of apophytochrome for a particular bilin, k_2.

2. For time-based measurements, samples were excited with 570 nm light with 2 nm bandpass. Fluorescence emission data was collected at 586 nm with 16 nm bandpass. Data was collected with 1-second integration for 15 to 30 minutes. Saturated fluorescence intensity (i.e., 100% assembly) is determined in parallel by incubating a control sample of phytochrome with a very large excess of PEB (>500-fold molar excess) for more than 1 hour.
3. The equations outlining the analysis of data for the standard kinetic assay with PEB and apophytochrome are outlined in Scheme 1, and example data is shown in Figure 3. Raw fluorescence data is transformed using Equation 3 in Scheme 1. When data is replotted on a semilog graph, k_{app} values for each assembly reaction are determined from the slope of the line. According to Equation 5 (Scheme 1), $1/k_{app}$ values for the different assemblies are then plotted versus 1/(PEB). The x- and y-intercepts for this data provide the K_{bilin} and k_{cat}, or k_2, respectively, for PEB.

Variations:

4. The analysis of data for the competitive assay using a reversible inhibitor of PEB-phytochrome formation such as BV (see Scheme 2), is carried out in much the same manner as outlined for the standard assay above. The raw fluorescence data is transformed using Equation 8 (Scheme 2). This data is graphed on a semilog plot to obtain the k_{app} as before. The K_I for BV, or K_{BV}, is estimated using the x-intercept of the plot of the $1/k_{app}$ versus the BV concentration (Equation 10 in Scheme 2).
5. The analysis of the data when using an irreversible inhibitor of PEB-apophytochrome adduct formation, such as PΦB or PCB, is much different from the previous examples (Scheme 3). Experimentally, the amount of competitor bilin is estimated from the degree of fluorescence inhibition relative to the control reaction with no inhibitor. Since both the concentration of PEB-phytochrome (PC) adduct and the k_{app} for PEB-PC formation are known, the k^I_{app} can be calculated using Equation 13 (Scheme 3). The K_I^{bilin} is obtained following a double reciprocal plot of the k^I_{app} versus the bilin concentration as shown in Equation 15 (Scheme 3).

4. BILIN AND APOPHYTOCHROME SPECIFICITY FOR THIOETHER LINKAGE FORMATION

4.1. Bilin Specificity for Thioether Linkage Formation

The question of bilin chromophore precursor specificity for holophytochrome assembly has been addressed using zinc blot analysis, difference spectroscopy, and fluorescence spectroscopy (15,37,40). The requirements for assembly are an A-ring ethylidene at the C3 position, as is present in PΦB, PCB, and PEB (37), and a C10 methine bridge. The former conclusion is based on in vivo feeding of BVs IIIα, IXα, and XIIIα; BV IXα and XIIIα feeding restored levels of spectrally active phytochrome, while BV IIIα had no apparent effect (15). Assembly of a bilin with an ethylidene at the C2 position cannot be ruled out, but this compound is not readily available, and based on BV IIIα feeding experiments, it is probably not biologically relevant. The requirement for a C10 methine bridge is based on the inability of rubins, including those possessing an A-ring ethylidene moiety, such as phycocyanorubin, to assemble with apophytochrome (61). The observations that the D-ring can be modified including 18-vinyl reduction,

$$\mathrm{BV{:}apoPC} \underset{k_{-3}}{\overset{k_3}{\rightleftharpoons}} \mathrm{BV + apoPC + PEB} \underset{k_{-1}}{\overset{k_1}{\rightleftharpoons}} \mathrm{PEB{:}apoPC} \xrightarrow{k_2} \mathrm{PEB\text{-}PC}$$

$$\frac{-d[\mathrm{apoPC}]_T}{dt} = k_2[\mathrm{PEB:apoPC}] \qquad \textbf{[Eq. 6]}$$

where $[\mathrm{apoPC}]_T = [\mathrm{apoPC}] + [\mathrm{PEB{:}apoPC}] + [\mathrm{BV{:}apoPC}]$

$$K_{PEB} = \frac{[\mathrm{PEB}][\mathrm{apoPC}]}{[\mathrm{PEB:apoPC}]} \qquad K_{BV} = \frac{[\mathrm{BV}][\mathrm{apoPC}]}{[\mathrm{BV:apoPC}]}$$

Combining the above equations,

$$\frac{-d[\mathrm{apoPC}]_T}{dt} = k_2\left[\frac{[\mathrm{PEB}]K_{BV}}{K_{PEB}K_{BV} + K_{BV}[\mathrm{PEB}] + K_{PEB}[\mathrm{BV}]}\right][\mathrm{apoPC}]_T \qquad \textbf{[Eq. 7]}$$

When [PEB] and [BV] >> [apoPC],

$$= k_{app}[\mathrm{apoPC}]_T$$

Integrating,

$$\ln\left[\frac{[\mathrm{apoPC}]_t}{[\mathrm{apoPC}]_0}\right] = -k_{app}t \qquad \textbf{[Eq. 8]}$$

where

$$k_{app} = k_2\left[\frac{[\mathrm{PEB}]K_{BV}}{K_{PEB}K_{BV} + K_{BV}[\mathrm{PEB}] + K_{PEB}[\mathrm{BV}]}\right] \qquad \textbf{[Eq. 9]}$$

Rearranging,

$$\frac{1}{k_{app}} = \left[\frac{K_{PEB}}{K_{BV}k_2[\mathrm{PEB}]}\right][\mathrm{BV}] + \left[\frac{K_{PEB} + [\mathrm{PEB}]}{k_2[\mathrm{PEB}]}\right] \qquad \textbf{[Eq. 10]}$$

Scheme 2. Reversible competitive inhibition of PEB phytochrome assembly. A kinetic model for PEB adduct formation in the presence of a reversible competitive inhibitor such as BV. The kinetics of PEB adduct formation should be pseudo-first-order as predicted by Equation 7. The raw fluorescence data is transformed and replotted as described by Equation 8. The slopes of these semilog replots yield k_{app} values. These values are used to construct the plot described by Equation 10. The x-intercept of Equation 10 yields an estimate of the equilibrium dissociation constant for the reversible competitive inhibitor.

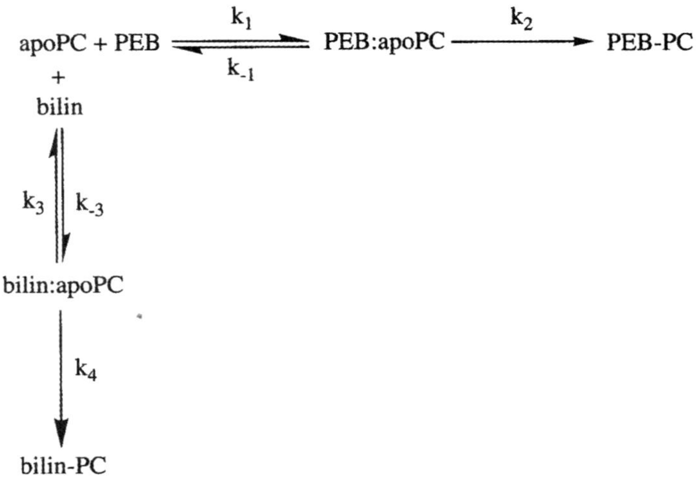

$$K_{PEB} = \frac{[\text{PEB}][\text{apoPC}]}{[\text{PEB}:\text{apoPC}]} \qquad Ki = \frac{[\text{bilin}][\text{apoPC}]}{[\text{bilin}:\text{apoPC}]}$$

When [PEB] >> [apoPC] and [bilin] >> [apoPC],

$$\frac{d[\text{PEB}-\text{PC}]}{dt} = k_2\left[\frac{[\text{PEB}]}{K_{PEB}+[\text{PEB}]}\right][\text{apoPC}]_T = k_{app}[\text{apoPC}]_T \qquad \textbf{[Eq. 11]}$$

$$\frac{d[\text{bilin}-\text{PC}]}{dt} = k_2\left[\frac{[\text{bilin}]}{K_i+[\text{bilin}]}\right][\text{apoPC}]_T = k^{I}_{app}[\text{apoPC}]_T \qquad \textbf{[Eq. 12]}$$

Therefore,

$$\frac{[\text{PEB}-\text{PC}]}{[\text{bilin}-\text{PC}]} = \frac{k_{app}}{k^{I}_{app}} \qquad \textbf{[Eq. 13]}$$

where

$$k^{I}_{app} = k_4\left[\frac{[\text{bilin}]}{K_i+[\text{bilin}]}\right] \qquad \textbf{[Eq. 14]}$$

Rearranging,

$$\frac{1}{k^{I}_{app}} = \frac{1}{[\text{bilin}]}\frac{K_i}{k_4} + \frac{1}{k_4} \qquad \textbf{[Eq. 15]}$$

Scheme 3. Irreversible inhibition of PEB phytochrome assembly. A kinetic model for PEB adduct formation in the presence of an irreversible competitive inhibitor such as PΦB. The formation of both PEB-phytochrome and competitor bilin–phytochrome are described by Equations 11 and 12. In the presence of large molar excesses of all bilins, these equations are first-order expressions. The $k_{app}{}^{i}$ values are calculated using Equation 13, then plotted as a function of the competitive inhibitor according to Equations 14 and 15. A plot of Equation 15 yields the dissociation constant (K_i) and the catalytic rate constant (k_4) for the competitive inhibitor of fluorescent adduct formation.

switching of the C17 and C18 methyl and vinyl moieties and elaboration of the C18 side chain reveal that the bilin binding pocket of apophytochrome is not very discerning with regard to the C18 substituent (15,37,40). The C15 methine bridge is not required for assembly, as demonstrated by the binding of PEB to apophytochome. BVs and bilirubins, which lack the A-ring ethylidene moiety, also do not form covalent adducts with phytochrome, although the former are capable of noncovalent

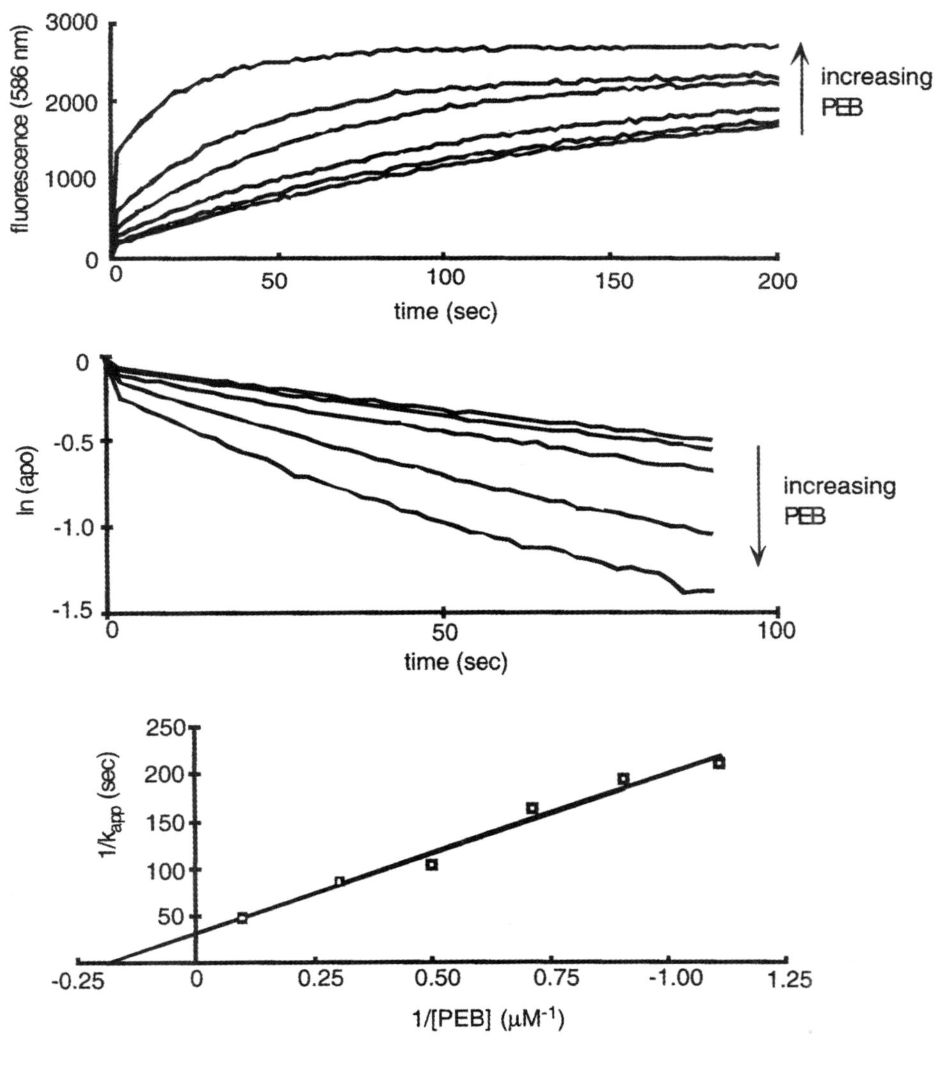

Figure 3. Fluorescence assay for holophytochrome assembly. Representative data for standard fluorescence analysis of PEB attachment to recombinant oat phytochrome A, after Li et al. (39). The upper panel shows raw fluorescence kinetic data as a function of increasing PEB concentration. The middle panel is a replot of the same data according to Equation 3 (Scheme 1), from which k_{app} values were estimated. The bottom panel depicts a replot of $1/k_{app}$ versus 1/(PEB) according to Equation 5 (Scheme 1). See text and Reference 39 for details.

interaction with phytochrome and, therefore, can act as reversible competitive inhibitors as discussed above. The requirement for both propionic acid moieties has been established by the inability of bilin–esters to bind to apophytochrome (3).

4.2. Apophytochrome Specificity for Thioether Linkage Formation

While bilin specificity has been actively addressed, less is known about the regions of the apophytochrome required for bilin attachment. Thus far, the only unequivocal requirement is the conserved cysteine, through which the bilin forms its thioether linkage (cys-321 in the case of oat *PHYA3*) (3,4). Much effort has been directed at trying to determine other amino acid residues or regions of the protein that are involved either directly or indirectly with the lyase activity. Deletion analysis of phytochrome and expression of the truncated proteins in either *E. coli* or yeast have established that neither the first 68 amino acids nor the entire C-terminal domain are required for the autocatalytic assemble of recombinant phytochromes (10,20). With recombinant rice apophytochrome however, the deletion of the first 80 residues abolished bilin binding (62). Site-directed mutagenesis of the region surrounding the conserved cysteine attachment site has been undertaken by several groups (3,9,49,58). These experiments have so far failed to identify other residues essential for bilin assembly, although an important role for the histidine residue adjacent to the conserved cysteine (i.e., H324 in pea PHYA) has been proposed based of the loss-of-function of site-directed mutants of this histidine residue (3,9,49). Interestingly, for the Cph1-related bacteriophytochrome BphP from *Deinococcus radiodurans*, which lacks the conserved cysteine residue, this adjacent histidine appears to be the site of bilin binding (8). Whether this histidine represents a catalytically important residue is presently unresolved. However, the observed bilin lyase activity recombinant pea apophytochrome mutants, in which this histidine residue was changed to a glutamine or arginine residue, suggest otherwise (3,58). Ongoing studies to identify catalytically important residues for bilin lyase activity will take advantage of the growing family of phytochrome-related proteins in cyanobacteria, in which deletions, insertions, and amino acid substitutions, which influence bilin ligation, can be assessed.

4.3. Assembly of Recombinant Phytochromes In Vivo

Recently, recombinant expression of phytochrome led to a novel application (11,28,38,67). Phytochrome expressed in a heterologous system such as *Saccharomyces cerevisiae* could be assembled in vivo if the chromophore was supplied exogenously. The key stumbling block was getting the chromophore into living cells. This could be accomplished by dissolving the chromophore precursor in DMSO, which was added to a minimal buffer medium at a final concentration of 50 μM (38). The dissolved chromophore was then diluted in the appropriate buffer to no more than 10% (vol/vol). The cells were able to take up the bilin that assembled with the recombinant phytochrome, while the cells remained viable. The ability to reconstitute holophytochromes in living cells provides a powerful tool for structure–function analysis of this photoreceptor family in nonplant cell systems and has also led to the development of a new family of apophytochrome-based fluorescent probes called phytofluors (43).

4.4. Phytofluors: A New Class of Fluorescent Protein Probe

Phytofluors are intensely orange fluorescent adducts that are formed spontaneously upon co-incubation of apophytochrome with PEB (see Figure 1B and Reference 43). The intense molar absorption coefficient of PEB-apophytochrome adducts and its spectrofluorometric properties (i.e., photostability, very sharp excitation, and emission maxima at 576 and 586 nm, respectively) make phytofluors ideal candidates as in vivo fluorescent protein tags. PEB can be fed to organisms that are expressing an apophytochrome gene. PEB is taken up by plant cells and autocatalytically assembles with apophytochrome to produce a fluorescent adduct that can be detected by techniques such as confocal microscopy (43). No central method has been developed for phytofluors that is broadly applicable to all possible uses of this novel fluorescent label. There are 2 requirements for the use of phytofluors: (*i*) ligation-competent apophytochrome, and (*ii*) PEB. Transgenic expression of various phytochromes in a variety of bacteria, yeast, and mammalian cells has been demonstrated. The key limitations for the application of this technique at present are PEB uptake and catabolism by different types of cells and the commercial availability of free PEB. In all the examples of the phytofluor technology, PEB has been supplied exogenously in a buffered Me_2SO solution. One goal for further development of this technology is the coexpression of bilin biosynthetic enzymes with apophytochrome. Research toward this end is ongoing in a number of laboratories.

ACKNOWLEDGMENTS

We thank Beronda Montgomery, Nicole Frankenberg, and Jihong Wang for helpful comments regarding this manuscript. We also gratefully acknowledge the support from the United States Department of Agriculture Competitive Research Grant No. AMD-9801768 to J.C.L.

ABBREVIATIONS

ALA, 5-aminolevulinic acid; BV, biliverdin IXα; HKRD, histidine kinase-related domain; Me_2SO, dimethyl sulfoxide; PCB, phycocyanobilin; PEB, phycoerythrobilin; PΦB, phytochromobilin; Pr, red light absorbing form of phytochrome; Pfr, far-red light absorbing form of phytochrome; PRD, PAS-related domain.

REFERENCES

1. **Auche, A.** 1908. Comptes Rendus De Societe Biologie *64*:297-298.
2. **Berkelman, T.R. and J.C. Lagarias.** 1986. Visualization of bilin-linked peptides and proteins in polyacrylamide gels. Anal. Biochem. *156*:194-201.
3. **Bhoo, S.H., T. Hirano, H.Y. Jeong, J.G. Lee, M. Furuya, and P.S. Song.** 1997. Phytochrome photochromism probed by site-directed mutations and chromophore esterification. J. Am. Chem. Soc. *119*:11717-11718.
4. **Boylan, M. and P. Quail.** 1989. Oat Phytochrome is biologically active in transgenic tomatoes. Plant Cell *1*:765-773.
5. **Clack, T., S. Mathews, and R.A. Sharrock.** 1994. The phytochrome apoprotein family in *Arabidopsis* is encoded by five genes—the sequences and expression of PhyD and PhyE. Plant Mol. Biol. *25*:413-427.
6. **Cornejo, J. and S.I. Beale.** 1997. Phycobilin biosynthetic reactions in extracts of cyanobacteria. Photosynth. Res. *51*:223-230.
7. **Cornejo, J., S.I. Beale, M.J. Terry, and J.C. Lagarias.** 1992. Phytochrome assembly—the structure and biological activity of 2(R),3(E)-phytochromobilin derived from phycobiliproteins. J. Biol. Chem. *267*:14790-14798.
8. **Davis, S.J., A.V. Vener, and R.D. Vierstra.** 1999. Bacteriophytochromes: phytochrome-like photoreceptors from nonphotosynthetic eubacteria. Science *286*:2517-2520.
9. **Deforce, L., M. Furuya, and P.S. Song.** 1993. Mutational analysis of the pea phytochrome a chromophore pocket—chromophore assembly with apophytochrome A and photoreversibility. Biochemistry *32*:14165-14172.
10. **Deforce, L., K.I. Tomizawa, N. Ito, D. Farrens, P.S. Song, and M. Furuya.** 1991. *In vitro* assembly of apophytochrome and apophytochrome deletion mutants expressed in yeast with phycocyanobilin. Proc. Natl. Acad. Sci. USA *88*:10392-10396.
11. **Eichenberg, K., T. Kunkel, T. Kretsch, V. Speth, and E. Schafer.** 1999. In vivo characterization of chimeric phytochromes in yeast. J. Biol. Chem. *274*:354-359.
12. **Elich, T.D. and J.C. Lagarias.** 1988. 4-Amino-5-hexynoic acid—a potent inhibitor of tetrapyrrole biosynthesis in plants. Plant Physiol. *88*:747-751.

13.**Elich, T.D. and J.C. Lagarias.** 1989. Formation of a photoreversible phycocyanobilin-apophytochrome adduct *in vitro* . J. Biol. Chem. *264*:12902-12908.

14.**Elich, T.D. and J.C. Lagarias.** 1987. Phytochrome chromophore biosynthesis. Both 5-aminolevulinic acid and biliverdin overcome inhibition by gabaculine in etiolated *Avena sativa* L. seedlings. Plant Physiol. *84*:304-310.

15.**Elich, T.D., A.F. McDonagh, L.A. Palma, and J.C. Lagarias.** 1989. Phytochrome chromophore biosynthesis. Treatment of tetrapyrrole-deficient *Avena* explants with natural and non-natural bilatrienes leads to formation of spectrally active holoproteins. J. Biol. Chem. *264*:183-189.

16.**Fodor, S.P.A., J.C. Lagarias, and R.A. Mathies.** 1990. Resonance Raman analysis of the Pr and Pfr forms of phytochrome. Biochemistry *29*:11141-11146.

17.**Furuya, M.** 1993. Phytochromes—their molecular species, gene families, and functions. Ann. Rev. Plant Physiol. Plant Mol. Biol. *44*:617-645.

18.**Grimm, R., G.K. Donaldson, SM Vandervies, E. Schafer, and A.A. Gatenby.** 1993. Chaperonin-mediated reconstitution of the phytochrome photoreceptor. J. Biol. Chem. *268*:5220-5226.

19.**Gross, J., M. Seyfried, L. Fukshansky, and E. Schaefer.** 1984. In vivo spectrophotometry, p. 131-158. *In* H. Smith and M.G. Holmes (Eds.), Techniques in Photomorphogenesis. Academic Press, New York.

20.**Hill, C., W. Gartner, P. Towner, S.E. Braslavsky, and K. Schaffner.** 1994. Expression of phytochrome apoprotein from *Avena sativa* in *Escherichia coli* and formation of photoactive chromoproteins by assembly with phycocyanobilin. Eur. J. Biochem. *223*:69-77.

21.**Hughes, J., T. Lamparter, F. Mittmann, E. Hartmann, W. Gartner, A. Wilde, and T. Borner.** 1997. A prokaryotic phytochrome. Nature *386*:663-663.

22.**Jones, A.M., C.D. Allen, G. Gardner, and P.H. Quail.** 1986. Synthesis of phytochrome apoprotein and chromophore are not coupled obligatorily. Plant Physiol. *81*:1014-1016.

23.**Jones, A.M., and M.D. Edgerton.** 1994. The anatomy of phytochrome, a unique photoreceptor in plants. Sem. Cell Biol. *5*:295-302.

24.**Jones, A.M., R.D. Vierstra, S.M. Daniels, and P.H. Quail.** 1985. The role of separate domains in the structure of phytochrome from etiolated *Avena sativa* L. Planta. *164*:505-506.

25.**Kehoe, D.M. and A.R. Grossman.** 1996. Similarity of a chromatic adaptation sensor to phytochrome and ethylene receptors. Science *273*:1409-1412.

26.**Kendrick, R.E. and G.H.M. Kronenberg (Eds.).** 1994. Photomorphogenesis in Plants, 2nd ed. Martinus Nijhoff Publishers, Dordrecht, The Netherlands.

27.**Kolukisaoglu, H.U., S. Marx, C. Wiegmann, S. Hanelt, and H.A.W. Schneider-Poetsch.** 1995. Divergence of the phytochrome gene family predates angiosperm evolution and suggests that Selaginella and Equisetum arose prior to Psilotum. J. Mol. Evol. *41*:329-337.

28.**Kunkel, T., V. Speth, C. Buche, and E. Schafer.** 1995. In vivo characterization of phytochrome-phycocyanobilin adducts in yeast. J. Biol. Chem. *270*:20193-20200.

29.**Kunkel, T., K. Tomizawa, R. Kern, M. Furuya, N.H. Chua, and E. Schafer.** 1993. *In vitro* formation of a photoreversible adduct of phycocyanobilin and tobacco apophytochrome B. Eur. J. Biochem. *215*:587-594.

30.**Lagarias, D.M., S.H. Wu, and J.C. Lagarias.** 1995. Atypical phytochrome gene structure in the green alga *mesotaenium caldariorum*. Plant Mol. Biol. *29*:1127-1142.

31.**Lagarias, J.C., J.M. Kelly, K.L. Cyr, and W.O. Smith, Jr.** 1987. Comparative photochemical analysis of highly purified 124 kilodalton oat and rye phytochromes in vitro. Photochem. Photobiol. *46*:5-13.

32.**Lagarias, J.C. and D.M. Lagarias.** 1989. Self assembly of synthetic phytochrome holoprotein in vitro. Proc. Natl. Acad. Sci. USA *86*:5778-5780.

33.**Lagarias, J.C. and F.M. Mercurio.** 1985. Structure function studies on phytochrome. Identification of light-induced conformational changes in 124-kDa *Avena* phytochrome in vitro. J. Biol. Chem. *260*:2415-2423.

34.**Lagarias, J.C. and H. Rapoport.** 1980. Chromopeptides from phytochrome. The structure and linkage of the Pr form of the phytochrome chromophore. J. Am. Chem. Soc. *102*:4821-4828.

35.**Lamparter, T., F Mittmann, W. Gartner, T. Borner, E. Hartmann, and J. Hughes.** 1997. Characterization of recombinant phytochrome from the cyanobacterium Synechocystis. Proc. Natl. Acad. Sci. USA *94*:11792-11797.

36.**Lapko, V.N. and P.S. Song.** 1995. A simple and improved method of isolation and purification for native oat phytochrome. Photochem. Photobiol. *62*:194-198.

37.**Li, L. and JC Lagarias.** 1992. Phytochrome assembly—defining chromophore structural requirements for covalent attachment and photoreversibility. J. Biol. Chem. *267*:19204-19210.

38.**Li, L. and J.C. Lagarias.** 1994. Phytochrome assembly in living cells of the yeast *Saccharomyces cerevisiae*. Proc. Natl. Acad. Sci. USA *91*:12535-12539.

39.**Li, L., J.T. Murphy, and J.C. Lagarias.** 1995. Continuous fluorescence assay of phytochrome assembly *in vitro*. Biochem. *34*:7923-7930.

40.**Lindner, I., B. Knipp, S.E. Braslavsky, W. Gartner, and K. Schaffner.** 1998. A novel chromophore selectively modifies the spectral properties of one of the two stable states of the plant photoreceptor phytochrome. Angew. Chem., Int. Ed. *37*:1843-1846.

41.**Litts, J.C., J.M. Kelly, and J.C. Lagarias.** 1983. Structure-function studies on phytochrome. Preliminary characterization of highly purified phytochrome from *Avena sativa* enriched in the 124-kilodalton species. J. Biol. Chem. *258*:11025-11031.

42.**McDowell, M.T. and J.C. Lagarias.** Purification and properties of phytochromobilin synthase from etiolated oat seedlings. Plant Physiol. (In press).

43.**Murphy, J.T. and J.C. Lagarias.** 1997. The phytofluors: a new class of fluorescent protein probes. Curr. Biol. *7*:870-876.

44.**Murphy, J.T. and J.C. Lagarias.** 1997. Purification and characterization of recombinant affinity peptide-tagged oat phytochrome A. Photochem. Photobiol. *65*:750-758.

45. **Pratt, L.H., M.M. Cordonnier-Pratt, P.M. Kelmenson, G.I. Lazarova, T. Kubota, and R.M. Alba.** 1997. The phytochrome gene family in tomato (*Solanum lycopersicum* L). Plant Cell Environ. *20*:672-677.
46. **Pratt, L.H., J.E. Wampler, and E.S. Rich, Jr.** 1984. An automated dual-wavelength spectrophotometer optimized for phytochrome assay. Anal. Instrum. *13*:269-287.
47. **Quail, P.H., M.T. Boylan, B.M. Parks, T.W. Short, Y. Xu, and D. Wagner.** 1995. Phytochromes: photosensory perception and signal transduction. Science *268*:675-680.
48. **Raps, S.** 1990. Differentiation between phycobiliprotein and colorless linker polypeptides by fluorescence in the presence of $ZnSO_4$. Plant Physiol. *92*:358-362.
49. **Remberg, A., P. Schmidt, S.E. Braslavsky, W. Gartner, and K. Schaffner.** 1999. Differential effects of mutations in the chromophore pocket of recombinant phytochrome on chromoprotein assembly and Pr-to-Pfr photoconversion. Eur. J. Biochem. *266*:201-208.
50. **Rudiger, W., T. Brandlmeier, I. Blos, A. Gossauer, and J.P. Weller.** 1980. Isolation of the phytochrome chromophore. The cleavage reaction with hydrogen bromide. Z. Naturforsch. *35*:763-769.
51. **Rudiger, W. and F. Lopez-Figueroa.** 1992. Photoreceptors in algae. Photochem. Photobiol. *55*:949-954.
52. **Rudiger, W., F. Thummler, E. Cmiel, and S. Schneider.** 1983. Chromophore structure of the physiologically active form (Pfr) of phytochrome. Proc. Natl. Acad. Sci. USA *80*:6244-6248.
53. **Sage, L.C.** 1992. Pigment of the Imagination: A History of Phytochrome Research. Academic Press, San Diego.
54. **Schiff, J.A.** 1972. A green safelight for the study of chloroplast development and other photomorphogenetic phenomena. Methods Enzymol. *24B*:321-322.
55. **Schmidt, P., U.H. Westphal, K. Worm, S. Braslavsky, W. Gartner, and K. Schaffner.** 1996. Chromophore-protein interaction controls the complexity of the phytochrome photocycle. J. Photochem. Photobiol. B. Biol. *34*:73-77.
56. **Schneider-Poetsch, H.A.W., B. Braun, S. Marx, and A. Schaumburg.** 1991. Phytochromes and bacterial sensor proteins are related by structural and functional homologies—hypothesis on phytochrome-mediated signal-transduction. FEBS Lett. *281*:245-249.
57. **Schneider-Poetsch, H.A.W., S. Marx, H.U. Kolukisaoglu, S. Hanelt, and B. Braun.** 1994. Phytochrome evolution—phytochrome genes in ferns and mosses. Physiol. Plant. *91*:241-250.
58. **Song, P.S., M.H. Park, and M. Furuya.** 1997. Chromophore: apoprotein interactions in phytochrome A. Plant Cell Environ. *20*:707-712.
59. **Taylor, B.L. and I.B. Zhulin.** 1999. PAS domains: internal sensors of oxygen, redox potential, and light. Microbiol. Mol. Biol. Rev. *63*:479-506.
60. **Terry, M.J. and J.C. Lagarias.** 1991. Holophytochrome assembly—coupled assay for phytochromobilin synthase *in organello*. J. Biol. Chem. *266*:22215-22221.
61. **Terry, M.J., M.D. Maines, and J.C. Lagarias.** 1993. Inactivation of phytochrome-chromophore and phycobiliprotein-chromophore precursors by rat liver biliverdin reductase. J. Biol. Chem. *268*:26099-26106.
62. **Tomizawa, K., J. Stockhaus, N.H. Chua, and M. Furuya.** 1995. Spectrophotometric and molecular properties of mutated rice phytochrome A. Plant Cell Physiol. *36*:511-516.
63. **Vierstra, R.D.** 1993. Illuminating phytochrome functions. Plant Physiol. *103*:679-684.
64. **Wada, M., T. Kanegae, K. Nozue, and S. Fukuda.** 1997. Cryptogam phytochromes. Plant Cell Environ. *20*:685-690.
65. **Wahleithner, J.A., L. Li, and J.C. Lagarias.** 1991. Expression and assembly of spectrally active recombinant holophytochrome. Proc. Natl. Acad. Sci. USA *88*:10387-10391.
66. **Wilde, A., Y. Churin, H. Schubert, and T. Borner.** 1997. Disruption of a *Synechocystis* sp. PCC 6803 gene with partial similarity to phytochrome genes alters growth under changing light qualities. FEBS Lett. *406*:89-92.
67. **Wu, S.H. and J.C. Lagarias.** 1996. The methylotrophic yeast synthesizes a functionally active chromophore precursor of the plant photoreceptor phytochrome. Proc. Natl. Acad. Sci. USA *93*:8989-8994.
68. **Wu, S.H., M.T. McDowell, and J.C. Lagarias.** 1997. Phycocyanobilin is the natural chromophore precursor of phytochrome from the green alga *Mesotaenium caldariorum*. J. Biol. Chem. *272*:25700-25705.
69. **Yeh, K.C. and J.C. Lagarias.** 1998. Eukaryotic phytochromes: light-regulated serine/threonine protein kinases with histidine kinase ancestry. Proc. Natl. Acad. Sci. USA *95*:13976-13981.
70. **Yeh, K.C., S.H. Wu, J.T. Murphy, and J.C. Lagarias.** 1997. A cyanobacterial phytochrome two-component light sensory system. Science *277*:1505-1508.

14 Analysis and Reconstitution of Phycobiliproteins: Methods for the Characterization of Bilin Attachment Reactions

Wendy M. Schluchter[1] and Donald A. Bryant[2]
[1]Department of Biological Sciences, University of New Orleans, New Orleans, LA, and [2]Department of Biochemistry and Molecular Biology, The Pennsylvania State University, University Park, PA, USA

1. INTRODUCTION

Phycobiliproteins are a homologous family of light-harvesting accessory proteins present in cyanobacteria (25,51), red algae (25), cryptomonads (36,52), and some species of prochlorophytes (41,48). The blue, violet, red, or yellow colors of the phycobiliproteins are due to linear tetrapyrrole chromophores called bilins that are covalently attached at cysteine residues (25). These water-soluble proteins are composed of α and β subunits. The $\alpha\beta$ monomers form $(\alpha\beta)_3$ trimers which further stack into $(\alpha\beta)_6$ hexamers. These disc-shaped trimers and hexamers can be stabilized or organized into larger structures by linker proteins. Through the association of several types of phycobiliproteins with these linker proteins [69), the large light-harvesting complex called the phycobilisome is formed (51,63). Cryptomonad phycobiliproteins have a different composition and structural organization and will not be discussed further in this chapter (for reviews on cryptomonad phycobiliproteins, see References 36, 52, 53, and 73).

There are three major types of phycobiliproteins, each having unique spectroscopic properties: (*i*) phycoerythrins (PEs; λ_{max} approximately 565 nm); (*ii*) phycocyanins (PCs; λ_{max} approximately 620 nm); and (*iii*) allophycocyanins (APs; λ_{max} approximately 650 nm) (27). These three proteins differ in both the numbers and the types of bilins that are associated with each $\alpha\beta$ monomer. Cyanobacterial phycobiliproteins are formed by the interaction of the apoprotein subunits with one or more of four different types of isomeric bilins: phycourobilin (PUB), phycoerythrobilin (PEB), phycobiliviolin (PXB), and phycocyanobilin (PCB) (see Figure 1 and Reference 27). Although cyanobacteria

Heme, Chlorophyll, and Bilins: ***Methods and Protocols***
Edited by A.G. Smith and M. Witty

have been shown to contain proteins similar to eukaryotic phytochrome (44,47, 64,77,78), definitive evidence for the occurrence of phytochromobilin (PφB) in cyanobacteria has not yet been obtained.

The phycobilisome is composed of an AP core that is surrounded by rods containing PC that radiate outwards from this core. In some organisms, PE is also a component of these peripheral rods and is found distal to the PC (28). In other cyanobacteria, phycoerythrocyanin (PEC) is present in small to moderate amounts under low light-intensity conditions and is likewise found distal to phycocyanin in the peripheral rods (9,12).

The α and β subunits that compose each phycobiliprotein share amino acid sequence similarity to each other and to the subunits of other phycobiliproteins, and this observation supports the hypothesis that this family of proteins evolved through gene duplication (63). Indeed, it is the β subunit of phycoerythrin that is thought to be the ancestral phycobiliprotein from which all others evolved (36, 71). The three-dimensional structures of at least one member of each of the major spectroscopic classes of phycobiliproteins have been determined (8,13,18,19,23,24, 59–62,65), and these structures show that the amino acid similarity translates into remarkable structural conservation (5). The subunit structure for this family of proteins resembles that of members of the globin family with a predominance of α helices and the complete absence of β-pleated sheets (61). The unique spectroscopic properties of each phycobiliprotein are believed to be due to the type(s) of bilin(s) attached, to the immediate electrical charge and polarity of the environment of the chromophore, and to the way by which the phycobiliprotein subunits hold the bilins in a stretched conformation (26,35,63). Linker proteins also affect the spectroscopic properties of the phycobiliproteins (26,35,63,65). Recently, the X-ray structure of the AP trimers carrying the core linker polypeptide was solved (65). This structure shows that this AP linker (and probably the related rod linker that interacts with phycocyanin) modifies the spectroscopic properties of the phycobiliprotein with which it is associated by causing slight shifts in bilin conformation as well as by bringing two bilins closer together within the trimer. The linker protein is located between two of the three β-AP subunits in the trimer and directly interacts with the PCBs of these two subunits (65). Approximately half of the surface of the linker protein is located within the cavity of the trimer (65).

In some strains of marine cyanobacteria, three different bilins may occur on their phycobiliproteins (55), whereas in other strains, such as *Synechococcus* sp. PCC 7002 and *Synechocystis* sp. PCC 6803, only PCB is present. Even in those strains which only contain PCB, two different stereoisomers occur on the C-phycocyanin β subunit at the C3′ of the bilin: the *R* configuration is found for the PCB attached at cysteine β-82 and the *S* configuration is found for the PCB attached at cysteine β-153 (62). The biochemical basis for how the biosynthesis of the phycobiliproteins is controlled, such that the correct bilin is attached to the proper cysteine residue with the appropriate stereochemistry, is a fascinating but incompletely understood process. Since autocatalytic reactions with apophycobiliproteins and free bilins have yielded nonnatural products (2,4,20), all evidence currently indicates that bilins are enzymatically attached to the appropriate apophycobiliprotein.

Phycobiliproteins have been studied for more than a century and a half now, and they have captured the imaginations of many scientists because of their brilliant colors. These proteins are relatively easy to isolate and purify because they comprise such a large proportion of the total protein in many cells. Much is known about their structure and function, but much less is known about the biosynthesis of the individual proteins and the assembly of the macromolecular phycobilisome. Most approaches toward understanding how these proteins are synthesized have been made in the attempts to reconstitute them. Most of these reconstitution studies have taken place in the last 10 years when recombinant DNA technology has allowed one to overproduce the apoproteins for such studies and to generate mutants. The majority of work done on phycobiliprotein reconstitution has been performed using cyanobacterial proteins. Therefore, this chapter will summarize some of the many methods that have been developed for analyzing and reconstituting phycobiliproteins from these organisms. However, it is hoped that this information will serve as a good starting point for researchers who are interested in studying the reconstitution and biosynthesis of phycobiliproteins from red algae, cyanobacteria, or cryptomonads. Also, since the last review on phycobiliprotein purification was written (27), a new method for the separation, characterization, and quantitation of phycobiliproteins utilizing reverse-phase high-performance liquid chromatography (HPLC) was developed (66). This method has been extensively used in the characterization of phycobiliproteins from newly discovered organisms (29) and from mutants defective in phycobiliprotein biosynthesis (42,68). The methods necessary for the reconstitution of phycobiliproteins are summarized below.

Nomenclature

The nomenclature for phycobiliproteins can be somewhat confusing and reflects, in part, historical developments in the study of phycobiliproteins. PCs and PEs were all originally given the prefix C- or R- to designate whether they were purified from cyanophytes (cyanobacteria) or rhodophytes (red algae). The designation B- was later introduced for a distinct type of PE from the red alga *Smithora naiadum,* which is a member of the order Bangiales (1,37). The three major types of Class I PE (those that contain five bilins per αβ monomer; see below) differ in their absorbance properties due to the types of

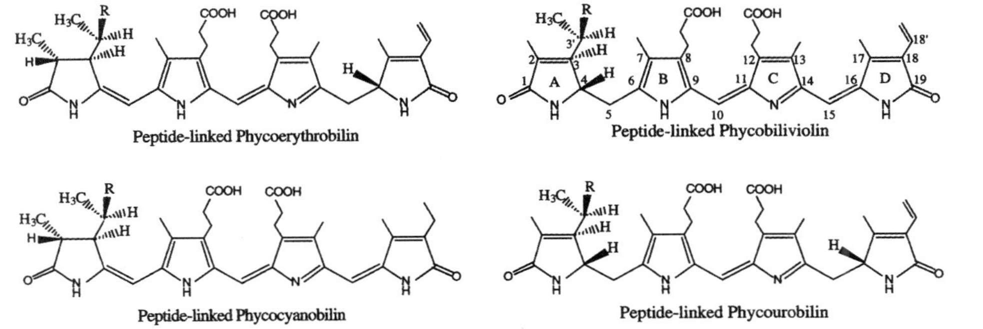

Figure 1. Structures of the four singly-linked peptide-linked bilins present in the phycobiliproteins of cyanobacteria. The numbering scheme for the carbon atoms is indicated.

bilins present on their αβ subunits. These proteins exhibit one, two, or three distinct peaks in the visible region of the spectrum and are called C-phycoerythrin (C-PE; containing only PEB), B-phycoerythrin (B-PE), and R-phycoerythrin (R-PE), respectively, regardless of the group from which they have been isolated (References 33, 37, and references therein). These designations seemed sufficient until marine unicellular cyanobacteria were shown to contain two forms of PE, PE I, and PE II, in the rods of their phycobilisomes (55, 67). PE I was less abundant than PE II. PE II has an extra bilin (PUB) on the α subunit (α-75) and contains both PEB and PUB. Thus far, only two *Synechococcus* strains (WH8020 and WH8103) have been shown to contain a PE with six chromophores per αβ monomer (55). Therefore, these two PEs are members of a new class of PE, dubbed Class II PE. However, PE I is more like other PEs that have been characterized, in that it contains only five bilins per αβ monomer and contains either only PEB or both PEB and PUB (55). Thus far, no red algal PE has been shown to be a member of Class II PE. B-PE, R-PE, and PE II complexes carry bilins on their associated linker protein, called γ (34, 45). A recently discovered red algal PE (from *Audouinella macrospora*) contains a PE with PCB, PEB, and PUB chromophores and is more like B- and R-PEs in that it contains 5 bilins per αβ monomer and has bilins present on its γ subunit (29). This PE is a Class I member, but is not by definition an R-PE, which have been shown to contain PUB and PEB that contribute to the three absorption peaks in the visible region.

Four major types of PC have been characterized (15,63), and all PC types contain PCB as the terminal acceptor bilin at cysteine β-82. C-PC contains PCB at all three cysteines (25,46,60). R-PC-I, present in some red algae including *Porphyridium cruentum*, contains PEB at cysteine β-155 and PCB at the other two positions (33). R-PC II, isolated from several unicellular marine cyanobacterial strains, contains PEB at cysteines α-84 and β-155 and PCB at cysteine β-82 (55,56). R-PC-III was isolated from *Synechococcus* sp. WH7805 and has a PCB:PEB ratio of 2:1, but differences in the absorption properties of this PC suggest that the chromophores are distributed differently than in R-PC-I (57). The fourth form of PC, R-PC-IV, was isolated from *Synechococcus* sp. WH8501 and was found to contain PUB attached at α-84 and PCB at the other two positions on the β subunit (67). Finally, PEC is structurally more similar to PCs than to PEs, but is found distal to PC in the phycobilisome rods of some cyanobacterial strains. PEC carries PXB at α-84 and PCB at both positions on the β subunit (9,12). Spectroscopic variants of AP (which contains one PCB on each subunit) have not yet been identified. Thus, the nomenclature for biliproteins devised previously has been rather haphazard and confusing. A new form of nomenclature has been suggested (51), but has not been widely used thus far.

2. HOLO-PHYCOBILIPROTEINS

2.1. Phycobiliprotein Purification

Phycobiliproteins may exist in different aggregation states depending upon the individual type of biliprotein, the organism from which it was isolated, the composition of the solution containing it (pH, ionic strength), and such factors as temperature and protein concentration. The purification of individual phycobiliproteins has been summarized previously (27). A few minor improvements have been introduced using fast protein, peptide, and polynucleotide liquid chromatography (FPLC) (Mono Q®; Amersham Pharmacia

Biotech, Piscataway, NJ, USA) (66), but for the most part, the conventional chromatographic methods are still widely used today. Therefore, this chapter will primarily summarize the methods for separation and purification of individual α and β subunits.

2.2. Storage and Recovery of Purified Phycobiliproteins

Phycobiliproteins are very stable when stored in phosphate buffer at pH 7.0 in the presence of a reducing agent [1–5 mM β-mercaptoethanol or dithiothreitol (DTT)] and sodium azide (1 mM) in the dark at 4°C. For long-term storage, ammonium sulfate may be added to 65% saturation at 4°C. When sealed at 4°C in the dark, such slurries–precipitates can be stored indefinitely. The phycobiliproteins can be recovered by centrifugation at 27 000× *g* for 15 minutes or by centrifugation in a microcentrifuge at 13 000× *g* for 15 minutes. The phycobiliprotein pellet should be resuspended in 5 mM phosphate buffer, pH 7.0, 1 mM β-mercaptoethanol, and dialyzed against the same buffer at 4°C prior to use.

2.3. Concentration Determination

Because the absorption properties of the phycobiliproteins are highly dependent on the aggregation state, pH, ionic strength, and protein concentration (26), the most reliable method to determine the concentration of phycobiliprotein solutions is to measure the absorption spectrum of the peptide-bound bilins by dissolving an aliquot of the protein in 8 M urea, pH 1.9, or in 10 mM TFA (trifluoroacetic acid); its concentration can then be determined by using the extinction coefficients given in Table 1 (Reference 27 and references therein). For phycobiliproteins that contain 2 or more different bilins per subunit, the contributions of each bilin type at various wavelengths must be considered. The contributions of the various chromophores at different wavelengths are listed in Table 1. In some cases, the concentration of the phycobiliprotein sample may be limiting; for example, this is often the case when isolating phycoerythrins from field samples of red algae. The spectra for PEs in 20% acetic acid (vol/vol) have been determined to be identical to those for the same protein dissolved in 8 M urea, pH 3.0 (37). This was also found to be true for AP and PC (A.N. Glazer, personal communication).

2.4. Purification of Individual Subunits by Conventional Chromatographic Methods

The first method for the separation and purification of phycobiliprotein subunits was developed by Glazer and Fang and was based upon methods used for the separation of the subunits of hemoglobin (30,31). All methods described thus far are performed under denaturing and acidic conditions which limit oxidation and other side reactions that can modify the bilins. Each procedure described will include the cyanobacterial source for the phycobiliprotein. Most of these conditions have been shown to work successfully for the separation of phycobiliproteins from a wide variety of sources, but some optimization of the method may be required if the user is attempting to adapt the method to the purification of subunits from a different phycobiliprotein (see Procedure 1). Each subunit, when renatured without its partner, is much less stable and tends to aggregate over time when in solution. Procedures 2 and 3 describe renaturation of phycocyanin subunits. Both renaturation procedures, followed by the last diethylaminoethyl (DEAE) chromatography step, have a recovery rate for renatured protein of approximately 25% (32). In contrast, when the α and β subunits are renatured together in a 1:1 molar ratio, the yield of reconstituted phycocyanin is between 40%–60% (31). The absorption spectra of the renatured α and β subunits purified from the PC of *Synechococcus* sp. PCC 6301 are shown in Figure 2.

❖ Procedure 1. Separation of PC Subunits

1. Prepare 25 to 50 mg of *Anabaena* sp. PCC 6411 PC in 100 mM Na-phosphate buffer, pH 7.0 2. See section 2.1.
2. Adjust pH to 3.0 by addition of glacial acetic acid with reductant present (10 mM β-mercaptoethanol).
3. Apply this mixture to a column of Bio-Rex®70 resin (weak cation exchange, minus 400 mesh, 2.2 × 13 cm; Bio-Rad Laboratories, Hercules, CA, USA) that has previously been equilibrated with 0.4% acetic acid, pH 3.0 (31). The PC subunits should adsorb to the top of the column.
4. Wash the column extensively with 2 M urea, 10 mM β-mercaptoethanol, pH 3.0.
5. Elute PC subunit by a stepwise increase in urea concentration (approximately one column volume each of 4.0 M, 6.0 M, 8.0 M, and 9.0 M urea, pH 3.0, and 10 mM β-mercaptoethanol). The α subunit typically elutes with 8.0 M urea, while elution of the β subunit requires 9.0 M urea.
6. For long-term storage, subunits can be dialyzed extensively against water, lyophilized, and subsequently renatured using either Procedure 2 or 3 below.

❖ Procedure 2. Renaturing Fresh Phycocyanin Subunits (31)

1. Dilute PC subunits from procedure 1 to 0.1 to 0.4 mg/mL protein with 8 M

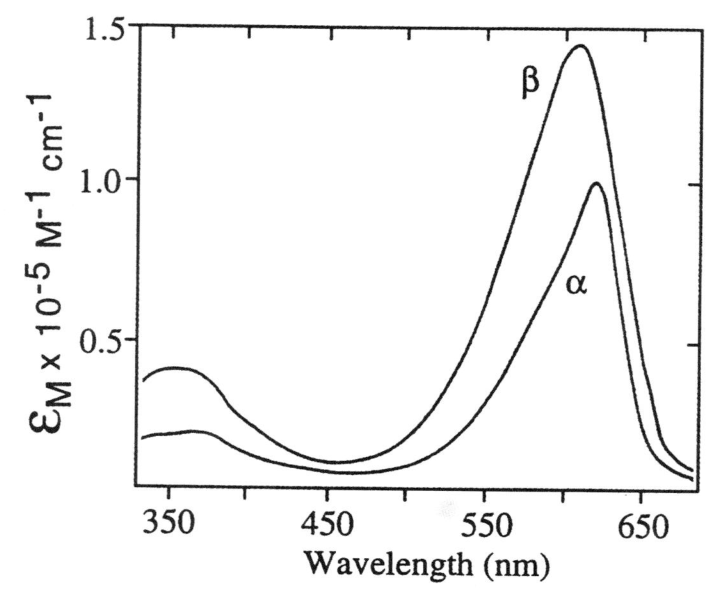

Figure 2 Absorption spectra of the individual renatured α and β subunits of phycocyanin purified from *Synechococcus* sp. PCC 6301. Spectra were determined in 50 mM Na phosphate, 1 mM β-mercaptoethanol, pH 7.0, and at α protein concentration of 1.05×10^{-5} M for the α-subunit and 1.11×10^{-5} M for the β subunit. The λ_{max} was 620 nm for the α subunit and 608 nm for the β subunit. This figure was modified with permission from Reference 32.

Table 1. Millimolar Extinction Coefficients of Peptide-Linked Bilins[a]

Bilin	ε^{495}	ε^{495}	ε^{495}	ε^{495}
PUB	94.0	0	0	0
PEB	18.3	53.7	8.5	0
PXB	6.8	28.4	38.6	0
PCB	1.45	6.0	16.2	35.4

[a]Extinction coefficients are $mM^{-1}cm^{-1}$ at the wavelength indicated. The absorption spectra of peptide-linked bilins were measured in 10 mM TFA or 8 M urea, pH 1.9. These values are taken from References 7, 33, and 44.

urea, 5 mM β-mercaptoethanol, pH 8.0.

2. Dialyzed against 3 M urea, 5 mM β-mercaptoethanol, 6 mM Na-phosphate, pH 6.7, at 4°C.
3. Dialyze against two changes of 10 mM Na-phosphate, 5 mM β-mercaptoethanol, pH 6.5, at 4°C.
4. Dialyze against 5 mM Na-phosphate, pH 7.0, at 4°C.

❖ Procedure 3. Renaturing Lyophilized Phycocyanin Subunits (32)

1. Dissolve lyophilized PC in 5 mM Na-phosphate, 1 mM β-mercaptoethanol, pH 7.0, and allow to stand overnight at 4°C.
2. Remove insoluble material by centrifugation.
3. Loaded the protein solution onto a DEAE cellulose DE-52 column (0.5 × 5 cm; Whatman, Clifton, NJ, USA) equilibrated in 5 mM Na-phosphate, 1 mM β-mercaptoethanol, pH 7.0.
4. Subunits can be eluted immediately with 200 mM Na-phosphate, 1 mM β-mercaptoethanol, pH 7.0. A large proportion of blue nonfluorescent material is typically retained at the top of this column. This extra step ensures that properly folded subunits are recovered.

Sedimentation analyses of α and β subunits at neutral pH indicate that each purified subunit has a tendency to dimerize at higher protein concentrations (greater than 0.2 mg/mL). Protein concentrations can be calculated from the absorption at 662.5 nm in 8 M urea, pH 1.9, using the molar extinction coefficient values of 33.2 $mM^{-1}cm^{-1}$ for the α subunit and 69.5 $mM^{-1}cm^{-1}$ for the β subunit (32).

The first separation method for AP subunits was developed by Gysi and Zuber for the protein from the thermophilic cyanobacterium *Mastigocladus laminosus* (40) and is described in Procedure 4.

❖ Procedure 4. Separation of Allophycocyanin from *Mastigocladus laminosus*

1. Prepare 10 mg of allophycocyanin in 20 mM phosphate buffer, 8.0 M urea, 10 mM β-mercaptoethanol, pH 8.0, and allowed to incubate for 2.5 hours at 37°C. See section 2.1.
2. Apply this mixture to a DEAE Sephadex® A-50 column (2.5 × 45 cm; Amersham Pharmacia Biotech) at room temperature, equilibrated in the same buffer.
3. Elute the AP subunits with a linear gradient (400 mL) of KCl (50 to 300 mM) in 20 mM phosphate, 8.0 M urea, 10 mM β-mercaptoethanol, pH 8.0. The β subunit of AP elutes first followed by the α subunit.
4. Fractions containing these purified subunits should be pooled and dialyzed exhaustively against 20 mM Na-phosphate buffer, pH 8.0.

This procedure was slightly modified for the purification of AP subunits from *Syne-*

chococcus sp. PCC 6301 and *Synechocystis* sp.; (22663; ATCC, Manassas, VA, USA) also called *Microcystis aeruginosa* (14) (Procedure 5).

❖ Procedure 5. Separation of Allophycocyanin Subunits from *Synechococcus* sp. PCC 6301 and *Synechocystis* sp.

1. Dissolve 325 mg purified and lyophilized AP in 50 mL of 10 mM K-phosphate, 8 M urea, 10 mM β-mercaptoethanol, pH 8.0, and equilibrate for 1 hour at room temperature. See section 2.1.
2. Load the material onto a DEAE Sephadex A-50 column (3.5 × 12 cm) and wash with equilibration buffer.
3. Use 200 mL of equilibration buffer plus 50 mM KCl to elute the elute the β subunit of AP.
4. Residual β subunit is eluted by repeated washes with 150 mL equilibration buffer plus 80 mM KCl.
5. Use equilibration buffer plus 180 mM KCl to elute the α subunit of AP.
6. Pool fractions of each subunit from steps 3 and 5 for dialysis against 25 mM ammonium acetate, pH 6.8, and concentration by ultrafiltration using an Amicon® cell with a 10 000 MWCO membrane (Millipore, Bedford, MA, USA).

A method similar to the one developed for the separation of the PC subunits was successfully used in the separation of the α, β, and γ subunits of phycoerythrin from *P. cruentum* (34). The only significant difference was in the development of the column. The γ subunit was eluted with 7.4 M urea, the α subunit with 8.0 M urea, and the β subunit with 9.0 M urea. Similar conditions were used to separate the subunits of phycoerythrin II (PE II) from the cyanobacterium *Gloeobacter violaceus* (10). The Bio-Rex 70 column (1.5 × 15 cm) with the PE subunits adsorbed was washed with 15 mL of 2.0 M urea, 30 mL of 4.0 M urea, and 50 mL of 6.0 M urea before development with a linear gradient of 6.0 to 10.0 M urea, pH 3.0 (20 mL total volume). The α subunit eluted first followed by the β subunit. Subunits were renatured by exhaustive dialysis against 50 mM K-phosphate buffer at pH 7.0 at room temperature.

Separation of the subunits of *Anabaena variabilis* PEC was first demonstrated by Bryant et al. using a modification of the method developed for the separation of PC subunits described above (12). The Bio-Rex 70 column (3.9 × 51 cm) was subjected to incremental step gradients of acidic urea as described previously, followed by elution of the α subunit by addition of 7.4 M urea, pH 3.0. Once the elution of the α subunit was complete, elution of the β subunit was accomplished by addition of 9.0 M urea, pH 3.0. Subunits were dialyzed against 50 mM ammonium acetate, pH 6.8.

2.5. Purification of Phycobiliproteins by HPLC

In 1987, HPLC was used to verify the purity of PC and AP preparations from *M. aeruginosa* (58); however, the method also showed that the AP and PC subunits could be separated on a C_{18} reverse-phase column. In 1990, Swanson and Glazer introduced a method for separation of phycobiliprotein subunits using C_4 reverse-phase HPLC (66). These HPLC methods have several advantages over the conventional chromatographic methods. They are more rapid and require much less starting material. When used in conjunction with a photodiode array detector, these methods also give immediate spectroscopic information about bilin content and subunit stoichiometry. The method of Swanson and Glazer has also successfully been used to separate phycobiliproteins obtained directly from purified phycobilisomes, giving quantitative information regarding phycobiliprotein stoichiometry and content in these mixtures (39). When both HPLC methods were compared, the method of Swanson and Glazer gave better resolution of phycobiliproteins isolated from *Arthrospira maxima* (38,39). The use of reverse-phase HPLC is clearly a better choice than conventional chromatographic procedures for determining stoichiometric information when the amount of starting material and speed are primary concerns.

The method of Swanson and Glazer uses a C_4 reverse-phase analytical column (250 × 10 mm) and a solvent system consisting of 0.1% TFA in water (Buffer A) and a 2:1 acetonitrile: isopropanol mixture (Buffer B). This purification procedure has been very successful in the separation and resolution of diverse types and mixtures of phycobiliproteins. The purified phycobiliprotein or phycobiliprotein mixture, typically 100 to 1500 μg in 200 to 500 μL in 5 mM Na-phosphate, pH 7.0, 1 mM β-mercaptoethanol is combined with an equal volume of 9.0 M urea, pH 2.0 (freshly prepared), and subjected to centrifuga-

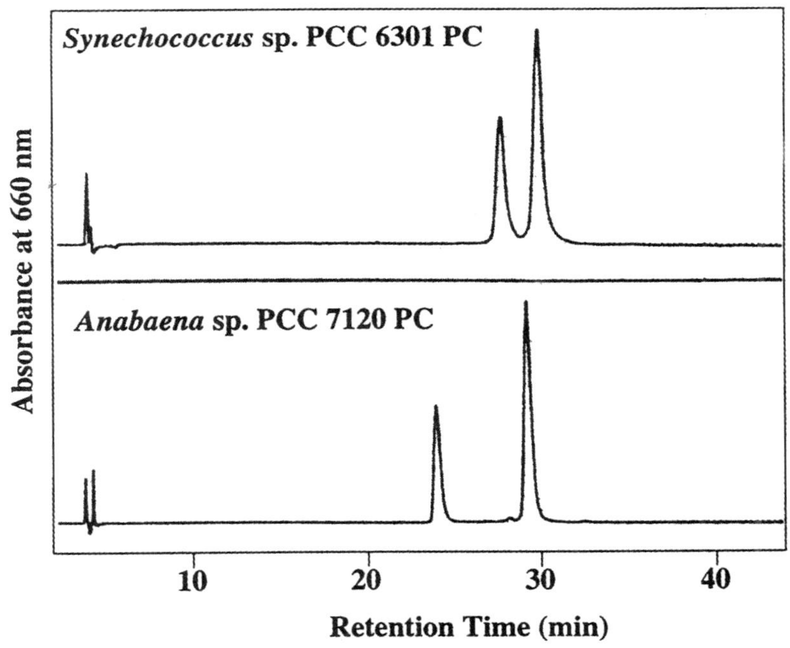

Figure 3. HPLC separation of cyanobacterial C-PC subunits. Purified PC from *Synechococcus* sp. PCC 6301 (top panel) or *Anabaena* sp. PCC 7120 (bottom panel) was separated on a C_4 reverse-phase HPLC column as described in the text. Elution of subunits was monitored at 660 nm in order to follow the absorbance of peptide-linked PCB. In each case, the α subunit elutes prior to the β subunit. This figure was modified with permission from Reference 66.

tion in a microcentrifuge for 5 minutes prior to injection on the column. A Hi-Pore® RP304 column (Bio-Rad Laboratories) equilibrated in 65% Buffer A and 35% Buffer B (1.5 mL/min) has typically been used. After injection of the sample, proteins are eluted with a linear gradient to 30% Buffer A and 70% Buffer B over 35 to 40 minutes depending upon the source of the phycobiliprotein (see Figure 3). With a few alterations of the elution gradient profile, this method has been successfully employed in the separation of a wide variety of phycobiliproteins from cyanobacteria, red algae, and cryptomonads (17,29,39,66,74). In fact, researchers have had success in the separation of phycobiliproteins from phycobilisome samples taken directly from sucrose gradients (after dialysis against 5 mM Na-phosphate, pH 7.0 followed by combination with an equal volume of 9.0 M urea, pH 1.9, prior to injection) (see Figure 4). Toole et al. combined the phycobilisomes taken directly from sucrose gradients (in sucrose–phosphate) with an equal volume of 8.4 M guanidine hydrochloride, pH 6.4 (followed by centrifugation), prior to loading on the C_4 column (Vydac/The Separations Group, Hesperia, CA, USA) using the gradient conditions described above (72). This method has also been successfully used to characterize the linker polypeptide and phycobiliprotein stoichiometry in phycobilisomes from *A. maxima* (38,39).

Some technical considerations to keep in mind for each separation include the need to use higher concentrations of urea to solubilize phycobiliprotein mixtures that may contain any given apophycobiliprotein. It has been observed that apophycobiliprotein subunits often do not bind as well as holo-subunits under these conditions, but that addition of urea to at least 6 M final concentration in the solution to be injected greatly increases the yield of apophycobiliprotein material (22). It is also very important to wash the column extensively between injections using a linear gradient to 100% Buffer B over 5 minutes followed by at least 5 to 10 minutes of washing the column with 100% Buffer B. The β subunits of phycobiliproteins are sometimes retained on the column, and these will usually be eluted by this treatment. If careful quantitation of a sample is required, it is wise to perform a blank injection between each run with samples in order to insure that the column is entirely free of residual phycobiliprotein subunits.

Preparative separation of phycobiliprotein subunits can be accomplished using this method in conjunction with a semi-preparative C_4 reverse-phase column (or by employing multiple runs on an analytical column). Subunits can be collected as they elute from the column, and the solvents can be removed by rotary evaporation. The aqueous subunits can then be diluted 2:9 with 9.0 M urea, pH 2.0, 10 mM β-mercaptoethanol, followed by dialysis against 50 mM Na-phosphate, pH 7.0 (22).

2.6. Methods for Analyzing the Quality of the Renatured Subunits

The best method to analyze the quality of renatured subunits is to compare the absorption spectrum of a dilute solution containing the subunit with the fluorescence excitation spectrum of the same solution. In order to obtain an accurate excitation spectrum, the absorbance at the long-wavelength maximum should be less than 0.05 OD so that reabsorption of emitted light will be minimized. If the two spectra differ significantly, then it is likely that the renatured subunit is not folded properly or that the chromophore(s) may have been chemically modified during purification and renaturation. If the majority of the protein has been oxidized, it is unlikely that the sample will be a good source of bilin in bilin transfer assays.

3. APOPHYCOBILIPROTEINS

In order to understand how phycobiliproteins are biosynthesized, one must have an effective assay system. One such system has successfully been developed and shown to be effective for the reconstitution of the α subunit of phycocyanin as described below. However, the overproduction of various apophycobiliproteins has been successfully accomplished, and this information is also described below.

3.1. Overproduction of Apophycobiliprotein Subunits

3.1.1. Apophycocyanin

The first successful overproduction of apophycobiliproteins was accomplished with the α and β subunits of phycocyanin (11). The *cpcBcpcA* genes encoding the β and α subunits, respectively, from the cyanobacterium *Synechococcus* sp. PCC 7002 were cloned into a vector and expressed in *Escherichia coli* using their native promoter (2,11). Both subunits were produced at a low level throughout growth of the *E. coli* culture. A lower level of expression of phycocyanin and allophycocyanin subunits throughout the growth of the culture typically seems to yield proteins that are properly folded. When the T7/pET vector system was used for the expression of the *cpcA* gene in BL21 DE3 pLysS cells, a significant proportion of apo-α-PE was present in inclusion bodies (W.M. Schluchter and A.N. Glazer, unpublished observations). Although the apo-α-PC subunit could be renatured from these inclusion bodies, the soluble subunit produced in *E. coli* expression cultures was always a better substrate for in vitro addition reactions than the product of these renaturation experiments (W.M. Schluchter and A.N. Glazer, unpublished results).

When the α and β subunits of PC are produced together, a high yield of αβ monomer is produced (2,11). After removal of unbroken cells and large cell membrane fragments by centrifugation (31 000× *g* for 30 min), apo-αβ-PC can be precipitated by addition of ammonium sulfate to 38% saturation. Following centrifugation at 18 000 × *g*, the pellet should be resuspended in a large volume of 50 mM Na-phosphate buffer, pH 7.0 (approximately three times the initial volume of the cell-free supernatant). This mixture should be immediately loaded on a DEAE cellulose DE-52 column, and the flow-through should be collected and pooled after rinsing with two column volumes of the phosphate buffer (2). The apo-αβ-PC subunits can be precipitated with ammonium sulfate added to 50% saturation. The pellet from this precipitation should be resuspended in a small volume of phosphate buffer with 2 mM β-mercaptoethanol. This mixture can be desalted and further purified by loading onto a gel filtration column (Sephadex G-100) run at room temperature. Fractions should be collected and monitored for purity by sodium dodecyl sulfate polyacrylamide gel electrophoresis (SDS-PAGE). These subunits should be stored under N_2 or degassed in Na-phosphate buffer at 4°C containing a reducing agent (10 mM β-mercaptoethanol or DTT) to prevent oxidation of cysteine residues.

3.1.2. Apophycoerythrin

Expression of *Calothrix* sp. PCC 7601 *cpeAcpeB* genes in *E. coli* resulted in the production of large amounts of insoluble apo-αβ-PE subunits, which were found in inclusion bodies (20). These proteins could be successfully solubilized in acid urea (9 M urea-HCl, 10 mM DTT, pH 2.5). After dialysis against 3 M urea-HCl, 10 mM β-mercaptoethanol, pH 2.5 at 4°C, insoluble material was removed by ultracentrifugation at 100 000× *g* for 30 minutes. The supernatant containing both subunits was applied to a Bio-Gel® P100 gel filtration column (5 × 75 cm; Bio-Rad Laboratories) with 3 M urea-HCl, 10 mM β-mercaptoethanol, pH 2.5, as the buffer at room temperature. The β subunit eluted first, followed by fractions containing both α and β subunits, and finally followed by fractions containing only the α subunit. Attempts to renature the β subunit were unsuccessful. However, the α subunit could be renatured as long as the protein concentration remained below 0.1 mg/mL. Dialysis against 50 mM Na-phosphate, 1.0 mM DTT, pH 7.0, and 0.1 mM NaN_3 resulted in renaturation of some apo-α-PE.

3.1.3. Producing Apophycobiliproteins as Fusions

Several different phycobiliprotein structural genes have been successfully fused with the genes encoding other proteins, and this has allowed the purification procedure to be simplified to a single affinity chromatography step (Y.A. Cai, W.M. Schluchter, and A.N. Glazer, unpublished results). The maltose binding protein has been employed in such fusions, as well as a domain of 24 amino acids containing 6 contiguous histidine residues that has usually been fused to the N termini of several phycobiliprotein subunits (including α-PC, β-PC, α-AP, and β-AP) from several different cyanobacteria. Following the manufacturer's procedures for purification of the fusion proteins, high yields of products were generally obtained. An important factor to remember is to add reductant throughout the purification procedure in order to keep the cysteines reduced. It is also best to purify only as much protein as is needed in the next week. Within 2 weeks at 4°C, these proteins tend to oxidize and begin denaturing. As a matter of practice, it is much easier to store frozen *E. coli* cells containing the overproduced apophycobiliprotein fusion and purify small batches of protein when one needs it. This insures that the substrate for in vitro addition reactions is properly folded and contains fully reduced cysteines.

3.1.4. Attaching Apophycobiliproteins to Agarose Beads

The covalent attachment of apo-α-PC to agarose beads greatly facilitated reconstitution studies because it was possible to perform addition reactions in a small microcentrifuge tube, to wash away excess bilin after the reaction was terminated if necessary, and then to measure the fluorescence of the sample after this process (21,22). The apoprotein in 50 mM Na-phosphate, pH 7.0, 5 mM EDTA was mixed with Affi-Gel® 15 (Bio-Rad Laboratories) at 1 mg of protein per mL of beads (22). The covalent attachment of the protein to the beads continued for 30 minutes at 4°C until the reaction was stopped by the addition of 0.05 volumes of 1 M glycylglycine, pH 7.0 (incubated for 1 hour at 4°C). To remove excess unbound protein, the beads were washed with 50 mM Na-phosphate, pH 7.0, 5 mM EDTA, 0.5 M NaCl, followed by 50 mM Na-phosphate, pH 7.0, 5 mM EDTA. The air was evacuated out of the flask containing the beads, and the beads were stored at 4°C in the same buffer with the addition of 5 mM DTT.

4. RECONSTITUTION OF HOLOPHYCOBILIPROTEINS

4.1. Nonenzymatic Assays

The first evidence that enzymes might be required for bilin addition to phycobiliproteins was revealed through the experiments of Arciero et al. with apo-PC

(2–4). When either PCB or PEB was added to apo-αβ-PC, covalent addition took place at the α-84 and β-82 sites, but not at the β-153 site. The primary products of those nonenzymatic additions were bilins at a higher oxidation state, with an extra double bond between C_2-C_3 of ring A (see Figure 1 for numbering scheme). Mesobiliverdin (MBV) was the product when PCB was added, and 15,16 dihydrobiliverdin was the product when PEB was added.

Nonenzymatic addition reactions have also been performed with apo-α-PC (20) and with apoallophycocyanin subunits (W.M. Schluchter and A.N. Glazer, unpublished results). In all cases, a phycobiliprotein adduct is formed, and there is no discrimination between the bilin isomers observed in such in vitro addition experiments. Such discrimination clearly must take place in vivo in organisms which contain more than one bilin attached to phycobiliproteins.

4.1.1. Assay Conditions

The single most important factor in these assays is that the apoprotein be fully reduced prior to addition of the bilin substrate. This is accomplished by using freshly purified apoprotein, adding DTT to 10 mM, and incubating this mixture for 30 minutes at room temperature (22). The DTT should be removed by gel filtration prior to addition of bilin. It has been observed that bilins will react with DTT when this compound is present at high concentrations (W.M. Schluchter and A.N. Glazer, unpublished results) (20). Generally, apophycobiliproteins are very stable when present in 5 to 50 mM Na-

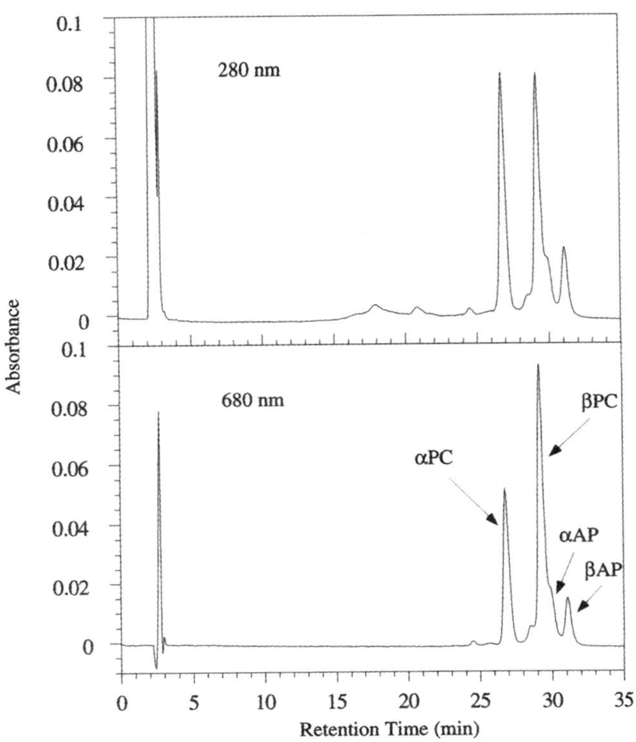

Figure 4. HPLC separation of phycobiliproteins present in the phycobilisomes purified from *Synechocystis* sp. PCC 6803. Phycobilisomes from sucrose gradients were dialyzed extensively against 5 mM Na-phosphate, pH 7.0, prior to injection on a C_4 reverse-phase column (see text for details). The elution of polypeptides was monitored at 280 nm (upper panel) and 680 nm (lower panel). The α-AP subunit is poorly resolved from the β-PC subunit.

phosphate, pH 7.0, or 10 to 50 mM Tris-HCl, pH 8.0, 75 mM NaCl, and therefore these conditions have been used in non-enzymatic assays.

❖ Procedure 6. Nonenzymatic Assay of Adduct Formation

1. The bilin, after dissolution in dimethyl sulfoxide (DMSO), is added to a final concentration of 10 to 50 μM to the reduced apoprotein that is present at a similar concentration (10–50 μM). Generally, the majority of the bilin combines with the apoprotein within 1 hour (2).
2. The reaction should be protected from the light at room temperature, but reaction mixtures can be purged with N_2 and left overnight at room temperature in the dark.
3. At the end of the incubation period, the phycobiliprotein should be separated from unreacted bilin, and this can be accomplished by one of several methods.
4. In instances in which nonaffinity tagged apoprotein is used, the reaction mixture should be loaded onto a small Sephadex G-25 column equilibrated with the same buffer used in the reaction. Bilins will bind to the surface of the resin [and require 10% (vol/vol) acetic acid to be released; the resin can usually be regenerated by standard procedures for reuse], while the phycobiliprotein will elute immediately (2).
5. If the phycobiliprotein has been affinity-tagged, one can proceed directly with the procedure for purification recommended by the manufacturer.
6. If the phycobiliprotein is covalently attached to agarose beads, the beads can be washed exhaustively to remove any trace bilins from the protein.

4.2. Enzymatic Assays

The evidence that enzymes were involved in bilin attachment to phycobiliproteins came from the characterization of the products of two genes, *cpcE* and *cpcF*, that occur downstream of *cpcBA*, the structural genes for the β and α subunits of phycocyanin, in *Synechococcus* sp. PCC 7002 (68,80). Insertional inactivation of either gene affected only PCB addition to the α subunit of PC. Fairchild et al. later showed that these two proteins acted together as a heterodimeric PC α subunit PCB lyase (22). Other *cpcE* and/or *cpcF* mutants have been characterized in *Synechococcus* sp. PCC 7942 (6), *Anabaena* sp. PCC 7120 (W.M. Schluchter and A.N. Glazer, unpublished results), and in *Calothrix* sp. PCC 7601 (70). In all of these cases, the mutants produce significantly reduced amounts of PC. Jung et al. (42) showed that a mutation in one or both of the *pecE* and *pecF* genes of *Anabaena* sp. PCC 7120, whose products show a high degree of sequence similarity with CpcE and CpcF, affected the level of the PEC holo-α-subunit. The PEC α subunit that could be purified from a *pecEF* mutant was found to contain a PCB adduct instead of the PXB (see Figure 1) chromophore that is normally present in wild-type cells. These results suggest that PecE and PecF form a heterodimeric PEC α subunit PXB lyase, and that in the absence of PecE and PecF, another lyase, possibly CpcE and CpcF, recognizes this site and adds PCB to the α-PEC subunit (42). Very recently, Zhao et al. have shown that PecE and PecF from *M. laminosus* act together to attach and isomerize PCB to PXB to the α subunit of PEC (79). This reaction required the presense of both subunits, because when one or both PecE and PecF were absent, the only product was MBV-α-PEC. In the cyanobacterium *Fremyella diplosiphon*, a mutation in *cpeY*, the product of which shows limited sequence similarity to the family of putative lyases including CpcE and which is located downstream of the structural genes encoding PE, produced markedly lower levels of PE. These observations suggest that CpeY is a lyase subunit as well (43).

The activities of only a few lyases have been tested in vitro, and to date, the most extensive analyses have been performed using *Synechococcus* sp. PCC 7002 CpcE and CpcF. So far, no lyase that can specifically attach bilins to the β subunit of any phycobiliprotein has been identified. Methods for assaying these enzymes will be summarized here in the hopes that this will encourage additional research in this area.

4.2.1. CpcE CpcF Expression

Recombinant CpcE and CpcF were produced in both soluble form and in the form of inclusion bodies in *E. coli*. However, Fairchild et al. showed that corenaturation of these two proteins in a 1:1 ratio led to the most activity (22).

❖ Procedure 7. Purification of Recombinant CpcE and CpcF

1. The *cpcE* and *cpcF* genes overexpressed using a T7/pET vector system and the majority of the recombinant proteins are found in inclusion bodies.
2. The inclusion bodies are collected by low-speed centrifugation after the cells have been lysed by passage through a French pressure cell. The inclusion bodies appear as a chalky white pellet and are easily differentiated from unbroken cells which usually appear more tan or brownish in color.
3. The inclusion bodies should be washed extensively using the following solutions: 50 mM Tris-HCl, 5 mM EDTA, pH 8.0; 50 mM Tris-HCl, pH 8.0, 1% Triton® X-100; 50 mM Tris-HCl, pH 8.0. Washing entails full resuspension, preferably using a tissue homogenizer, followed by centrifugation at 8000× *g*; the inclusion bodies containing CpcE/CpcF will be in the pellet fraction.
4. The inclusion body proteins are solubilized with 9.0 M urea-HCl, pH 1.9, 1 mM DTT. The concentrations of each protein should be determined spectrophotometrically using the $\varepsilon^{280\ \mathrm{nm}}$ for each protein (calculated from the Trp [ε = 5540 $M^{-1}cm^{-1}$] and Tyr [ε = 1480 $M^{-1}cm^{-1}$] content of the proteins) (54).
5. This estimate should be compared with the staining intensities of diluted aliquots of each urea-solubilized protein on SDS-PAGE. The $\varepsilon^{280\ \mathrm{nm}}$ for *Synechococcus* sp. PCC 7002 CpcE and CpcF under denaturing conditions are 35 640 $M^{-1}cm^{-1}$ and 20 220 $M^{-1}cm^{-1}$, respectively (22).
6. These proteins should be mixed in a 1:1 molar ratio at a concentration of 0.15 to 0.3 mg/mL prior to renaturation.

Several methods have been used successfully to renature these proteins. A concentrated mixture can be diluted approximately 1:10 with 50 mM Tris-HCl, 75 mM NaCl, pH 8.0; the dilution is followed by extensive dialysis against the same buffer at 4°C. This procedure yielded renatured heterodimeric CpcECpcF, but direct dialysis of the diluted proteins in 9.0 M urea against the same Tris-NaCl buffer produced similar results. In both cases, the yield of renatured CpcECpcF was approximately 50%. The extinction coefficients for native CpcE and CpcF were calculated (from the Trp and Tyr content of each protein) to be 38 440 $M^{-1}cm^{-1}$ and 21 060 $M^{-1}cm^{-1}$, respectively. After filter sterilization through a 0.2 μm membrane to prevent microbial growth, these proteins were stable for weeks at 4°C. Although other purification procedures have been utilized for preparations of proteins for more rigorous kinetic analyses (21), the procedure described above yielded a preparation of enzyme with high activity.

4.2.2. Bilin Donors

PEB and PCB can be cleaved from holophycobiliproteins and purified as described elsewhere in this volume (see Chapter 8) and in References 2 and 22. There is presently no reported method for the purification of the precursor of peptide-linked PUB or of the doubly-linked forms of PEB and PUB. However, it has been shown that CpcECpcF from *Synechococcus* sp. PCC 7002 (13) and *Anabaena* sp. PCC 7120 (C.F. Chan, W.M. Schluchter, and A.N. Glazer, unpublished results) will transfer the bilin from a holo-α-PC sub-

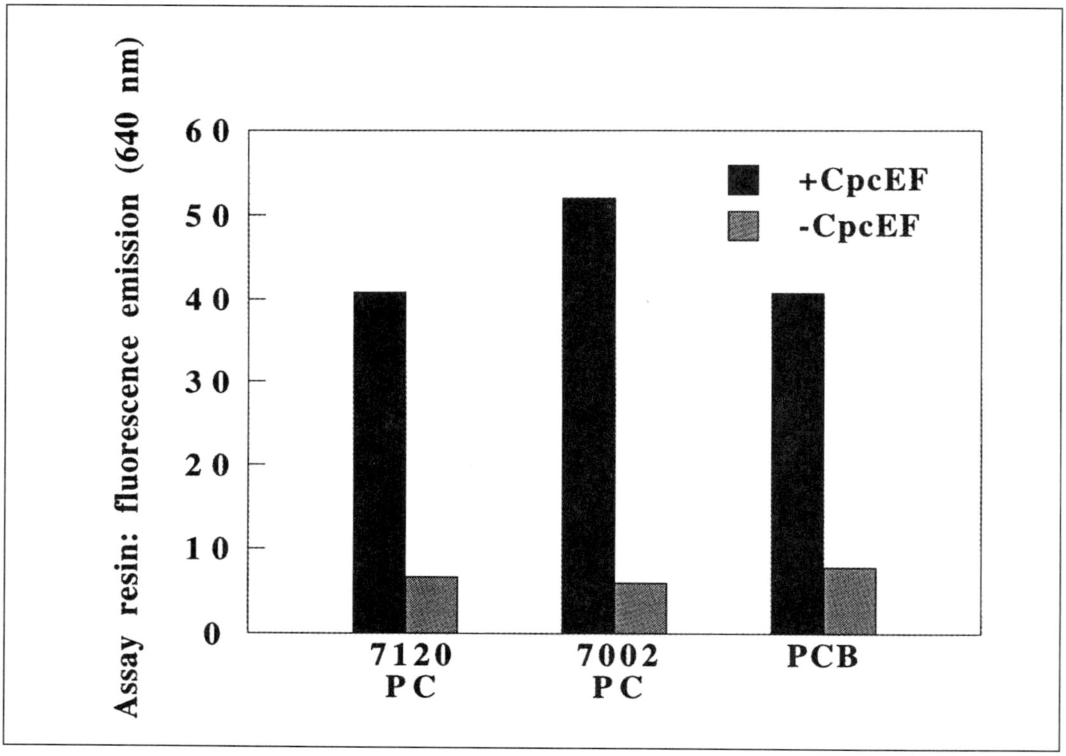

Figure 5. Bilin addition assays with *Anabaena* sp. PCC 7120 apo-α-PC resin. Assay conditions were as follows. Approximately 300 μL of settled resin (containing *Anabaena* sp. PCC 7120 apo-α-PC subunit covalently attached as described in Reference 22 was in a 1.5-mL microcentrifuge tube containing 0.8 mL of reaction assay buffer (50 mM Tris-HCl, pH 8.0, 75 mM NaCl, 1 mM $MgCl_2$, 1 mM Na pyrophosphate, 1 mM thioglycollate). The enzyme to be tested was *Anabaena* sp. PCC 7120 CpcECpcF (overproduced and purified as described in this chapter; W.M. Schluchter, C. Chan, and A.N. Glazer, manuscript in preparation). In assays where the enzyme was added (+CpcEF), *Anabaena* sp. PCC 7120 CpcECpcF were present at 0.25 μM. In control assays, the same volume of reaction assay buffer was added in place of CpcECpcF (-CpcEF). The reaction was initiated by the addition of the bilin donor. After incubation at 37°C in the dark for 1 hour, the resin was washed extensively as described in the text to remove any remaining donor bilin. The fluorescence emission of the resin present in each assay was measured at 640 nm because this is the peak of fluorescence emission for the native holo-α-PC. The donor bilin was purified PCB (11.6 μM; labeled as PCB), purified holophycocyanin from *Anabaena* sp. PCC 7120 (0.92 μM; labeled as 7120 PC), or purified holophycocyanin from *Synechococcus* sp. PCC 7002 (1.0 μM; labeled as 7002 PC). The *Anabaena* sp. PCC 7120 CpcECpcF lyase catalyzed the addition of free PCB to *Anabaena* sp. PCC 7120 apo-α-PC. However, this enzyme also catalyzed the reverse reaction by transferring bilin from the α-PC subunit (purified either from *Anabaena* sp. PCC 7120 or from *Synechococcus* sp. PCC 7002; W.M. Schluchter, C. Chan, and A.N. Glazer, unpublished results).

unit to an apo-α-PC subunit. It is unknown whether all lyases have this transfer activity. However, it is possible that many of these enzymes also serve as repair enzymes or as part of the phycobiliprotein degradation pathway under nutrient starvation conditions (16).

4.2.3. Enzyme Assay Conditions

The first assays performed to test the activity of *Synechococcus* sp. PCC 7002 CpcECpcF used apo-α-PC bound to resin as the substrate (22). This greatly facilitated the removal of unreacted bilins or the holo-α-PC substrate and the enzyme prior to the measurement of the incorporation of PCB onto the α-PC resin. However, affinity-tagged apophycobiliproteins have been successfully used as substrates in these same reactions (W.M. Schluchter and A.N. Glazer, unpublished results).

❖ Procedure 8. Enzymatic Assay of Adduct Formation

1. The fully prereduced apophycobiliprotein is added to a microcentrifuge tube. If the subunit is affinity tagged, approximately 0.3 to 0.6 mg is used. However, if the subunit is attached to a solid support, an aliquot corresponding to approximately 300 to 500 μL of settled beads is added.
2. The buffer conditions (as determined for optimal activity of the *Synechococcus* sp. PCC 7002 CpcECpcF) are 50 mM Tris-HCl, pH 8.0, 75 mM NaCl, 1 mM $MgCl_2$, 1 mM disodium pyrophosphate, and 1 mM thioglycolate.
3. The enzyme to be tested should be added to a final concentration of 0.1 to 0.4 μM.
4. The reaction is usually initiated by the addition of the bilin substrate. Free bilin should be dissolved in DMSO at concentrations between 0.8 to 2 mM and added to the reaction to a final concentration of 10 to 20 μM. If the source of the bilin is to be a holophycobiliprotein, then this protein should be added to a final concentration of 1 to 10 μM.
5. The reaction should be incubated in the dark at 37°C for 1 hour.

4.3. Methods for Analysis: Detection of Covalent Products

Enzymatic bilin addition reactions should always be compared with control nonenzymatic reactions using one or more of the following methods. If holophycobiliproteins were the source of bilin for the addition reaction, care must be taken to insure that all residual holophycobiliprotein has been removed. This is most easily accomplished when the apophycobiliprotein is attached to a solid support. The beads are washed extensively with 9.0 M urea-HCl, pH 2.5, followed by 50 mM Na-phosphate, pH 7.0. When the apo-subunit has been affinity tagged, it is very likely that any holo-phycobiliprotein added as a source of bilin can dimerize with either affinity-tagged apo-subunits or affinity-tagged enzyme-mediated bilin adducts and will copurify with the affinity-tagged subunit. Therefore, purification of the affinity-tagged protein must be performed according the manufacturer's procedure under denaturing conditions whenever possible. If this is not possible, then another method for the detection and separation of these two subunits should be used (see HPLC separation below).

4.3.1. Absorbance

Absorbance is the easiest and most straightforward method to detect an addition product. Unfortunately, this is the method that gives one the least amount of information about the product. Although it is a good starting point, this method should never be used as the only indicator of which product(s) is present. When PCB is added to apo-α-PC in the absence of CpcECpcF, the unnatural MBV adduct predominates and can be easily distinguished from the PCB product. The absorbance maximum of MBV attached to the native PC subunit occurs at 647 nm, whereas the absorbance maximum for the proper PCB adduct occurs at 622 nm (2). However, in cases in which multiple products may be attached at the same site, the absorbance spectrum of the addition product will usually be difficult to interpret (20). The absorbance of the peptide-bound bilins present can be determined by denaturation of the addition product using one of the methods described above. For PEB addition experiments, the nonenzymatically favored product, DBV, was found to accumulate (4,20). DBV exhibits characteristic absorbance maxima at 606 and 330 nm in native proteins (74); for denatured subunits in acidic urea solutions, a 330 nm absorbance peak is diagnostic of peptide-linked DBV, whereas a 308 nm peak is characteristic of peptide-linked PEB (33,74).

4.3.2. Fluorescence

Free bilins exhibit little fluorescence in solution but become highly fluorescent once they have been covalently attached to phycobiliproteins, because they are rigidly held in a stretched conformation that does not facilitate nonradiative decay of the excited state (22). Therefore, the fluorescence emission spectrum of both control and enzymatic reactions can be measured as a way of monitoring the products of the reaction (see Figure 5). The MBV product of nonenzymatic PCB addition to apophycocyanin is easily distinguished from the natural PCB product, because both the absorbance and fluorescence are red-shifted relative to the PCB product. The MBV adduct, with a fluorescence emission maximum at 668 nm, is much less fluorescent than the PCB adduct, which has a fluorescence emission maximum at 643 nm (3,22). Additionally, the extinction coefficients for the long wavelength absorbing species of MBV peptides in 10 mM TFA were determined to be 40% lower than those of the naturally occurring PCB-bearing peptides (2).

Much less is known about the fluorescence properties of the unnatural DBV adduct formed when PEB is added to apo-PC or apo-α-PE (4,20). The use of fluorescence to monitor product accumulation with putative lyases that attach PEB may be complicated by the fact that multiple products accumulate in nonenzymatic reactions. Therefore, absorbance and fluorescence spectroscopy may not work as well as one of the following methods for the characterization of enzymatic bilin addition to apo-PE.

4.3.3. HPLC Separation and Detection

If the holo- and apo-subunits, which might be produced or used as substrates in an enzymatic reaction, can be separated by C_4 reverse-phase chromatography as described above, then this method provides an excellent way to detect the transfer of bilin from a holophycobiliprotein to an apo-subunit. Such separations are usually best achieved if the source of the holo-subunit is from another organism. The transfer reaction of PCB from *Anabaena* sp. PCC 7120 holo-α-PC to *Synechococcus* sp. PCC 7002 apo-α-PC mediated by *Synechococcus* sp. PCC 7002 CpcECpcF was detected using this method (22). *Synechococcus* sp. PCC 7002 CpcECpcF proteins can also transfer a bilin from *Synechococcus* sp. PCC 7002 holo-PC to *Anabaena* sp. PCC 7120 apo-α-PC subunit (see Figure 6).

4.3.4. Characterization of the Product by Tryptic Digestion

This is the most quantitative method of characterization of the bilin product (2,3,20). The addition product is cleaved using trypsin, and tryptic peptides are separated on a C_{18} reverse-phase column (45). Tryptic peptides can be collected, their absorption spectra in 10 mM TFA determined, and their composition evaluated by amino acid analysis or sequencing to show rigorously which bilin was added to a particular site(s) on the apophycobiliprotein subunit. If multiple products are present, this is the best method to determine how many products have been formed and to quantitate their relative amounts. Keep in mind that for each phycobiliprotein, digestion by more than one protease may be required to obtain a fragment sufficiently small to allow its isolation and characterization. Digestion procedures for each type of phycobiliprotein have been published (7,49,50,55,67,74–76), and it is recommended that the user refer to the optimized procedure for the particular phycobiliprotein with which he/she is working. The procedure described below was used successfully on C-PC and R-PE (2,45).

The addition product should be separated from unreacted bilin by chromatography on Sephadex G-25. The phycobiliprotein should then be fully denatured by acidification with 1 N HCl to pH 2.0 and stored under N_2 for 45 minutes. Trypsin (TCPK-treated; Worthington Biochemical, Lakewood, NJ, USA), dissolved in 1 mM HCl at 5 mg/mL concentration, is added to 2% (wt/wt) to the denatured phycobiliprotein in HCl. This mixture is titrated to pH 7.5 with 1 N NaOH after the addition of ammonium bicarbonate to 100 mM. After incubation of this mixture for 2 hours at 30°C in the dark, an additional aliquot of trypsin is added, and the incubation is continued for another 2 hours under the same conditions. The reaction is stopped by the addition of glacial acetic acid to 30% (vol/vol). If a large amount of protein is being digested, then fractionation on Sephadex G-50 in 30% acetic acid (vol/vol) is a good method to separate undigested material from tryptic peptides. If the amount of material is scaled for analytical purposes, then the colored material can be collected and loaded directly onto a SepPak C_{18} cartridge. The cartridge can be washed with 0.1% TFA followed by elution by 60% acetonitrile, 40% 0.1% TFA. The eluate should be collected, dried under N_2, and redissolved in 10 mM TFA prior to HPLC separation. However, if the amount of material is scaled for preparative purposes, the colored material in the eluate from the gel exclusion chromatography in 30% acetic acid should then be concentrated under N_2 before dilution with 50 mM Na-phosphate, pH 2.5. The mixture should then be fractionated on an ion-exchange column (SP-Sephadex G-25, 2 × 6.5 cm) and eluted with a linear gradient of 0 to 0.6 M NaCl in 50 mM Na-phosphate, pH 2.5. Fractions containing colored material should be collected, desalted on the SepPak C_{18} cartridge as described above, before separation by HPLC.

The conditions used for separating the tryptic peptides of phycocyanin follow. However, for each phycobiliprotein, different gradient conditions may be required, and optimization of these conditions should be pursued prior to preparative-scale analyses. For the phycocyanin of *Synechococcus* sp. PCC 7002, a C_{18} reverse-phase analytical column (5 µm, 4.6 × 250 mm) should be used for separation of tryptic peptides (see Figure 7). The solvent system is 0.1 M Na-phosphate, pH 2.1 (Buffer A) and acetonitrile (Buffer B) with flow rates of 1.5/mL min. Peptides are loaded at 20% Buffer B (80% Buffer A)

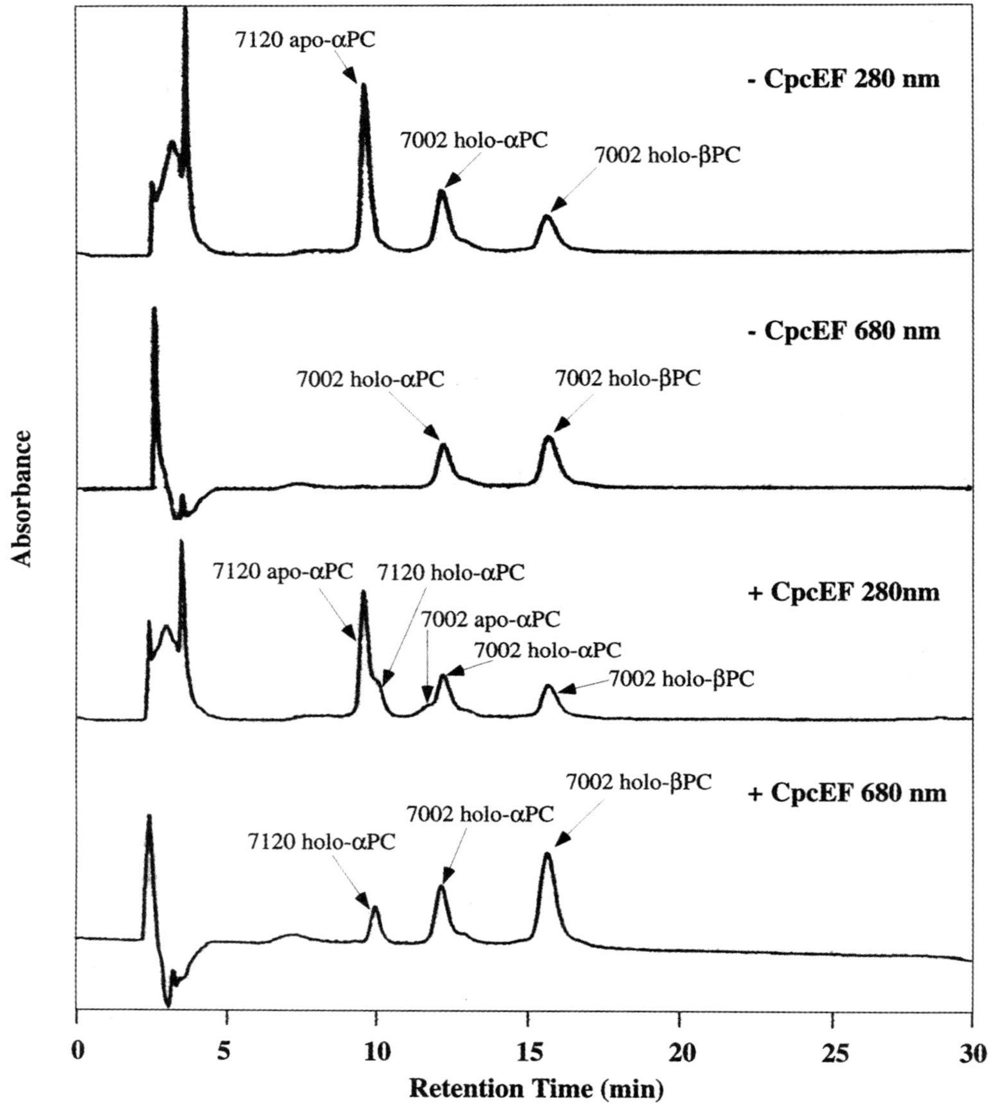

Figure 6. Monitoring the transfer of bilin from *Synechococcus* sp. PCC 7002 PC to *Anabaena* sp. PCC 7120 apo-α-PC by C_4 reverse-phase HPLC. Each assay contained 100 µg of *Anabaena* sp. PCC 7120 apo-α-PC, 75 µg of *Synechococcus* sp. PCC 7002 PC, 0.2 µM *Synechococcus* sp. PCC 7002 CpcECpcF (if present) in a volume of 400 µL (reaction assay buffer conditions are as described in Figure 5). Reactions were allowed to proceed for 16 hours at room temperature in the dark. Each reaction was combined with 800 µL of 9 M urea, pH 1.9, mixed, and centrifuged prior to injection on the C_4 column (as described in this chapter). After injection, buffer conditions (buffers are those from Swanson and Glazer; Reference 66) are as follows: 2 minutes at 35% Buffer B (65% Buffer A), a 1-minute linear gradient to 53% Buffer B (47% Buffer A), followed by a linear gradient to 63% Buffer B over 20 minutes (22). Each assay was monitored at 280 nm (reflecting protein content) and 680 nm (reflecting bilin content). Retention times for various components are as follows: *Anabaena* sp. PCC 7120 apo-α-PC, 9.5 minutes; *Anabaena* sp. PCC 7120 holo-α-PC, 10 minutes; *Synechococcus* sp. PCC 7002 apo-α-PC, 11.7 minutes; *Synechococcus* sp. PCC 7002 holo-α-PC, 12.2 minutes; *Synechococcus* sp. PCC 7002 holoβ-PC, 15.8 minutes. *Synechococcus* sp. PCC 7002 CpcECpcF is capable of transferring bilin from 7002 holo-α-PC to *Anabaena* sp. PCC 7120 apo-α-PC (W.M. Schluchter and A.N. Glazer, unpublished results).

and eluted with a linear gradient to 40% Buffer B (60% Buffer A) over 20 minutes (2).

5. CONCLUDING REMARKS

Although phycobiliproteins were among the first proteins to be characterized and much is known about their structures, relatively little is still known concerning the details of chromophore attachment to this large and highly diverse protein family. This situation has not improved dramatically in spite of the availability of the complete genomic sequence of the cyanobacterium *Synechocystis* sp. PCC 6803. It is hoped that the procedures described above for the production of substrate proteins and for the characterization of bilin attachment reactions will aid other researchers interested in the characterization of new phycobiliproteins or in the characterization of the biosynthesis of phycobiliproteins.

ACKNOWLEDGMENTS

We thank Dr. Alexander N. Glazer for helpful comments. This research was supported in part by United States Public Health Service (USPHS) Grant No. GM-31625 (to D.A.B.), a National Research Service Award Grant No. GM-16935 (to W.M.S.), and the LA Board of Regents Grant No. LEQSF(1999-2002)-RD-A-45 (to W.M.S.).

ABBREVIATIONS

AP, allophycocyanin; DBV, 15,16 dihydrobiliverdin; DTT, dithiothreitol; EDTA, ethylenediamine tetraacetate; HPLC, high-performance liquid chromatography; MBV, mesobiliverdin; PC, phycocyanin; PCB, phycocyanobilin; PE, phycoerythrin; PEB, phycoerythrobilin; PEC, phycoerythrocyanin; PUB, phycourobilin; PXB, phycobiliviolin; PϕB, phytochromobilin; TFA, trifluoroacetic acid.

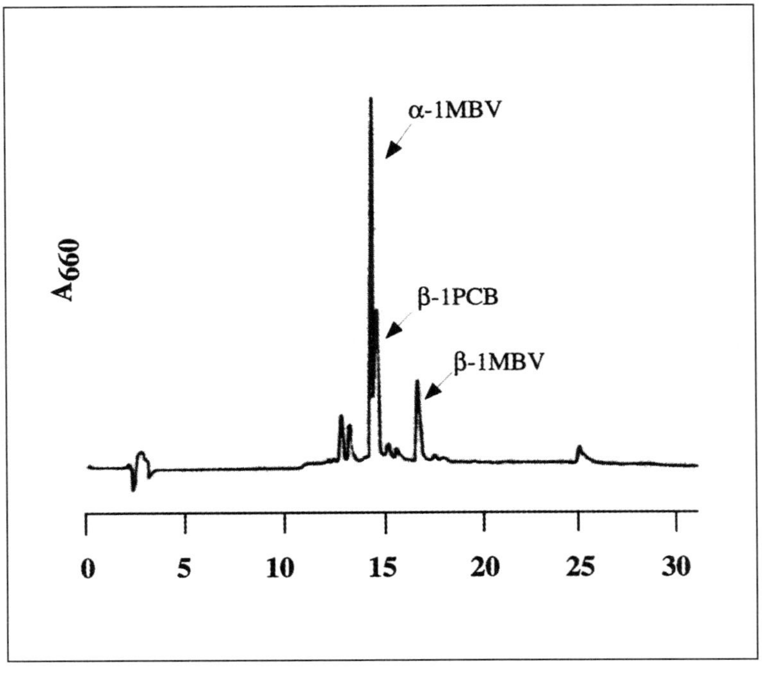

Figure 7. HPLC elution profile from a C_{18}-reverse phase column of a tryptic digest of a preparation of *Synechococcus* sp. PCC 7002 apophycocyanin after reaction with free PCB, in the absence of enzymes. The major products are MBV at the α-84 (α-1MBV) and β-82 (β-1MBV) sites with some PCB forming at the β-82 site (β-1PCB; the amount of this product is variable). The elution of bilin-linked peptides was monitored at 660 nm. This figure was modified with permission from Reference 2.

REFERENCES

1. **Airth, R.L. and L.R. Blinks.** 1956. A new phycoerythrin from *Porphyra naiadum*. Biol. Bull. *111*:321-327.
2. **Arciero, D.M., D.A. Bryant, and A.N. Glazer.** 1988. In vitro attachment of bilins to apophycocyanin I: specific covalent adduct formation at cysteinyl residues involved in phycocyanobilin binding in C-phycocyanin. J. Biol. Chem. *263*:18343-18349.
3. **Arciero, D.M., J.L. Dallas, and A.N. Glazer.** 1988. In vitro attachment of bilins to apophycocyanin: II: determination of the structures of tryptic bilin peptides derived from the phycocyanobilin adduct. J. Biol. Chem. *263*:18350-18357.
4. **Arciero, D.M., J.L. Dallas, and A.N. Glazer.** 1988. In vitro attachment of bilins to apophycocyanin: III: properties of the phycoerythrobilin adduct. J. Biol. Chem. *263*:18358-18363.
5. **Betz, M.** 1997. One century of protein crystallography: the phycobiliproteins. Biol. Chem. *378*:167-176.
6. **Bhalerao, R.P., L.K. Kind, and P. Gustafsson.** 1994. Cloning of the *cpcE* and *cpcF* genes from *Synechococcus* sp. PCC 6301 and their inactivation in *Synechococcus* sp. PCC 7942. Plant Mol. Biol. *26*:313-326.
7. **Bishop, J.E., H. Rapoport, A.V. Klotz, C.F. Chan, A.N. Glazer, P. Guglistaller, and H. Zuber.** 1987. Chromopeptides from phycoerythrocyanin: structure and linkage of the three bilin groups. J. Am. Chem. Soc. *109*:875-881.
8. **Brejc, K., R. Ficner, R. Huber, and S. Steinbacher.** 1995. Isolation, crystallization, crystal structure analysis and refinement of allophycocyanin form the cyanobacterium *Spirulina platensis* at 2.3 Å resolution. J. Mol. Biol. *249*:424-440.
9. **Bryant, D.A.** 1982. Phycoerythrocyanin and phycoerythrin: properties and occurrence in cyanobacteria. J. Gen. Microbiol. *128*:835-844.
10. **Bryant, D.A., G. Cohen-Bazire, and A.N. Glazer.** 1981. Characterization of the biliproteins of *Gloeobacter violaceus*: chromophore content of a cyanobacterial phycoerythrin carrying phycourobilin chromophore. Arch. Microbiol. *129*:190-198.
11. **Bryant, D.A., J.M. Dubbs, P.I. Fields, R.D. Porter, and R. de Lorimier.** 1985. Expression of phycobiliprotein genes in *Escherichia coli*. FEMS Microbiol. Lett. *29*:343-349.
12. **Bryant, D.A., A.N. Glazer, and F.A. Eiserling.** 1976. Characterization and structural properties of the major phycobiliproteins of *Anabaena* sp. Arch. Microbiol. *110*:61-75.
13. **Chang, W., T. Jiang, Z. Wan, J. Zhang, Z. Yang, and D. Liang.** 1996. Crystal structure of R-phycoerythrin from *Polysiphonia urceolata* at 2.8 Å resolution. J. Mol. Biol. *262*:721-731.
14. **Cohen-Bazire, G., S. Béguin, S. Rimon, A.N. Glazer, and D. M. Brown.** 1977. Physico-chemical and immunological properties of allophycocyanins. Arch. Microbiol. *111*:225-238.
15. **de Lorimier, R., S.M. Wilbanks, and A.N. Glazer.** 1993. Genes of the R-phycocyanin II locus of marine *Synechococcus* spp. and comparison of protein-chromophore interactions in phycocyanins differing in bilin composition. Plant Mol. Biol. *21*:225-237.
16. **Dolganov, N. and A.R. Grossman.** 1999. A polypeptide with similarity to phycocyanin α-subunit phycocyanobilin lyase involved in degradation of phycobilisomes. J. Bacteriol. *181*:610-617.
17. **Ducret, A., W. Sidler, G. Frank, and H. Zuber.** 1994. The complete amino acid sequence of R-Phycocyanin-I α and β subunits from the red alga *Porphyridium cruentum*: structural and phylogenetic relationships of the phycocyanins within the phycobiliproteins families. Eur. J. Biochem. *221*:563-580.
18. **Duerring, M., R. Huber, W. Bode, R. Ruembeli, and H. Zuber.** 1990. Refined three-dimensional structure of phycoerythrocyanin from the cyanobacterium *Mastigocladus laminosus* at 2.7 Å. J. Mol. Biol. *211*:633-644.
19. **Duerring, M., G.B. Schmidt, and R. Huber.** 1991. Isolation, crystallization, crystal structure analysis and refinement of constitutive C-phycocyanin from the chromatically adapting cyanobacterium *Fremyella diplosiphon* at 1.66 Å resolution. J. Mol. Biol. *217*:577-592.
20. **Fairchild, C.D., and A.N. Glazer.** 1994. Nonenzymatic bilin addition to the a-subunit of an apophycoerythrin. J. Biol. Chem. *269*:28988-28996.
21. **Fairchild, C.D., and A.N. Glazer.** 1994. Oligomeric structure, enzyme kinetics, and substrate specificity of the phycocyanin α subunit phycocyanobilin lyase. J. Biol. Chem. *269*:8686-8694.
22. **Fairchild, C.D., J. Zhao, J. Zhou, S.E. Colson, D.A. Bryant, and A.N. Glazer.** 1992. Phycocyanin a subunit phycocyanobilin lyase. Proc. Natl. Acad. Sci. USA *89*:7017-7021.
23. **Ficner, R. and R. Huber.** 1993. Refined crystal structure of phycoerythrin from *Porphyridium cruentum* at 0.23 nm resolution and localization of the g subunit. Eur. J. Biochem. *218*:103-106.
24. **Ficner, R., K. Lobeck, G. Schmidt, and R. Huber.** 1992. Isolation, crystallization, crystal structure analysis and refinement of B-phycoerythrin from the red alga *Porphyridium sordidum* at 2.2 Å resolution. J. Mol. Biol. *228*:935-950.
25. **Glazer, A.N.** 1989. Light guides: directional energy transfer in a photosynthetic antenna. J. Biol. Chem. *264*:1-4.
26. **Glazer, A.N.** 1981. Photosynthetic accessory proteins with bilin prosthetic groups, p. 51-96. *In* M.D. Hatch and N.K. Boardman (Eds.), The Biochemistry of Plants, vol. 8. Academic Press, New York.
27. **Glazer, A.N.** 1988. Phycobiliproteins. Methods Enzymol. *167*:291-303.
28. **Glazer, A.N.** 1988. Phycobilisomes. Methods Enzymol. *167*:304-312.
29. **Glazer, A.N., C.F. Chan, and J.A. West.** 1997. An unusual phycocyanobilin-containing phycoerythrin of several bluish-colored, arcochaetioid, freshwater red algal species. J. Phycol. *33*:617-624.
30. **Glazer, A.N. and S. Fang.** 1973. Chromophore content of blue-green algal phycobiliproteins. J. Biol. Chem. *248*:659-662.
31. **Glazer, A.N. and S. Fang.** 1973. Formation of hybrid proteins from the α and β subunits of phycocyanins of unicellular and filamentous blue-green algae. J. Biol. Chem. *248*:663-671.

32.**Glazer, A.N., S. Fang, and D.M. Brown.** 1973. Spectroscopic properties of C-phycocyanin and of its α and β subunits. J. Biol. Chem. *248*:5679-5685.
33.**Glazer, A.N. and C.S. Hixson.** 1975. Characterization of R-phycocyanin: chromophore content of R-phycocyanin and C-phycoerythrin. J. Biol. Chem. *250*: 5487-5495.
34.**Glazer, A.N. and C.S. Hixson.** 1977. Subunit structure and chromophore composition of rhodophytan phycoerythrins: *Porphyridium cruentum* B-phycoerythrin and b-phycoerythrin. J. Biol. Chem. *252*: 32-42.
35.**Glazer, A.N., D.J. Lundell, G. Yamanaka, and R.C. Williams.** 1983. The structure of a "simple" phycobilisome. Ann. Microbiol. (Inst. Pasteur). *134B*:159-180.
36.**Glazer, A.N. and G.J. Wedemayer.** 1995. Cryptomonad biliproteins—an evolutionary perspective. Photosyn. Res. *46*:93-105.
37.**Glazer, A.N., J.A. West, and C. Chan.** 1982. Phycoerythrins as chemotaxonomic markers in red algae: a survey. Biochem. Sys. Ecol. *10*:203-215.
38.**Gómez-Lojero, C., B. Pérez-Gómez, D.W. Krogmann, and A. Peña-Diaz.** 1997. The tricylindrical core of the phycobilisome of the cyanobacterium *Arthrospira (Spirulina) maxima*. Int. J. Biochem. Cell Biol. *29*:959-970.
39.**Gómez-Lojero, C., B. Pérez-Gómez, G. Prado-Flores, D.W. Krogmann, A. Cárabez-Rejo, and A. Peña-Diaz.** 1997. The phycobilisomes of the cyanobacterium *Arthrospira (Spirulina) maxima*. Int. J. Biochem. Cell Biol. *29*:1191-1205.
40.**Gysi, J. and H. Zuber.** 1974. Isolation and characterization of allophycocyanin from the thermophilic blue-green alga *Mastigocladus laminosus* cohn. FEBS Lett. *48*:209-213.
41.**Hess, W.R., F. Partensky, G.W.M. van der Staay, J.M. Garcia-Fernandez, T. Börner, and D. Vaulot.** 1996. Coexistence of phycoerythrin and a chlorophyll α/β antenna in a marine prokaryote. Proc. Natl. Acad. Sci. USA *93*:11126-11130.
42.**Jung, L.J., C.F. Chan, and A.N. Glazer.** 1995. Candidate genes for the phycoerythrocyanin α subunit lyase: biochemical analysis of *pecE* and *pecF* interposon mutants. J. Biol. Chem. *270*:12877-12884.
43.**Kahn, K., D. Mazel, J. Houmard, N. Tandeau de Marsac, and M.R. Schaefer.** 1997. A role for cpeYZ in cyanobacterial phycoerythrin biosynthesis. J. Bacteriol. *179*: 998-1006.
44.**Kehoe, D. and A. Grossman.** 1996. Similarity of a chromatic adaptation sensor to phytochrome and ethylene receptors. Science *273*:1409-1412.
45.**Klotz, A.V. and A.N. Glazer.** 1985. Characterization of the bilin attachment sites in R-phycoerythrin. J. Biol. Chem. *260*:4856-4863.
46.**Lagarias, J.C., A.V. Klotz, J.L. Dallas, A.N. Glazer, J.E. Bishop, J.F. O'Connell, and H. Rapoport.** 1988. Exclusive A-ring linkage for singly attached phycocyanobilins and phycoerythrobilins in phycobiliproteins. J. Biol. Chem. *263*:12977-12985.
47.**Lamparter, T., F. Mittmann, W. Gartner, T. Börner, E. Hartmann, and J. Hughes.** 1997. Characterization of recombinant phytochrome from the cyanobacterium *Synechocystis*. Proc. Natl. Acad. Sci. USA *94*:11792-7.
48.**Lokstein, H., C. Steglich, and W.R. Hess.** 1999. Light-harvesting antenna function of phycoerythrin in *Prochlorococcus marinus*. Biochim. Biophys. Acta. *1410*:97-98.
49.**Lundell, D.J. and A.N. Glazer.** 1981. Allophycocyanin B: α common β subunit in *Synechococcus* allophycocyanin B (λ_{max} 670 nm) and allophycocyanin (λ_{max} 650 nm). J. Biol. Chem. *246*:12600-12606.
50.**Lundell, D.J., A.N. Glazer, R.J. DeLange, and D.M. Brown.** 1984. Bilin attachment sites in the α and β subunits of B-phycoerythrin: amino acid sequence studies. J. Biol. Chem. *259*:5472-5480.
51.**MacColl, R.** 1998. Cyanobacterial phycobilisomes. J. Struct. Biol. *124*:311-334.
52.**MacColl, R., L.E. Eisele, M. Dhar, J.-P. Ecuyer, S. Hopkins, J. Marrone, R. Barnard, H. Malak, and A.J. Lewitus.** 1999. Bilin organization in cryptomonad biliproteins. Biochemistry *38*:4097-4105.
53.**MacColl, R. and D. Guard-Friar.** 1987. Phycobiliproteins. CRC Press, Boca Raton.
54.**Mach, H., C.R. Middaugh, and R.V. Lewis.** 1992. Statistical determination of the average values of the extinction coefficients of tryptophan and tyrosine in native proteins. Anal. Biochem. *200*:74-80.
55.**Ong, L.J. and A.N. Glazer.** 1991. Phycoerythrins of marine unicellular cyanobacteria I: bilin types and locations and energy transfer pathways in *Synechococcus* sp. phycoerythrins. J. Biol. Chem. *266*:9515-9527.
56.**Ong, L. J. and A.N. Glazer.** 1987. R-phycocyanin II, a new phycocyanin occurring in marine *Synechococcus* species. Identification of the terminal energy acceptor bilin in phycocyanins. J. Biol. Chem. *262*:6323-6327.
57.**Ong, L.J. and A.N. Glazer.** 1988. Structural studies of phycobiliproteins in unicellular marine cyanobacteria, p. 102-121. *In* S.E. Stevens, Jr. and D.A. Bryant (Eds.), Light-Energy Transduction in Photosynthesis: Higher Plant and Bacterial Models. American Society of Plant Physiologists, Rockville, MD.
58.**Padgett, M.P. and D.W. Krogmann.** 1987. Large scale preparation of pure phycobiliproteins. Photosyn. Res. *11*:225-235.
59.**Reuter, W., G. Wiegand, R. Huber, and M.E. Than.** 1999. Structural analysis at 2.2 Å of orthorhombic crystals presents the asymmetry of the allophycocyanin-linker complex, AP-$L_C^{7.8}$, from phycobilisomes of *Mastigocladus laminosus*. Proc. Natl. Acad. Sci. USA *96*:1363-1368.
60.**Schirmer, T.,W. Bode, and R. Huber.** 1987. Refined three-dimensional structures of two cyanobacterial C-phycocyanins at 2.1 and 2.5 Å resolution. J. Mol. Biol. *196*:677-695.
61.**Schirmer, T., W. Bode, R. Huber, W. Sidler, and H. Zuber.** 1985. X-ray crystallographyic structure of the light-harvesting biliprotein C-phycocyanin from the thermophilic cyanobacterium *Mastigocladus laminosus* and its resemblance to globin structures. J. Mol. Biol. *184*:257-277.
62.**Schirmer, T., R. Huber, M. Schneider, W. Bode, M. Miller, and M.L. Hackert.** 1986. Crystal structure analysis and refinement at 2.5 Å of hexameric C-phycocyanin from the cyanobacterium *Agmenellum quadruplicatum*: the molecular model and its implications from light-harvesting. J. Mol. Biol. *188*:651-676.

63.**Sidler, W.A.** 1994. Phycobilisome and phycobiliprotein structure, p. 139-216. *In* D.A. Bryant (Ed.), The Molecular Biology of Cyanobacteria, Vol. 1. Kluwer Academic, Dordrecht, The Netherlands.

64.**Sineshchekov, V., J. Hughes, E. Hartmann, and T. Lamparter.** 1998. Fluorescence and photochemistry of recombinant phytochrome from the cyanobacterium *Synechocystis*. Photochem. Photobiol. *67*:263-267.

65.**Stec, B., R.F. Troxler, and M.M. Teeter.** 1999. Crystal structure of C-phyocyanin from *Cyanidium caldarium* provides a new perspective on phycobilisome assembly. Biophys. J. *76*:2912-2921.

66.**Swanson, R.V. and A.N. Glazer.** 1990. Separation of phycobiliprotein subunits by reverse-phase high pressure liquid chromatography. Anal. Biochem. *188*:295-299.

67.**Swanson, R.V., L.J. Ong, S.M. Wilbanks, and A.N. Glazer.** 1991. Phycoerythrins of marine unicellular cyanobacteria II: characterization of phycobiliproteins with unusually high phycourobilin content. J. Biol. Chem. *266*:9528-9534.

68.**Swanson, R.V., J. Zhou, J.A. Leary, T. Williams, R. de Lorimier, D.A. Bryant, and A.N. Glazer.** 1992. Characterization of phycocyanin produced by *cpcE* and *cpcF* mutants and identification of an intergenic suppressor of the defect in bilin attachment. J. Biol. Chem. *267*:16146-16154.

69.**Tandeau de Marsac, N. and G. Cohen-Bazire.** 1977. Molecular composition of cyanobacterial phycobilisomes. Proc. Natl. Acad. Sci. USA *74*:1635-1639.

70.**Tandeau de Marsac, N., D. Mazel, V. Capuano, T. Damerval, and J. Houmard.** 1990. Genetic analysis of the cyanobacterial light-harvesting antenna complex, p. 143-153. *In* D. Drews and E.A. Dawes (Eds.), Molecular Biology of Membrane-Bound Complexes in Phototropic Bacteria. Plenum, New York.

71.**Thomas, J.-C. and C. Passquet.** 1999. Characterization of a phycoerythrin without α-subunits from a unicellular red alga. J. Biol. Chem. *274*:2472-2482.

72.**Toole, C.M., T.L. Plank, A.R. Grossman, and L.K. Anderson.** 1998. Bilin deletions and subunit stability in cyanobacterial light-harvesting proteins. Mol. Microbiol. *30*:475-486.

73.**Wedemayer, G.J., D.G. Kidd, and A.N. Glazer.** 1996. Cryptomonad biliproteins: bilin types and locations. Photosyn. Res. *48*:163-170.

74.**Wedemayer, G.J., D.G. Kidd, D.E. Wemmer, and A.N. Glazer.** 1992. Phycobilins of cryptophycean algae: occurrence of dihydrobiliverdin and mesobiliverdin in cryptomonad biliproteins. J. Biol. Chem. *267*:7315-7331.

75.**Wemmer, D.E., G.J. Wedemayer, and A.N. Glazer.** 1993. Phycobilins of cryptophycean algae: novel linkage of dihydrobiliverdin in a phycoerythrin 555 and a phycocyanin 645. J. Biol. Chem. *268*:1658-1669.

76.**Wilbanks, S.M. and A.N. Glazer.** 1993. Rod structure of a phycoerythrin II-containing phycobilisome II: complete sequence and bilin attachment site of a phycoerythrin g subunit. J. Biol. Chem. *268*:1236-1241.

77.**Wilde, A., Y. Churin, H. Schubert, and T. Börner.** 1997. Disruption of a *Synechocystis* sp. PCC 6803 gene with partial similarity to phytochrome genes alters growth under changing light qualities. FEBS Lett. *406*: 89-92.

78.**Yeh, K. C., S.-H. Wu, J.T. Murphy, and J.C. Lagarias.** 1997. A cyanobacterial phytochrome two-component light sensory system. Science *277*:1505-1508.

79.**Zhao, K.H., M.G. Deng, M. Zheng, M. Zhou, A. Parbel, M. Storf, M. Meyer, B. Strohmann, and H. Scheer.** 2000. Novel activity of a phycobiliprotein lyase: both the attachment of phycocyanobilin and the isomerization to phycobiliviolin are catalyzed by the proteins PecE and PecF encoded by the phycoerythrocyanin operon. FEBS Lett. *469*:9-13.

80.**Zhou, J., G.E. Gasparich, V.L. Stirewalt, R. de Lorimier, and D.A. Bryant.** 1992. The *cpcE* and *cpcF* genes of *Synechococcus* sp. PCC 7002: construction and phenotypic characterization of interposon mutants. J. Biol. Chem. *267*:16138-16145.

Index